PITT PRESS SERIES

D. IVNII IVVENALIS

SATVRAE XIV

FOURTEEN SATIRES OF JUVENAL

EDITED BY

J. D. DUFF

WITH A NEW INTRODUCTION BY

MICHAEL COFFEY

University College London

CAMBRIDGE
AT THE UNIVERSITY PRESS
1970

Published by the Syndics of the Cambridge University Press
Bentley House, 200 Euston Road, London N.W.1
American Branch: 32 East 57th Street, New York, N.Y.10022

© Cambridge University Press 1970

Library of Congress Catalogue Card Number: 71–129933

Standard Book Number: 521 07370 7

First Edition 1898
Reprinted 1900
1904 with corrections
1909 with corrections
1914 with corrections
1925 with corrections
1929
1932
1940
1948
1951
1955
1957
1962
Reset and reprinted 1970 with a new introduction

Printed in Great Britain
at the University Printing House, Cambridge
(Brooke Crutchley, University Printer)

PREFACE

Perhaps it is reason enough for adding another to the many English editions of Juvenal, that all our recent editors have excluded the Sixth Satire, the most brilliant in detail and by far the longest of Juvenal's poems. The present edition includes 530 lines of this celebrated piece. It may also be noted that the text of Juvenal has been materially affected by recent discoveries, of which full account has been taken here.

The Introduction deals first with the Life of Juvenal. In this section the different sources from which our information is derived, are considered in turn; and the dates assigned by Friedländer to the different Books of Satires are accepted generally as proved. The next section contains a sketch of *satura* as treated by Juvenal's predecessors in this kind of writing: here (pp. xxiv–xxix) I follow closely the late Professor Nettleship's Essay on the subject now reprinted in his *Lectures and Essays* (second series). When writing my Introduction, I was not aware that this Essay had been reprinted in an easily accessible form; or I should have been content to refer to it. The same volume contains (p. 117) an Essay on Juvenal's Life and Poems, published originally in the *Journal of Philology* (vol. xvi): this contains the best criticism of Juvenal I have ever read, and I have frequently quoted from it in different parts of my book. My third section deals with Juvenal himself—his relation to his predecessors, his characteristics, moral and literary, and his motives for writing satire. The two remaining sections contain a brief account of the Manuscripts and Scholia, based upon personal study of the material.

The Text is based upon F. Bücheler's last edition (1893), which will always be memorable as giving the first full and trustworthy account of the readings of *P* (the *codex*

Pithoeanus), incomparably the best of the multitude of extant MSS. Yet *P* has many errors; and in a number of passages, especially in the later satires, where Bücheler retains its reading, I have felt unable to follow him, while I have occasionally retained the reading of *P* where he discards it.

My knowledge of *P* is derived entirely from the labours of others—Bücheler's edition, R. Beer's *Spicilegium*, and the two facsimile pages, one added by Beer to his book, the other published by E. Chatelain in his *Paléographie des Classiques Latins* (*livraison* 10). Much may be learned by study of these two pages.

For the interpolated MSS. (ω), to which class all complete MSS. except *P* belong, I have studied all the collations accessible to me, especially those contained in Jahn's larger edition (1851) and in Hosius' *Apparatus Criticus ad Iuvenalem*; and, besides examining other specimens of this class, I have collated three early MSS. which had not previously been used by editors.

The first of these, of the ninth century, is in the British Museum (15600 Add.). It is a good specimen of its class, with blunders of its own but agreeing in certain cases with *P* against other MSS.: thus it reads *fictus* (v. 70); *subito* (vi. 65); *Est pretium curae* (vi. 474); *hicte tricę* (vi. 565); *Haec* (vii. 41); *Nullo quippe modo* (vii. 100); *quid do* (vii. 165); *pravam* (viii. 33); *effudit* (x. 78); *pallidus mi* (x. 82); *sapiat qui* (xi. 81); *currunt* (xii. 77). These are the chief exceptions to the rule that the readings inserted by the corrector in the text or margin of *P*, or those found in ω (the two are very often identical), are found also in this MS. It has an important variant in Sat. viii. 148 where see note.

The second MS., also of the ninth century, is in the Library of Trinity College, Cambridge (O, 4, 11). It is imperfect, a quire of eight having been lost which contained from Sat. vi. 455 to Sat. vii. 95. This is a worse

MS. than the first, adhering still more closely to the inter-
polated tradition. A few marginal variants from some good
source (e.g. *auditor* for *adiutor* III. 322; *torrentis* for
torpentis IV. 43; *figuli* for *tegulae* IV. 135; *Iunco* for *Iunio*
XV. 27) are its most interesting feature; but a complete
collation of it would add little to Hosius' material.

The third MS., of the tenth century, is preserved in the
same Library (O, 4, 10), and is a document of much greater
value and interest: indeed I believe that, though vastly
inferior to *P*, it ranks among the very best of the other
MSS. *T*, as I have called it, gives for the first time MS.
authority for the true reading in Sat. VI. 13, and in a
number of cases, recorded in the *apparatus* to this edition,
confirms the reading of *P*. It obviously belongs to the same
family as *V* and *B* of Hosius but is much better than either
of them, and perfect while they are both mutilated. I hope
soon to publish a full collation elsewhere.

No emendation of my own is printed in the text, but a
certain number are suggested in the notes: a list will be
found in the Index. They are not mere guesses, but based
on some peculiarity in the reading of *P*, or on the agree-
ment in error of *P* and *T*, or on some hint contained in the
ancient Scholia. The punctuation has been dealt with
freely; but when an alteration has been made that affects
the sense, the reasons for alteration have been explained
in a note.

As to the Commentary, my endeavour has been to make
it explanatory rather than illustrative: there was no need
to do again what had been done once for all by Professor
Mayor. Nevertheless, as Professor Mayor has not yet
published his collections on the Sixth Satire, I allow my-
self some freedom of citation there; also, as I print, after
Bücheler, many new readings from *P*, these admit of, or
require, confirmation from other writers; and, apart from
these special cases, every scholar knows that an apt illus-
tration is often the best of comments. In order to under-

stand my author as fully as possible, I have read through a large part of the extant literature between Plautus and Juvenal; but I have tried not to quote from other authors more than was necessary. In the case of a writer so difficult and obscure as Juvenal often is, much space is required merely for explanation. Whatever seemed to me to need explanation, I have tried to explain. But every commentator must feel the truth of what Johnson says in his *Preface to Shakespeare*: 'it is impossible for an expositor not to write too little for some and too much for others: he can only judge what is necessary by his own experience'. Johnson speaks elsewhere of 'the commentator's rage for saying something where there is nothing to be said': but this foible of editors is perhaps more pardonable than the opposite practice, of saying nothing where it is quite certain that something ought to be said, if it amounts to no more than a confession of ignorance.

I have frequently called attention to those restrictions of metre which forced Juvenal, like other Latin poets, to substitute some metrical equivalent for what he really wished to say. The ancient critics—Seneca, for instance, and Quintilian—often make excuse for their poets on this account; but I doubt whether modern critics have attended sufficiently to the point. The natural Latin for 'sons' in the accusative is *filios*: it is obvious that Juvenal could not use this word: so, where *natos* is inappropriate, *iuvenes* has to do the business as best it can. But when a poet refrains, as Juvenal does, from using *filio*, or even *filium*, then various devices are necessary: apostrophe, the use of plural for singular and of diminutives which have no diminutive sense. Virgil uses elision, synizesis, and hiatus to force cretics into his verse, but Juvenal uses none of them for this purpose, the only exceptions being *plurimum* and *quantulum*, each used once by elision.

Some of the books which I have most frequently used without giving references to them, are as follows: the seven volumes of Smith's Dictionaries; Mommsen's History; Mommsen and Marquardt's *Handbuch*, especially Marquardt's *Privatleben*, a perfect model of what such a book should be; Friedländer's *Sittengeschichte Roms*; the same author's editions of Martial and the *Cena Trimalchionis*; Hirschfeld's *Untersuchungen*; Cagnat's *Épigraphie Latine*; Bouché-Leclercq's *Institutions Romaines*. In matters of syntax, and especially with regard to silver-age peculiarities, I have made constant use of Riemann's *Syntaxe Latine* and of the admirable *Études sur Tite Live* by the same author.

It remains to speak of works which deal specially with Juvenal. The chief of these is Professor J. E. B. Mayor's Commentary. I owe more to this than to all other sources put together; nor have I been able always to indicate, as clearly as I should wish, the amount of my obligation. For twenty years I have used his book constantly; and some years ago, when preparing a course of lectures on Juvenal, I wrote, together with other material, extracts from Professor Mayor's notes on the earlier Satires, in an interleaved Teubner text; this formed the nucleus of the present commentary. I also owe much to Professor L. Friedländer's edition, but not more, I think, than to his other works mentioned above, especially the *Sittengeschichte*: the Indices, also, to his editions of Martial and Juvenal are a real boon. When his edition was published, I had prepared more than half my commentary for the Press; but from that time onward I constantly kept his book before me, and was also able to make some changes in the proof-sheets of what I had already written. Weidner's edition, which I had been consulting before with little advantage, I discarded almost entirely when I had Friedländer before me. The only other edition which I have regularly used is that of the late J. D. Lewis: its great merits are good sense and power of apt quotation. I have studied the papers on Juvenal in Madvig's *Opuscula* and, in almost every case, accepted his conclusions. I must also mention a large number of articles and notes by different scholars in the philological Reviews, English and German—especially the series of papers by Bücheler in the *Rheinisches Museum*. My obligations to all these authorities are, I believe, acknowledged on any passage where I accept their views.

Finally, I owe a special debt of gratitude to Mr W. T. Lendrum, Fellow of Gonville and Caius College. He has read nearly the whole of the commentary in proof; and almost every page bears marks of his fine scholarship and exact knowledge of Roman institutions under the Empire.

The corrections introduced into successive reprints are mainly due to the edition of Professor Housman.

J. D. DUFF

CONTENTS

INTRODUCTION

I

LIFE OF JUVENAL

Of the life of Decimus Junius Juvenalis very little is certainly known. The materials for a biography are scanty and are mainly derived from sources of doubtful character. Again, when the source happens to be above suspicion, it is by no means certain that the information thus supplied really refers to Juvenal himself. We know that Juvenal was of Italian birth, lived at Rome, and wrote satire during the first thirty years of the second century. But when more precise detail is asked for, it is necessary to distinguish between ascertained fact and more or less plausible hypothesis.

Our knowledge of the lives of Latin authors, apart from the evidence of their own writings, and incidental notices in contemporary or later literature, is drawn from fragments of a lost work by Suetonius. This was a biographical history of Roman literature, entitled *De Viris Illustribus*. The excellent lives of Terence and Horace preserved in some of their manuscripts, are taken from this source; and the meagre notices of other writers, such as Lucretius, including little more than the name and dates of birth and death, are extracts from the same work, which were added by Jerome to his Latin translation of the Greek Chronicle of Eusebius. This Chronicle was compiled about A.D. 328 and translated fifty years later.

Juvenal lived too late to have Suetonius for his biographer. It is probable that the *De Viris Illustribus* was published before A.D. 114. For the younger Pliny, who died in that year, was not one of the authors included in the work; and it is highly improbable that Suetonius, if his book was written after this date, would not have mentioned so distinguished a personage and so intimate a friend as

Pliny. But we know that Juvenal was still living fourteen years later. Further, Juvenal tells us little about himself; and he is mentioned by none of his contemporaries except Martial. (The information gleaned from his own satires and from Martial is considered below.) The next mention of him occurs in Lactantius, two hundred years later. Nor is this long silence surprising; for classical literature came to an end with his death. But it is surprising that his name never occurs in the Letters of Pliny. These letters appeared between A.D. 97 and 110, to which period Juvenal's first book must be assigned, and they contain many references to literary events. Yet Juvenal is never mentioned. It is perhaps not uncharitable to suppose that if Juvenal had praised Pliny, the passage would have been preserved in a letter, together with some information about the satirist. Their silence may be due to mutual want of sympathy; for it is clear that Pliny belonged to the Lord Chesterfields and Sir William Temples of literature, while Juvenal was one of the Johnsons and Swifts.

The Ancient Biographies

As we have no biography from Suetonius, we must gather what information we can from other sources. The first of these consists in the anonymous biographies attached to many of the interpolated manuscripts of Juvenal.

These Lives are very numerous: at least twelve have been preserved. But they are generally of slight authenticity and value: it is easy to see how this or that detail is drawn from no other source than a passage, perhaps misunderstood, of Juvenal himself. One Life, lately discovered, asserts that he was born in A.D. 55, and gives the names of his parents; but the inference to be drawn from this unsupported statement in a manuscript of the fifteenth century, is not that the writer had access to any special information, but that he felt the need of it and was willing to invent it himself. In spite of some variations in detail between the different Lives, only one need be taken into account, as it is certain that all the others are derived from it.[1] It runs as follows:

VITA D. IUNII IUVENALIS. Iunius Iuvenalis, libertini locupletis incertum est filius an alumnus, ad mediam fere aetatem declamavit animi magis causa quam quod se scholae aut foro praepararet. deinde paucorum versuum satura non absurde composita in Paridem pantomimum poetamque eius semenstribus militiolis tumentem genus scripturae industriose excoluit. et tamen diu ne modico quidem auditorio quicquam committere est ausus. mox magna frequentia magnoque successu bis ac ter auditus est, ut ea quoque quae prima fecerat inferciret novis scriptis [VII. 90]

> quod non dant proceres, dabit histrio. tu Camerinos
> et Bareas, tu nobilium magna atria curas?
> praefectos Pelopea facit, Philomela tribunos.

erat tum in deliciis aulae histrio multique fautorum eius cottidie provehebantur. venit ergo Iuvenalis in suspicionem, quasi tempora figurate notasset, ac statim per honorem militiae quamquam octogenarius urbe summotus est missusque ad praefecturam cohortis in extrema parte tendentis Aegypti. id supplicii genus placuit, ut levi atque ioculari delicto par esset. verum intra brevissimum tempus angore et taedio periit.

This Life is itself probably a compilation dating from the fourth century: it was, if G. Valla may be trusted, prefixed to the commentary of Probus which is discussed below in connection with the Scholia. It is remarkable that a document professing to be a biography of Juvenal does not give us the date of his birth and death. The allusion to Paris is the only reference to contemporary persons or events. For solving the problem of Juvenal's chronology, the document offers no assistance; rather it increases the difficulties. However, the first sentence is superior to the rest in style, and also conveys information which cannot have been extracted from the satires. There is therefore no reason to reject this tradition, that Juvenal was 'the son, or adopted son, of a rich freedman, and practised rhetoric till

[1] Bücheler (ed. 1893), p. 234.

about middle age, more as an amusement than as a serious preparation for teaching rhetoric or pleading in the courts'. The latter statement is abundantly confirmed by the rhetorical tone which pervades his compositions. Such an occupation, though strange to us, would be natural enough to a Roman. The practice of speaking in the rhetorical schools was continued by many throughout life as an intellectual pastime, and even by men who had no intention of putting their skill to practical use in the courts. It was one of the accomplishments of a gentleman to have this power.[1] Thus we learn from M. Seneca that Ovid used constantly to practise speaking in the schools, and was considered a skilful debater.[2] A man must, of course, have had some means, in order to spend his time in a pursuit which paid him nothing. But the Biography makes it probable that Juvenal had some expectations of wealth, though it seems likely, as we shall see below, that they were disappointed.

The remainder of the Life is inferior in style; and the statements there made are suspicious. They seem to rest on conclusions drawn from the satires,[3] combined with more or less credible traditions. One exception, however, must be made. Juvenal himself never alludes to his banishment; and a tradition to this effect is found in all the Lives, in several Scholia, and also in later writers, e.g. Sidonius Apollinaris in the fifth century.[4] His punishment is

[1] Cf. Mart. II. 7. I *declamas belle, causas agis, Attice, belle.*

[2] M. Seneca *Controv.* II. 2 (10). 9 *tunc autem cum studeret habebatur bonus declamator* [*Naso*]. Notes of an argument by Ovid then follow.

[3] Thus compare *semenstribus militiolis tumentem* with Juv. VII. 88, 89.

[4] *Carm.* IX. 269 (Ovid's fate is compared to Juvenal's)

> *non qui tempore Caesaris secundi*
> *aeterno incoluit Tomos reatu,*
> *nec qui consimili deinde casu*
> *ad vulgi tenuem strepentis auram*
> *irati fuit histrionis exul.*

ascribed to different emperors, and the place of his exile differs, in the different authorities. Some say he was banished by Trajan to Scotland, others by Domitian to Egypt; according to most he died in exile; one tells us that he returned to Rome and died of grief on finding that his friend Martial had gone home to Spain. But all agree that he was banished, and that the cause of offence was his mention of the actor Paris in Sat. VII. 87–92.

Now the evidence of the satires, which we shall consider presently, makes it incredible that Juvenal spent a great part of his later life at a distance from Rome; and common sense tells us that a man of eighty cannot have been sent to an important military post on a distant frontier. But it is also difficult to believe that a tradition so universal is not based upon some fact. That Juvenal does not mention it himself, is not surprising. On the other hand, the statement deserves the more credit, because it cannot be extracted from any passage in the satires. There is some independent evidence, as we shall see, that Juvenal served in the army; and it is possible that he was appointed, long before he was eighty, to some very distant command, and that this gave rise to the story of his exile. But it is possible, too, that he was really banished for a short time during Domitian's reign, and also that the cause assigned is the true one. But, if this be so, the offending verses of the seventh satire must have been written many years before the satire itself was published in its present form.[1] Such a practice on Juvenal's part is borne out to some extent by the Life quoted above; but it is certain that both Paris and Domitian were dead long before the satire was published. If, then, we are not justified in rejecting the tradition of Juvenal's banishment, still the date and place of his exile must be considered as quite uncertain.

One other statement of the anonymous biographer has just been referred to—that which implies that in the satires,

[1] See the following section.

as we have them, there are passages inserted from earlier and less successful attempts. If this means (and apparently it must) that Juvenal wrote and published, or at least recited, satires before the death of Domitian, the chronological problem, obscure enough already, becomes further complicated. If the seventh satire refers to Domitian as a patron of literature (this is Nettleship's opinion), while the fourth satire refers to his death as past history, no safe inferences can be drawn from the order in which the satires now stand.

But the statement is ill-authenticated, and involves very serious difficulties. On what subject was it safe to write satire under Domitian? To attack the Emperor, or his courtiers, or public institutions, was out of the question. Is it credible that Juvenal stooped to flatter like Martial? Again, how is it that Martial, an intimate friend, writing to Juvenal from Spain about A.D. 101, while describing generally his friend's course of life, says not a word of satires? The biographer's statement is probably a fiction. It is dealt with incidentally in the two following sections.

The Evidence of the Satires

We get upon surer ground, when we turn to the satires themselves, and examine how far we can ascertain their dates by internal evidence.[1]

They were not all published together: no book of the ancient form could have held them all. Nor were they published singly; but like many works of Latin authors, such as the satires and epistles of Horace and the epigrams of Martial, they were published in 'books' (*libelli*) at different dates, each book containing somewhat less than a thousand lines.[2] This division into books is preserved in

---

[1] The dates here given are those of Friedländer.

[2] Juvenal speaks in the first satire (l. 86) of *noster libellus*: the word could not be used either of the first satire or of the whole sixteen, but is quite appropriate of the first five together.

many manuscripts; and the grammarians quote Juvenal by it: e.g. *Iuvenalis in secundo* refers to the sixth satire. Hence, if we can date any satire, the date will hold good approximately of the whole book which includes it.

The first book contains five satires; and the first satire (which, serving as a preface, was probably the last written) contains an allusion[1] to the condemnation of Marius Priscus for misgovernment in Africa. Now we know from Pliny, who was prosecutor in the case,[2] that the date of this trial was A.D. 100. It follows that the book was published after this date, though how much later is, within limits, a matter of conjecture. Thus Domitian had been dead at least four years before Juvenal's first book was published; and, further, the 'dancer' Paris was put to death by Domitian in the year 83; so that neither emperor nor actor was in a position to resent allusions to them occurring in a satire of much later date. The second and fourth satires of this book contain attacks on Domitian,[3] which could not conceivably have been circulated, even privately, before 96, the year of Domitian's death; and, in fact, a passage in the latter[4] refers plainly to his assassination.

The second book contains only one satire, but that a very long one, the sixth. In it there are mentioned, as topics of the day, a comet boding disaster to the Kings of Armenia and Parthia, and a great earthquake in the East accompanied by inundations. Now a conspicuous comet was visible at Rome in November of 115;[5] and Trajan began his campaign against Parthia in the following year. (It is true that the Armenian campaign was over in 114; but this slight inaccuracy on Juvenal's part need not disturb us.) Also, there was a famous earthquake at Antioch in December of 115, in which Trajan himself had a narrow escape

[1] Ll. 49, 50.   [2] Pliny *Epp.* II. 11 and 12.
[3] II. 29–33; IV. 37, 73, 84 etc.
[4] IV. 153.
[5] This date depends on the observations of Chinese astronomers: see Friedl. p. 9.

from death. It seems a certain inference, that Juvenal's second book was published not earlier, and not much later, than A.D. 116.

The third book contains the seventh, eighth, and ninth satires. The first of these begins with the statement that the prospects of literature, especially of poetry, depend entirely upon the patronage of 'Caesar'. The question thus arises, and has been much discussed, which of the emperors is spoken of. Now Trajan never returned alive from his Eastern campaigns but died in Cilicia in 117. What more probable than that this compliment was paid, on his accession, to Hadrian, the first ruler since Claudius who was 'a man of parts' and took a serious interest in literature? Indeed, he was a poet himself, and his 'dying address to his soul' has been famous in all ages.[1] Some think Trajan is referred to, in spite of the obvious chronological difficulty, and appeal to a passage in Pliny,[2] where it is said that certain *studia* have got a new lease of life under Trajan. But, as Friedländer shows, the context proves that Pliny has in mind, not poetry, but philosophy and rhetoric. Others go further back than Trajan for this imperial patron of literature. Nettleship[3] argues for Domitian. If this were so, it would show that the satires are not arranged in anything like chronological order: the seventh must have been written long before the fourth. It is probable that the emperor addressed, whoever he was, had just succeeded to the throne; for, in the days of patronage, men of letters clung with strange persistency to the belief

[1]
> *animula vagula, blandula,*
> *hospes comesque corporis,*
> *quae nunc abibis in loca,*
> *pallidula, rigida, nudula?*
> *nec ut soles dabis iocos.*

[2] *Paneg.* c. 47 *quem honorem dicendi magistris, quam dignationem sapientiae doctoribus habes! ut sub te spiritum et sanguinem et patriam receperunt studia!*

[3] *Journal of Philology*, XVI, 55 ff.

that each new monarch must prove a Maecenas. Thus
George the Third was certainly a bad judge of literature,
yet we read in Boswell (II. 229, ed. 1874): 'the accession of
George the Third to the throne of these kingdoms opened
a new and brighter prospect to men of literary merit'.
But Domitian became emperor in A.D. 81: so the question
arises, whether the evidence is sufficient to establish, what
is in itself highly improbable, that the seventh satire was
written at least twenty years before the first. Nettleship's
arguments are not successful in proving this.[1] There is
really no reason to doubt that the seventh satire was
addressed to Hadrian, and that the third book was pub-
lished soon after the new emperor arrived at Rome from the
East in August A.D. 118.

The fourth book contains the tenth, eleventh, and twelfth
satires; but no allusion occurs in them which serves to fix
the date of publication.

The fifth book, containing the rest of the satires, has two
passages, in the thirteenth[2] and fifteenth[3] satires, which
prove that the former was written in 127 and the latter at
least a year later.

These dates show that Juvenal was writing satire be-
tween A.D. 100 and 130. Friedländer dates the first book
between 112 and 116, thinking it improbable that there
was a much longer interval between the first two than
between any other consecutive books. But a somewhat
earlier date seems to be suggested by the incessant re-
ferences to the reign of Domitian. The muzzle was re-
moved in 96; and it seems unlikely that Juvenal refrained
from speech for nearly twenty years. Pliny and Tacitus lost

[1] Thus he argues from *novorum* | *fatorum* (VII. 189) that Juv.
speaks of Quintilian's good fortune as quite recent; but
novorum there means 'strange'. The argument based on
Martial's intimacy with Juvenal is considered below. For the
anachronisms of *Sat.* VII, see p. xxxiv. ff.

[2] L. 17: see note there.

[3] L. 27: see note there.

little time in liberating their souls about Domitian: the *Panegyricus* was spoken in A.D. 100; the *Agricola* was published in 98. The Biography told us that Juvenal began writing satire about middle age. A man reaches middle age soon after forty. Hence we may conclude that Juvenal was born between A.D. 60 and 72. He himself in his first satire[1] speaks of his youth as already past, and in the eleventh[2] uses the language of an old man. The date of his death is quite uncertain, except that it cannot have been earlier than 128.

When we turn to the satires themselves, not for dates, but for facts in Juvenal's life, we find that there is little to be learned. In this respect, as in many others, he does not follow the example of previous satirists. The satires of Lucilius, as we know from Horace,[3] gave a complete picture of their author's life; and Horace's own satires and epistles are full of autobiography. But Juvenal's manner is solemn and didactic, not chatty and anecdotic. We learn from himself little more than this: that he had the usual education of the higher classes (I. 15–17); that he lived from early youth at Rome but went for holidays to Aquinum, with which he had some connection (III. 319); that he inherited a small estate (VI. 57) and had a small farm at Tibur, probably distinct from the other (XI. 66); that he had also a house in Rome where he entertained his friends in a modest way (XI. 190); that he was not a professed follower of any philosophic school (XIII. 121–3); and that he had been in Egypt (XV. 45).

Two tolerably certain inferences may also be drawn from what he says or leaves unsaid: that he was unmarried, and that he understood from experience the hardships of poverty and dependence. It appears that he did not

---

[1] L. 25 *quo tondente gravis inveni mihi barba sonabat.*

[2] L. 203 *nostra bibat vernum contracta cuticula solem.*

[3] Hor. *Sat.* II. 1, 32 ff. *quo fit ut omnis | votiva pateat veluti descripta tabella | vita senis* (the motto on the title-page of Boswell's *Johnson*).

inherit much of the wealth of the *libertinus* of Aquinum. The circumstances of Martial's life were very similar: he also was a man of letters living in Rome and probably unmarried; and, though the owner of a small estate, he has never done complaining of his poverty and the hardships of a client's life.

The Inscription at Aquinum

It was said above that Juvenal may have served in the army. He had seen something of the world, as a Roman officer naturally would. He had certainly visited Egypt;[1] and there are references to Britain,[2] from which it has been inferred that he had seen our island. The Biographies also speak of him as banished to a distant country to hold a military command. But those who believe in his military service, rely chiefly on an inscription found at Aquinum. The inscription was probably intended for an altar and was engraved on a marble stone, which is now lost. When it was sought for in 1846, the inhabitants could only show the place where it had been. However, it had been previously copied by more than one competent person of credit and is accepted as genuine by the best judges. It runs as follows: the letters in italics are supplied by editors (*Corp. Inscr. Lat.* x. 5382):

Ce*re*ri sacrum
*D. Iu*nius Iuvenalis
trib. coh. *I* Delmatarum
II vir quinq. flamen
divi Vespasiani
vovit dedica*vit*que
sua pec.

---

[1] xv. 45.
[2] ii. 160 (the conquest of the Orkneys in 84); xiv. 196 (the forts of the Brigantes); iv. 141 (the oysters of Richborough); x. 14 (the whales of the Atlantic); ii. 161 (the short nights in Britain).

'This offering was vowed and dedicated to Ceres, at his own charges, by Junius Juvenalis, tribune of the first cohort of Dalmatians, *duumvir quinquennalis*,[1] *flamen*[2] of the deified emperor Vespasian.'

Does this inscription refer to Juvenal himself or to a member of his family? We have seen that he mentions (III. 319) his connection with Aquinum, and the goddess to whom the offering was made. Also, if the son of a rich freedman, he was a likely person to hold the highest municipal magistracies. Again, the mention of the worship of Vespasian points to the reign of the Flavian dynasty, i.e. the end of the first century. It is certainly possible that Juvenal served in the army, and reached the rank of tribune (which carried with it the privileges of the equestrian order) in his younger days before he turned his attention to satire. And this hypothesis has generally been accepted.

It is, however, beset with difficulties. The first, and best, sentence of the Biography says nothing of a military career but rather seems to exclude it by the account there given of Juvenal's occupations until middle age. Again, Juvenal, like Persius, does not generally, and especially in the satire devoted to the subject, speak with favour of a military life. Again, if Juvenal was rich enough to fill these municipal offices, which were a considerable burden upon their occupants,[3] how can we account for the complaints of poverty, whether of clients or literary men, with which the

---

[1] The *duumviri* in the provincial towns corresponded to the consuls at Rome; the *quinquennales* were the *duumviri* of every fifth year, and had duties like those of the censors, having to draw up a list of the local senators (*decuriones*) and citizens.

[2] Julius Caesar was the first mortal man who had a *flamen* to conduct his worship: this was Mark Antony: cf. Cic. *Phil.* II. 110; Suet. *Jul.* 76. The distinction was granted to all the deified Emperors.

[3] These magistrates were not only unpaid, but had to pay a considerable sum, called *honorarium*, as a contribution to the town treasury. In the second century, it became difficult to find candidates for the offices.

satires abound? Again, the evidence of the satires goes to prove that Juvenal lived habitually in Rome, at least after the end of the first century: is it not surprising that the first satire, which represents the author as familiar with all the conspicuous figures in the crowded streets of Rome and with all the scandals attached to them, should be written by a country gentleman with a dignified position and a comfortable fortune? The language of the satires is the language of a poor and disappointed man. Lastly, it must be noticed that the inscription did not contain the poet's *praenomen*, which is known from the Lives and Scholia to have been *Decimus*.

The conclusion is that here again certainty is beyond our reach. My own belief is that the local magnate of Aquinum was not the poet himself, but a kinsman by blood or adoption. But it is possible that the fact is otherwise; and there are allusions in the satires which are most easily explained by the hypothesis that Juvenal was for a time an officer in the Roman army.

The Evidence of Martial

We shall consider next what light is thrown on Juvenal's life and occupations by the poet Martial. There is no positive proof that Martial's Juvenal is our Juvenal; but their identity is highly probable and has generally been accepted by scholars as a matter of course. Though Juvenal never mentions Martial, directly[1] or indirectly, Martial speaks of Juvenal as a very intimate friend and addresses two epigrams to him personally. One wonders what the satirist thought of Martial's flattery of Domitian. Now the 'books' of Martial's epigrams can be accurately dated by internal evidence. The seventh, in which Juvenal is twice mentioned,[2] was published in the autumn of 92; and the twelfth, which contains the last mention of the satirist,[3]

[1] Metre alone would make it impossible for Martial's name to occur in Juvenal's verse.

[2] VII. 24 and 91.

[3] XII. 18.

was published in 101 or 102 after Martial had returned to Spain.

The epigrams prove that Juvenal was living, and had been living for some time past, at Rome in the year 92, and that he was again at Rome in 101 or 102. Thus it is possible that he may have been in exile between these dates: he is not mentioned during the interval by Martial; and it is known that Nerva recalled many who had been exiled by Domitian.

Two other points are to be noticed in the epigrams. In 92 Martial applies to Juvenal the epithet *facundus*. Does the epithet show that Juvenal was known to him as a poet, or as a rhetorician? Neither inference can be drawn with certainty, as Martial elsewhere applies the word both to Virgil and to Cicero. But, if Juvenal had written no satires at that time and was only known to Martial as a student of rhetoric, the epithet is perfectly appropriate. On the other hand, if Juvenal had written satires before this date, or even before 101, it is inconceivable that Martial, so ready to praise far humbler literary efforts of his friends, should not mention them. This is another proof that Juvenal did not publish satires till after the death of Domitian.

In the second place, the last epigram (XII. 18) represents Juvenal's life very much as he represents it himself: Martial contrasts his own easy days and restful nights in Spain with the annoyances Juvenal continues to undergo while toiling up the hilly streets, in the noise and heat of Rome, on his way to pay court to the rich and powerful.

It may be added here that there are some obvious imitations or reminiscences of Martial in the satires, especially the earlier ones, and also that there is a remarkable correspondence between the work of the two writers, 'not only in their views of literature, but in the subjects they treat, the persons they mention, their language and expression, and their general tone'.[1] Ample evidence of this

---

[1] Nettleship, *J. Philol.* XVI, 47.

will be found in the notes to this edition; but it seems
unnecessary, in order to account for this likeness, to
suppose, as Nettleship does, that they were in the habit
of working together. The facts which have been stated
above, go to prove that Martial ceased to write about the
time when Juvenal began; and the resemblance will not
seem more than can be accounted for, if we believe that
Juvenal, having already a thorough knowledge of Martial's
epigrams, began to direct his satires against the same period
and persons whom Martial had already riddled with his
lighter artillery.

A caution may be useful against the practice, which was
carried too far by the older editors, of supposing that,
wherever the same name occurs in the two poets, they
both refer to the same person. This is a mistake. For
Martial, as we know from himself,[1] consistently used
fictitious names in his satirical epigrams. Juvenal's practice
was quite different. He only attacked persons who were not
in a position to resent it—those who were no longer living,
or had been condemned by a judicial sentence, or were of
no social importance.[2] Hence he mentions many persons
of whom we know nothing from any other source; but
there is no reason to believe that he uses any fictitious
names, with the possible exception of the correspondents
(Postumus, Ponticus, Calvinus, Gallius) to whom some of
the later satires are addressed.

[1] Mart. I. praef.; II. 23; IX. 95; X. 33. For the practice of
combining real and fictitious names in satirical writing, cf.
Pliny, *Epp.* VI. 21. 5 (of Vergilius Romanus who had composed
a play in imitation of the Old Comedy) *insectatus est vitia,
fictis nominibus decenter, veris usus est apte.*

[2] See below, p. xxxiv ff.

II

THE *SATURA* BEFORE JUVENAL

It was the boast of Roman writers and critics[1] that satire was a genuine national creation, that they had invented for themselves, and not borrowed from Greece, at least one important kind of literature. The claim of originality must be allowed, although satire was much influenced, at more than one period of its history, by Greek example.

Roman satire, however, has nothing to do with the Satyric drama of Athens. The word *satura* probably means 'medley', being a feminine noun derived from *satur*, like *dira* 'a curse' and *noxia* 'a hurt'. It seems to have been applied originally to a dish, containing various ingredients, and, by metaphor, to a law, comprising miscellaneous enactments. It is generally supposed that the term, by a similar metaphor, was applied to a form of literature, treating of various subjects and written in a mixture of prose and verse; and it is apparently for this reason that Juvenal speaks[2] of *nostri farrago libelli*.

The word itself, however, is much older than the kind of literature to which it became eventually restricted. For the early history of the word, our chief guide is a passage in Livy,[3] where, tracing the rise of scenic representations at Rome, he incidentally refers to *saturae*, as one of many native Italian forms of the drama. He describes it as an improvised dialogue of rude and unpolished verse, which was first raised to the rank of a dramatic representation, by union with certain Etruscan performances, consisting of music and dancing only. In this improved form, *saturae*, supplemented by music, continued to hold the stage at Rome, until they were superseded by the Greek play introduced by Livius Andronicus about 240 B.C. Livy's

---

[1] Horace, *Sat.* I. 10. 66: *Graecis intacti carminis*; Quintil. x. 1. 93: *satura quidem tota nostra est.*

[2] I. 86.   [3] VII. 2. 4–13.

words seem to imply that a *satura* differed from a regular play in having no plot. Dialogue it possessed from the beginning; but the scenes, of which it was composed, had no connection with each other. It consisted, apparently, of a succession of scenes, drawn at random from common life, in which the Fescennine spirit of rude and offensive banter disported itself with little pretensions to art and less to decorum. Supplemented by musical accompaniment borrowed from Etruria, and appropriate gesticulation, this form of art held its ground until it was banished from the stage by the higher form introduced by Livius Andronicus from Greece. Thus, the essential features of the original *satura* were dialogue and the absence of a plot; and these characteristics it still preserved, when it was driven from the stage and transferred to paper by men of letters.

Of these, the earliest was Ennius (239–169 B.C.). His *saturae* have perished; but we may gather, from certain allusions to them, that in his hands Greek influences added new features to the originally dramatic *satura*. Quintilian tells us that they contained a dialogue between Life and Death; and Gellius quotes two lines from a *satura* in which Ennius versified the fable of the lark and her young.[1] The former notice suggests the influence of Greek popular philosophy or moralising which is so characteristic a feature of later satire; and the latter reminds us of the fable, introduced by Horace into one of his satires, of the town and country mouse.[2] Thus Ennius, under the influence of Greek studies, seems to have added two new ingredients to the *farrago* of satire—philosophy and the fable, which remained more or less permanent features of the later forms of this branch of literature.

In the hands of Lucilius (180?–103 B.C.), the *satura* did not lose its character as a brief narrative or picture of life, with an element of dialogue, tinged here and there with philosophical reflection, and full of autobiographical detail.

---

[1] Quintil. IX. 2. 36; Gellius, II. 29.   [2] *Sat.* II. 6. 80–117.

But it further underwent an entirely new influence, that
of the Old Comedy of Athens. Lucilius was the first writer,
who impressed upon satire that character of invective
which ultimately, in the hands of Horace, Persius, and
Juvenal, became most essential to it. It is probable that
the preponderance, which invective assumed over the
other more kindly ingredients of satire in the work of
Lucilius, was largely due to the character of his age. He
wrote at a time when the corruption and incapacity of the
governing class were growing every year more flagrant;
and his own sympathies lay with the more moderate
section of the reforming party, represented by Scipio and
Laelius. Thus, as written by Lucilius, satire becomes
mainly the scourge of incapacity in high places: 'Lucilius
flogged the town', says Persius.[1] He is never tired of
deploring the decay of old Roman virtue, and the growth
of luxury, avarice, and selfish ambition. It is in this sense
that Horace declares Lucilius to depend entirely on the
Old Comedy.[2] We know, however, that his satires dealt
with many different subjects and were written in many
different metres. The fragments still extant would not give
a high idea of his genius: but we know that his country-
men thought him the chief master in this style, and that
even in Quintilian's time he was considered by some to be
not only the greatest satirist but the greatest poet who had
used the Latin language.[3]

The next writer of *saturae*, whom we must mention, is
Marcus Terentius Varro (116–28 B.C.), the learned
antiquary and friend of Cicero. He wrote 150 books of
saturae Menippeae, so called because they were an imita-
tion of the works of Menippus, a cynic philosopher of
uncertain date, who reappears in the pages of Lucian.
Here again we see the influence of Greek literature. Of
Varro's *saturae* only scanty fragments survive: but these
are sufficient to show that our loss is great. An interesting

[1] I. 114.   [2] *Sat.* I. 4. 6.   [3] Quintil. X. 1. 93.

and appreciative account of them is given by Mommsen,[1] who finds so little to praise in Latin literature. They were, as Quintilian says,[2] *saturae* of the genuine old kind, a medley of both prose and verse, giving a series of pictures of contemporary life at Rome, whether of social, moral, literary, or philosophical interest. We can see also that there was a decided personal note in them, an element of autobiography.

The last statement is eminently true also of the satires of Horace (65–8 B.C.). But here there is no prose and no variety of metres; there is also a considerable modification of tone. Political invective was impossible for a man in Horace's position: it remained for him to attack with reproof, or more often with ridicule, the social mistakes or ethical shortcomings of his contemporaries. But invective is by no means the staple of these satires. Apparently Horace thought that Lucilius had gone too far in this direction. The form is often dramatic, thus following the tradition of the ancient *satura*: indeed in the second and maturer book there is nothing which is not either a scene or a conversation.

The next name in the history of satire is that of Persius (A.D. 34–62). But, though there is a highly peculiar and personal flavour in his strange genius, he was still in the imitative stage at the time of his early death, and his form is entirely derived from his predecessors, especially Horace.

The *saturae* of Petronius, whose author is believed to be the Petronius who died in A.D. 66, are of a very different character. The work originally consisted of a number of books; the extant portions are extracts from the fifteenth and sixteenth. The manner of Varro is to some extent revived, prose alternating with verse in various metres. But the chief innovation is that the *satura* has now assumed the form of a connected narrative, describing the adventures

---

[1] *Roman History*, IV, 594 ff.   [2] Quintil. X. I. 95.

of a Greek freedman in various towns of Italy, related by himself. The fact that only certain episodes are preserved, makes it impossible to trace exactly the evolution of the story. The work of Petronius is highly dramatic; and the pictures of life and character are drawn by a master's hand. For easy humour and graphic realism, there is nothing in Latin, nothing even in Greek, literature that can compare with the 'Dinner of Trimalchio', which is the subject of the longest and most important fragment now extant. It would hardly be extravagant to compare the genius of Petronius with that of Sterne, alike in its excellences and also in the stains by which it is disfigured.

We see then that the name *satura* is applied to compositions of very different kinds. There seems to be little in common between the poems of Lucilius and the romance of Petronius. Yet, as Nettleship points out, there are two characteristics which reappear in each successive writer of *satura*, and which are not entirely lost even in Petronius or Juvenal. First, there is always a marked personal element: the writer addresses the general public or an imaginary companion; hence the frequent occurrence of dialogue. Secondly, the *satura* never contains a regular plot: it consists of talk, flowing at will and not bound by the laws of more formal literature.

Of the specimens still extant, Varro's fragments show most clearly the mixture of verse and prose: from the same *satura* there are fragments in prose and in half-a-dozen different metres. In Petronius the narrative is written in prose; and the poetry inserted in the narrative is generally placed in the mouth of a speaker with some formal introduction. For variety of subject, the fragments of Lucilius are perhaps the best example. His work might have been called 'The Life and Opinions of Gaius Lucilius'. He simply put on paper the thoughts passing through his mind; and it is an accident, due to the times in which he lived, that his thoughts turned largely on public affairs and the misconduct of public men. The true modern parallel to Lucilius is to be found in such works of self-revelation as Montaigne's *Essays*. Beside the romance of Petronius we may set Borrow's *Lavengro*: it also has no plot, and consists of a succession of isolated scenes and dialogues. In both books the narrator is unchanged; and so a certain degree of continuity is secured. Of satire proper, the form which *satura* ultimately assumed, the classical English authors are Dryden, Pope, and Swift. But in English literature the development has been different; and a more faithful image of the ancient *satura*, its discursive, personal, and humorous character, and its freedom from rhetoric, will be found in the passages where Thackeray drops his narrative for a time and addresses himself directly to the reader.

III

JUVENAL AS A SATIRIST

The development of *satura* has now been traced from its first appearance as a rudimentary form of drama down to the romance of Petronius. It remains to consider how it was manipulated by Juvenal, the last and the most powerful of the Roman satirists.

But, before discussing Juvenal's literary method and ethical purpose, it is desirable to point out the remarkable difference between his earlier and later satires. The dividing line comes after the ninth satire, which is the last of the third book. This difference may be due partly to advancing years and failing powers; but, as there was no long interval of time between the third and fourth books, it is probable that Juvenal deliberately discarded some of the means on which he had formerly relied to produce his most striking effects.

The difference is seen both in form and substance. The

satires in the last two books are really letters, and not
satires at all: each is addressed to a friend; there is no
dialogue and little dramatisation. The style is different,
much less abrupt and elliptical. The sentences are longer
and more complicated; there is far more repetition. Nor is
the contrast less striking, when we consider the substance
of the later satires. They are moral essays: like the letters
of Seneca, they are nominally addressed to a correspondent
but deal in a general way with such questions as 'the
value of prayer', 'the desire of revenge', 'the influence of
parents', and so on. Illustrations, which are frequent, are
taken, not from the Roman streets, but from Greek history
and mythology. The first nine satires present a wonderfully
vivid picture of life at Rome at the end of the first century;
the last seven have a different object, and the notices of
contemporary events are merely incidental. Read the
fifteenth satire after the first, and the difference will seem
astonishing.

This unlikeness has given rise to a theory, originated
half in jest by Ribbeck,[1] that a number of what pass for
Juvenal's satires, were really written by an anonymous
forger of somewhat later date. If this *jeu d'esprit* were
worthy of serious refutation, it would be easy to show that,
in spite of general unlikeness, there are many minute
resemblances, which a forger would hardly have hit upon,
between the earlier and later satires. But a theory, which
would have us believe that a nameless bookseller's hack
wrote the tenth satire, carries its own refutation. It is
enough to say that authors, either ancient or modern, are
not confined to a single manner. If Juvenal had chosen to
call his later poems epistles, it is certain that Ribbeck's
theory would never have been given to the world. Juvenal
followed, perhaps consciously, the example of Horace,
who, in later life, discarded satire for epistle. The style of
Horace does not, indeed, like that of Juvenal, become more

[1] *Der echte und der unechte Juvenal* (Berlin, 1865).

prolix with increasing years; but the unlikeness between Horace's earliest satires and his epistles is as marked as that between the early and late satires of Juvenal. The latter preferred to call all his poems by the name of satires. Now the peculiar excellence of the *satura* was just this, that it was not bound, either in form or matter, by rigid rules; yet it was a decided innovation to apply the name to epistles.

Our next business is to consider how far Juvenal, especially in his first three books, conforms to the practice of his predecessors. He does not pose as an innovator on the established methods. In the preface to his work, he refers to both Lucilius and Horace, and professes to be carrying on their tradition.[1] Nevertheless, his treatment of the *satura* is, in many respects, peculiar to himself. Indeed, it is doubtful whether the words 'satire' and 'satirical' would bear exactly their present meaning if Juvenal had never written. Much in the same way Martial's genius has given to the word 'epigram' a shade of meaning which it did not necessarily convey before his time.

In the hands of Juvenal the prose and mixed metres of Varro and Petronius have entirely disappeared. He writes in hexameter verse, a metre employed largely by Lucilius and exclusively by Horace. But his management of the metre is quite different from theirs. Instead of the easy, conversational verse of Horace, we find the powerful but monotonous measure of Lucan and the later Roman Epic.[2] This change is of great significance: Juvenal rightly felt that the frequent elisions and irregular pauses, the roughness and occasional oddness of Horace's *Sermones* would be out of place in his own satires.

[1] I. 20 and 51.
[2] Two lines of Turnus, a contemporary of Martial's, who had a high reputation as a satirist, are quoted by the Scholiast on Juv. I. 71: *ex quo Caesareas suboles Lucusta cecidit | horrida cura sui verna nota Neronis.* The second verse is corrupt; but the first sounds quite like a verse of Juvenal.

Further, the element of dialogue which we have seen to be essential to the older *satura*, has almost disappeared. There is a fragment of dialogue in the first satire; the whole of the ninth is a real dialogue; and the third retains something of the ancient form, though it is in fact a monologue of three hundred lines put into the mouth of Umbricius. But generally Juvenal himself is the only speaker.

Some characteristics, however, of the *satura* still survive. It was originally a succession of isolated scenes; and what better description could be given of much of Juvenal's work? In the sixth satire, especially, the want of connection and even of consistency between the different topics of which it is composed, is so striking that some editors suppose that each paragraph must have been written independently and the whole strung together with little regard for sequence or coherence. It must be allowed that Juvenal is remarkably indifferent to what Quintilian calls *oeconomia*, the proper arrangement and disposition of his matter. Hardly any satire, except the third, has a consistent and satisfactory framework; in many he seems to have forgotten the beginning before he reaches the end; and in nearly all, the attentive reader is puzzled by irrelevant digressions, sudden transitions, and the unexplained absence of topics which he has promised to treat of. He is a marked example of the fault of his age, the tendency to sacrifice the whole to the parts. It is possible, however, that he would have pleaded in excuse for these faults the traditional freedom of *satura*.[1]

Thus there are great differences of form between the satires of Juvenal and those of his predecessors. But in matter the difference is not less marked. Lucilius 'lashed

---

[1] Some of the minor external characteristics of *satura*, e.g. the use of diminutives, are fully preserved in Juvenal. He also allows much greater variety at the end of the line than the epic poets do.

the town'; but he also discoursed on innumerable subjects,
from his own journeys to the anomalies of Latin grammar.
And Horace, who duly carried out his own precepts,
declares that the satirist must be terse, but above all
versatile—must change from grave to gay, must be now an
orator, now a wit, now a poet, and must reserve his
strength.[1] It will be seen at once how little this description
suits Juvenal. He is the most diffuse and the least versatile
of writers, and uses the force of a steam-hammer to crack
a nut. Again, Horace, as a satirist, does not stand on his
dignity: we are allowed, and expected, to laugh at him when
he is lectured by his slave or button-holed by the bore
on the Sacred Way. But this attitude of self-depreciation,
and all the other touches that gave humour, gaiety, and
kindliness to the *satura*, have disappeared in Juvenal.
The one thing we find in those poems which are really
satires, is vigorous and even violent declamation against
vices and vicious persons. The fact is that Juvenal was a
born rhetorician, who had cultivated his natural gift by
the practice of half a lifetime; and continuous rhetoric is
entirely opposed to the spirit of the genuine *satura*. To
sum up: while the later satires of Juvenal are not really
satires at all, he has, in his earlier work, discarded almost
entirely the peculiarities of metre, treatment, and tone,
which had been characteristic of this kind of literature and
may be seen surviving in the *Sermones* of Horace. The
'medley' is no longer a medley; for the invective, im-
ported into it almost accidentally by Lucilius, has over-
whelmed the other ingredients and leavened the whole
lump.

Juvenal probably felt, indeed he hints as much himself,
that he had not by nature such powers as would enable him
to rival his predecessors in their own line. To sketch living
types was the object of their art; and this was quite out of
his reach. 'From Juvenal we hear what people on particular

---

[1] *Satt.* I. 10. 9–14.

occasions have done; but we know nothing of their personality; he cannot draw a character, he cannot laugh. Think of Juvenal's Virro and then of Petronius' Trimalchio; the one is a figure cut out in paper, the other a living man.'[1] His motive also, at least his ostensible motive, was different: he professes no wish to please or to instruct, hardly even to reform; he tells us repeatedly in his introductory satire that his only source of inspiration is his burning anger against vice and crime. He was quite conscious that he was making satire do a work it had never done before—that, to use his own words, 'he was ignoring the limits and objects of former satirists, and that satire was putting on the high buskin' of tragedy or epic.[2] His innovations could not be better described.

Now moral indignation must command respect, especially when it is expressed with fearless indifference to consequences. But some critics have maintained that Juvenal's 'indignation' is the mere literary convention of a man with a gift for satire and a resolve to write it; and that his brave words exposed him to no risks, because the abuses which he attacked had ceased to exist. This raises two distinct questions which must be briefly considered. First, was Juvenal a hero like the Hebrew prophets, who rebuked wicked kings to their faces? Secondly, was Juvenal an honest man, or is his moralising attitude a mere pretence?

In seeking to answer these questions, we are again confronted by the difficulties of chronology; probability must be our guide. In the first place, Juvenal himself tells us that he intends to turn his satire against the dead, because it is too dangerous to attack the living.[3] And this is what he seems actually to have done. Nettleship truly says that 'Juvenal's manner is at times so unreal that it is impossible for the reader to be sure whether the poet is

[1] Nettleship, *J. Philol.* XVI. 65. [2] VI. 634 ff.
[3] I. 170 ff.

referring to contemporary events or only professing to do so'. This is just the reason that weakens Nettleship's argument for the early date of the seventh satire, based upon the persons there mentioned. Juvenal's idea of historical perspective is very peculiar. Thus in Sat. XIII (l. 157) he speaks of Gallicus as if he were then prefect of the city; but the satire was written about A.D. 127, and we know from Statius[1] that Gallicus was prefect in the year 89 and was then more than sixty years old. In the first satire there is still worse confusion: the mention of Marius's condemnation proves the date to be not earlier than A.D. 100; but, wishing to show by an example the danger of satirising living criminals, he actually chooses Sofonius Tigellinus, a notorious figure in Nero's reign, who was forced to commit suicide in the year 69. In view of such eccentricities as these, which might have been intended to mystify commentators then unborn, the difficulty of which Nettleship speaks is a real one. Yet the evidence of the satires, already considered,[2] makes it probable, if not certain, that all the emperors attacked by Juvenal, Domitian as well as the whole list from Tiberius to Otho, were dead before the attacks were published; and also that the living persons whom he assailed were for various reasons powerless to take revenge. But if there was little danger in this method, there was also little glory. Juvenal certainly did not borrow it from Lucilius or from Lucilius's master Aristophanes, who, after Cleon's death, refuses to trample any more on his familiar victim.

But it may be argued that, although Juvenal speaks of the dead as if they were still living, and of events of forty years before as if they were passing before his eyes, still his descriptions of the corrupt state of Roman society are true in the main of his own day. It is said that Juvenal's account of

---

[1] *Silv.* I. 4: the book was published in 89. For Gallicus's age, see l. 53.

[2] See pp. xiv–xvii.

Roman life agrees substantially with that given by Tacitus. But here we must remember that the historical works of Tacitus do not touch the second century; he intended to deal with the reigns of Nerva and Trajan in his old age,[1] but we do not know whether he lived to do so. If we turn for information to Pliny's Letters, which were written between A.D. 97 and A.D. 110, just when the earlier and fiercer satires of Juvenal were appearing, we find ourselves in a different world from that scourged by the satirist. There are still traces of the bad old times; but the notorious figures of Domitian's reign, who still survive, the Reguluses and Veientos, must blush to find themselves still living in such an era of public and private virtue. The emperor, Trajan, is a model of manly worth; the empress sets an example to all women. The rising generation is moral and industrious, with a taste for literature and a proper respect for Pliny. As the other picture was all shade, so this is all light. But if there is an element of convention in Juvenal's ferocity, there is certainly not less in Pliny's rose-colour.

However that may be, no sober critic will maintain that the social condition of Rome in the first century tended to maintain a high standard of morals. We see a city with no conceivable rival in size, wealth, and importance; a free population supported by the state, and finding their business in amusements, often of a brutal and inhuman kind; a multitude of slaves brought together from all parts of the world and avenging themselves unconsciously on society by the corruption of domestic life; and a number of freedmen combining enormous wealth with the tastes and habits of slaves. It is impossible to suppose that a virtuous emperor, even aided by a staff of industrious officials, could reform such a society, and that the vices attacked by the satirist existed only in his own imagination. Yet it is certain that the worst scandals which had disgraced the first century were banished, at least from public life, before

[1] Tac. *Hist.* I. I.

Juvenal published his first book: vice and folly, lust and cruelty no longer sat upon the throne; robbery and murder were no longer protected, and even encouraged, by the laws under which the informers had plied their infamous trade. But, in spite of this, the emperors and the informers are, next to the aristocracy, the chief objects of Juvenal's satire.

It must, then, be admitted that Juvenal is, in some degree, tilting against windmills. A man may do this in all sincerity, but it is not the business of a hero. I have often thought that some light is thrown on the position of Juvenal, when he began to write, by a passage in Pliny's Letters, where the writer explains his motives for attacking a notorious 'informer' in the senate. The passage runs: 'On the death of Domitian I reflected that here was a signal and glorious opportunity to punish guilt, to avenge misfortune, and to *bring oneself into notice*.'[1] It is certain that Pliny thought all these motives creditable, as the main object of his letters, which were carefully edited before publication, is to put his own conduct in a favourable light. Pliny was a wealthy senator and a distinguished pleader; Juvenal seems to have been needy and obscure. Yet the same motives may well have inspired Juvenal's fourth satire and the eloquent harangue which Pliny hurled across the Senate-house against a foe no longer to be feared, and which he afterwards published with additions.[2] That satire, under such circumstances, must fall very flat, is true of satire generally; but it is not true of Juvenal: we read him still.

But, if Juvenal was not a hero, it by no means follows that he was dishonest and insincere. We shall understand him best if we believe that his eye is fixed throughout on the reign of Domitian, and the horrors, political and social,

[1] Pliny, *Epp.* IX. 13. 2: *occiso Domitiano statui mecum ac deliberavi esse magnam pulchramque materiam insectandi nocentes, miseros vindicandi, se proferendi.*

[2] *Ibid.* § 23.

which he must himself have witnessed when he was of
full age to appreciate them. Like Tacitus and Pliny, he had
to hide his feelings at the time and play at oratory in the
schools; but all the while he was sharpening his weapons,
and when freedom of speech and better government re-
turned in the time of Trajan, his resentment rushed forth
all the hotter for its long suppression. He speaks of
Domitian just as Tacitus does in the *Agricola* (A.D. 98) and
Pliny in the *Panegyricus* (A.D. 100): only they admit that
with Trajan a better era began. But Juvenal never once
acknowledges the better times, but for which he could
never have written with such freedom. Was his anger too
hot? or was the rhetorician determined not to relieve the
gloom of his picture?

We have said that society at Rome was exceedingly
corrupt, and we admit that Juvenal's pictures of private
morality are substantially true; yet exception must be
taken to the principle on which he distributes his censures.
His hottest wrath seems to be reserved for the most venial
faults; improprieties and breaches of social convention
are, in his eyes, worse than crimes. When a Gracchus
fought in the arena as a gladiator, this is expressly stated
to be a greater disgrace to his order than the most horrible
vices.[1] Lateranus is fond of horses, and drives himself,
though not by day, along the public roads: he is placed in
the same list as the forger, the adulterer, and the spoiler of
the provinces.[2] A list of Nero's crimes is given: his
murders come first, and his appearances on the stage form
the climax.[3] Nettleship aptly compares the paradox in De
Quincey's *Art of Murder*: 'if once a man indulges himself
in murder, very soon he comes to think little of robbing;
and from robbing he comes next to drinking and Sabbath-
breaking, and from that to incivility and procrastination'.
The philosophy which Juvenal despised would have sup-
plied him with a truer ethical standard, and also been of

[1] II. 143 ff.   [2] VIII. 146–82.   [3] VIII. 217–23.

service to him in other ways. He says, incorrectly, that Zeno taught men to abstain from cannibalism;[1] the followers of Zeno could have taught him a more useful and more difficult lesson, to give up national antipathies. For Juvenal's patriotism is of that narrow type which considers it a virtue to hate foreigners: all orientals come under his ban, Jews, Greeks, and, above all, Egyptians.

Thus it is impossible to claim for Juvenal that he was willing, in his own words, 'to sacrifice his life for the truth'; or that he had a delicate appreciation of shades of moral delinquency; or that he was a man of wide sympathies. It has been argued from the great licence of language which he allows himself in attacking vice, that his own morals were not above reproach; but such evidence is quite insufficient. It is probable that he is here following the tradition of *satura*. Even Persius, of whose blameless life there is ample evidence, uses language of this kind; and very inappropriate it seems in his pages.

What then are the qualities which have gained for Juvenal his immense reputation? He had great gifts, both of character and intellect. He was evidently a man of stern and serious temperament, an ardent admirer of the old Roman *gravitas*, that combination of force and dignity on which the world-wide Empire was based. This is the root whence springs his hatred of luxury, effeminacy, and indecorum on the part of public men. He has no passages more pleasing than the short digressions, where he describes the simple habits and plain meals of a Fabricius or a Curius,[2]

These names of men so poor,
Who could do mighty things.

Stern as he is, he shows a marked liking and sympathy for children,[3] differing in this respect from another satirist, with whose temperament he has a good deal in common;

[1] xv. 106. [2] xi. 77 ff.; xiv. 161–88.
[3] v. 138–45; ix. 60 and 61; xi. 145–55; xiv. 44–50; xv. 135–40.

but the *saeva indignatio* of Swift did not relent even before
the innocence of childhood. Yet, in his attacks on women,
Juvenal carries his habitual exaggeration to its furthest
pitch. The excessive asperity, which characterises his
earlier work, gives place, in the later satires, to a more
attractive strain of moral exposition, often diffuse, but
grave and simple, where 'the reader is carried along from
point to point with sweetness and dignity'.[1]

His chief literary qualities are his power of painting
lifelike scenes, and his command of brilliant epigrammatic
phrase. He cannot draw a character, but his pictures of
external life are admirably real and vivid. One of the very
best is the account of Sejanus's fall in the tenth satire; but
it is in his earlier poems that they must chiefly be sought.
The third satire is a long succession of such pictures,
excellent in themselves and put together with an artistic
skill which elsewhere he either disdains or cannot reach.
It is described as follows by Mr Mackail:[2] 'In this elabo-
rate indictment of the life of the capital, put into the mouth
of a man who is leaving it for a little sleepy provincial
town, he draws a picture of the Rome he knew, its social
life and its physical features, its everyday sights and
sounds, that brings it before us more clearly and sharply
than even the Rome of Horace or Cicero. The drip of the
water from the aqueduct that passed over the gate from
which the dusty, squalid Appian Way stretched through its
long suburb; the garret under the tiles, where, just as now,
the pigeons sleeked themselves in the sun and the rain

---

[1] Nettleship, *loc. cit.* p. 65. Cf. FitzGerald, *Letters*, p. 299.
'I have been reading Juvenal with Translation etc. in my Boat.
Nearly the best things seem to me what one may call Epistles
rather than Satires: viii. To Ponticus: xi. To Persicus: and
xii, xiii and xiv to several others; and, in these, leaving out the
directly satirical Parts. Satires iii and x are prostituted by
Parliamentary and vulgar use, and should lie by for a while.
One sees Lucretius, I think, in many parts.'

[2] *Latin Literature*, p. 223.

drummed on the roof; the narrow, crowded streets, half choked with the builders' carts, ankle-deep in mud, and the pavement ringing under the heavy military boots of guardsmen; the tavern waiters trotting along with a pyramid of hot dishes on their heads; the flower-pots falling from high window-ledges; night, with the shuttered shops, the silence broken by some sudden street brawl, the darkness shaken by a flare of torches as some great man, wrapped in his scarlet cloak, passes along from a dinner-party with his long train of clients and slaves: these scenes live for us in Juvenal, and are perhaps the picture of ancient Rome that is most abidingly impressed on our memory.'

Even in the narrow compass of a single line, Juvenal can draw a picture, perfectly expressive of his meaning and indelible in the memory of the reader. Let us take an example from what is, on the whole, the weakest of the satires, the fifteenth. The phenomenon which we express by saying 'the schoolmaster is abroad', is put thus by Juvenal:

de conducendo loquitur iam rhetore Thyle.

It is obvious that the Latin is immensely superior in dramatic force and a kind of humour. But the position of this line should be noticed. In the previous lines, the same thing has been already said twice, but with much less point and vigour. This artifice is common in Juvenal. He has the practised rhetorician's eye for a climax; and, sure of the effect which he has reserved for the end, he refuses to spoil it by any premature coruscations. But to reach such a point, how far Latin literature has travelled from the art of Virgil!

Indeed, Juvenal's greatest gift is his power of coining phrases, a power which he had cultivated by the practice of declamation and which must have gained him thunders of applause in the rhetorical schools, where these *sententiae*, as they were called, were prized to excess. The thought

may sometimes be a commonplace, but the form is so
perfect that posterity, in despair of finding any better
expression of the familiar idea, has constantly adopted his
foreign phrase. Tacitus has a share of this peculiar gift;
so has Quintilian; but no Roman writer, not even Horace
himself, has left so many phrases and lines, of which all are
familiar proverbs among educated Europeans, and some
are habitually quoted even by those who have never heard
the name of Juvenal. Such are the lines:

si natura negat, facit indignatio versum.

quis tulerit Gracchos de seditione querentes?

dat veniam corvis, vexat censura columbas.

nemo repente fuit turpissimus.

nil habet infelix paupertas durius in se
quam quod ridiculos homines facit.

haut facile emergunt quorum virtutibus obstat
res angusta domi.

plurima sunt quae
non audent homines pertusa dicere laena.

rara avis in terris nigroque simillima cycno.

hoc volo, sic iubeo: sit pro ratione voluntas.

sed quis custodiet ipsos
custodes?

occidit miseros crambe repetita magistros.

summum crede nefas animam praeferre pudori
et propter vitam vivendi perdere causas.

dum serta, unguenta, puellas
poscimus, obrepit non intellecta senectus.

cantabit vacuus coram latrone viator.

expende Hannibalem: quot libras in duce summo
invenies?

ingenui vultus puer ingenuique pudoris.

maxima debetur puero reverentia.

In some of these phrases a common thought is expressed with great epigrammatic skill; but they are by no means all commonplaces. The last may have sounded even paradoxical to readers who thought that reverence should be reserved for old age; but the paradox is admirable in its truth and force. Again, there is a bitter, and not obvious, truth in the saying, that the worst sting of poverty is that it makes the poor man ridiculous.

One of the characters in Victor Hugo's novel, *Les Misérables*, had learned Latin for the sole purpose of reading Juvenal, who represented in his eyes all that was worth reading in the language. This was a hasty judgement; for Juvenal is not one of the greatest names in Latin literature. But his great literary gift and his honest purpose have made his fame secure. For our own age his satire ought to have a special interest; as there are considerable resemblances between modern society in great cities and the busy life that surged before Juvenal's eyes in Rome just eighteen centuries ago.

IV

THE MANUSCRIPTS

Juvenal, in his character as a moralist, was much read in the Middle Ages: there were probably few large monastic libraries which did not contain a copy of the satires; and some contained more than one. Two MSS., one of the ninth, the other of the tenth century, now in the Library of Trinity College, Cambridge, both belonged formerly to St Augustine's Monastery at Canterbury. Hence the number of MSS. still extant is very large; there is probably no classical author of whom we possess so many. Beer had counted 120 in the libraries of Austria, Germany, and France. No complete MS. is older than the ninth century; and there are few of that antiquity. They are all divided into two classes, according to the recension of the text they contain. But the better class is now represented by only one complete MS. This is preserved in the Library of the Medical School at Montpellier. It is sometimes called *Codex Montepessulanus*, but more often *Pithoeanus*, from the fact that it once belonged to Pierre Pithou whose name is written upon it. Pithou was a French scholar and jurist, who published in 1585 an edition of Juvenal and Persius based upon his MS. The abbreviation *P* is commonly used to denote this MS. and its readings. It dates from the ninth century; and though little, if at all, older than some of the other MSS., it has no rival in merit and occupies a class by itself. All the other MSS. belong to the second and inferior class: they (and their readings) are denoted in critical editions by the symbol ω. The earliest editions of Juvenal, as those of Ferrara (1474) and Louvain (1475), were printed from MSS. of the interpolated class; nor was the value of *P* fully understood until the edition of Otto Jahn, published in 1851.

P has been corrected at different dates by several different hands, which cannot be distinguished with certainty. The readings of the correctors are indicated by the symbol *p*. The endeavour of the corrector was generally to substitute ω for *P*, the worse reading for the better, as he used a MS. of the inferior class to correct by. Many passages, however, have escaped his mistaken zeal; and, even where the change has been made, the erasure does not always entirely conceal the original reading.

There is also another source from which the true reading has in many cases been retrieved, and which throws light on the surpassing merit of *P*.

This is afforded by the ancient Scholia, which are preserved in *P*. The corrector, while altering and corrupting the text, did not trouble to alter the Scholium, or note in the margin, which still remains to prove the corruption. Thus in Sat. 1. 2 all MSS. read *Codri*; but the Scholium in *P*, immediately opposite the text, gives *Cordi*. Turning to the text of *P*, we see the word is written *Co dri*: the inference is clear that the true form of the name is *Cordi*, and that it was originally found in *P*, before the corrector erased the *r* and changed *i* to *ri*, as he could easily and neatly do. The corrector was haunted by the recollection of *Codrus* in Sat. III. 203 ff., and made up his mind that Juvenal meant the same person here. By another ill-timed exercise of memory he changed *tradentur tubera* in Sat. V. 116 to *raduntur t.*, which gives no sense; but *radere tubera* of Sat. XIV. 7 was in his mind. Take another instance of help from the Scholia. In Sat. VIII. 155 all MSS. read *torvumque iuvencum*; but in the Scholium of *P* we find *robumque i.* We next find that in *P* the first letters of the adjective are a correction for something else. We infer that the corrector, innocent of evil intention, struck out *rob-* and substituted *torv-*, which he found in his MS. of the inferior class, *torvum* having been originally a 'gloss' or explanation of the rare word *robum*. And the reading has in this case external confirmation; as it is found in the *Florilegium Sangallense*, a St Gall MS. containing selections from Juvenal; for even in the Middle Ages the satirist had to submit to expurgation, and there are many such *Florilegia*. If this last MS. gave the whole text of the satires, it would be a worthy rival of *P* itself: it is one source of the reading *mulio consul* in Sat. VIII. 148.

These instances give some idea of the superiority of *P* over the other MSS., because even where it now gives a corrupt reading, we see that in many cases this is the fault of the corrector. Next let us give some of the many instances where the corrector has, no doubt by oversight, failed to alter the true reading to what is found in ω.

In Sat. VII. 99 all MSS. but *P* read *petit hic plus temporis atque olei plus*; *P* gives *perit* for *petit*. The first reading is good Latin; but the second is obviously right. It is a version of the common proverb *perdere oleum*, which no copyist could have hit upon *suo Marte*. Again, in Sat. XV. 27

P rightly reads *Iunco* where nearly all the other MSS. read *Iunio*. In the first place, there is no instance in the text of Juvenal of a word like *Iunio* scanning as a spondee: see notes to Satt. VI. 82; VII. 185. Next, we know that an Aemilius Iuncus was consul in October A.D. 127, a date which suits the context perfectly. Lastly, the oldest authority for any part of Juvenal's text, the Bobio palimpsest in the Vatican, written in capitals, happens to contain this line with fifty-one others, and reads *Iunco*. Many similar instances will be found in the *apparatus criticus*.

But it must not be supposed that the corrector did nothing but corrupt the text. There were many mistakes in *P*. First, there were many omissions of letters, words, and parts of lines: all these are filled in by the corrector. Further, certain letters were constantly confused in *P*: *l* is often written for *i*, *t* for *i*, *v* for *b*: thus it gives *locari* for *iocari* (III. 40) and conversely *ioculis* for *loculis* (I. 89): *despictat* for *despiciat* (V. 82): *vervum* for *verbum* (I. 161). These are harmless blunders, which can easily be detected. Yet they have sometimes led to more serious corruptions, of which there is a striking example in Sat. VIII. 148. All MSS. there read *multo sufflamine consul*; yet the true reading, preserved by the Scholiast and the St Gall *Florilegium*, is *s. mulio consul*. The process of corruption is this: in the archetype of our MSS. *mulio* had become *multo*; and then the serious corruption was made by the reader who, to save the scansion, transposed the two words, and thereby made restoration almost hopeless. But the artifices of the writer of *P* are generally of a transparent kind: when he read *iumenta* as *tumenta* (VII. 180), and *summula* as *summuia* (VII. 174), he tried to make the words look like Latin by writing *tumentia* and *summa uia*. The lines, as he wrote them, neither scan nor construe; but it is just his supreme merit that he did not try to make them do so. To write *tumenta* for *iumenta* is a blunder; to change *tumenta* to *tumentia* is a venial interpolation; but to transpose the words *sufflamine multo* is a grave interpolation. Now the difference between *P* and ω is just this, that grave interpolations are rare in *P* and common in ω. In Sat. VIII. 148, Bücheler's report makes it highly probable that *P* read *sufflamine multo consul*, and that the corrector transposed the words.

There are other letters which are apt to be confused in *P*: we find final *s* for final *t*, and vice versa, e.g. *lucebit* for *lucebis* (I. 155), *deducit* for *deducis* (I. 157), *stetis* for *stetit* (I. 149). We may infer that in some ancestor of *P*, as in *P* itself, the two letters were written much alike. The readings *necabis* for *negabis* (III. 168), *ignium* for *Ionium* (VI. 93), are remarkable: as *c* and *g*, *g* and *o* are quite unlike in any cursive hand, these errors are probably derived from some MS. written in capitals or uncials. *P* has another instructive error in Sat. VI. 159, reading *observant ubi festa nudo pede sabbata reges*: this is of course unmetrical, and a less honest copyist would have 'corrected' it. The true reading *mero pede* is given by the corrector and by all other MSS. *Mero* is a very rare and strange equivalent for *nudo*, which was accordingly written above as a 'gloss'; and the copyist of *P* wrote down the gloss in the text by mistake. Again *P* makes Sat. X. 221 a heptameter by adding *tutor* to the line: this word is a gloss, explanatory of *Hirrus* in the line below, which has been added by an error to the text. These two mistakes prove that the MS. from which *P* was copied had glosses between the lines: it was probably written in uncials soon after the sixth century, with no divisions between the words.

Again, it is now clear that *P* once contained the lost part of the last book. Line 60 of the 16th satire is the last line of the last page of what is now the last quire of the MS. But the concluding formula (*explicit liber quintus*) is not found, though there is room for it and it is appended to all the other books. The appearance of the MS. proves that it is mutilated, and once contained more than it does now; no one who has held it in his hand can doubt this, says Bücheler. But none of the other MSS. give any indication of a loss: they subjoin the usual formula after l. 60, as if the satire and book had come to their natural conclusion. This is another proof of the unique value of *P*.

Lastly, good spelling is a sure sign of a good MS.; and the spelling of *P* is generally good and classical. Bücheler notes, for its rarity, a single instance of the spelling *solatia* for *solacia*. The other MSS. are full of barbarous spellings, especially in proper names, compared with which *solatia* is a venial offence.

It was said above that *P* is the only entire MS. which represents the better tradition of the text. But there are various fragments which belong to the same class. Of these the *Florilegium* of St Gall has already been mentioned. At Aarau five leaves of a similar MS. with Scholia have been found: they had been used to bind other books: and Scholia like those of *P* are preserved at St Gall, though the MS. from which they were copied, is no longer extant.

With these exceptions, all the MS. authority for Juvenal's text belongs to the interpolated class. All these MSS. contain a large number, not of mere blunders, but of interpolations and false readings. In many cases, these give a sufficiently good sense; but the readings peculiar to *P*, and the other marks of superiority already mentioned, have convinced all scholars that the text of *P* must be preferred in every case where it is not debarred by the sense. Thus in Sat. I. 143 *crudus* (so *P*) and *crudum* (so ω) both give good sense; yet the first must be preferred. On the other hand there are not a few cases where the reading of ω must be preferred on grounds of meaning or language. Thus in Sat. III. 61 all editors read, with ω, *quota portio faecis Achaei*, where *P* has *Achaeae*. It is certain that a nominative is wanted; so that, if *Achaeae* is an adj. agreeing with *faecis*, it is wrong. Yet it is possible that *Achaeae*

stands for *Achaea ē* (i.e. *est*), which gives good sense: such
contractions are used especially at the end of a long line.
In some cases it is very difficult to decide whether the
reading of *P* or ω should be preferred. Thus in Sat. XIII.
65 *mirandis sub aratro* is dull and flat, while *miranti sub
aratro* is vigorous and pointed; yet the worse word has
the better authority. So this question arises: is it more
likely that Juvenal missed an obvious point which a copyist
hit upon, or that the reading of *P* is corrupt? On the
whole, the text of Juvenal would not have lost much if no
other MS. except *P* (including the supplements of *p*)
had been preserved; but it would have lost a great deal if
all the other existing MSS. had been saved and *P* alone
lost.

Though the reading of this class of MSS. is indicated
by ω, it must not be understood that they all agree precisely
everywhere. Some resemble *P* more than others, having
fewer interpolations. Some of them have probably been
corrected by the aid of a MS. of the *P* class. Still the
resemblance is so general that a reading is attributed to ω,
where it is found in a large number, if not in all, of this
class.

A collation of eleven select MSS. belonging to this
class—seven containing the whole of the satires, and four,
extracts from them—has been published by C. Hosius,[1]
by which their divergences from *P* and from one another
can be easily seen. A MS. which agrees with the reading of
P more often than any of these, except perhaps *A*, is now
in the Library of Trinity College; but, in spite of this
comparative merit, it undoubtedly belongs to the inferior
class, and is valuable mainly for its confirmation of readings
of *P*.

The relation of ω to *P* is a problem which has never been
solved and is perhaps insoluble. It is impossible that they
can be corrupt copies of *P*, if only because of the early

---

[1] *Apparatus criticus ad Iuvenalem* (Bonn, 1888).

date of some of them. One Vienna MS. is considered by experts to be even older than *P*. Bücheler, the highest authority on the subject, is of opinion that both classes are derived from a common original, which contained, together with the text, a great number of various readings and glosses; and that the two different recensions arose from the fact that the copyists sometimes adopted the text, sometimes the variants of the original. But this does not account for the essential superiority of *P*. How did it come about that the copyist of *P* stuck to the text in so many cases where the other copyists departed from it? The great antiquity of the inferior recension is proved by the fact that it is followed, at least as often as *P*, in quotations from Juvenal, occurring in Fathers and grammarians of the fourth and fifth centuries. Also the Bobio palimpsest, already spoken of, though it preserves one true reading, *Iunco*, clearly belongs to the interpolated class of MSS.; yet the character in which it is written shows that it must be some centuries earlier than *P*.

The readings of *P* are now certainly known, as Bücheler collated it himself throughout for his edition of 1893. Unless another complete MS. of the same class turns up, it is not likely that the text will be materially improved; for conjectural emendation has done little for this author. But the text of Juvenal is probably in a fair state of preservation; though, when we think how many generations of scholars saw no difficulty in *multo sufflamine consul* (VIII. 148), no wise critic will deny that there may be many similar passages which contain corruptions no less serious and as little suspected.

The *apparatus* given in this edition is a selection of readings which have been chosen on purpose to let the reader know where the text is doubtful, and also to illustrate the superiority of *P*, based partly upon the readings peculiar to it, partly on the nature of its mistakes. Yet, as interpolations do occur in *P*, instances of these are given too. Emendations printed in the text are referred to their authors; and passages where the true reading is preserved by the Scholiast are indicated. The more important readings of the Trinity College MS., mentioned above, are also recorded. It is hoped that the selected readings will be sufficient to afford an object lesson in the main matter, the difference between blunders and interpolations.

V

THE SCHOLIA

The Scholia, or ancient commentaries preserved in the MSS., are divided, like the MSS. themselves, into two classes. The better class are known as the *Scholia Pithoeana*, being contained in *P* and in none of the interpolated MSS. The inferior Scholia are traditionally attributed to Cornutus, who is also the reputed author of a commentary on Persius preserved in MSS. of that writer. Who Cornutus was, and when he lived, is far from certain. However, in a note on Sat. IX. 37, he refers to *Magister Heiricus*; hence Jahn suggested that he was a pupil of Heiric who was born in 843 at Auxerre and was distinguished as a teacher and writer. It is certain that Annaeus Cornutus, the friend and executor of Persius, can have had nothing to do with these Scholia; but it is possible that the medieval scholar may have assumed the name, in order to give weight to his annotations on Persius. His commentary on Juvenal must be of great antiquity, as it is found even in MSS. of the ninth century, written in a contemporary hand. Occasional traces of ancient learning are found in this collection: thus the allusion to Catullus in Sat. VI. 7— not a very recondite allusion—is correctly explained: but the great mass of the annotations are merely dull and superfluous, and some are remarkable for an extraordinary depth of ignorance and folly. To substantiate this charge, two explanations are copied here from a Cambridge MS. of the tenth century. They are chosen for their absurdity; but though less dull, they are not really less instructive than most of the others. On Sat. i. 3 the Scholium runs: '*togatas*' *vero feminino genere vocavit propter luxuriam illius temporis, ut Virgilius* '*o vere Phrygiae neque Phryges*', *id est, non viri sed feminae.* This shows some knowledge, however applied: the next is due to mere native brilliance. On Sat. i. 75 *criminibus debent...stantem extra pocula caprum*, Cornutus explains: CAPRUM: *Caper fuit philosophus abstemius, unde dicitur extra pocula.* But he does not go on to explain how the wicked owe to their guilt an abstemious philosopher. Now compare the *Scholia Pithoeana* on these two passages. On *togatas* we find, '*togatae sunt comoediae Latinae, quales Afranius fecit*': on *caprum,* '*dicit emblematum opus*'. The comparison shows indeed, as Jahn says, *quid distent aera lupinis.*

The Scholia of Cornutus are very copious, but a detailed account of them is superfluous. By their dim light the Middle Ages studied their corrupt text of Juvenal; they were repeatedly copied and circulated in MSS. from the ninth century onwards, and are generally found with little variation in the interpolated MSS.[1] They were also written in a later hand on the margin of *P*—the same hand which added the Life at the end of the MS.—but have been crossed out by some judicious reader (Pithou himself, according to Beer), whose pen has spared the good Scholia.

These latter are found in *P*—hence their name of *Pithoeana*—and also were included in all the fragmentary MSS. which preserve the same recension as *P*; and, as was said above, they are preserved without a text in a MS. at St Gall. The great importance of these Scholia,

[1] Jahn's inference (*Prolegom. ad Persium*, p. cxvi) that this commentary was printed in the Louvain edition of Juvenal (1475), is apparently incorrect. There is a copy of this rare book in Peterhouse Library; it has no printed commentary.

and their unique value for settling the text of Juvenal, have
been shown incidentally in the previous section: in many
cases their *lemmata* give a reading older and better than
that of *P* itself. It is clear that, as they now stand, they
are the work of more than one hand; but the more
valuable annotations were probably taken from a com-
mentary on Juvenal, attributed to a scholar called Probus
on the authority of G. Valla, who, in his edition of
Juvenal published at Venice in 1486, gave extracts from
a MS., now lost, containing '*Probi grammatici in Iuvenalem
commentarii*'. These notes went no further than Sat. VIII.
198. Valla treated his materials, after the fashion of his
time, with some freedom; but a comparison of his quota-
tions from Probus with the Scholia of *P* makes it certain
that both come from an identical source, or rather that the
Scholia are extracts from the commentary which Valla had
before him in a completer form. Chronology prevents us
from identifying this Probus with M. Valerius Probus of
Berytus, a famous critic of the first century, of whom a
short life by Suetonius is extant and to whom Martial
alludes.[1] The commentary of this unknown Probus was
probably written towards the end of the fourth century,
when it is known on the evidence of Ammianus[2] that
Juvenal was much read. That the commentary cannot be
earlier than this is made probable by a Scholium on Sat.
X. 24, which speaks of a Cerealis as *praefectus urbi*: it is
known that Neratius Cerealis filled that office in A.D. 352.
The author was a learned man, with considerable know-
ledge of the ancient poets and historians. In this way there
are preserved some topographical notices of ancient
Rome; some facts, chiefly biographical, of notable persons;
and some fragments of ancient literature. Among these

---

[1] Suet. *de Gramm.* 24; Mart. III. 2. 12.

[2] XXVIII. 4. 14 *quidam detestantes ut venena doctrinas, Iuve-
nalem et Marium Maximum curatiore studio legunt, nulla volu-
mina praeter haec in profundo otio contrectantes.*

last are: an epigram of Martial, not included in his own MSS.; four verses of Statius' lost epic *De Bello Germanico*; two verses of the satirist Turnus, and two of the poetess Sulpicia. Both of the last authors are mentioned with praise by Martial.

NEW INTRODUCTION

BY

MICHAEL COFFEY

J. D. Duff's edition of Juvenal shows both wide learning
and common sense. The reference by A. E. Housman to
its 'candour and clear perception' is an unusual tribute
from a great scholar to whom censure came more easily
than praise.[1] The sections that follow indicate the main
developments in scholarly work on Juvenal since the
appearance of Duff's edition. Detailed references to
books and articles mentioned briefly in the notes are
given in the bibliography (pp. lxxxiv–lxxxix).

I
THE LIFE OF JUVENAL

There is no new external evidence of substance for
Juvenal's life. Prosopographical research has shown that
many holders of the *gens* name Iunius were of Spanish
origin. Some wealthy men from Spain settled at the
fashionable and expensive resort of Tibur (Tivoli), where
Juvenal seems to have owned a farm (XI. 65). One of
the consuls of A.D. 81, C. Iulius Iuvenalis, had the same
cognomen as the poet, but his gentile name was different.
In general the *cognomen* Iuvenalis, which may origin-
ally have been foreign, suggests humble birth; two of
its bearers were freedmen living in the neighbourhood
of Aquinum, a town near Monte Cassino mentioned
by Juvenal in an apparently autobiographical context

[1] *Iuvenalis Saturae* ed. 2 preface, p. xxix.

(III. 318–21).[1] While the names are not a safe guide to Juvenal's status, they show the possibility at least that the poet who championed pristine Roman values was himself of non-Italian origin.

It must remain an open question whether the Iunius Iuvenalis who dedicated the inscription to Ceres was the poet or a near kinsman.[2] The dedicator was an affluent and respected magistrate of Aquinum, whose duties included petty jurisdiction and censorial authority as well as routine council business; he was also a local priest of the cult of a deified emperor.[3] His rank of commander of auxiliaries carried with it equestrian status. Such military tribunates were sometimes of short duration and a prelude to an equestrian career of imperial service, but there is no evidence that a Iunius Iuvenalis received any such preferment.[4] Whether or not Juvenal the poet was of equestrian rank, Martial in a poem written c. A.D. 102[5] describes him as leading the restless life of a *cliens* in Rome, in contrast to his own idyllic life of retirement in Spain. But too much should not be made of the contrast, for it is clear from Martial's disillusioned indictment of provincial narrowness in the prose preface to the twelfth book that in the poem to Juvenal he

[1] On the significance of the names see Syme, *Tacitus*, pp. 774–6, also p. 602.

[2] See Knoche, *Die römische Satire*, p. 89 and Syme, *Tacitus*, p. 775.

[3] On the status and duties of the *duumvir quinquennalis* and *flamen* of the imperial house in a *colonia* see Meiggs, *Roman Ostia*, pp. 174–80.

[4] We do not know how Juvenal obtained his personal experience of Egypt (xv. 45); he may have seen Britain also (Duff, p. xix), but it should not be supposed that Juvenal had visited every place that he describes. For equestrian careers see Highet, *Juvenal*, pp. 34 f.

[5] XII. 18; on the dating of Martial xii see Sherwin-White on Pliny, *Epp.* III. 21. 2.

deliberately overstates his own contentment; it is possible that his description of the quality of Juvenal's life in Rome is likewise exaggerated.[1] Pliny's silence on Juvenal, referred to briefly by Duff (p. x), is perhaps not without significance. Pliny liked Martial personally and helped him financially.[2] As Juvenal and Martial were friends, it is unlikely that Juvenal was not known to Pliny. For Pliny was anxious to act as patron to writers and to mention in his letters the literary celebrities of the times,[3] and even though Juvenal's first book may not have been published until after Pliny's departure for Bithynia, which was possibly as early as A.D. 109, his talent will have been recognised earlier, especially in an age in which poets were accustomed to recite their works in public.[4] Pliny's failure to mention him suggests dislike of Juvenal himself or of his manner of writing. Nor does Juvenal mention Pliny, but it is possible, though by no means certain, that his satires contain some innuendos against Pliny and his family.[5]

The ancient Life of Juvenal from which all the others derive without independent value (Duff, p. xi) is itself suspect; compared with ancient Lives of other Roman poets it is deficient in factual information and for the most part ill-written. It is possible, as Duff believed, that the statements in the opening sentence deserve credence,

[1] Highet, *Juvenal*, p. 18, perhaps takes this peom too seriously.

[2] Plin. *Epp.* III. 21.

[3] See, for example, the help he gave Suetonius the biographer (*Epp.* I. 24. 1 and Sherwin-White *a.l.*); see also Syme, *Historia* 9 (1960), 364 ff. The death of Silius Italicus, to whom Martial had addressed poems (e.g. VII. 63), was the occasion for an obituary (*Epp.* III. 7); Pliny had not the occasion to mention the esteemed Statius, on whom see Juv. VII. 82 ff., because his death had probably taken place earlier than the years of the Letters.

[4] On the date of Pliny's departure from Rome see Sherwin-White, p. 80, but also J. Crook, *Class. Rev.* N.S. 17 (1967), 313 f.

[5] Highet, *Juvenal*, pp. 291–3.

and they are at least plausible. But even here scepticism
is possible, for the facts of Horace's parentage and
remarks about declamation in *Sat.* I may have inspired
the fabrication of data about Juvenal's earlier life. The
main point of the Life, the poet's exile as punishment for
an oblique reference to an old imperial scandal that was
construed as having a present application, though accep-
ted cautiously by Duff (p. xiii), is very dubious. Exile
during the last years of Domitian, the earliest possible
date, is most improbable, for there is abundant evidence
that during the reign of terror so powerfully described
by Tacitus (*Agr.* 45) offences that reflected on the majesty
of the emperor were punished with death.[1] Lines on an
imperial favourite of a previous dynasty written by a
poet of no political or social importance are unlikely to
have appeared offensive to either Trajan or Hadrian,
busy military administrators who spent many years away
from Rome. Further Juvenal would not have inserted
into the seventh satire, which begins with a vague hope
for the patronage of the emperor (almost certainly
Hadrian), anything that could conceivably have appeared
to be politically tendentious. Nor would the tense
atmosphere at the beginning of the reign of Hadrian
immediately after the execution of the four consulars
have seemed a prudent time for political innuendo.[2] But
Hadrian's reign was for the most part moderate, and it
is not even plausible that an emperor assiduous for the
proper government of the boundaries of the empire
should, as the Life states, send a senile man to a distant
military post as a kind of practical joke. The whole
story of an exile is flimsy and was probably invented in
the fourth century, when there was no genuine bio-

---

[1] Note particularly Suet. *Dom.* 10; see Coffey, *Lustrum* 8
(1963), 167.

[2] Script. Hist. Aug. *Hadr.* VII. 1–3; Syme, *Tacitus*, pp. 244 f.

graphical tradition about Juvenal. If it is genuine, it must refer to an event of which we know neither the circumstances nor the date.

The story of exile under Domitian has been made the most important item in some modern accounts of the life of Juvenal, notably that by Gilbert Highet, who believes that Juvenal the poet made the dedication at Aquinum at the outset of a promising career that was destroyed by exile and loss of property during the period A.D. 93–96: the broken pauper returned embittered and turned to savage denunciation until success or patronage mellowed his later years.[1] A similar pattern of events is postulated by Peter Green, whose discussion of the ancient biographical evidence is none the less circumspect.[2] But without discounting the possibility or even likelihood of some disappointment in Juvenal's own life, it is arguable that *saeva indignatio* may also be a product of temperament irrespective of personal wealth or status and that the conventions of poetry of the Silver age offered a highly developed rhetoric for indulging in such emotions.

A few more pieces of chronological evidence may be added to Duff's discussion. The date of Juvenal's birth is uncertain. Juvenal may have referred sixty years later to Fonteius Capito, the consul of A.D. 67, an unmemorable person, because that was the year of his own birth; it is a slender indication, but the date is intrinsically plausible.[3] It has been suggested that the second satire

---

[1] Highet, *Juvenal*, pp. 4–39; summary, pp. 40 f.

[2] P. Green, *Juvenal*, pp. 10–22.

[3] XIII. 16 f.; Syme, *Tacitus*, pp. 774 f. Some scholars argue for an earlier birth date: Highet, *Juvenal*, pp. 5 and 233, favours *c.* A.D. 60, interpreting I. 25 strictly as written by a man of more than forty-five years of age (see Censorinus *de die nat.* XIV. 2); Green, *Juvenal*, p. 11, advocates A.D. 55, an unsupported date given in one of the later ancient lives.

refers to Tacitus' *Histories*,[1] first known of in A.D. 105 and completed possibly by the end of A.D. 109. The effeminate behaviour of the emperor Otho is described as:

> res memoranda novis annalibus atque recenti
> historia. (II. 102–3).
>
> (matter to be related in the latest *Annals* and a newly
> published *History*).

This merely gives an indeterminate *terminus post quem;* the word *recenti* may imply something fresh in the memory, not necessarily of very recent appearance. If *novis annalibus* refers in an imprecise way to Tacitus, the date for the publication of Juvenal's first book is unlikely to have been much earlier than A.D. 115. In the eighth and tenth satires Juvenal may have drawn on the appropriate parts of the *Annals*, a work perhaps not completed until A.D. 120 or later.[2] As Duff points out (on XV. 27) the reference to a consulship of A.D. 127 as something recent is vague, and so the fifteenth satire may have been composed some years later. It is therefore possible that Juvenal was still alive after the death of Hadrian in A.D. 138.

II

THE *SATURA* BEFORE JUVENAL

There was speculation in antiquity about the original meaning of the word *satura*. Diomedes, a Latin grammarian of the fourth century A.D., whose views are based on those of the late Republican polymath Varro, offers four explanations, two of which, the derivation of *satura* from the Greek satyr play or from a legal phrase,

---

[1] Highet, *Juvenal*, pp. 11 f.

[2] On the chronology of Tacitus see Syme, *Tacitus*, pp. 117–20, 473; in relation to Juvenal, p. 777; and on the title *Annales*, pp. 253, n. 1.

may safely be rejected. The other two are plausible: *satura* as a literary term was a metaphor based either on a full dish of varied first-fruits offered to the gods or on a kind of sausage stuffed with many ingredients.[1] Both of these metaphors suggest abundance and variety of contents. Many modern scholars have accepted the culinary rather than the cult metaphor as better suited to the subject matter of *satura*.[2] In describing the themes of his satire as *nostri farrago libelli* Juvenal (i. 86) alludes to what may have become standard text-book theory, but he does so disrespectfully, for *farrago* is always used of mixed meal for cattle, not of food for men. Some modern scholars have been dissatisfied with the ancient theories, and an attempt has been made to connect *satura* with Etruscan *satir*, which is said to mean 'speak' or 'declare'.[3] The theory has attractions, as it is appropriate for the conversation and discourse of *satura* as represented by Horace's term *sermo* and does not presuppose a dramatic source for any hypothetical Etruscan influence. But there are difficulties. The meaning of the Etruscan word is perhaps not free from uncertainty; it is also unlikely that an Etruscan loan word would have been chosen for a genre which contained much Greek material. It seems likely that Ennius chose the title *satura* for a collection of informal verse, and that, if the explanation of a metaphor from a cult offering or a recipe is accepted, coined it in the manner of Posidippus'

[1] Diomedes, *Gramm. Lat.* i. 485 f. K. The text and a translation of this passage will be found in Van Rooy, *Studies in Classical Satire*, pp. xii f., who discusses (pp. 1–29) the word *satura*.

[2] E.g. Knoche, *Röm. Sat.* p. 12, and Weinreich, *Röm. Sat.* p. xi.

[3] P. Meriggi, *Stud. Etrusc.* 11 (1937), 157 and 197. This etymology is discussed by Walde-Hofmann, *Lat. etym. Wörterb.* s.v. *satura* and accepted without discussion by L. R. Palmer, *Lat. Lang.* p. 48.

Soros (heap of winnowed grain) and other fanciful metaphorical Greek titles from the Hellenistic period.[1]

This conclusion depends on the assumption that there was no *satura* in Rome or Italy before Ennius. The tradition found in Livy that there was a pre-literary dramatic *satura*, though accepted by Duff (p. xxiv) and some other modern scholars, is highly suspect.[2] His vagueness and the length of the chronological stages of his account raise grave suspicions, nor does he mention the fundamental fact of Latin literary history that it was a translated Greek drama that Livius Andronicus introduced to the Roman stage.[3] This tendentiously patriotic omission makes Livy's account of Livius Andronicus incredible. Neither does his account of a dramatic *satura*, which has no independent corroboration, deserve credence.[4] It has long been suspected that Livy's account is not a statement of a true chronological development but an artificial schematic construction. Just as his dubious account of Fescennines was necessary to explain the origin of the dialogue (*diverbia*) of comedy without reference to Greek plays, so also a Roman dramatic lyric without plot was a necessary hypothesis to account for the *cantica* of Roman plays. It thus seems very likely that Livy's chronological sequence is groundless speculation and that there never was a dramatic

[1] On metaphorical Hellenistic titles see T. B. L. Webster, *Hellenistic Poetry and Art*, pp. 45 f. and on Posidippus also H. Lloyd-Jones, *Journ. Hell. Stud.* 83 (1963), 75–90.

[2] Scholars accepting the tradition of a dramatic *satura* include B. L. Ullman, *Class. Phil.* 9 (1914), 1–23; O. Weinreich, *Röm, Sat.* pp. xiv–xvii; J. Wight Duff, *Lit. Hist. Rome*, I, 61.

[3] Contrast Livy's imprecise *post aliquot annis* with the reliable date of 240 B.C. given by Cicero, *Brut.* 72, for Livius Andronicus' first production in Rome. Duff's account may mislead, for it seems to combine what Livy said with what he ought to have said.

[4] Val. Max. II. 4. 4 is modelled on Livy or his immediate source.

satura. Livy's chapter owes much to the Peripatetic formulation of the development of Greek drama from its beginnings to its maturity, as is found in such treatises as Aristotle's *Poetics*, and it is possible that a theorist postulated a Roman dramatic entity that corresponded to a satyr play and by a piece of linguistic opportunism gave it a title that sounded similar.[1] This is not to deny the possible existence in Rome of pre-literary dramatic pieces, though we have no knowledge of their nature nor even their name.[2] But there is no connexion between any such hypothetical pieces and Roman literary *satura.* Quasi-dramatic elements in *satura* may be attributed to the influence of Comedy, of the prose Dialogue such as that of Plato and even, in the introduction of an imaginary objector, to the devices of rhetoric.

For Ennius, author of Rome's first great national epic, the four books of *saturae*, poems written in a variety of metres, were minor works. It is clear from the scanty remains that he wrote personal pieces on various topics in an informal manner; but an element of censoriousness seems relatively unimportant in the work of the first creator of *satura.*[3] It was Lucilius to whom Juvenal looked back as the early master who gave the genre its

[1] O. Jahn's view, *Hermes* 2 (1867), 225–51, was amplified by Fr. Leo, *Hermes* 24 (1889), 67–84, and G. L. Hendrickson, *Am. Journ. Phil.* 15 (1894), 1–30; in more recent times Knoche, *Röm. Sat.* p. 9 and E. Burck in Kiessling-Heinze, *Hor. Sat.* p. 370 are sceptical about the existence of a dramatic *satura.* The arguments of Leo and Hendrickson for Peripatetic influence have not been substantially affected by any later criticism; for the suggestion that Livy's *satura* was given its name on the basis of theorizing about satyr play as at Ar. *Poet.* 1449 a 20, see Leo, p. 77.

[2] See Duckworth, *Nat. Rom. Com.* pp. 8 ff.

[3] An edition and translation of the fragments will be found in Warmington, *Rem. Old Lat.* I. 382–95. The standard edition is that of Vahlen.

most famous characteristics, outspoken criticism and a censorious temper. He wrote nothing but *saturae*; after early experiments with dramatic metres he chose the hexameter, which became the standard medium for all later verse *saturae*. The content of his work was wide in its range. He retailed personal experience without inhibition, attacked contemporary politicians including some of the most powerful, castigated moral standards, showed some interest in philosophy and also offered criticism and sometimes systematic exposition of grammar and literary principles.[1] Lucilius was a wealthy knight, who lived in Republican times when speech was relatively free. Horace, the son of a freedman, wrote in an age of political crisis and would not have been able to denounce the influential by name, even if he had wished to do so. But though no contemporary of consequence is ridiculed in his *saturae*, none the less he makes forceful attacks on wickedness and cant. He writes discreetly about himself, discusses the function of his *satura*, and moralizes without pretension. Less ebullient than Lucilius, he uses an urbane irony that was suited to a more fastidious age.[2] The six *Satires* of Persius are permeated with reminiscences of the *Satires* and *Epistles* of Horace. This precocious satirist, experienced only in books and ideas, is original in his involuted image-laden language and in his doctrinaire

[1] There is an edition and translation of the fragments in Warmington, *Rem. Old Lat.* III. The standard edition is that of Fr. Marx.

[2] There is a good edition of Horace's *Satires* by A. Palmer; the best modern text is that of Fr. Klingner. N. Rudd, *The Satires of Horace*, gives a full interpretation and shows (pp. 258–73) that the frequently stated contrast between Juvenal who denounced wickedness and Horace who gently mocked folly and 'foibles' is fallacious. Ed. Fraenkel, *Horace*, provides a distinguished study of the satires as part of Horace's developing creative personality.

commitment to the Stoic sect.[1] Some satirists of the Flavian period (A.D. 70–96) were esteemed by Quintilian (X. I. 94) but their work has disappeared almost without trace. Such was the tradition of verse satire inherited by Juvenal.

The alternative convention of *satura*, the so-called Menippean satire, written in a mixture of verse and prose, though sharing some themes and a stylistic level with verse *satura*, is in many ways a separate tradition. Varro's Menippeans have some of the topics of Lucilius; though lacking his vehement denunciation of individuals they are devoted to discussions of personal and political morality and also literary problems. There is extensive treatment of the views of the philosophical sects. Varro is particularly concerned about the decline in old Roman Republican moral values. But in form he is immediately dependent on Menippus, a Greek-speaking Cynic of the third century B.C. There is also a fantastic invention in his *saturae*, which is unusual in Latin literature.[2] Menippean satire became in the first century A.D. a medium of great flexibility. In A.D. 54 Seneca wrote a work in a mixture of prose and verse entitled *Apocolocyntosis* ('Pumpkinification'), a skit on the apotheosis of the recently deceased emperor Claudius, in which vindictive ridicule of a dead enemy contains a measure of informed political criticism and is expressed with a masterly

[1] The *Oxf. Class. Text* of Juvenal, ed. Clausen, also contains an excellent text of Persius. The edition of Conington-Nettleship provides a translation as well as text and commentary. See also the essay on Persius by R. G. M. Nisbet, in (ed.) Sullivan, *Crit. Essays, Sat.* pp. 39–71.

[2] Cic. *Ac. post.* I. 8 makes Varro's immediate dependence on Menippus explicit. There is no English edition or translation of the approximately six hundred fragments of Varro's Menippeans. The standard edition is to be found in Bücheler-Heraeus, *Petronii Saturae*, pp. 177–250.

variation and economy of style.[1] A few years later
Petronius in his *Satyricon* turned the prose part of Menipp-
pean satire into a long novel and extended some verse
parts to be independent display pieces, and in his tale of
the bizarre adventures of rascals attained an ironical
mockery of contemporary Roman values.[2]

III

JUVENAL AS A SATIRIST

The dominant characteristic of Juvenal's satire is
rhetoric. The topics of much of his work and their treat-
ment are permeated by the prevailing mode of thought
and expression in the post-Augustan period. From the
beginning Latin poetry contained elements of rhetoric.
The *Aeneid* has many marks of a carefully organised
rhetoric in the shaping of speeches and in the well
placed general maxim (*sententia*). But in the first century
A.D. an intensified declamatory practice had affected all
aspects of literary composition. For the minor talents
this was merely the adroit manipulation of 'stock re-
sponses', but a few major writers mastered the resources
of an all-pervasive technique. Lucan sometimes lapses
into an exaggerated verbal facility and an irrelevant
sensationalism but in his clamorous depiction of a de-
moralised Rome subjected to the tyranny of Caesars and
in the fervent denunciation of corruption and sensuality
he has much in common with Juvenal. Juvenal claims

[1] Text and transl. by W. H. D. Rouse (together with
Petronius) Loeb Classical Library. The standard text is that
of Bü
cheler-Heraeus. On problems of interpretation see
M. Coffey, *Lustrum* 6 (1961), 239–71 and 309–11.

[2] Text and translation by M. Heseltine, Loeb Classical
Library. The best modern text is that of K. Müller. There is
an introduction to the translation of Petronius by J. Sullivan.

that *indignatio* impels him to write (1. 79); this word is used of the synthetic indignation of the declaimer as well as of a genuine personal anger.[1] A proper assessment of Juvenal's rhetoric is relevant to the understanding of all aspects of his work, both its general import and the interpretation of particular passages.

In the first satire, in many ways an introductory programme to the first book (Satires 1–5), Juvenal offers with an insistent series of angry rhetorical questions a syllabus of wickedness to be attacked either by giving brief pictures of typical sinners such as the murderer, swindler and adulterer or by describing at greater length the misdeeds and humiliations arising from avarice, the cardinal vice of Roman moralising; he reviews a skilfully varied succession of vices that his rhetorical training (1. 15 ff.) has equipped him to attack.[2] Towards the end after an imaginary interlocutor warns of the danger of denouncing individuals by name, Juvenal promises that his attacks will be directed at those already dead. Under the principate no individual could risk unrestricted free speech. Though by the time Juvenal published the tyranny of Domitian was in the past, sinister informers[3]

---

[1] On *indignatio* see Quint. XI. 3. 61 and Anderson, *Yale Class. Stud.* 17 (1961), 3–93. On indignation as a force animating writers in the first century A.D. see B. Otis, *Ovid*, p. 340. Kenney, *Latomus* 22 (1963), 704–20, has sensible general remarks on the role of rhetoric in Juvenal. For a detailed study of the influence of the practices of the rhetorical schools on Juvenal see De Decker, *Juvenalis Declamans* (Ghent, 1913).

[2] On *Sat.* I see Kenney, *Proc. Camb. Philol. Soc.* 8 (1962), 29–40; Anderson, *Yale Class. Stud.* 15 (1957), 34–45.

[3] Pliny *Epp.* IV. 7; see M. Winterbottom, *Journ. Rom. Stud.* 54 (1964), 92–6; see also Highet, *Juv.* pp. 54 ff. There was some continuity of government: Titinius Capito was *ab epistulis* under Domitian, Nerva and Trajan: see Syme, *Tacitus*, p. 93.

(*delatores*) survived even under the liberal administration
of Trajan. But the fear that an emperor whose reign
started tolerantly might turn tyrant would have deterred
a satirist from gaining the reputation of a universal
castigator. Further there were dangers from the civil law
of libel. In fact Juvenal does not criticise imperial policy.
He also avoids direct attacks on important contem-
poraries, even such figures as the ostentatious and wealthy
sophists who on a modern view seem to have been fitting
objects of satire.[1] Juvenal's failure to attack them may be
attributed not merely to prudence but also to a lack of
interest in portraying an individual except as an illus-
tration of a general pattern of behaviour; for such
moralising an example from the past would suffice. But
an older example may sometimes suggest a more modern
reference. The trial of Marius Priscus in A.D. 100 is men-
tioned by Juvenal as a recent example of provincial
maladministration, but the reader may also have thought
of the disgrace in *c.* A.D. 107 of Vibius Maximus, prefect
of Egypt.[2] Juvenal takes as his example of the foreigner
who achieves high success in an equestrian career
Tiberius Iulius Alexander, prefect of Egypt (A.D. 67–70)
and probably also praetorian prefect. The violence of his
language may be explained by the fact that Alexander
was a supporter of the Flavians, but he may also be
thinking of the career of more recent upstarts.[3] Though
Juvenal usually took his illustrative examples from the
past not in order to refer covertly to the present but

[1] Syme, *Tacitus*, pp. 511, 499.

[2] For Marius Priscus see I. 49 and VIII. 120; on the downfall
of Vibius Maximus see *Pap. Oxy.* 471, Syme, *Historia* 6 (1957),
480–7, and Sherwin-White on Plin. *Epp.* IX. 1.

[3] Juv. I. 129 ff. On the career of Ti. Iulius Alexander see
Furneaux on Tac. *Ann.* xv. 28. 4 and on his probable tenure of
the praetorian prefecture *Pap. Hib.* 215 and E. G. Turner,
Journ. Rom. Stud. 44 (1954), 54–64.

because he was concerned with a general mode of behaviour, it is perhaps better not to insist on too rigid a uniformity of procedure.[1]

Three satires from the first book are devoted to a general topic. The theme of *Sat.* II is the degradation attendant on homosexual practices and that of *Sat.* V the humiliation endured by the indigent client at his patron's dinner party. An underlying unity of theme explains the structure of these works. In *Sat.* III the structure is based on the contrast between the honest Umbricius and the series of unpleasant and dangerous experiences that make life in the great city intolerable.[2] The fourth satire is without parallel in Juvenal's writings, for it describes what purports to be a single historical event, a meeting of Domitian's imperial council to deliberate concerning the monstrous turbot that had been presented to the emperor. It is political satire in that Domitian's chief political advisers are present and that for all the prejudiced exaggeration it provides what is probably an accurate account of the procedure of an imperial council.[3] But it is primarily a moral study of a degrading situation, in which, along with the sycophant and the sinister, honest imperial administrators such as Acilius Glabrio, father and son, were compelled to grovel in adulation of the tyrant. There is a further dimension of satire, for the description of the council contains parody of a laudatory poem by Statius on Domitian's German

[1] Great caution is necessary in the handling of Juvenal's proper names; see the criticisms by Syme, *Tacitus*, p. 778, of Highet's interpretations, *Juvenal*, esp. pp. 289–94.

[2] On the structure of *Sat.* III see Anderson, *Yale Class. Stud.* 15 (1957), 55–68.

[3] See J. Crook, *Consilium principis* (Cambridge, 1955) and Syme, *Tacitus*, pp. 5 f. and 636. On the naive political judgment IV. 153 f. see M. P. Charlesworth, *Journ. Rom. Stud.* 27 (1937), 60 ff.

war. The preliminary attack on Crispinus before the main narrative is not of structural relevance and is best explained as an inconsequential prelude in the manner of Horace.[1]

The second book of Satires contains a single work of almost 700 lines, *Sat.* VI, an enormous indictment of deplorable characteristics and activities of women, ranging from lechery and sadism to a literary dilettantism. It seems impossible to impose a convincing scheme of structure on material that is presented in a succession of scenes and descriptions. An attack in catalogue form on the irksome characteristics of women was traditional, and a rhetorical poet's virtuoso treatment of the topic should not be condemned by irrelevant standards of structural balance.[2] The satires of the third book have as their main theme some aspect of degeneracy or debasement. In *Sat.* VII Juvenal complains that the serious writer and rhetorician looks in vain for patronage and cannot live well without some loss of integrity; the initial hope for imperial support sounds hollow and forlorn.[3] In *Sat.* VIII he attacks the degeneracy of the nobility and the irresponsibility of imperial administrators in contrast to Republican new men such as Cicero and Marius, and in *Sat.* IX uses the dialogue form to

[1] E.g. Hor. *Sat.* I. 2. W. C. Helmbold and E. N. O'Neil, *Am. Journ. Philol.* 77 (1956), 68–73, and Anderson, *Yale Class. Stud.* 15 (1957), 68–80, argue for a unity of structure, but see Highet, *Juvenal*, pp. 76 and 256 f.

[2] For an early Greek example see Semonides 7 D (a text and transl. will be found in Edmonds, *Elegy and Iambus* II. 216–225, Loeb Classical Library). It is not possible to separate the themes of attack on women and discussion of marriage. For well-argued attempts to find unity see Highet, *Juvenal*, pp. 91 ff. and 267 f. and Anderson, *Class. Philol.* 51 (1956), 73–94. The value judgments by Nettleship quoted in Duff's notes at the beginning of some of the satires are unhelpful.

[3] On *Sat.* VII see Anderson, *Class. Philol.* 57 (1962), 154–8.

present the furthest extreme of degradation in the relationship of patron and client.[1]

No scholar would take seriously nowadays the view of Ribbeck that the Satires of the remaining books are spurious. But some changes of technique may be noted. In *Sat.* x Juvenal expands with extended use of traditional *exempla* the thesis of the futility of the prayers of the generality of men that had already been treated succinctly in Persius' second satire. In *Sat.* xi homily precedes a personal invitation to a morally blameless meal. The dedicatees of this and other satires are not known historical persons; some or all may have been imaginary. *Sat.* xii also combines what purports to be a personal theme, thanksgiving for a friend's escape from a storm at sea, with ironical treatment of the legacy-hunter. Irony is also present in *Sat.* xiii, the first poem of the fifth book, a consolation offered to a man who had been cheated of some money.[2] In *Sat.* xiv the theme of parental example passes into that of avarice; unity of structure has been defended on the grounds that avarice is particularly easy to teach.[3] Many of the arguments in the satires of books four and five are expounded without denunciatory rancour, but *indignatio* is at its most forceful in the narrative of cannibalism in Egypt in *Sat.* xv and the incomplete *Sat.* xvi[4] expresses from the view-

[1] *Sat.* ix is sometimes highly esteemed by modern scholars for its ironical understatement, e.g. by Kenney, *Latomus* 22 (1963), 716, but the poem does not retain this quality consistently. Green, *Juvenal*, p. 13, regards it as of poor quality.

[2] See Anderson, *Class. Philol.* 57 (1962), 149 ff. as against Highet, *Juvenal*, pp. 140–4, who does not allow for ironical mockery.

[3] On *Sat.* xiv see O'Neil, *Class. Philol.* 55 (1960), 251 ff.

[4] On problems concerning *Sat.* xvi see Housman, preface p. lvii and Highet, *Juvenal*, pp. 287 ff. On some further questions of theme and structure in Juvenal, see Coffey, *Lustrum* 8 (1963), 202–9.

point of a civilian under-dog an almost obsessive sense
of grievance at the privileges of military life.

Juvenal was not a wide-ranging critic of contemporary
administration, nor will sociological reliability be ex-
pected of an angry rhetorical satirist. Some of his themes
seem to belong to the past. By his time the old nobility
that he attacks in *Sat.* VIII hardly existed at all. What was
left of the Republican aristocracy became degenerate in
the Julio-Claudian period and demonstrated its final
political impotence in the brief reign of Galba.[1] Nobility
that fell short of its obligations was a stock topic of
ancient declamation.[2] Juvenal's attack could have had a
contemporary reference in so far as it included the
immoral and unseemly behaviour of the holder of high
office. Often Juvenal's attitude towards a section of
society or an institution varies according to his rhetorical
thesis. He is indignant at the exploitation of provincials
by Roman pro-magistrates (e.g. VIII. 87–139) and regards
the moral standards of the inhabitants of some of the
farthest territories as superior to those of the officials
who corrupt them (II. 159–70), but he also expresses
contempt for their aspirations to education.[3] He has a
vague sense of the dignity of Rome's past, but in *Sat.* X
the traditional pomp of a magistrate's procession is an
object of derision, for it reveals the futility of power
(x. 36–46).

It has been argued that Juvenal's examples of sophisti-

[1] Syme, *Tacitus*, p. 778; also pp. 566–84.

[2] E.g. the examples given by Valerius Maximus (III. 4), a
rhetorician of the time of Tiberius; see Highet, *Juvenal*,
p. 272.

[3] On provincial oratory see xv. 110 ff. and J. R. C. Martyn,
Hermes 92 (1964), 121–3. Tacitus' attitude is similarly ambi-
guous: admiration for the unspoiled simplicity of the Germans
in *Germ.* and contempt for the Britons' imitation of Roman life
as *pars servitutis* (*Agr.* 21).

cated vice come from the age of Nero and that the recently deceased old lady mentioned by Pliny who owned her private dancing troupe reflected the tastes of those times.[1] No doubt many of the new governing class of the late first and early second centuries had higher standards of morality than their immediate predecessors, but Pliny, it may be surmised, does not give a representative view of contemporary morals, as he seems anxious to portray himself as the friend of agreeable people. But Juvenal, according to his brief in *Sat.* VI. 434-56, would have found the intellectual women praised by Pliny worse than tedious.[2] Juvenal hated wealthy freedmen, particularly those of oriental origin (VII. 14 ff.). Inscriptions from Rome's port Ostia illustrate that to some extent Syrian Orontes did flow into the Tiber (III. 62) and that some of the freedmen were very rich. But they also show that the relationship between patron and freedman was often one of friendship.[3] Juvenal expresses the angry jealousy felt by the supine indigent, who despised the energetic enterprise and commercial success of the lowborn and regarded patronage as the sole source of material well-being.[4] For the slave except the arrogant flunkey

---

[1] See Sherwin-White on Plin. *Epp.* I. 14. 4; VII. 24. 3.

[2] Sherwin-White on Plin. *Epp.* V. 16. 3. Sherwin-White's commentary gives valuable information on many topics treated by Juvenal. Duff (pp. xxxiv f.) warned against accepting without reservation Pliny's presentation of the morality of his times. A. Serafini, *Stud. sat. Giovenale*, p. 16, regards Juvenal's picture of morality in Rome as substantially true; see Coffey, *Lustrum* 8 (1963), 164 f.

[3] Some of the names are of Semitic origin; see on this Meiggs, *Roman Ostia*, p. 224, and on the role of freedmen pp. 217-24. Meiggs's valuable study of the life of a Roman town supplements and corrects literary sources on Italian life. The best compact study of Roman society is J. Carcopino, *Daily life in ancient Rome.*

[4] On this attitude see Green, *Juvenal*, pp. 25-35.

Juvenal shows compassion,[1] and there is from time to time in references to the poor and the humiliated a humane attitude that is rare in Latin literature.[2]

Though Juvenal's impassioned plea for human pity (xv. 131–50) has the colouring of organised philosophy, in general he regards philosophers as eccentrics or hypocrites. As he asserts in the manner of Horace that he is not committed to any sect (XIII. 120–3), it seems unwarranted to claim his adherence to any philosophical school.[3] His moral fervour is not that of the doctrinaire reformer.

It has already been shown that the traditional *exempla* of the rhetorical schools are an essential part of Juvenal's exposition of his themes. Some aspects of his technique may be noted. The melting down of the effigies of a fallen tyrant or favourite by an angry populace to make domestic utensils is a rhetorician's commonplace which Juvenal incorporates in a description of the downfall of Sejanus (x. 56–107) that may derive some of its power from the narrative of Tacitus. But for Juvenal perhaps real experience rejuvenated a *locus communis*, as immediately after the assassination of Domitian he may

---

[1] E.g. I. 92 f.; VI. 219 f.; VI. 475–95; the arrogant slave, v. 59–75. Juvenal's attitude accords with the humanitarianism of some of the decisions attributed to Hadrian on the treatment of slaves; see, for example, Script. Hist. Aug. *Hadr.* XVIII. 7f.

[2] E.g. pity for the man killed on a street accident (III. 257–67), and for the pauper who loses in a fire what little he has (III. 200–9). Sometimes sympathy is combined with contempt as in the humiliations of the husband in *Sat.* VI or of the *cliens* in *Sat.* V.

[3] Cf. Hor. *Epp.* I. 1. 14. On Pythagoreans as a comic *exemplum* see III. 229 and for the philosopher as hypocritical betrayer see III. 116–18. Highet, *Trans. Proc. Am. Phil. Ass.* 80 (1949), 254–70 asserts that the later satires show signs of a 'conversion' to Epicureanism, but occasional arguments based on philosophical doctrines, e.g. the Stoicism of Zeno in *Sat.* XV, belonged also to the intellectual equipment of the uncommitted, particularly the rhetorician.

have witnessed the destruction of his statues.[1] The celebrated description of Messalina's erotic excesses (VI. 115–32) owes many garish details to a standard practice case of declamation that had nothing to do with a historical person, but the unreality of the schools is combined with a historical plausibility to create a sensational paradigm of aristocratic impurity.[2] Sometimes Juvenal gives an unexpected pejorative application to a usually edifying historical example. Cornelia, mother of the Gracchi, a traditional model of Republican maternal virtue, is to Juvenal a type of female arrogance to be compared directly to Niobe, who in turn is compared to the white sow of Alba.[3]

While there is much bathos that is often both comic and bitter, Juvenal's language departs from that of his predecessors in its occasional approximation to the grand rhetorical manner of Lucan and sometimes for a few lines, as in the description of early Rome (XI. 111–16), attains a level akin to high poetry. Juvenal seems conscious of an intensification of manner that may appear to violate the traditions of the genre of satire, but evades the charge of stylistic pretentiousness by asserting that the heinous misdeeds of figures of tragic poetry are of everyday occurrence in contemporary Rome (VI. 634–40). His range of vocabulary is wide with many colloquialisms, but for the most part he avoids both the verbal

[1] See Plut. *Praec. rep. ger.* XXVII. 820 E, Diog. Laert. v. 77, and for the historical reality Plin. *Pan.* 52; on the *locus communis* see B. Lavagnini, *Athenaeum* 25 (1947), 83–8.

[2] Sen. Rhet. *Contr.* I. 2; de Decker, *Juv. Declamans*, p. 29, n. 2, who in his useful study shows the extent of Juvenal's indebtedness to the matter and manner of the corpus of Latin rhetorical declamations of the Elder Seneca and ps.-Quintilian but does not discuss Juvenal's creative contribution to traditional material.

[3] Juv. VI. 167–77; contrast Val. Max. IV. 4 *proem.*

obscenities that were probably considered unsuitable for the language of public recitation and also at the other extreme words that belong exclusively to epic and tragedy.[1] His imagery shows a similar variation of manner.[2] Like most satirists he has a penchant for the incongruous or even grotesque simile to illustrate unseemly behaviour, as when the intemperate woman athlete who vomits is compared to a snake that falls into a vat (VI. 431 f.). Elsewhere a tone of disgust is mingled with pathos, as when in the context of the futility of a prayer for long life Juvenal compares the helpless paralytic to a baby swallow with beak wide open for food (x. 228–32). A few lines later he uses the epic manner to compare the death of Priam to that of a sacrificed ox (x. 258–71). The Virgilian associations add pathos to the compassionate description; he condemns the fault but pities the victim. Juvenal's satires have an intense rhetoric and an enhanced level of language that explain Jos. Scaliger's description of them as *satirae tragicae*,[3] and like the historical writings of Tacitus they display supreme skill in portraying with a savage irony the vices and pretensions of imperial Rome.

IV

THE MANUSCRIPTS

Since the first publication of Duff's edition, three important critical texts of Juvenal have appeared. A. E. Housman's edition was the first to offer a critical apparatus in which there is a balanced appraisal of readings from

[1] On Juvenal's range of style see Highet, *Juvenal*, pp. 295 f.

[2] Anderson discusses aspects of Juvenal's imagery *Am. Journ. Phil.* 81 (1960), 243–60; see Coffey, *Lustrum* 8 (1963), 212.

[3] Referred to by Knoche, *Röm. Sat.* p. 94.

manuscripts of both classes. His knowledge of the manuscripts was limited but his critical acumen of the highest distinction; his apparatus contains some exegetical notes, including defence of his own conjectures. In the preface of the first edition he discusses problems of recension, conjecture and interpolation and in that of the second details of interpretation. His lucid prefaces are marred by an arrogant contempt for other scholars, even those without whose work his own edition could not have been made. Knoche published in 1950 an edition based on a wide knowledge of the manuscripts of the so-called interpolated class, which he divides into four main groups with contamination; in his critical apparatus he uses an elaborate system of *sigla* in order to present economically the readings of an unusually large number of sources and reports extensively the suggestions of modern scholars. His edition is invaluable for the amplitude of the information it provides. Clausen in his text (1959) sets aside any grouping of the enormous mass of interpolated manuscripts and uses for the most part only those manuscripts of this class which formed the basis of Housman's apparatus together with some items mentioned by Housman and Knoche as of occasional interest. He made a thorough inspection of P and so offers a reliable report of its readings such as has hitherto not been available. Clausen's edition is judicious and is the most convenient modern critical text. But even though the critical study of Juvenal in the twentieth century has achieved much, Duff's text may still be accepted as basically sound, for his judgment was good and he was unwilling to reject a reading merely because it was not found in the *optimus codex*.[1]

---

[1] For a review of Housman's first edition see W. M. Lindsay, *Class. Rev.* 19 (1905), 463–5 and on the second U. Knoche, *Gnomon* (1933), pp. 242–54. For a review of Knoche's edition

Some more recently discovered evidence for the tradition akin to P has been made available in the manuscript Parisinus 8072 of the tenth century (Clausen's R); citations of Juvenal in the metrical florilegium compiled in A.D. 825 by Mico, a monk of St Riquier near Corbie, belong to the same tradition.[1] Three fragments not known to Duff are of some interest. A sixth century Ambrosian fragment of uncertain textual affiliation preserves the following lines of *Sat.* XIV: 250–6, 268–19, 303–19.[2] A leaf of a papyrus codex written at about A.D. 500 containing *Sat.* VII. 149–98 was found at Antinoë in Egypt. The relatively uncorrupted text offers nothing new of value but has some readings found only in a few manuscripts. Critical signs in the papyrus seem to indicate that l. 192 was regarded as in some way noteworthy.[3] Fragments of Juvenal (II. 32–89 and III. 35–93) were found as the impress of writing left behind on the surface of the inside of the covers of a Carolingian school textbook in Orléans after the leaf glued to the cover had been stripped away. The original Juvenal leaves were of a date no later than the middle of the ninth century and

see S. Prete, *Gnomon* 24 (1952), 328–31 and on Clausen's see J. G. Griffith, *Class. Rev.* N.S. 11 (1961), 51–8, and Anderson, *Am. Journ. Philol.* 82 (1961) 434–41. As an example of Duff's judgment note his acceptance at VII. 177 of Jahn's conjecture *scindes* as a 'certain correction'; it has since been confirmed by manuscript evidence (see Clausen *a.l.*).

[1] On R see C. E. Stuart, *Class. Quart.* 3 (1909), 1–7 and Housman, 2nd ed. p. xlii; on Mico see Knoche, *Handschr. Grundl. Philologus Supplbd* 33, 1 (1940), 44 and Courtney, *Bull. Inst. Class. Stud.* 14 (1967), 45; see also Clausen (ed.) p. ix.

[2] On the readings of the Ambrosian fragment see Housman, 2nd ed., p. lv.

[3] On the Antinoe papyrus see C. H. Roberts, *Journ. Egypt. Arch.* 21 (1935), 199–209; l. 192 had been regarded by Jahn as interpolated but is probably genuine.

the text akin to that of P but also showing traces of the interpolated tradition.[1] Readings of importance from these fragments are reported in Clausen's critical apparatus. Scholars have given some weight to a Vienna manuscript of late ninth-century date which holds a intermediate position between the two traditions.[2]

It was discovered at the end of the last century that a late eleventh century manuscript of Juvenal in Oxford contained 36 lines of *Sat.* VI (after 365 and 373) that were not found in any other manuscript and that the last five lines of the main piece of 34 lines appeared as three (VI. 346–8) in the rest of the tradition.[3] Housman, who did much to elucidate these difficult and obscene lines, believed them genuine, concluding that while VI. 346–8 were an accidental truncation made after the loss of 1–29 of the Oxford lines, their transposition to follow 345 was deliberate.[4] Opinion on the genuineness of these lines has been divided, but nowadays it seems that a majority of scholars would accept them as authentic.[5] Assuming the lines to be genuine the disquieting lesson

[1] On the Orléans manuscript (cod. Aurelianensis 295) see A. P. McKinlay and E. K. Rand, *Harv. Stud. Class. Phil.* 49 (1938), 229–63.

[2] On Vindobonensis 107 see Clausen (ed.), p. ix.

[3] The manuscript, Oxoniensis bibl. Bodleianae Canonicianus Class. Lat. 41, was first reported on by E. O. Winstedt, *Class. Rev.* 13 (1899), 201–5. The lines are usually referred to as Juv. VI. 365, O 1–34 and VI. 373 A and B.

[4] Housman, 2nd ed., pp. xxix and xlviii f.; see also *Class. Rev.* 13 (1899), 266 f., 15 (1901), 263–6; Courtney, *Mnem.* Ser. 4, 15 (1962), 266 believes probably rightly that the truncation was made deliberately.

[5] Against authenticity: U. Knoche, *Philologus* 93 (1938), 196–217 and B. Axelson, ΔΡΑΓΜΑ M. P. Nilsson, pp. 41–55; in favour: Highet, *Juvenal*, pp. 335 f., Clausen (ed.), p. xiii, E. Courtney, *Mnem.* Ser. 4, 15 (1962), 262–6, J. G. Griffith, *Hermes* 91 (1963), 104–14; see also Coffey, *Lustrum* 8 (1963), 179–84.

of the discovery is not merely the possibility of a lacuna at a hitherto unsuspected point but also the degree of falsification to which a defective passage could be subjected in late antiquity.

One of the most vexatious editorial problems in Juvenal has been that of interpolations, not the limited verbal alterations associated with one part of the tradition, but the postulated intrusion of single lines or even small groups. Duff's attitude was of a characteristic good sense and moderation.[1] Housman, while arguing for the expulsion of a number of verses, believed that, though some lines in Juvenal could be regarded as flat and unnecessary, that was not in itself grounds for declaring them spurious.[2] A more radical approach in the tradition of Jahn is advocated by Jachmann, who argues that the detection of spurious lines is the principal critical problem in Juvenal.[3] A few of his cleverly argued examples are convincing, e.g. the deletion of XII. 50–1, already attacked by Bentley (on Hor. *AP* 337), but some of the difficulties that he attempts to solve by the deletion of one or more lines are better approached by postulating a limited area of corruption, e.g. at VIII. 6–8, where, in contrast to Jachmann's expulsion of the three lines, Housman had supposed that the first half of l. 7 was corrupt and read *pontifices posse ac*, not a certain correction but right in method. Jachmann regards Juvenal as a succinct writer who never wastes a word and so sometimes mistakes the characteristic rhetorical expansiveness of the poet himself for the work of an interpolator. His approach has left his mark on later editions: Knoche

---

[1] See v. 91 n.

[2] 2nd ed. pp. xxi–xxxvi; Housman's discussion of the interpolation problem is masterly.

[3] G. Jachmann, *Studien zu Juv.* (Nachr. Göttingen, 1943), pp. 187–266.

brackets more than a hundred lines and Clausen about forty. While much may be learnt from Jachmann's critical acumen, the answers to the problems he rightly raises will only rarely be found in excision.[1]

It is possible that Juvenal may have altered the original version of some parts of his poems, but it is not necessary to assume that he made a thorough revision.[2] After his death or even in his lifetime he became an unfashionable author; his style would have seemed unattractive to the archaizing taste of an age that ridiculed the rhetoric of Lucan (Fronto 157N.) and some of his themes tendentious at a time when emperors behaved like some of his worst miscreants.[3] Texts of Juvenal, it may be supposed, did not disappear altogether, but he is not cited by grammarians earlier than Servius, not even Donatus.[4] In the later part of the fourth century[5] there was a revival

[1] Some scholars have reacted against Jachmann's approach, e.g. J. G. Griffith, *Class. Rev.* N.S. 1 (1951), 138–42; *Festschr. B. Snell*, pp. 101–11.

[2] On possible author-variants see Griffith, *Festschrift B. Snell*, pp. 101–11. Fr. Leo, *Hermes* 44 (1909), 600–17, proposed the theory of a second edition by the author, a view regarded as an open question by Housman, 2nd ed. pp. xxxviii–xli and Clausen (ed.), p. xiii.

[3] On knowledge of Juvenal in the Late Empire, the Dark Ages, Mediaeval and later times see the excellent treatment by Highet, *Juvenal*, pp. 180–232. Highet's book has rightly been criticised for the excessively discursive and sensational description of the individual satires (see, for example, Kenney, *Class. Rev.* N.S. 5 (1955), 278–81), but the fully documented and lucid discussions of the main problems of interpretation make it a work of considerable value and importance.

[4] A possible exception is the unimportant grammarian Phocas, who cites Juvenal; on Strzelecki's dating (*RE* 20, 1 (1941), 318 f.) he is earlier than Donatus.

[5] A. D. E. Cameron, *Hermes* 92 (1964), 363–77, suggests that a renaissance in Juvenal studies should be attributed to the middle rather than to the end of the fourth century; see also idem, *Journ. Rom. Stud.* 54 (1964), 15–28.

of interest. He is quoted frequently by Servius and according to the historian Ammianus Marcellinus (XXVIII. 4. 14) was in vogue among Roman aristocrats who despised almost all other literature. About the same time grammarians elucidated his text and its subject-matter. Some of the worst corruptions in the text of Juvenal probably antedate this period.[1] A manuscript subscript by Nicaeus states that he emended the text of Juvenal in the house of Servius, but such activity does not necessarily imply that a thorough critical edition was made.[2] The ancestor of P and of its associated manuscripts belongs to this era; while containing some corruptions not found in the rest of the tradition it was comparatively free from arbitrary alterations.[3] At about the same time another version of the text was produced, in which, in order to turn Juvenal into a writer more easily assimilable by the relatively unlettered, many unintelligent changes were made that often turned his genius into a normalised mediocrity. This recension was the origin of the majority of the extant manuscripts, which represent an ever continuing process of falsification in varying degrees; but it is impossible to trace with confidence the vicissitudes of transmission that befell an increasingly contaminated mass of manuscripts of a popular author, of whom more than one copy survived

[1] See Knoche, *Handschr. Grundl. Philologus Supplbd* 33, 1. (1940), 68.

[2] Note the sceptical remarks by Housman (ed.), *Lucan*, pp. xvi ff.

[3] Clausen (ed.), p. xii, n. 2, gives a list of passages where P alone preserves the truth. For a discussion of places where the other tradition is to be preferred see Housman, 2nd ed., pp. xvii f. One example of a reading of P accepted by Duff but rejected by later editors will suffice: at v. 169 they rightly prefer the more positively significant *tacetis* of the interpolated manuscripts to the *iacetis* of P (see Duff *a.l.*).

the Dark Ages into Carolingian times.[1] We also owe to the scholarship of the fourth-century revival the short notes that make up the transmitted corpus of the ancient scholia of Juvenal.[2] The accompanying citations of the text (lemmata) are important witness of the tradition represented by P, but there are indications that the notes were originally composed for a more inter-polated text.[3]

V

TRANSLATIONS AND COMMENTARIES

There is a good translation with scholarly introduction and notes on interpretation by Peter Green, and a read-able but not always scholarly version by Rolf Humphries. A commentary on the whole of Juvenal, including those parts omitted by Duff, will be found in L. Friedlaender's edition.

[1] The most important study of the complicated transmission of the text is that of U. Knoche, *Handschr. Grundl.*, esp. ch. 1. See also Courtney, *Bull. Inst. Class. Stud.* 14 (1967), 38–50.

[2] There is an excellent edition of the ancient scholia by P. Wessner. On some characteristics of the scholia see Highet, *Juvenal*, p. 299. On the value of the text and notes used by G. Valla (Duff, pp. l–li) see W. Anderson, *Traditio* 21 (1965), 383–424.

[3] Courtney, *Bull. Inst. Class. Stud.* 14 (1967), 40–3, has argued that Housman, 2nd ed., p. xxviii, overestimated the value of readings given in the scholia. On the original text from which the scholia were made see Wessner, (ed.) *Schol.* pp. xvii f.

VI

BIBLIOGRAPHY

A very useful bibliography with descriptive and evalua-
tive comments is given by Highet, *Juvenal*, 339–46. A select
bibliography including work later than that available to Highet
will be found in P. Green, *Juvenal*, 299–303. For extensive
bibliographical reports see the Section on Juvenal in the general
accounts of work on the Roman satirists in Bursian-Jahres-
berichten, the last of which was by R. Helm, 282 (1943),
15–37. For the period 1941–61 see M. Coffey, *Lustrum* 8 (1963),
pp. 161–215 and 268–70. In the following bibliography a few
particularly attractive suggestions on the text have been singled
out for specific mention; this does not imply adverse criticism
of others.

W. S. Anderson, Juvenal 6, A problem in structure, *Class.
Philol.* 51 (1956), 73–94.

Studies in Book I of Juvenal, *Yale Class, Stud.* 15 (1957),
31–90.

Imagery in the Satires of Horace and Juvenal, *Am. Journ.
Philol.* 81 (1960), 225–60.

Juvenal and Quintilian, *Yale Class. Stud.* 17 (1961), 3–93.

The programs of Juvenal's later books, *Class. Philol.* 57 (1962),
145–60.

Anger in Juvenal and Seneca, *Univ. California Publ. in
Class. Philol.* 19, 3 (1964), 127–96.

Valla, Juvenal and Probus, *Traditio* 21 (1965), 383–424.

B. Axelson, A problem of genuineness in Juvenal, ΔΡΑΓΜΑ
M. P. Nilsson Dedicatum (Lund, 1939), 41–55.

D. R. S. Bailey, Seven emendations, *Class. Rev.* N.S. 9 (1959),
201 f. [XIII. 153 for *solitus* read *solidum*].

A. T. von S. Bradshaw, *Glacie aspersus maculis*: Juv. v. 104,
Class, Quart. N.S. 15 (1965), 121–5.

F. Bücheler-G. Heraeus, *Petronii Saturae*, ed. 6 (Berlin, 1922).

A. D. E. Cameron, The Roman friends of Ammianus, *Journ.
Roman Stud.* 54 (1964), 15–28.

Literary Allusions in the Historia Augusta, *Hermes* 92 (1964),
363–77.

J. Carcopino, *Daily Life in Ancient Rome*, rev. ed. (Penguin
books, Harmondsworth, 1956).

M. P. Charlesworth, Flaviana. IV. Juvenal, Domitian and the
'Cerdones', *Journ. Roman Stud.* 27 (1937), 60–2.

W. V. Clausen, Two notes on Juvenal, *Class. Rev.* N.S. 1 (1951), 73 f. [A discussion of II. 168 and X. 311 ff.]

Silva coniecturarum, *Am. Journ. Philol.* 76 (1955), 52–60. [A discussion of II. 132 f.; V. 103–6 (reading *aut glaucis sparsus maculis*); XI. 106 f.]

A. Persi Flacci et D. Iuni Iuvenalis Saturae (Oxford, 1959).

M. Coffey, Seneca, Apocolocyntosis 1922–1958, *Lustrum* 6 (1961), 239–71. and 309–11

Juvenal. Report for the Years 1941–61, *Lustrum* 8 (1963), 161–215 and 268–70.

J. Colin, 'Galerus': pièce d'armament du gladiateur ou coiffure de prêtre salien?
Les Études Classiques 23 (1955), 409–15 [discusses VIII. 198–210].

S. T. Collins, Notes on Juvenal, Apuleius, etc., *Class. Quart.* 3 (1909), 279 f. [XVI. 25 reads *adsit*, a certain correction.]

J. Conington-H. Nettleship, *The Satires of A. Persius Flaccus* ed. 3 (Oxford, 1893).

E. Courtney, *Vivat ludatque cinaedus, Mnemosyne* Ser. 4, 15 (1962), 262–6.

Juvenaliana, *Bulletin of the Institute of Classical Studies* 13 (1966), 38–43, [discussion of I. 37–41; I. 142 f.; VI. 116–21; VI. 133–5; VIII. 108–22; X. 311–3; XI. 145–8; XII. 24–33; XIII. 34–7; XIII. 247–9 (suggests *Drusum for surdum*.]

The Transmission of Juvenal's Text, *Bulletin of the Institute of Classical Studies* (London) 14 (1967), 38–50.

J. A. Crook, *Consilium principis* (Cambridge, 1955).

J. De Decker, Juvenalis Declamans, *Étude sur la rhétorique déclamatoire dans les satires de Juvénal* (Ghent, 1913).

G. E. Duckworth, *The Nature of Roman Comedy* (Princeton, 1952).

J. Wight Duff, *A literary history of Rome from the origins to the close of the Golden Age*, rev. ed. by A. M. Duff (London, 1960).

A literary history of Rome in the Silver Age from Tiberius to Hadrian, rev. ed. by A. M. Duff (London, 1964).

D. E. Eichholz, The art of Juvenal and his Tenth Satire, *Greece and Rome*, N.S. 3 (1956), 61–9.

G. B. A. Fletcher, Alliteration in Juvenal, *Durham Univ. Journal* 36 (1943–4), 59–64.

E. Fraenkel, *Horace* (Oxford, 1957).

L. Friedlaender, *D. Junii Juvenalis Saturarum Lib. V mit erklärenden Anmerkungen* (Leipzig, 1895).

    Roman Life and Manners under the Early Empire, trans. J. H. Freese and L. A. Magnus (4 vols., London, 1908–13).

G. Giangrande, Juvenalian Emendations and Interpretations, *Eranos* 63 (1965), 26–41 [discussion of I. 155 ff.; VI. 50–1; x. 196 f. (for *ille alio* read *ardalio*); XI. 145–51 (for *in magno* read *Inachio*); XII. 10–16.]

    Textkritische Beiträge zu lateinischen Dichtern, *Hermes* 95 (1967), 110–21 [discussion of III. 216–20; V. 103–6; VI. 451–6].

P. Green, *Juvenal: The Sixteen Satires*, translated with an introduction and notes (Penguin Books, Harmondsworth, 1967).

J. G. Griffith, Varia Iuvenaliana, *Class. Rev.* N.S. 1 (1951), 138–42 [discusses I. 2; VII. 15 f. (defence of the manuscript tradition)].

    Author-Variants in Juvenal: a reconsideration, *Festschrift Bruno Snell* (Munich, 1956), 101–11.

    A gerundive in Juvenal, *Class. Rev.* N.S. 10 (1960), 189–92 [at XII. 14 defends *a grandi*.]

    Juvenal and stage-struck patricians, *Mnemosyne*, Ser. 4, 15 (1962), 256–61 [on VIII. 192–6].

    The survival of the longer of the so-called 'Oxford' fragments of Juvenal's sixth satire, *Hermes* 91 (1963), 104–14.

    Frustula Iuvenaliana, *Class. Quart.* N.S. 19 (1969), 379–87.

    Juvenal, statius and the Flavian Establishment, *Greece and Rome*, 2nd ser. 16 (1969), 134–50.

E. L. Harrison, Neglected hyperbole in Juvenal, *Class. Rev.* N.S. 10 (1960), 99–101 [at III. 240, *Liburna* is the right reading.]

W. C. Helmbold, The Structure of Juvenal 1, *Univ. California Publ. Class. Philol.* 14 (1954), 47–59.

    and E. N. O'Neil, The structure of Juvenal 4, *Amer. Journ. Philol.* 77 (1956), 68–73.

    and E. N. O'Neil, The Form and purpose of Juvenal's seventh satire, *Class. Philol.* 54 (1959), 100–8.

G. L. Hendrickson, The dramatic *Satura* and the Old Comedy at Rome, *Amer. Journ. Philol.* 15 (1894), 1–30.

M. Heseltine, transl. *Petronius, The Satyricon*, Loeb Classical Library (London and New York, 1916).

G. Highet, The Philosophy of Juvenal, *Transactions of the American Philol. Assoc.* 80 (1949), 254–70.

    Juvenal the Satirist, (Oxford, 1954).

A. E. Housman, The new fragment of Juvenal, *Class. Rev.* 13 (1899), 266 f.

The new fragment of Juvenal, *Class. Rev.* 15 (1901), 263–6.

D. Iunii Iuvenalis Saturae, ed. 2 (Cambridge, 1938).

R. Humphries, *The satires of Juvenal* (Bloomington, Indiana, 1958).

G. Jachmann, Studien zu Juvenal, *Nachrichten d. Akademie der Wissenschaften Göttingen Phil.-Hist. Kl.* 6 (1943), 187–266.

O. Jahn, *Satura, Hermes* 2 (1867), 225–51.

E. J. Kenney, The first satire of Juvenal, *Proceedings of the Cambridge Philological Society* N.S. 8 (1962), 29–40.

Juvenal, Satirist or Rhetorician? *Latomus* 22 (1963), 704–20.

D. A. Kidd, Juvenal, 1. 149 and 10. 106–7, *Class. Quart.* N.S. 14 (1964), 103–8.

A. Kiessling-R. Heinze, *Q. Horatius Flaccus Satiren*, ed. 6 mit Nachwort und bibliographischen Nachträgen von Erich Burck (Berlin, 1957).

Fr. Klingner, *Horatius Opera*, ed. 3 (Leipzig, 1959).

U. Knoche, Ein Wort zur Echtheitskritik—[Juvenal] 6, 365, 1–34, *Philologus* 93 (1938), 196–217.

Handschriftliche Grundlagen des Juvenaltextes, *Philologus Supplementband* 33, 1 (1940).

Die römische Satire, ed. 2 (Göttingen, 1957).

D. Iunius Iuvenalis Saturae (Munich, 1950).

E. Laughton, Juvenal's other elephants, *Class. Rev.* N.S. 6 (1956), 201 [discussion of x. 148 ff.].

B. Lavagnini, Motivi diatribici in Lucrezio e in Giovenale, *Athenaeum* 25 (1947), 83–8.

Fr. Leo, Varro und die Satire, *Hermes* 24 (1889), 67–84; repr. in *Ausgewählte Kleine Schriften* 1 (Rome, 1960), pp. 283–300.

Doppelfassungen bei Juvenal, *Hermes* 44 (1909), 600–17.

H. Lloyd-Jones, The Seal of Posidippus, *Journ. Hellenic Stud.* 83 (1963), 75–99.

L. A. MacKay, Two notes on Juvenal, *Class. Rev.* 58 (1944), 45 f. [discusses III. 186; VII. 16].

A. P. McKinlay and E. K. Rand, A fragment of Juvenal in a manuscript of Orléans, *Harvard Stud. Class. Philol.* 49 (1938), 229–63.

J. R. C. Martyn, Juvenal on Latin Oratory, *Hermes* 92 (1964), 121–3.

Friedlaender's Essays on Juvenal (Amsterdam, 1969) (translation of most of the introduction to Friedlaender's edition of Juvenal).

Fr. Marx, *C. Lucilii Carminum Reliquiae* (2 vols., Leipzig, 1904 and 1905).

H. A. Mason, Is Juvenal a Classic? in (ed.) J. P. Sullivan, *Critical essays on Roman literature (Satire)* (London, 1963), 93–176.

R. Meiggs, *Roman Ostia* (Oxford, 1960).

P. Meriggi, Osservazioni sull' Etrusco, *Studi Etruschi*, 11 (1937), 157, 196 f. and n. 177.

K. Müller, *Petronii Arbitri Satyricon* (Munich, 1961).

R. G. M. Nisbet, review and discussion of Clausen, A. Persi Flacci et D. Iuni Iuvenalis Saturae, *Journ. Roman Stud.* 52 (1962), 233–8 [suggests emendations at VI. 107; XI. 112 (for *mediamque* read *tacitamque*); XIII. 23; XIII. 108 (for *vexare* read *vectare*); XIV. 269.]

Persius, in (ed.) J. P. Sullivan, *Critical essays on Roman literature (Satire)* (London, 1963), pp. 39–71.

E. N. O'Neil, The structure of Juvenal's fourteenth satire, *Class. Philol.* 55 (1960), 251–3.

B. Otis, *Ovid as an epic poet* (Cambridge, 1966).

A. Palmer, *The Satires of Horace* (London, 1883).

L. R. Palmer, *The Latin Language* (London, 1954).

C. H. Roberts, The Antinoë fragment of Juvenal, *Journ. Egyptian Archaeology* 21 (1935), 199–209.

D. S. Robertson, Juvenal 8, 241, *Class. Rev.* 42 (1928), 60 f. (read *quantum igni Leucade*).

Juvenal 5, 103–6, *Class. Rev.* 60 (1946), 19 f.

N. Rudd, *The Satires of Horace* (Cambridge, 1966).

A. Serafini, *Studio sulla satira di Giovenale* (Florence, 1957).

A. N. Sherwin-White, *The Letters of Pliny. A historical and social commentary* (Oxford, 1966).

B. Snell, Etrusco-Latina, *Studi Italiani di Filologia Classica*, 17 (1940), 251 f.

P. T. Stevens, *In summa*, *Class. Rev.* 63 (1949), 91 f. [discussion of III. 79 f.].

C. E. Stuart, An uncollated manuscript of Juvenal, *Class. Quart.* 3 (1909), 1–7.

J. P. Sullivan, *Petronius. The Satyricon and the Fragments* (Penguin Books, Harmondsworth, 1965).

The Satyricon of Petronius (London, 1968).

R. Syme, C. Vibius Maximus, Prefect of Egypt, *Historia* 6 (1957), 480–7.

Tacitus (2 vols., Oxford, 1958).

Pliny's less successful friends, *Historia* 9 (1960), 362–79.

A. Thierfelder, Juvenal 6, 57, *Hermes* 76 (1941), 317 f. [emends, probably rightly, *cedo* to *credo*].

E. G. Turner, Tiberius Iulius Alexander, *Journ. Roman Stud.* 44 (1954), 54–64.

B. L. Ullman, Dramatic Satura, *Class. Philol.* 9 (1914), 1–23.

J. Vahlen, *Ennianae Poesis Reliquiae*, ed. 3 (Leipzig, 1928).

C. A. Van Rooy, *Studies in Classical Satire and related literary theory* (Leiden, 1965).

A. Walde-J. B. Hofmann, *Lateinisches etymologisches Wörterbuch*, ed. 3 (2 vols., Heidelberg, 1938 and 1954).

E. H. Warmington, *Remains of Old Latin I, Ennius and Caecilius*, Loeb Classical Library, rev. ed. (Cambridge, Mass., and London, 1961).

Remains of Old Latin III, Lucilius; The Twelve Tables, Loeb Classical Library (Cambridge, Mass., and London, 1957).

T. B. L. Webster, *Hellenistic Poetry and Art* (London, 1964).

O. Weinreich, *Römische Satiren*, ed. 2 (Zürich and Stuttgart, 1962).

P. Wessner, *Scholia in Iuvenalem vetustiora* (Leipzig, 1931).

E. O. Winstedt, A Bodleian manuscript of Juvenal, *Class. Rev.* 13 (1899), 201–5.

M. Winterbottom, Quintilian and the *vir bonus*, *Journ. Roman Stud.* 54 (1964), 90–7.

I am very grateful to Mr E. Courtney, who read my typescript and made valuable suggestions. Errors are entirely my responsibility.

January 1970   M.C.

Centaur – a supposed monster of Thessaly, ½-man 7½ horse

ABBREVIATIONS

P	= The original reading of the *Codex Pithoeanus* (Montpellier 125).
p	= The reading substituted for *P* by later hands.
ω	= The reading of all, or a majority of, the other MSS.
S	= The reading preserved in the Scholia.
T	= The original reading of a tenth-century MS. in the Library of Trinity College, Cambridge (O, iv, 10).

Büch. = The reading of F. Bücheler's edition (1893).

Jahn = The reading of O. Jahn's larger edition (1851).

Letters erased in *P* or *T* are indicated by a corresponding number of asterisks.

myth. king of Aegina
father of Peleus 7 Telamon
Grand " " Achilles 7 Ajax

IVVENALIS

SATVRARVM

LIBER PRIMVS

SATVRA I

Semper ego auditor tantum? numquamne reponam
vexatus totiens rauci Theseide Cordi?
inpune ergo mihi recitaverit ille togatas,
hic elegos? inpune diem consumpserit ingens
Telephus aut, summi plena iam margine libri,   5
scriptus et in tergo necdum finitus Orestes?
nota magis nulli domus est sua, quam mihi lucus
Martis et Aeoliis vicinum rupibus antrum
Vulcani; quid agant venti, quas torqueat umbras
Aeacus, unde alius furtivae devehat aurum   10
pelliculae, quantas iaculetur Monychus ornos,
Frontonis platani convulsaque marmora clamant
semper et adsiduo ruptae lectore columnae.
expectes eadem a summo minimoque poeta.
et nos ergo manum ferulae subduximus, et nos   15
consilium dedimus Sullae, privatus ut altum
dormiret. stulta est clementia, cum tot ubique
vatibus occurras, periturae parcere chartae.
cur tamen hoc potius libeat decurrere campo,
per quem magnus equos Auruncae flexit alumnus,   20
si vacat ac placidi rationem admittitis, edam.
  cum tener uxorem ducat spado, Mevia Tuscum
figat aprum et nuda teneat venabula mamma,
patricios omnis opibus cum provocet unus
quo tondente gravis iuveni mihi barba sonabat,   25

2 Cordi *PS*: Codri *pω*

cum pars Niliacae plebis, cum verna Canopi
Crispinus, Tyrias umero revocante lacernas,
ventilet aestivum digitis sudantibus aurum
nec sufferre queat maioris pondera gemmae,
difficile est saturam non scribere. nam quis iniquae   30
tam patiens urbis, tam ferreus, ut teneat se,
causidici nova cum veniat lectica Mathonis
plena ipso, post hunc magni delator amici
et cito rapturus de nobilitate comesa
quod superest, quem Massa timet, quem munere palpat   35
Carus et a trepido Thymele summissa Latino?
quid referam quanta siccum iecur ardeat ira,   45
cum populum gregibus comitum premit hic spoliator
pupilli prostantis et hic damnatus inani
iudicio? quid enim salvis infamia nummis?
exul ab octava Marius bibit et fruitur dis
iratis, at tu victrix provincia ploras.   50
haec ego non credam Venusina digna lucerna?
haec ego non agitem? sed quid magis? Heracleas
aut Diomedeas aut mugitum labyrinthi
et mare percussum puero fabrumque volantem,
cum leno accipiat moechi bona, si capiendi   55
ius nullum uxori, doctus spectare lacunar,
doctus et ad calicem vigilanti stertere naso?
cum fas esse putet curam sperare cohortis
qui bona donavit praesepibus et caret omni
maiorum censu, dum pervolat axe citato   60
Flaminiam puer Automedon? nam lora tenebat
ipse, lacernatae cum se iactaret amicae.
nonne libet medio ceras inplere capaces
quadrivio, cum iam sexta cervice feratur
hinc atque inde patens ac nuda paene cathedra   65
et multum referens de Maecenate supino
signator falsi, qui se lautum atque beatum

67 falsi *P*: falso *pω*

exiguis tabulis et gemma fecerit uda?
occurrit matrona potens, quae molle Calenum
porrectura viro miscet sitiente rubetam 70
instituitque rudes melior Lucusta propinquas
per famam et populum nigros efferre maritos.
aude aliquid brevibus Gyaris et carcere dignum,
si vis esse aliquid. probitas laudatur et alget,
criminibus debent hortos praetoria mensas 75
argentum vetus et stantem extra pocula caprum.
quem patitur dormire nurus corruptor avarae,
quem sponsae turpes et praetextatus adulter?
si natura negat, facit indignatio versum
qualemcumque potest, quales ego vel Cluvienus. 80
    ex quo Deucalion nimbis tollentibus aequor
navigio montem ascendit sortesque poposcit
paulatimque anima caluerunt mollia saxa
et maribus nudas ostendit Pyrrha puellas,
quidquid agunt homines, votum timor ira voluptas 85
gaudia discursus, nostri farrago libelli est.
et quando uberior vitiorum copia? quando
maior avaritiae patuit sinus? alea quando
hos animos? neque enim loculis comitantibus itur
ad casum tabulae, posita sed luditur arca. 90
proelia quanta illic dispensatore videbis
armigero. simplexne furor sestertia centum
perdere et horrenti tunicam non reddere servo?
quis totidem erexit villas, quis fercula septem
secreto cenavit avus? nunc sportula primo 95
limine parva sedet turbae rapienda togatae.
ille tamen faciem prius inspicit et trepidat ne
suppositus venias ac falso nomine poscas:
agnitus accipies. iubet a praecone vocari
ipsos Troiugenas, nam vexant limen et ipsi 100
nobiscum. 'da praetori, da deinde tribuno.'

68 fecerit *PTS*: fecerat *p*   89 loculis *pω*: ioculis *P*

sed libertinus prior est. 'prior' inquit 'ego adsum.
cur timeam dubitemve locum defendere, quamvis
natus ad Euphraten, molles quod in aure fenestrae
arguerint, licet ipse negem? sed quinque tabernae   105
quadringenta parant. quid confert purpura maior
optandum, si Laurenti custodit in agro
conductas Corvinus oves, ego possideo plus
Pallante et Licinis?' expectent ergo tribuni,
vincant divitiae, sacro ne cedat honori   110
nuper in hanc urbem pedibus qui venerat albis,
quandoquidem inter nos sanctissima divitiarum
maiestas, etsi funesta pecunia templo
nondum habitat, nullas nummorum ereximus aras,
ut colitur Pax atque Fides Victoria Virtus   115
quaeque salutato crepitat Concordia nido.
sed cum summus honor finito conputet anno,
sportula quid referat, quantum rationibus addat,
quid facient comites quibus hinc toga, calceus hinc est
et panis fumusque domi? densissima centum   120
quadrantes lectica petit, sequiturque maritum
languida vel praegnas et circumducitur uxor.
hic petit absenti nota iam callidus arte
ostendens vacuam et clausam pro coniuge sellam.
'Galla mea est' inquit 'citius dimitte. moraris?   125
profer, Galla, caput. noli vexare, quiescet.'
   ipse dies pulchro distinguitur ordine rerum:
sportula, deinde forum iurisque peritus Apollo
atque triumphales, inter quas ausus habere
nescio quis titulos Aegyptius atque Arabarches,   130
cuius ad effigiem non tantum meiere fas est.
vestibulis abeunt veteres lassique clientes
votaque deponunt, quamquam longissima cenae

106 purpura maior p: purpuraemator P: purpurae amator
some mss   114 habitat P: habita* T: habitas pω
   126 quiescet P: quiescit pω

spes homini; caulis miseris atque ignis emendus.
optima silvarum interea pelagique vorabit　　　　135
rex horum vacuisque toris tantum ipse iacebit.
nam de tot pulchris et latis orbibus et tam
antiquis una comedunt patrimonia mensa.
nullus iam parasitus erit. sed quis ferat istas
luxuriae sordes? quanta est gula quae sibi totos　　　　140
ponit apros, animal propter convivia natum!
poena tamen praesens, cum tu deponis amictus
turgidus, et crudus pavonem in balnea portas.
hinc subitae mortes atque intestata senectus
et nova nec tristis per cunctas fabula cenas:　　　　145
ducitur iratis plaudendum funus amicis.

nil erit ulterius quod nostris moribus addat
posteritas, eadem facient cupientque minores:
omne in praecipiti vitium stetit. utere velis,
totos pande sinus. dices hic forsitan 'unde　　　　150
ingenium par materiae? unde illa priorum
scribendi quodcumque animo flagrante liberet
simplicitas?' cuius non audeo dicere nomen?
quid refert, dictis ignoscat Mucius an non?
'pone Tigellinum: taeda lucebis in illa,　　　　155
qua stantes ardent qui fixo pectore fumant,
et latum media sulcum deducis harena.'
qui dedit ergo tribus patruis aconita, vehatur
pensilibus plumis atque illinc despiciat nos?
'cum veniet contra, digito compesce labellum:　　　　160
accusator erit qui verbum dixerit "hic est."
securus licet Aeneam Rutulumque ferocem
committas, nulli gravis est percussus Achilles
aut multum quaesitus Hylas urnamque secutus:
ense velut stricto quotiens Lucilius ardens　　　　165

143 crudus P: crudum pω　　　150 dices P: dicas p
155 lucebis pω: lucebit PT　　　157 deducis p: deducit PT
161 verbum] vervum P: verum pω

infremuit, rubet auditor cui frigida mens est
criminibus, tacita sudant praecordia culpa.
inde irae et lacrimae. tecum prius ergo voluta
haec animo ante tubas, galeatum sero duelli
paenitet.' experiar quid concedatur in illos   170
quorum Flaminia tegitur cinis atque Latina.

SATVRA III

Quamvis digressu veteris confusus amici,
laudo tamen, vacuis quod sedem figere Cumis
destinet atque unum civem donare Sibyllae.
ianua Baiarum est et gratum litus amoeni
secessus. ego vel Prochytam praepono Suburae;   5
nam quid tam miserum, tam solum vidimus, ut non
deterius credas horrere incendia, lapsus
tectorum adsiduos ac mille pericula saevae
urbis et Augusto recitantes mense poetas?
sed dum tota domus raeda componitur una,   10
substitit ad veteres arcus madidamque Capenam.
hic ubi nocturnae Numa constituebat amicae—
nunc sacri fontis nemus et delubra locantur
Iudaeis quorum cophinus faenumque supellex,
omnis enim populo mercedem pendere iussa est   15
arbor et eiectis mendicat silva Camenis—
in vallem Egeriae descendimus et speluncas
dissimiles veris. quanto praesentius esset
numen aquis, viridi si margine cluderet undas
herba nec ingenuum violarent marmora tofum.   20

    hic tunc Umbricius 'quando artibus' inquit 'honestis
nullus in urbe locus, nulla emolumenta laborum,
res hodie minor est, here quam fuit, atque eadem cras

169 anim̄ ante tubas *P*: animante tuba *p*
III 16 eiectis *p*: electis *P*   18 praesentius *Heinsius*: pra
(*the rest erased*) *P*: praestantius *pω*

deteret exiguis aliquid, proponimus illuc
ire, fatigatas ubi Daedalus exuit alas,   25
dum nova canities, dum prima et recta senectus,
dum superest Lachesi quod torqueat, et pedibus me
porto meis nullo dextram subeunte bacillo.
cedamus patria. vivant Artorius istic
et Catulus, maneant qui nigrum in candida vertunt,   30
quis facile est aedem conducere flumina portus,
siccandam eluviem, portandum ad busta cadaver,
et praebere caput domina venale sub hasta.
quondam hi cornicines et municipalis harenae
perpetui comites notaeque per oppida buccae,   35
munera nunc edunt et verso pollice vulgus
cum iubet, occidunt populariter; inde reversi
conducunt foricas, et cur non omnia? cum sint,
quales ex humili magna ad fastigia rerum
extollit quotiens voluit Fortuna iocari.   40
quid Romae faciam? mentiri nescio; librum,
si malus est, nequeo laudare et poscere; motus
astrorum ignoro; funus promittere patris
nec volo nec possum; ranarum viscera numquam
inspexi; ferre ad nuptam quae mittit adulter,   45
quae mandat, norunt alii; me nemo ministro
fur erit, atque ideo nulli comes exeo tamquam
mancus et extinctae, corpus non utile, dextrae.
quis nunc diligitur nisi conscius et cui fervens
aestuat occultis animus semperque tacendis?   50
nil tibi se debere putat, nil conferet umquam,
participem qui te secreti fecit honesti:
carus erit Verri, qui Verrem tempore quo vult
accusare potest. tanti tibi non sit opaci
omnis harena Tagi quodque in mare volvitur aurum,   55
ut somno careas ponendaque praemia sumas

37 cum iubet *P*: quem iubet *T*: quem lubet *pω*   38 non?
omnia *Büch.*   40 iocari *pω*: locari *P*

tristis et a magno semper timearis amico.
    quae nunc divitibus gens acceptissima nostris,
et quos praecipue fugiam, properabo fateri,
nec pudor opstabit. non possum ferre, Quirites,   60
Graecam urbem; quamvis quota portio faecis Achaei?
iam pridem Syrus in Tiberim defluxit Orontes
et linguam et mores et cum tibicine chordas
obliquas nec non gentilia tympana secum
vexit et ad circum iussas prostare puellas; ʼ   65
ite, quibus grata est picta lupa barbara mitra.
rusticus ille tuus sumit trechedipna, Quirine,
et ceromatico fert niceteria collo.
hic alta Sicyone, ast hic Amydone relicta,
hic Andro, ille Samo, hic Trallibus aut Alabandis,   70
Esquilias dictumque petunt a vimine collem,
viscera magnarum domuum dominique futuri.
ingenium velox, audacia perdita, sermo
promptus et Isaeo torrentior. ede quid illum
esse putes. quemvis hominem secum attulit ad nos: 75
grammaticus rhetor geometres pictor aliptes
augur schoenobates medicus magus, omnia novit
Graeculus esuriens; in caelum, iusseris, ibit.
in summa non Maurus erat neque Sarmata nec Thrax
qui sumpsit pinnas, mediis sed natus Athenis.   80
horum ego non fugiam conchylia? me prior ille
signabit fultusque toro meliore recumbet,
advectus Romam quo pruna et cottona vento?
usque adeo nihil est, quod nostra infantia caelum
hausit Aventini baca nutrita Sabina?   85
quid quod adulandi gens prudentissima laudat
sermonem indocti, faciem deformis amici,
et longum invalidi collum cervicibus aequat
Herculis Antaeum procul a tellure tenentis,
miratur vocem angustam, qua deterius nec   90

        61 Achaei ω: Achaeae P: *perhaps* Achaea est

ille sonat quo mordetur gallina marito?
haec eadem licet et nobis laudare, sed illis
creditur. an melior, cum Thaida sustinet aut cum
uxorem comoedus agit vel Dorida nullo
cultam palliolo? mulier nempe ipsa videtur,   95
non persona, loqui.
    nec tamen Antiochus nec erit mirabilis illic
aut Stratocles aut cum molli Demetrius Haemo:
natio comoeda est. rides, maiore cachinno   100
concutitur; flet, si lacrimas conspexit amici,
nec dolet; igniculum brumae si tempore poscas,
accipit endromidem; si dixeris "aestuo," sudat.
non sumus ergo pares: melior, qui semper et omni
nocte dieque potest aliena sumere vultum   105
a facie, iactare manus, laudare paratus,
si bene ructavit, si rectum minxit amicus,
si trulla inverso crepitum dedit aurea fundo.
scire volunt secreta domus atque inde timeri.
et quoniam coepit Graecorum mentio, transi
gymnasia atque audi facinus maioris abollae.   115
stoicus occidit Baream, delator amicum,
discipulumque senex, ripa nutritus in illa,
ad quam Gorgonei delapsa est pinna caballi.
non est Romano cuiquam locus hic, ubi regnat
Protogenes aliquis vel Diphilus aut Hermarchus,   120
qui gentis vitio numquam partitur amicum,
solus habet. nam cum facilem stillavit in aurem
exiguum de naturae patriaeque veneno,
limine summoveor, perierunt tempora longi
servitii; nusquam minor est iactura clientis.   125
    quod porro officium, ne nobis blandiar, aut quod
pauperis hic meritum, si curet nocte togatus
currere, cum praetor lictorem impellat et ire
praecipitem iubeat dudum vigilantibus orbis,

117 discipulumque] discipulamque *Ritter*

ne prior Albinam et Modiam collega salutet? 130
divitis hic servo cludit latus ingenuorum
filius; alter enim quantum in legione tribuni
accipiunt, donat Calvinae vel Catienae.
da testem Romae tam sanctum, quam fuit hospes
numinis Idaei, procedat vel Numa vel qui
servavit trepidam flagranti ex aede Minervam:
protinus ad censum, de moribus ultima fiet 140
quaestio. "quot pascit servos? quot possidet agri
iugera? quam multa magnaque paropside cenat?"
quantum quisque sua nummorum servat in arca,
tantum habet et fidei. iures licet et Samothracum
et nostrorum aras, contemnere fulmina pauper 145
creditur atque deos dis ignoscentibus ipsis.
quid quod materiam praebet causasque iocorum
omnibus hic idem, si foeda et scissa lacerna,
si toga sordidula est et rupta calceus alter
pelle patet, vel si consuto vulnere crassum 150
atque recens linum ostendit non una cicatrix?
nil habet infelix paupertas durius in se,
quam quod ridiculos homines facit. "exeat" inquit
"si pudor est, et de pulvino surgat equestri
cuius res legi non sufficit, et sedeant hic 155
lenonum pueri quocumque ex fornice nati,
hic plaudat nitidi praeconis filius inter
pinnirapi cultos iuvenes iuvenesque lanistae"
sic libitum vano, qui nos distinxit, Othoni.
quis gener hic placuit censu minor atque puellae 160
sarcinulis impar? quis pauper scribitur heres?
quando in consilio est aedilibus? agmine facto
debuerant olim tenues migrasse Quirites.
haut facile emergunt quorum virtutibus opstat
res angusta domi, sed Romae durior illis 165
conatus: magno hospitium miserabile, magno

131 servo P: servi $p\omega$ 141 agri $p\omega$: agros P

servorum ventres, et frugi cenula magno.
fictilibus cenare pudet, quod turpe negabis
translatus subito ad Marsos mensamque Sabellam
contentusque illic veneto duroque cucullo.   170
pars magna Italiae est, si verum admittimus, in qua
nemo togam sumit nisi mortuus. ipsa dierum
festorum herboso colitur si quando theatro
maiestas tandemque redit ad pulpita notum
exodium, cum personae pallentis hiatum   175
in gremio matris formidat rusticus infans,
aequales habitus illic similesque videbis
orchestram et populum; clari velamen honoris
sufficiunt tunicae summis aedilibus albae.
hic ultra vires habitus nitor, hic aliquid plus   180
quam satis est interdum aliena sumitur arca.
commune id vitium est, hic vivimus ambitiosa
paupertate omnes. quid te moror? omnia Romae
cum pretio. quid das, ut Cossum aliquando salutes?
ut te respiciat clauso Veiento labello?   185
ille metit barbam, crinem hic deponit amati;
plena domus libis venalibus. "accipe et istud
fermentum tibi habe." praestare tributa clientes
cogimur et cultis augere peculia servis.
    quis timet aut timuit gelida Praeneste ruinam   190
aut positis nemorosa inter iuga Volsiniis aut
simplicibus Gabiis aut proni Tiburis arce?
nos urbem colimus tenui tibicine fultam
magna parte sui; nam sic labentibus obstat
vilicus, et veteris rimae cum texit hiatum,   195
securos pendente iubet dormire ruina.
vivendum est illic ubi nulla incendia, nulli
nocte metus. iam poscit aquam, iam frivola transfert
Ucalegon, tabulata tibi iam tertia fumant:
tu nescis; nam si gradibus trepidatur ab imis,   200

ultimus ardebit quem tegula sola tuetur
a pluvia, molles ubi reddunt ova columbae.
lectus erat Codro Procula minor, urceoli sex
ornamentum abaci nec non et parvulus infra
cantharus, et recubans sub eodem marmore Chiro;   205
iamque vetus Graecos servabat cista libellos,
et divina opici rodebant carmina mures.
nil habuit Codrus, quis enim negat? et tamen illud
perdidit infelix totum nihil. ultimus autem
aerumnae cumulus, quod nudum et frusta rogantem   210
nemo cibo, nemo hospitio tectoque iuvabit.
si magna Asturici cecidit domus, horrida mater,
pullati proceres, differt vadimonia praetor.
tum gemimus casus urbis, tunc odimus ignem.   215
ardet adhuc, et iam accurrit qui marmora donet,
conferat inpensas; hic nuda et candida signa,
hic aliquid praeclarum Euphranoris et Polycliti,
haec Asianorum vetera ornamenta deorum,
hic libros dabit et forulos mediamque Minervam,
hic modium argenti. meliora ac plura reponit   220
Persicus orborum lautissimus et merito iam
suspectus, tamquam ipse suas incenderit aedes.
si potes avelli circensibus, optima Sorae
aut Fabrateriae domus aut Frusinone paratur,
quanti nunc tenebras unum conducis in annum.   225
hortulus hic, puteusque brevis nec reste movendus
in tenuis plantas facili diffunditur haustu.
vive bidentis amans et culti vilicus horti,
unde epulum possis centum dare Pythagoreis.
est aliquid, quocumque loco, quocumque recessu   230
unius sese dominum fecisse lacertae.
    plurimus hic aeger moritur vigilando, sed ipsum

210 est cumulus *P*: est *struck out by p, omitted in* ω
218 haec Asianorum *PS*: fecasianorum *p*ω
227 diffunditur *p*: defunditur *P*

languorem peperit cibus inperfectus et haerens
ardenti stomacho; nam quae meritoria somnum
admittunt? magnis opibus dormitur in urbe.   235
inde caput morbi. raedarum transitus arto
vicorum inflexu et stantis convicia mandrae
eripient somnum Druso vitulisque marinis.
si vocat officium, turba cedente vehetur
dives et ingenti curret super ora Liburna   240
atque obiter leget aut scribet vel dormiet intus;
namque facit somnum clausa lectica fenestra.
ante tamen veniet: nobis properantibus opstat
unda prior, magno populus premit agmine lumbos
qui sequitur; ferit hic cubito, ferit assere duro   245
alter, at hic tignum capiti incutit, ille metretam.
pinguia crura luto, planta mox undique magna
calcor, et in digito clavus mihi militis haeret.
nonne vides quanto celebretur sportula fumo?
centum convivae, sequitur sua quemque culina.   250
Corbulo vix ferret tot vasa ingentia, tot res
inpositas capiti, quas recto vertice portat
servulus infelix et cursu ventilat ignem.
scinduntur tunicae sartae modo, longa coruscat
serraco veniente abies, atque altera pinum   255
plaustra vehunt, nutant alte populoque minantur.
nam si procubuit qui saxa Ligustica portat
axis et eversum fudit super agmina montem,
quid superest de corporibus? quis membra, quis ossa
invenit? obtritum vulgi perit omne cadaver   260
more animae. domus interea secura patellas
iam lavat et bucca foculum excitat et sonat unctis
striglibus et pleno componit lintea guto.
haec inter pueros varie properantur, at ille
iam sedet in ripa taetrumque novicius horret   265

238 Druso *Pω*: surdo *Speyer*   240 Liburna *PT*: Liburno *p*
246 tignum *P*: lignum *pω*

guagés: troubled water

porthmea nec sperat caenosi gurgitis alnum   *ship of alder wood*
infelix nec habet quem porrigat ore trientem.   *coin?*
respice nunc alia ac diversa pericula noctis: *consider*
quod spatium tectis sublimibus unde cerebrum   *brain*
testa ferit, quotiens rimosa et curta fenestris   *cracked* *defective*   270
vasa cadant, quanto percussum pondere signent   *weight* *mark*
et laedant silicem. possis ignavus haberi   *sluggish* *regard*
et subiti casus improvidus, ad cenam si   *coming* *so many*
intestatus eas: adeo tot fata, quot illa   *for indeed*
nocte patent vigiles te praetereunte fenestrae.   *pass by*   275
ergo optes votumque feras miserabile tecum,   *you must pray* *vow*
ut sint contentae patulas defundere pelves.   *extended pour down* *basins*
ebrius ac petulans qui nullum forte cecidit,   *drunk* *violent*
dat poenas, noctem patitur lugentis amicum   *is punished*
Pelidae, cubat in faciem, mox deinde supinus.   *a moment after* *lie face upward*   280
ergo non aliter poterit dormire? quibusdam   *otherwise*
somnum rixa facit. sed quamvis improbus annis   *quarrel*
atque mero fervens cavet hunc, quem coccina laena   *scarlet* *cloak*
vitari iubet et comitum longissimus ordo,
multum praeterea flammarum et aenea lampas.   *besides*   285
me, quem luna solet deducere vel breve lumen   *escort*
candelae cuius dispenso et tempero filum,   *weigh out* *regulate*
contemnit. miserae cognosce prohoemia rixae,   *brawl*
si rixa est ubi tu pulsas, ego vapulo tantum.   *I am flogged*
stat contra starique iubet: parere necesse est;   290
nam quid agas, cum te furiosus cogat et idem
fortior? "unde venis?" exclamat "cuius aceto,
cuius conche tumes? quis tecum sectile porrum   *cut* *leek*
sutor et elixi vervecis labra comedit?   *cobbler* *boiled sheep head* *eat up*
nil mihi respondes? aut dic aut accipe calcem.   *make known* *heel*   295
ede ubi consistas, in qua te quaero proseucha?"   *have your stand* *synagogue*
dicere si temptes aliquid tacitusve recedas,
tantumdem est: feriunt pariter, vadimonia deinde   *bail?*
irati faciunt. libertas pauperis haec est: *just as much*
288 prohoemia P: praemia pω   293 conche PS: concha pω

poenas dare to be punished.

pulsatus rogat et pugnis concisus adorat   300
ut liceat paucis cum dentibus inde reverti.
nec tamen haec tantum metuas. nam qui spoliet te
non derit clausis domibus, postquam omnis ubique
fixa catenatae siluit compago tabernae.
interdum et ferro subitus grassator agit rem;   305
armato quotiens tutae custode tenentur
et Pomptina palus et Gallinaria pinus,
sic inde huc omnes tamquam ad vivaria currunt.
qua fornace graves, qua non incude catenae?
maximus in vinclis ferri modus, ut timeas ne   310
vomer deficiat, ne marrae et sarcula desint.
felices proavorum atavos, felicia dicas
saecula quae quondam sub regibus atque tribunis
viderunt uno contentam carcere Romam.
    his alias poteram et pluris subnectere causas.   315
sed iumenta vocant et sol inclinat, eundum est;
nam mihi commota iam dudum mulio virga
adnuit. ergo vale nostri memor, et quotiens te
Roma tuo refici properantem reddet Aquino,
me quoque ad Helvinam Cererem vestramque Dianam   320
converte a Cumis. saturarum ego, ni pudet illas,
auditor gelidos veniam caligatus in agros.'

SATVRA IV

Ecce iterum Crispinus, et est mihi saepe vocandus
ad partes, monstrum nulla virtute redemptum
a vitiis, aegrae solaque libidine fortes
deliciae: viduas tantum spernatur adulter.
quid refert igitur, quantis iumenta fatiget   5

320 vestramque ω: vestamque P   322 auditor P (probably):
adiutor pω

IV 3 aegrae] aegra P: aeger pω   4 deliciae viduas P: delicias
viduae p   spernatur P: aspernatur ω

porticibus, quanta nemorum vectetur in umbra,
iugera quot vicina foro, quas emerit aedes?
nemo malus felix, minime corruptor et idem
incestus, cum quo nuper vittata iacebat
sanguine adhuc vivo terram subitura sacerdos.   10
sed nunc de factis levioribus. et tamen alter
si fecisset idem, caderet sub iudice morum;
nam quod turpe bonis Titio Seioque, decebat
Crispinum. quid agas, cum dira et foedior omni
crimine persona est? mullum sex milibus emit,   15
aequantem sane paribus sestertia libris,
ut perhibent qui de magnis maiora loquuntur.
consilium laudo artificis, si munere tanto
praecipuam in tabulis ceram senis abstulit orbi;
est ratio ulterior, magnae si misit amicae,   20
quae vehitur cluso latis specularibus antro.
nil tale expectes: emit sibi. multa videmus
quae miser et frugi non fecit Apicius. hoc tu
succinctus patria quondam, Crispine, papyro?
hoc pretio squamae? potuit fortasse minoris   25
piscator quam piscis emi; provincia tanti
vendit agros, sed maiores Apulia vendit.
qualis tunc epulas ipsum gluttisse putamus
induperatorem, cum tot sestertia, partem
exiguam et modicae sumptam de margine cenae,   30
purpureus magni ructarit scurra Palati,
iam princeps equitum, magna qui voce solebat
vendere municipes fracta de merce siluros.
incipe, Calliope. licet hic considere, non est
cantandum, res vera agitur. narrate, puellae   35
Pierides. prosit mihi vos dixisse puellas.
    cum iam semianimum laceraret Flavius orbem
ultimus et calvo serviret Roma Neroni,

9 vittata *Pω*: vitiata *S*   25 pretio squame *P*: pretium squame
pω   33 fracta *p*: facta *PT*: *perhaps* Pharia

incidit Adriaci spatium admirabile rhombi
ante domum Veneris, quam Dorica sustinet Ancon,   40
implevitque sinus; nec enim minor haeserat illis
quos operit glacies Maeotica ruptaque tandem
solibus effundit torrentis ad ostia Ponti
desidia tardos et longo frigore pingues.
destinat hoc monstrum cumbae linique magister   45
pontifici summo. quis enim proponere talem
aut emere auderet, cum plena et litora multo
delatore forent? dispersi protinus algae
inquisitores agerent cum remige nudo,
non dubitaturi fugitivum dicere piscem   50
depastumque diu vivaria Caesaris, inde
elapsum veterem ad dominum debere reverti.
si quid Palfurio, si credimus Armillato,
quidquid conspicuum pulchrumque est aequore toto,
res fisci est, ubicumque natat. donabitur ergo,   55
ne pereat. iam letifero cedente pruinis
autumno, iam quartanam sperantibus aegris,
stridebat deformis hiems praedamque recentem
servabat. tamen hic properat, velut urgueat auster.
utque lacus suberant, ubi quamquam diruta servat   60
ignem Troianum et Vestam colit Alba minorem,
obstitit intranti miratrix turba parumper.
ut cessit, facili patuerunt cardine valvae;
exclusi spectant admissa obsonia patres.
itur ad Atriden. tum Picens 'accipe' dixit   65
'privatis maiora focis. genialis agatur
iste dies, propera stomachum laxare sagina
et tua servatum consume in saecula rhombum.
ipse capi voluit.' quid apertius? et tamen illi
surgebant cristae; nihil est quod credere de se   70
non possit cum laudatur dis aequa potestas.

sed derat pisci patinae mensura. vocantur
ergo in consilium proceres, quos oderat ille,
in quorum facie miserae magnaeque sedebat
pallor amicitiae. primus, clamante Liburno   75
'currite, iam sedit,' rapta properabat abolla
Pegasus, attonitae positus modo vilicus urbi.
anne aliud tum praefecti? quorum optimus atque
interpres legum sanctissimus omnia quamquam
temporibus diris tractanda putabat inermi   80
iustitia. venit et Crispi iucunda senectus,
cuius erant mores qualis facundia, mite
ingenium. maria ac terras populosque regenti
quis comes utilior, si clade et peste sub illa
saevitiam damnare et honestum adferre liceret   85
consilium? sed quid violentius aure tyranni,
cum quo de pluviis aut aestibus aut nimboso
vere locuturi fatum pendebat amici?
ille igitur numquam derexit bracchia contra
torrentem, nec civis erat qui libera posset   90
verba animi proferre et vitam inpendere vero.
sic multas hiemes atque octogensima vidit
solstitia, his armis illa quoque tutus in aula.
proximus eiusdem properabat Acilius aevi
cum iuvene indigno quem mors tam saeva maneret   95
et domini gladiis tam festinata; sed olim
prodigio par est in nobilitate senectus,
unde fit ut malim fraterculus esse gigantis.
profuit ergo nihil misero, quod cominus ursos
figebat Numidas Albana nudus harena   100
venator. quis enim iam non intellegat artes
patricias? quis priscum illud miratur acumen,
Brute, tuum? facile est barbato inponere regi.
nec melior vultu quamvis ignobilis ibat
Rubrius, offensae veteris reus atque tacendae,   10

83 regenti *pω*: gerenti *P*

et tamen inprobior saturam scribente cinaedo.
Montani quoque venter adest abdomine tardus,
et matutino sudans Crispinus amomo
quantum vix redolent duo funera, saevior illo
Pompeius tenui iugulos aperire susurro,   110
et qui vulturibus servabat viscera Dacis
Fuscus marmorea meditatus proelia villa,
et cum mortifero prudens Veiento Catullo,
qui numquam visae flagrabat amore puellae,
grande et conspicuum nostro quoque tempore monstrum,   115
caecus adulator dirusque a ponte satelles,
dignus Aricinos qui mendicaret ad axes
blandaque devexae iactaret basia raedae.
nemo magis rhombum stupuit; nam plurima dixit,
in laevum conversus, at illi dextra iacebat   120
belua. sic pugnas Cilicis laudabat et ictus
et pegma et pueros inde ad velaria raptos.
non cedit Veiento, sed ut fanaticus oestro
percussus, Bellona, tuo divinat et 'ingens
omen habes' inquit 'magni clarique triumphi.   125
regem aliquem capies, aut de temone Britanno
excidet Arviragus. peregrina est belua, cernis
erectas in terga sudes?' hoc defuit unum
Fabricio, patriam ut rhombi memoraret et annos.
'quidnam igitur censes? conciditur?' 'absit ab illo   130
dedecus hoc' Montanus ait; 'testa alta paretur,
quae tenui muro spatiosum colligat orbem.
debetur magnus patinae subitusque Prometheus.
argillam atque rotam citius properate, sed ex hoc
tempore iam, Caesar, figuli tua castra sequantur.'   135
vicit digna viro sententia. noverat ille
luxuriam inperii veterem noctesque Neronis
iam medias aliamque famem, cum pulmo Falerno
arderet. nulli maior fuit usus edendi

135 figuli *P*: tegulae *pω*

tempestate mea; Circeis nata forent an   140
Lucrinum ad saxum Rutupinove edita fundo
ostrea callebat primo deprendere morsu,
et semel aspecti litus dicebat echini.
surgitur et misso proceres exire iubentur
consilio, quos Albanam dux magnus in arcem   145
traxerat attonitos et festinare coactos
tamquam de Chattis aliquid torvisque Sycambris
dicturus, tamquam et diversis partibus orbis
anxia praecipiti venisset epistula pinna.
atque utinam his potius nugis tota illa dedisset   150
tempora saevitiae, claras quibus abstulit urbi
inlustresque animas impune et vindice nullo.
sed periit postquam cerdonibus esse timendus
coeperat. hoc nocuit Lamiarum caede madenti.

SATVRA V

Si te propositi nondum pudet atque eadem est mens,
ut bona summa putes aliena vivere quadra,
si potes illa pati quae nec Sarmentus iniquas
Caesaris ad mensas nec vilis Gabba tulisset,
quamvis iurato metuam tibi credere testi.   5
ventre nihil novi frugalius. hoc tamen ipsum
defecisse puta, quod inani sufficit alvo:
nulla crepido vacat? nusquam pons et tegetis pars
dimidia brevior? tantine iniuria cenae,
tam ieiuna fames, cum possit honestius illic   10
et tremere et sordes farris mordere canini?
   primo fige loco, quod tu discumbere iussus
mercedem solidam veterum capis officiorum.
fructus amicitiae magnae cibus, inputat hunc rex,
et quamvis rarum tamen inputat. ergo duos post   15

148 tamquam et *PT*: tamquam ω: tamquam ex *Büch.*
v 4 Gabba *PT*: Galba *p*ω   10 possit *P*: possis *p*ω

si libuit menses neglectum adhibere clientem,
tertia ne vacuo cessaret culcita lecto,
'una simus' ait. votorum summa. quid ultra
quaeris? habet Trebius propter quod rumpere somnum
debeat et ligulas dimittere, sollicitus ne   20
tota salutatrix iam turba peregerit orbem,
sideribus dubiis aut illo tempore quo se
frigida circumagunt pigri serraca Bootae.
     qualis cena tamen. vinum quod sucida nolit
lana pati: de conviva Corybanta videbis.   25
iurgia proludunt, sed mox et pocula torques
saucius, et rubra deterges vulnera mappa,
inter vos quotiens libertorumque cohortem
pugna Saguntina fervet commissa lagona.
ipse capillato diffusum consule potat   30
calcatamque tenet bellis socialibus uvam
cardiaco numquam cyathum missurus amico;
cras bibet Albanis aliquid de montibus aut de
Setinis, cuius patriam titulumque senectus
delevit multa veteris fuligine testae,   35
quale coronati Thrasea Helvidiusque bibebant
Brutorum et Cassi natalibus. ipse capaces
Heliadum crustas et inaequales berullo
Virro tenet phialas: tibi non committitur aurum,
vel si quando datur, custos adfixus ibidem,   40
qui numeret gemmas, ungues observet acutos.
da veniam, praeclara illi laudatur iaspis;
nam Virro, ut multi, gemmas ad pocula transfert
a digitis, quas in vaginae fronte solebat
ponere zelotypo iuvenis praelatus Iarbae.   45
tu Beneventani sutoris nomen habentem
siccabis calicem nasorum quattuor ac iam

38 berullo *P*: berillos *pω*
                     mie
41 acutos *P*: *perhaps* observet. amico   42 illi *P*: illic *pω*

quassatum et rupto poscentem sulpura vitro.
si stomachus domini fervet vinoque ciboque,
frigidior Geticis petitur decocta pruinis: 50
non eadem vobis poni modo vina querebar?
vos aliam potatis aquam. tibi pocula cursor
Gaetulus dabit aut nigri manus ossea Mauri
et cui per mediam nolis occurrere noctem,
clivosae veheris dum per monumenta Latinae: 55
flos Asiae ante ipsum, pretio maiore paratus
quam fuit et Tulli census pugnacis et Anci
et, ne te teneam, Romanorum omnia regum
frivola. quod cum ita sit, tu Gaetulum Ganymedem
respice, cum sities. nescit tot milibus emptus   60
pauperibus miscere puer: sed forma, sed aetas
digna supercilio; quando ad te pervenit ille?
quando rogatus adest calidae gelidaeque minister?
quippe indignatur veteri parere clienti
quodque aliquid poscas et quod se stante recumbas.   65
maxima quaeque domus servis est plena superbis.
ecce alius quanto porrexit murmure panem
vix fractum, solidae iam mucida frusta farinae,
quae genuinum agitent, non admittentia morsum.
sed tener et niveus mollique siligine fictus   70
servatur domino. dextram cohibere memento,
salva sit artoptae reverentia. finge tamen te
inprobulum, superest illic qui ponere cogat:
'vis tu consuetis, audax conviva, canistris
impleri panisque tui novisse colorem?'   75
'scilicet hoc fuerat, propter quod saepe relicta
coniuge per montem adversum gelidasque cucurri
Esquilias, fremeret saeva cum grandine vernus
Iuppiter et multo stillaret paenula nimbo.'
    aspice quam longo distinguat pectore lancem   80
quae fertur domino squilla, et quibus undique saepta

70 fictus *P*: factus *p*   80 distinguat *P*: distendat *pω*

asparagis qua despiciat convivia cauda,
dum venit excelsi manibus sublata ministri.
sed tibi dimidio constrictus cammarus ovo
ponitur exigua feralis cena patella.   85
ipse Venafrano piscem perfundit: at hic qui
pallidus adfertur misero tibi, caulis olebit
lanternam; illud enim vestris datur alveolis quod
canna Micipsarum prora subvexit acuta,
propter quod Romae cum Boccare nemo lavatur,   90
quod tutos etiam facit a serpentibus atris.
mullus erit domini, quem misit Corsica vel quem
Tauromenitanae rupes, quando omne peractum est
et iam defecit nostrum mare, dum gula saevit,
retibus adsiduis penitus scrutante macello   95
proxima, nec patimur Tyrrhenum crescere piscem.
instruit ergo focum provincia, sumitur illinc
quod captator emat Laenas, Aurelia vendat.
Virroni muraena datur, quae maxima venit
gurgite de Siculo; nam dum se continet auster,   100
dum sedet et siccat madidas in carcere pinnas,
contemnunt mediam temeraria lina Charybdim:
vos anguilla manet longae cognata colubrae,
aut glacie aspersus maculis Tiberinus, et ipse
vernula riparum, pinguis torrente cloaca   105
et solitus mediae cryptam penetrare Suburae.
   ipsi pauca velim, facilem si praebeat aurem.
'nemo petit, modicis quae mittebantur amicis
a Seneca, quae Piso bonus, quae Cotta solebat
largiri; namque et titulis et fascibus olim   110
maior habebatur donandi gloria. solum
poscimus ut cenes civiliter. hoc face et esto,
esto, ut nunc multi, dives tibi, pauper amicis.'
   anseris ante ipsum magni iecur, anseribus par
altilis, et flavi dignus ferro Meleagri   115

spumat aper. post hunc tradentur tubera, si ver
tunc erit et facient optata tonitrua cenas
maiores. 'tibi habe frumentum' Alledius inquit
'o Libye; disiunge boves, dum tubera mittas.'
structorem interea, nequa indignatio desit,   120
saltantem spectes et chironomunta volanti
cultello, donec peragat dictata magistri
omnia; nec minimo sane discrimine refert,
quo gestu lepores et quo gallina secetur.
duceris planta velut ictus ab Hercule Cacus   125
et ponere foris, si quid temptaveris umquam
hiscere, tamquam habeas tria nomina. quando propinat
Virro tibi sumitve tuis contacta labellis
pocula? quis vestrum temerarius usque adeo, quis
perditus, ut dicat regi 'bibe'? plurima sunt quae   130
non audent homines pertusa dicere laena.
quadringenta tibi si quis deus aut similis dis
et melior fatis donaret homuncio, quantus
ex nihilo, quantus fieres Virronis amicus.
'da Trebio, pone ad Trebium. vis, frater, ab ipsis   135
ilibus?' o nummi, vobis hunc praestat honorem,
vos estis fratres. dominus tamen et domini rex
si vis tu fieri, nullus tibi parvolus aula
luserit Aeneas nec filia dulcior illo;
iucundum et carum sterilis facit uxor amicum.   140
sed tua nunc Mycale pariat licet et pueros tres
in gremium patris fundat semel, ipse loquaci
gaudebit nido, viridem thoraca iubebit
adferri minimasque nuces assemque rogatum,
ad mensam quotiens parasitus venerit infans.   145
    vilibus ancipites fungi ponentur amicis,
boletus domino, sed quales Claudius edit
ante illum uxoris, post quem nihil amplius edit.
    Virro sibi et reliquis Virronibus illa iubebit

116 spumat *P*: fumat *p*   tradentur *P*: radentur *T*: raduntur *pω*

poma dari, quorum solo pascaris odore,   150
qualia perpetuus Phaeacum autumnus habebat,
credere quae possis subrepta sororibus Afris:
tu scabie frueris mali, quod in aggere rodit
qui tegitur parma et galea metuensque flagelli
discit ab hirsuta iaculum torquere capella.   155
   forsitan inpensae Virronem parcere credas.
hoc agit ut doleas; nam quae comoedia, mimus
quis melior plorante gula? ergo omnia fiunt,
si nescis, ut per lacrimas effundere bilem
cogaris pressoque diu stridere molari.   160
tu tibi liber homo et regis conviva videris:
captum te nidore suae putat ille culinae,
nec male coniectat; quis enim tam nudus, ut illum
bis ferat, Etruscum puero si contigit aurum
vel nodus tantum et signum de paupere loro?   165
spes bene cenandi vos decipit. 'ecce dabit iam
semesum leporem atque aliquid de clunibus apri,
ad nos iam veniet minor altilis.' inde parato
intactoque omnes et stricto pane iacetis.
ille sapit qui te sic utitur. omnia ferre   170
si potes, et debes. pulsandum vertice raso
praebebis quandoque caput nec dura timebis
flagra pati, his epulis et tali dignus amico.

169 iacetis *P*: tacetis *pω*

LIBER SECVNDVS

SATVRA VI

Credo Pudicitiam Saturno rege moratam
in terris visamque diu, cum frigida parvas
praeberet spelunca domos, ignemque Laremque
et pecus et dominos communi clauderet umbra,
silvestrem montana torum cum sterneret uxor   5
frondibus et culmo vicinarumque ferarum
pellibus, haut similis tibi, Cynthia, nec tibi, cuius
turbavit nitidos extinctus passer ocellos,
sed potanda ferens infantibus ubera magnis
et saepe horridior glandem ructante marito.   10
quippe aliter tunc orbe novo caeloque recenti
vivebant homines, qui rupto robore nati
compositive luto nullos habuere parentes.
multa Pudicitiae veteris vestigia forsan
aut aliqua exstiterint et sub Iove, sed Iove nondum   15
barbato, nondum Graecis iurare paratis
per caput alterius, cum furem nemo timeret
caulibus et pomis, et aperto viveret horto.
paulatim deinde ad superos Astraea recessit
hac comite, atque duae pariter fugere sorores.   20
anticum et vetus est genium contemnere fulcri:
omne aliud crimen mox ferrea protulit aetas,
viderunt primos argentea saecula moechos.
conventum tamen et pactum et sponsalia nostra   25
tempestate paras, iamque a tonsore magistro
pecteris, et digito pignus fortasse dedisti.
certe sanus eras. uxorem, Postume, ducis?
dic, qua Tisiphone, quibus exagitare colubris?

VI 12 rupto] rupe et *Scholte*   13 compositive *T*: compositi**
P: compositique ω

ferre potes dominam salvis tot restibus ullam,   30
cum pateant altae caligantesque fenestrae,
cum tibi vicinum se praebeat Aemilius pons?
    sed placet Ursidio lex Iulia, tollere dulcem
cogitat heredem, cariturus turture magno
mullorumque iubis et captatore macello.   40
quid fieri non posse putes, si iungitur ulla
Ursidio? si moechorum notissimus olim
stulta maritali iam porrigit ora capistro,
quem totiens texit perituri cista Latini?
quid quod et antiquis uxor de moribus illi   45
quaeritur? o medici, nimiam pertundite venam.
delicias hominis. Tarpeium limen adora
pronus et auratam Iunoni caede iuvencam,
si tibi contigerit capitis matrona pudici.
    porticibusne tibi monstratur femina voto   60
digna tuo? cuneis an habent spectacula totis
quod securus ames quodque inde excerpere possis?
nupta senatori comitata est Eppia ludum
ad Pharon et Nilum famosaque moenia Lagi,
prodigia et mores urbis damnante Canopo.
inmemor illa domus et coniugis atque sororis   85
nil patriae indulsit, plorantesque improba natos,
utque magis stupeas, ludos Paridemque reliquit.
sed quamquam in magnis opibus plumaque paterna
et segmentatis dormisset parvula cunis,
contempsit pelagus; famam contempserat olim,   90
cuius apud molles minima est iactura cathedras.
Tyrrhenos igitur fluctus lateque sonantem
pertulit Ionium constanti pectore, quamvis
mutandum totiens esset mare. iusta pericli
si ratio est et honesta, timent pavidoque gelantur   95

82 ludum $P\omega$: ludium *some mss.*
92 sonantem] sonorum *Bentley*
93 ignium *changed to* Ionium *in P*

pectore nec tremulis possunt insistere plantis:
fortem animum praestant rebus quas turpiter audent.
si iubeat coniunx, durum est conscendere navem,
tunc sentina gravis, tunc summus vertitur aer:
quae moechum sequitur, stomacho valet. illa maritum   100
convomit, haec inter nautas et prandet et errat
per puppem et duros gaudet tractare rudentis.
qua tamen exarsit forma, qua capta iuventa
Eppia? quid vidit propter quod ludia dici
sustinuit? nam Sergiolus iam radere guttur   105
coeperat et secto requiem sperare lacerto;
praeterea multa in facie deformia, sicut
attritus galea mediisque in naribus ingens
gibbus et acre malum semper stillantis ocelli.
sed gladiator erat. facit hoc illos Hyacinthos,   110
hoc pueris patriaeque, hoc praetulit illa sorori
atque viro. ferrum est quod amant. hic Sergius idem
accepta rude coepisset Veiento videri.
   'optima sed quare Censennia teste marito?'
bis quingena dedit; tanti vocat ille pudicam.
nec pharetris Veneris macer est aut lampade fervet;
inde faces ardent, veniunt a dote sagittae.
libertas emitur. coram licet innuat atque   140
rescribat; vidua est, locuples quae nupsit avaro.
   'cur desiderio Bibulae Sertorius ardet?'
si verum excutias, facies, non uxor amatur.
tres rugae subeant et se cutis arida laxet,
fiant obscuri dentes oculique minores: 145
'collige sarcinulas' dicet libertus 'et exi.
iam gravis es nobis et saepe emungeris. exi
ocius et propera. sicco venit altera naso.'
interea calet et regnat poscitque maritum
pastores et ovem Canusinam ulmosque Falernas—   150
quantulum in hoc?—pueros omnes, ergastula tota,

137 quingena *P*: quingenta *pω*

quodque domi non est, sed habet vicinus, ematur.
mense quidem brumae, quo iam mercator Iaso
clausus, et armatis opstat casa candida nautis,
grandia tolluntur crystallina, maxima rursus   155
myrrhina, deinde adamans notissimus et Berenices
in digito factus pretiosior. hunc dedit olim
barbarus incestae, dedit hunc Agrippa sorori,
observant ubi festa mero pede sabbata reges
et vetus indulget senibus clementia porcis.   160
   'nullane de tantis gregibus tibi digna videtur?'
sit formosa decens dives fecunda, vetustos
porticibus disponat avos, intactior omni
crinibus effusis bellum dirimente Sabina,
rara avis in terris nigroque simillima cycno: 165
quis feret uxorem cui constant omnia? malo,
malo Venustinam quam te, Cornelia, mater
Gracchorum, si cum magnis virtutibus adfers
grande supercilium et numeras in dote triumphos.
tolle tuum, precor, Hannibalem victumque Syphacem   170
in castris et cum tota Carthagine migra.
'parce, precor, Paean, et tu, dea, pone sagittas;
nil pueri faciunt, ipsam configite matrem'
Amphion clamat; sed Paean contrahit arcum.
extulit ergo greges natorum ipsumque parentem,   175
dum sibi nobilior Latonae gente videtur
atque eadem scrofa Niobe fecundior alba.
quae tanti gravitas, quae forma, ut se tibi semper
imputet? huius enim rari summique voluptas
nulla boni, quotiens animo corrupta superbo   180
plus aloes quam mellis habet. quis deditus autem
usque adeo est, ut non illam quam laudibus effert,
horreat inque diem septenis oderit horis?
   quaedam parva quidem, sed non toleranda maritis.

159 mero *pω*: nudo *P*   167 Venustinam *Büch.*: Venusinam
Pω   172 dea pone *Graevius*: depone *Pω*

nam quid rancidius, quam quod se non putat ulla 185
formosam nisi quae de Tusca Graecula facta est,
de Sulmonensi mera Cecropis? omnia graece,
cum sit turpe magis nostris nescire latine;
hoc sermone pavent, hoc iram gaudia curas,
hoc cuncta effundunt animi secreta *Latinae*. 190

si tibi legitimis pactam iunctamque tabellis 200
non es amaturus, ducendi nulla videtur
causa, nec est quare cenam et mustacea perdas
labente officio crudis donanda, nec illud
quod prima pro nocte datur, cum lance beata
Dacicus et scripto radiat Germanicus auro. 205
si tibi simplicitas uxoria, deditus uni
est animus, summitte caput cervice parata
ferre iugum. nullam invenies quae parcat amanti;
ardeat ipsa licet, tormentis gaudet amantis
et spoliis; igitur longe minus utilis illi 210
uxor, quisquis erit bonus optandusque maritus.
nil umquam invita donabis coniuge, vendes
hac opstante nihil, nihil haec si nolet emetur.
haec dabit affectus: 'ille excludatur' amicus
iam senior, cuius barbam tua ianua vidit. 215
testandi cum sit lenonibus atque lanistis
libertas et iuris idem contingat harenae,
non unus tibi rivalis dictabitur heres.
'pone crucem servo.' 'meruit quo crimine servus
supplicium? quis testis adest? quis detulit? audi; 220
nulla umquam de morte hominis cunctatio longa est.'
'o demens, ita servus homo est? nil fecerit, esto:
hoc volo, sic iubeo, sit pro ratione voluntas.'
imperat ergo viro. sed mox haec regna relinquit
permutatque domos et flammea conterit, inde 22
avolat et spreti repetit vestigia lecti;
ornatas paulo ante fores, pendentia linquit

213 nolet] nollet *P*: nolit *T*: nollit *p*

vela domus et adhuc virides in limine ramos.
sic crescit numerus, sic fiunt octo mariti
quinque per autumnos, titulo res digna sepulcri.   230
   desperanda tibi salva concordia socru.
illa docet spoliis nudi gaudere mariti,
illa docet missis a corruptore tabellis
nil rude nec simplex rescribere, decipit illa
custodes aut aere domat. tunc corpore sano   235
advocat Archigenen onerosaque pallia iactat.
abditus interea latet et secretus adulter.
scilicet expectas ut tradat mater honestos
atque alios mores quam quos habet? utile porro   240
filiolam turpi vetulae producere turpem.
   nulla fere causa est in qua non femina litem
moverit. accusat Manilia, si rea non est.
conponunt ipsae per se formantque libellos,
principium atque locos Celso dictare paratae.   245
   endromidas Tyrias et femineum ceroma
quis nescit, vel quis non vidit vulnera pali?
quem cavat adsiduis rudibus scutoque lacessit
atque omnes implet numeros dignissima prorsus
Florali matrona tuba, nisi si quid in illo   250
pectore plus agitat veraeque paratur harenae.
quem praestare potest mulier galeata pudorem?
quale decus, rerum si coniugis auctio fiat,   255
balteus et manicae et cristae crurisque sinistri
dimidium tegimen, vel si diversa movebit
proelia, tu felix ocreas vendente puella.
hae sunt quae tenui sudant in cyclade, quarum
delicias et panniculus bombycinus urit?   260
aspice quo fremitu monstratos perferat ictus
et quanto galeae curvetur pondere, quanta
poplitibus sedeat quam denso fascia libro,
et ride positis scaphium cum sumitur armis.

248 rudibus] *udibus *P*: sudibus *pω*

dicite vos neptes Lepidi caecive Metelli   265
Gurgitis aut Fabii, quae ludia sumpserit umquam
hos habitus, quando ad palum gemat uxor Asyli.
    semper habet lites alternaque iurgia lectus
in quo nupta iacet, minimum dormitur in illo.
tunc gravis illa viro, tunc orba tigride peior,   270
cum simulat gemitus occulti conscia facti,
aut odit pueros, aut ficta paelice plorat,
uberibus semper lacrimis semperque paratis
in statione sua atque expectantibus illam,
quo iubeat manare modo: tu credis amorem,   275
tu tibi tunc, uruca, places fletumque labellis
exsorbes, quae scripta et quot lecture tabellas,
si tibi zelotypae retegantur scrinia moechae.
dic aliquem sodes hic, Quintiliane, colorem.   280
haeremus. dic ipsa. 'olim convenerat' inquit
'ut faceres tu quod velles, nec non ego possem
indulgere mihi. clames licet et mare caelo
confundas, homo sum.' nihil est audacius illis
deprensis; iram atque animos a crimine sumunt.   285
    unde haec monstra tamen vel quo de fonte, requiris?
praestabat castas humilis fortuna Latinas
quondam, nec vitiis contingi parva sinebant
tecta, labor somnique breves et vellere Tusco
vexatae duraeque manus ac proximus urbi   290
Hannibal et stantes Collina turre mariti.
nunc patimur longae pacis mala, saevior armis
luxuria incubuit victumque ulciscitur orbem.
nullum crimen abest facinusque libidinis, ex quo
paupertas Romana perit. hinc fluxit ad istos   295
et Sybaris collis, hinc et Rhodos et Miletos
atque coronatum et petulans madidumque Tarentum.
prima peregrinos obscaena pecunia mores

270 tunc gravis ω: cum gravis P
295 ad istos Pithou: ad Indos P: ad Histros ω

intulit, et turpi fregerunt saecula luxu
divitiae molles. quid enim venus ebria curat,   300
grandia quae mediis iam noctibus ostrea mordet,
cum perfusa mero spumant unguenta Falerno,
cum bibitur concha, cum iam vertigine tectum
ambulat et geminis exsurgit mensa lucernis?   305
i nunc et dubita, qua sorbeat aera sanna
Tullia, quid dicat notae collactea Maurae
Maura, Pudicitiae veterem cum praeterit aram.

   audio quid veteres olim moneatis amici: 346
'pone seram, prohibe.' sed quis custodiet ipsos
custodes? cauta est et ab illis incipit uxor.
iamque eadem summis pariter minimisque libido,
nec melior, silicem pedibus quae conterit atrum,   350
quam quae longorum vehitur cervice Syrorum.
ut spectet ludos, conducit Ogulnia vestem,
conducit comites sellam cervical amicas
nutricem et flavam cui det mandata puellam.
haec tamen argenti superest quodcumque paterni,   355
levibus athletis et vasa novissima donat:
multis res angusta domi, sed nulla pudorem
paupertatis habet nec se metitur ad illum
quem dedit haec posuitque modum. tamen utile quid sit,
prospiciunt aliquando viri, frigusque famemque   360
formica tandem quidam expavere magistra:
prodiga non sentit pereuntem femina censum.
ac velut exhausta redivivus pullulet arca
nummus et e pleno tollatur semper acervo,
non usquam reputant, quanti sibi gaudia constent.   365

   si gaudet cantu *mulier*, *sunt* organa semper   380
in manibus, densi radiant testudine tota
sardonyches, crispo numerantur pectine chordae.
quo tener Hedymeles operas dedit, hunc tenet, hoc se

306 i nunc et *p*: inunget *P*
365 usquam *Büch.*: nusquam *P*: umquam *pω*

solatur, gratoque indulget basia plectro.
quaedam de numero Lamiarum ac nominis Appi   38
et farre et vino Ianum Vestamque rogabat,
an Capitolinam deberet Pollio quercum
sperare et fidibus promittere. quid faceret plus
aegrotante viro, medicis quid tristibus erga
filiolum? stetit ante aram nec turpe putavit   39
pro cithara velare caput dictataque verba
pertulit, ut mos est, et aperta palluit agna.
dic mihi nunc quaeso, dic, antiquissime divum,
respondes his, Iane pater? magna otia caeli;
non est, quod video, non est quod agatur aput vos.   39
haec de comoedis te consulit, illa tragoedum
commendare volet, varicosus fiet haruspex.

   sed cantet potius quam totam pervolet urbem,
audax et coetus possit quae ferre virorum
cumque paludatis ducibus praesente marito   40
ipsa loqui recta facie siccisque mamillis.
haec eadem novit quid toto fiat in orbe,
quid Seres, quid Thraces agant, secreta novercae
et pueri, quis amet, quis diripiatur adulter.   40
instantem regi Armenio Parthoque cometem
prima videt, famam rumoresque illa recentis
excipit ad portas, quosdam facit; isse Niphatem
in populos magnoque illic cuncta arva teneri   41
diluvio, nutare urbes, subsidere terras
quocumque in trivio cuicumque est obvia, narrat.

   nec tamen id vitium magis intolerabile quam quae
vicinos humiles rapere et concidere loris
exorata solet. nam si latratibus alti   4
rumpuntur somni, 'fustes huc ocius' inquit
'adferte' atque illis dominum iubet ante feriri,

   385 Appi S: Ap* P: alti pω
   404 diripiatur P: decipiatur pω
   415 exorata ω: exortata P: *perhaps* experrecta

deinde canem. gravis occursu, taeterrima vultu
balnea nocte subit, conchas et castra moveri
nocte iubet, magno gaudet sudare tumultu,   420
cum lassata gravi ceciderunt bracchia massa.
convivae miseri interea somnoque fameque
urguentur. tandem illa venit rubicundula, totum   425
oenophorum sitiens, plena quod tenditur urna
admotum pedibus, de quo sextarius alter
ducitur ante cibum rabidam facturus orexim.

    illa tamen gravior, quae cum discumbere coepit,
laudat Vergilium, periturae ignoscit Elissae,   435
committit vates et comparat, inde Maronem
atque alia parte in trutina suspendit Homerum.
cedunt grammatici, vincuntur rhetores, omnis
turba tacet, nec causidicus nec praeco loquetur,
altera nec mulier; verborum tanta cadit vis.   440
tot pariter pelves ac tintinnabula dicas
pulsari. iam nemo tubas, nemo aera fatiget:
una laboranti poterit succurrere Lunae.
inponit finem sapiens et rebus honestis;
nam quae docta nimis cupit et facunda videri,   445
crure tenus medio tunicas succingere debet,
caedere Silvano porcum, quadrante lavari.
non habeat matrona, tibi quae iuncta recumbit,
dicendi genus, aut curvum sermone rotato
torqueat enthymema, nec historias sciat omnes,   450
sed quaedam ex libris et non intellegat. odi
hanc ego quae repetit volvitque Palaemonis artem
servata semper lege et ratione loquendi
ignotosque mihi tenet antiquaria versus
nec curanda viris. opicae castiget amicae   455
verba: soloecismum liceat fecisse marito.

    nil non permittit mulier sibi, turpe putat nil,
cum virides gemmas collo circumdedit et cum

    455 viris. opicae *Housman*: viris opicae *edd.*

auribus extentis magnos commisit elenchos;
intolerabilius nihil est quam femina dives.   4̇0̇
interea foeda aspectu ridendaque multo
pane tumet facies aut pinguia Poppaeana
spirat, et hinc miseri viscantur labra mariti:
ad moechum lota veniunt cute. quando videri
vult formonsa domi? moechis foliata parantur,   4̇5̇
his emitur, quidquid graciles huc mittitis Indi.
tandem aperit vultum et tectoria prima reponit,
incipit agnosci, atque illo lacte fovetur
propter quod secum comites educit asellas
exul Hyperboreum si dimittatur ad axem.   4̇
sed quae mutatis inducitur atque fovetur
tot medicaminibus coctaeque siliginis offas
accipit et madidae, facies dicetur an ulcus?

   est pretium curae penitus cognoscere toto
quid faciant agitentque die. si nocte maritus   4̇
aversus iacuit, periit libraria, ponunt
cosmetae tunicas, tarde venisse Liburnus
dicitur et poenas alieni pendere somni
cogitur, hic frangit ferulas, rubet ille flagello,
hic scutica; sunt quae tortoribus annua praestent.   4̇
verberat atque obiter faciem linit, audit amicas
aut latum pictae vestis considerat aurum
et caedit, longi relegit transversa diurni
et caedit, donec lassis caedentibus 'exi'
intonet horrendum iam cognitione peracta.   4̇
praefectura domus Sicula non mitior aula.
nam si constituit solitoque decentius optat
ornari et properat iamque expectatur in hortis
aut aput Isiacae potius sacraria lenae,
disponit crinem laceratis ipsa capillis
nuda umero Psecas infelix nudisque mamillis.
'altior hic quare cincinnus?' taurea punit

          466 huc ω: hic P   479 flagello P: flagellis p

continuo flexi crimen facinusque capilli.
quid Psecas admisit? quaenam est hic culpa puellae,
si tibi displicuit nasus tuus? altera laevum   495
extendit pectitque comas et volvit in orbem.
est in consilio materna admotaque lanis
emerita quae cessat acu; sententia prima
huius erit, post hanc aetate atque arte minores
censebunt, tamquam famae discrimen agatur   500
aut animae. tanta est quaerendi cura decoris;
tot premit ordinibus, tot adhuc conpagibus altum
aedificat caput. Andromachen a fronte videbis,
post minor est, credas aliam. cedo, si breve parvi
sortita est lateris spatium, breviorque videtur   505
virgine Pygmaea nullis adiuta cothurnis,
et levis erecta consurgit ad oscula planta.
   nulla viri cura interea nec mentio fiet
damnorum. vivit tamquam vicina mariti,
hoc solo propior quod amicos coniugis odit   510
et servos, gravior rationibus. ecce furentis
Bellonae matrisque deum chorus intrat et ingens
Gallus adest, cui rauca cohors, cui tympana cedunt,   515
plebeia et Phrygia vestitur bucca tiara.
grande sonat metuique iubet septembris et austri
adventum, nisi se centum lustraverit ovis
et xerampelinas veteres donaverit ipsi,
ut quidquid subiti et magni discriminis instat,   520
in tunicas eat et totum semel expiet annum.
hibernum fracta glacie descendet in amnem,
ter matutino Tiberi mergetur et ipsis
verticibus timidum caput abluet, inde superbi
totum regis agrum nuda ac tremibunda cruentis   525
erepet genibus; si candida iusserit Io,

493 crimen ω: crinem P
511 gravior rationibus *Büch.*: gravirationibus P: gravis rationi-
bus T: gravis est rationibus pω

ibit ad Aegypti finem calidaque petitas
a Meroe portabit aquas, ut spargat in aede
Isidis, antiquo quae proxima surgit ovili.
credit enim ipsius dominae se voce moneri: 530
en animam et mentem cum qua di nocte loquantur.
ergo hic praecipuum summumque meretur honorem,
qui grege linigero circumdatus et grege calvo
plangentis populi currit derisor Anubis.
illius lacrimae meditataque murmura praestant
ut veniam culpae non abnuat, ansere magno   540
scilicet et tenui popano corruptus, Osiris.
cum dedit ille locum, cophino faenoque relicto
arcanam Iudaea tremens mendicat in aurem,
interpres legum Solymarum et magna sacerdos
arboris ac summi fida internuntia caeli.   544
implet et illa manum, sed parcius; aere minuto
qualiacumque voles Iudaei somnia vendunt.
spondet amatorem tenerum vel divitis orbi
testamentum ingens calidae pulmone columbae
tractato Armenius vel Commagenus haruspex;   550
pectora pullorum rimabitur, exta catelli,
interdum et pueri; faciet quod deferat ipse.
Chaldaeis sed maior erit fiducia; quidquid
dixerit astrologus, credent a fonte relatum
Hammonis, quoniam Delphis oracula cessant   555
et genus humanum damnat caligo futuri.
praecipuus tamen est horum, qui saepius exul,
cuius amicitia conducendaque tabella
magnus civis obit et formidatus Othoni.
inde fides artis, sonuit si dextera ferro   560
laevaque, si longo castrorum in carcere mansit.
nemo mathematicus genium indemnatus habebit,
sed qui paene perit, cui vix in Cyclada mitti
contigit et parva tandem caruisse Seripho.

547 somnia *pω*: omnia *P*   551 rimabitur *PT*: rimatur et *pω*

consulit ictericae lento de funere matris,   565
ante tamen de te Tanaquil tua, quando sororem
efferat et patruos, an sit victurus adulter
post ipsam; quid enim maius dare numina possunt?
hae tamen ignorant quid sidus triste minetur
Saturni, quo laeta Venus se proferat astro,   570
quis mensis damnis, quae dentur tempora lucro:
illius occursus etiam vitare memento,
in cuius manibus ceu pinguia sucina tritas
cernis ephemeridas, quae nullum consulit et iam
consulitur, quae castra viro patriamque petente   575
non ibit pariter numeris revocata Thrasylli.
ad primum lapidem vectari cum placet, hora
sumitur ex libro; si prurit frictus ocelli
angulus, inspecta genesi collyria poscit;
aegra licet iaceat, capiendo nulla videtur   580
aptior hora cibo nisi quam dederit Petosiris.
si mediocris erit, spatium lustrabit utrimque
metarum et sortes ducet frontemque manumque
praebebit vati crebrum poppysma roganti.
divitibus responsa dabunt Phryx augur et Indae,   585
conductus dabit astrorum mundique peritus
atque aliquis senior qui publica fulgura condit:
plebeium in circo positum est et in aggere fatum;
quae nudis longum ostendit cervicibus aurum,
consulit ante falas delphinorumque columnas,   590
an saga vendenti nubat caupone relicto.
     hae tamen et partus subeunt discrimen et omnis
nutricis tolerant fortuna urguente labores,
sed iacet aurato vix ulla puerpera lecto.   594
transeo suppositos et gaudia votaque saepe
ad spurcos decepta lacus atque inde petitos
pontifices, salios Scaurorum nomina falso
corpore laturos. stat Fortuna inproba noctu   605
585 dabunt *P*: dabit *p*   Indae *Büch.*: inde *Pω*: Indus *some mss.*

adridens nudis infantibus, hos fovet omni
involvitque sinu, domibus tunc porrigit altis
secretumque sibi mimum parat; hos amat, his se
ingerit utque suos semper producit alumnos.

    hic magicos adfert cantus, hic Thessala vendit   610
philtra, quibus valeat mentem vexare mariti
et solea pulsare natis. quod desipis, inde est,
inde animi caligo et magna oblivio rerum
quas modo gessisti. tamen hoc tolerabile, si non
et furere incipias ut avunculus ille Neronis,   615
cui totam tremuli frontem Caesonia pulli
infudit; quae non faciet quod principis uxor?
ardebant cuncta et fracta conpage ruebant,
non aliter quam si fecisset Iuno maritum
insanum. minus ergo nocens erit Agrippinae   620
boletus, siquidem unius praecordia pressit
ille senis tremulumque caput descendere iussit
in caelum et longa manantia labra saliva,
haec poscit ferrum atque ignes, haec potio torquet,
haec lacerat mixtos equitum cum sanguine patres.   625
tanti partus equae, tanti una venefica constat.

    oderunt natos de paelice; nemo repugnet,
nemo vetet, iam iam privignum occidere fas est.
vos ego, pupilli, moneo, quibus amplior est res,
custodite animas et nulli credite mensae;   630
livida materno fervent adipata veneno.
mordeat ante aliquis quidquid porrexerit illa
quae peperit, timidus praegustet pocula papas.
fingimus haec altum satura sumente cothurnum
scilicet, et finem egressi legemque priorum   635
grande Sophocleo carmen bacchamur hiatu,
montibus ignotum Rutulis caeloque Latino?
nos utinam vani. sed clamat Pontia 'feci,
confiteor, puerisque meis aconita paravi,

       629 ego *pω*: equo (*apparently*) *P*: *perhaps* quoque

quae deprensa patent; facinus tamen ipsa peregi.'   640
tune duos una, saevissima vipera, cena?
tune duos? 'septem, si septem forte fuissent.'
credamus tragicis quidquid de Colchide torva
dicitur et Procne, nil contra conor. et illae
grandia monstra suis audebant temporibus, sed   645
non propter nummos. minor admiratio summis
debetur monstris, quotiens facit ira nocentes
hunc sexum: rabie iecur incendente feruntur
praecipites ut saxa iugis abrupta, quibus mons
subtrahitur clivoque latus pendente recedit: 650
illam ego non tulerim, quae conputat et scelus ingens
sana facit. spectant subeuntem fata mariti
Alcestim, et similis si permutatio detur,
morte viri cupiant animam servare catellae.
occurrent multae tibi Belides atque Eriphylae   655
mane, Clytaemestram nullus non vicus habebit.
hoc tantum refert, quod Tyndaris illa bipennem
insulsam et fatuam dextra laevaque tenebat,
at nunc res agitur tenui pulmone rubetae—
sed tamen et ferro, si praegustabit Atrides   660
Pontica ter victi cautus medicamina regis.

647 nocentes *PT*: nocentem *pω*   660 praegustabit *PS*: prae-
gustaret *pω*

LIBER TERTIVS

SATVRA VII

Et spes et ratio studiorum in Caesare tantum.
solus enim tristes hac tempestate Camenas
respexit, cum iam celebres notique poetae
balneolum Gabiis, Romae conducere furnos
temptarent, nec foedum alii nec turpe putarent   5
praecones fieri, cum desertis Aganippes
vallibus esuriens migraret in atria Clio.
nam si Pieria quadrans tibi nullus in umbra
ostendatur, ames nomen victumque Machaerae
et vendas potius, commissa quod auctio vendit   10
stantibus, oenophorum tripedes armaria cistas
Alcithoen Pacci, Thebas et Terea Fausti.
hoc satius quam si dicas sub iudice 'vidi'
quod non vidisti, faciant equites Asiani
quamquam et Cappadoces faciant equitesque Bithyni,   15
altera quos nudo traducit Gallia talo.
nemo tamen studiis indignum ferre laborem
cogetur posthac, nectit quicumque canoris
eloquium vocale modis laurumque momordit.
hoc agite, o iuvenes. circumspicit et stimulat vos   20
materiamque sibi ducis indulgentia quaerit.
siqua aliunde putas rerum speranda tuarum
praesidia atque ideo croceae membrana tabellae
impletur, lignorum aliquid posce ocius et quae
componis, dona Veneris, Telesine, marito,   25
aut clude et positos tinea pertunde libellos.
frange miser calamum vigilataque proelia dele,

VII 12 Alcithoen *some mss.*: Alcitheon P   16 Gallia *p*ω:
Gallica *P*   22 speranda *Housman*: spectanda *P*: expectanda ω
24 impletur *p*ω: implentur *P*

qui facis in parva sublimia carmina cella,
ut dignus venias hederis et imagine macra.
spes nulla ulterior; didicit iam dives avarus   30
tantum admirari, tantum laudare disertos,
ut pueri Iunonis avem. sed defluit aetas
et pelagi patiens et cassidis atque ligonis.
taedia tunc subeunt animos, tunc seque suamque
Terpsichoren odit facunda et nuda senectus.   35
    accipe nunc artes. ne quid tibi conferat iste
quem colis et Musarum et Apollinis aede relicta,
ipse facit versus, atque uni cedit Homero
propter mille annos. et si dulcedine famae
succensus recites, Maculonis commodat aedes.   40
haec longe ferrata domus servire iubetur,
in qua sollicitas imitatur ianua portas.
scit dare libertos extrema in parte sedentis
ordinis et magnas comitum disponere voces:
nemo dabit regum, quanti subsellia constant   45
et quae conducto pendent anabathra tigillo
quaeque reportandis posita est orchestra cathedris.
nos tamen hoc agimus tenuique in pulvere sulcos
ducimus et litus sterili versamus aratro.
nam si discedas, laqueo tenet ambitiosi   50
consuetudo mali; tenet insanabile multos
scribendi cacoethes et aegro in corde senescit.
sed vatem egregium, cui non sit publica vena,
qui nil expositum soleat deducere, nec qui
communi feriat carmen triviale moneta,   55
hunc, qualem nequeo monstrare et sentio tantum,
anxietate carens animus facit, omnis acerbi
inpatiens, cupidus silvarum aptusque bibendis
fontibus Aonidum. neque enim cantare sub antro
Pierio thyrsumque potest contingere maesta   60

35 facunda et nuda *pω*: facundae inunda *P*   40 Maculonis
P: Maculonus *pω*: maculosas *Heinrich*   41 haec *P*: ac *pω*

paupertas atque aeris inops, quo nocte dieque
corpus eget: satur est cum dicit Horatius 'euhoe.'
quis locus ingenio, nisi cum se carmine solo
vexant et dominis Cirrhae Nysaeque feruntur
pectora vestra duas non admittentia curas?   65
magnae mentis opus nec de lodice paranda
attonitae, currus et equos faciesque deorum
aspicere et qualis Rutulum confundat Erinys.
nam si Vergilio puer et tolerabile desset
hospitium, caderent omnes a crinibus hydri,   70
surda nihil gemeret grave bucina: poscimus ut sit
non minor antiquo Rubrenus Lappa cothurno,
cuius et alveolos et laenam pignerat Atreus?
non habet infelix Numitor quod mittat amico:
Quintillae quod donet habet, nec defuit illi   75
unde emeret multa pascendum carne leonem
iam domitum; constat leviori belua sumptu
nimirum et capiunt plus intestina poetae.
contentus fama iaceat Lucanus in hortis
marmoreis, at Serrano tenuique Saleio   80
gloria quantalibet quid erit, si gloria tantum est?
curritur ad vocem iucundam et carmen amicae
Thebaidos, laetam cum fecit Statius urbem
promisitque diem; tanta dulcedine captos
adficit ille animos tantaque libidine volgi   85
auditur; sed cum fregit subsellia versu,
esurit, intactam Paridi nisi vendit Agauen.
ille et militiae multis largitur honorem,
semenstri digitos vatum circumligat auro.
quod non dant proceres, dabit histrio. tu Camerinos   90
et Baream, tu nobilium magna atria curas?
praefectos Pelopea facit, Philomela tribunos.
haut tamen invideas vati quem pulpita pascunt.
quis tibi Maecenas, quis nunc erit aut Proculeius

79 iaceat] taceat P   80 Saleio S: Saleno P: salino ω

aut Fabius? quis Cotta iterum, quis Lentulus alter?   95
tunc par ingenio pretium, tunc utile multis
pallere et vinum toto nescire decembri.
  vester porro labor fecundior, historiarum
scriptores? perit hic plus temporis atque olei plus.
nullo quippe modo millensima pagina surgit   100
omnibus et crescit multa damnosa papyro;
sic ingens rerum numerus iubet atque operum lex.
quae tamen inde seges? terrae quis fructus apertae?
quis dabit historico quantum daret acta legenti?
  'sed genus ignavum, quod lecto gaudet et umbra.'   105
dic igitur quid causidicis civilia praestent
officia et magno comites in fasce libelli.
ipsi magna sonant, sed tum cum creditor audit
praecipue, vel si tetigit latus acrior illo
qui venit ad dubium grandi cum codice nomen;   110
tunc inmensa cavi spirant mendacia folles
conspuiturque sinus: veram deprendere messem
si libet, hinc centum patrimonia causidicorum,
parte alia solum russati pone Lacertae.
consedere duces, surgis tu pallidus Aiax   115
dicturus dubia pro libertate bubulco
iudice. rumpe miser tensum iecur, ut tibi lasso
figantur virides, scalarum gloria, palmae.
quod vocis pretium? siccus petasunculus et vas
pelamydum aut veteres, Maurorum epimenia, bulbi   120
aut vinum Tiberi devectum, quinque lagonae.
si quater egisti, si contigit aureus unus,
inde cadunt partes ex foedere pragmaticorum.
Aemilio dabitur quantum licet, et melius nos
egimus. huius enim stat currus aeneus, alti   125
quadriiuges in vestibulis, atque ipse feroci
bellatore sedens curvatum hastile minatur

99 perit *P*: petit *pω*   100 nullo quippe modo *P*: namque
oblita modi *pω*   114 Lacertae *pω*: lacernae *P*

eminus et statua meditatur proelia lusca.
sic Pedo conturbat, Matho deficit, exitus hic est
Tongilii, magno cum rhinocerote lavari   130
qui solet et vexat lutulenta balnea turba
perque forum iuvenes longo premit assere Maedos
empturus pueros argentum murrina villas;
spondet enim Tyrio stlattaria purpura filo.
et tamen est illis hoc utile. purpura vendit   135
causidicum, vendunt amethystina; convenit illi
et strepitu et facie maioris vivere census,
sed finem inpensae non servat prodiga Roma.
fidimus eloquio? Ciceroni nemo ducentos
nunc dederit nummos, nisi fulserit anulus ingens.   140
respicit haec primum qui litigat, an tibi servi
octo, decem comites, an post te sella, togati
ante pedes. ideo conducta Paulus agebat
sardonyche, atque ideo pluris quam Gallus agebat,
quam Basilus. rara in tenui facundia panno.   145
quando licet Basilo flentem producere matrem?
quis bene dicentem Basilum ferat? accipiat te
Gallia vel potius nutricula causidicorum
Africa, si placuit mercedem ponere linguae.

declamare doces? o ferrea pectora Vetti,   150
cum perimit saevos classis numerosa tyrannos.
nam quaecumque sedens modo legerat, haec eadem stans
perferet atque eadem cantabit versibus isdem;
occidit miseros crambe repetita magistros.
quis color et quod sit causae genus atque ubi summa   155
quaestio, quae veniant diversae forte sagittae,
nosse volunt omnes, mercedem solvere nemo.
'mercedem appellas? quid enim scio?' 'culpa docentis
scilicet arguitur, quod laevae parte mamillae
nil salit Arcadico iuveni, cuius mihi sexta   160

139 fidimus eloquio P: ut redeant veteres ω
156 diversae forte P: diversa parte pω

quaque die miserum dirus caput Hannibal inplet,
quidquid id est de quo deliberat, an petat urbem
a Cannis, an post nimbos et fulmina cautus
circumagat madidas a tempestate cohortes.
quantum vis stipulare et protinus accipe, quid do   165
ut totiens illum pater audiat?' haec alii sex
vel plures uno conclamant ore sophistae
et veras agitant lites raptore relicto;
fusa venena silent, malus ingratusque maritus,
et quae iam veteres sanant mortaria caecos.   170
ergo sibi dabit ipse rudem, si nostra movebunt
consilia, et vitae diversum iter ingredietur,
ad pugnam qui rhetorica descendit ab umbra,
summula ne pereat qua vilis tessera venit
frumenti. quippe haec merces lautissima. tempta,   175
Chrysogonus quanti doceat vel Polio quanti
lautorum pueros: artem scindes Theodori.
balnea sescentis et pluris porticus in qua
gestetur dominus quotiens pluit—anne serenum
expectet spargatque luto iumenta recenti?   180
hic potius, namque hic mundae nitet ungula mulae;
parte alia longis Numidarum fulta columnis
surgat et algentem rapiat cenatio solem.
quanticumque domus, veniet qui fercula docte
conponat, veniet qui pulmentaria condit.   185
hos inter sumptus sestertia Quintiliano,
ut multum, duo sufficient; res nulla minoris
constabit patri quam filius. 'unde igitur tot
Quintilianus habet saltus?' exempla novorum
fatorum transi: felix et pulcer et acer,   190
felix et sapiens et nobilis et generosus,
adpositam nigrae lunam subtexit alutae;

165 quid do *PT*: quod do *p*: quiddam *Merry*
174 summula *p*: summauia *P*   177 scindes *Jahn*: scindens *Pω*
180 iumenta] tumentia *P*

felix orator quoque maximus et iaculator,
et si perfrixit, cantat bene. distat enim quae
sidera te excipiant modo primos incipientem   195
edere vagitus et adhuc a matre rubentem.
si Fortuna volet, fies de rhetore consul;
si volet haec eadem, fiet de consule rhetor.
Ventidius quid enim? quid Tullius? anne aliud quam
sidus et occulti miranda potentia fati?   200
servis regna dabunt, captivis fata triumphum.
felix ille tamen corvo quoque rarior albo,
paenituit multos vanae sterilisque cathedrae,
sicut Lysimachi probat exitus atque Secundi
Carrinatis; et hunc inopem vidistis, Athenae,   205
nil praeter gelidas ausae conferre cicutas.
di, maiorum umbris tenuem et sine pondere terram
spirantisque crocos et in urna perpetuum ver,
qui praeceptorem sancti voluere parentis
esse loco. metuens virgae iam grandis Achilles   210
cantabat patriis in montibus et cui non tunc
eliceret risum citharoedi cauda magistri,
sed Rufum atque alios caedit sua quemque iuventus,
Rufum, quem totiens Ciceronem Allobroga dixit.

    quis gremio Celadi doctique Palaemonis adfert   215
quantum grammaticus meruit labor? et tamen ex hoc
quodcumque est, minus est autem quam rhetoris aera,
discipuli custos praemordet acoenonoetus,
et qui dispensat, frangit sibi. cede, Palaemon,
et patere inde aliquid decrescere, non aliter quam   220
institor hibernae tegetis niveique cadurci,
dummodo non pereat mediae quod noctis ab hora
sedisti, qua nemo faber, qua nemo sederet
qui docet obliquo lanam deducere ferro;
dummodo non pereat totidem olfecisse lucernas,   225
quot stabant pueri, cum totus decolor esset

Flaccus et haereret nigro fuligo Maroni.
rara tamen merces quae cognitione tribuni
non egeat. sed vos saevas inponite leges,
ut praeceptori verborum regula constet,   230
ut legat historias, auctores noverit omnes
tamquam ungues digitosque suos, ut forte rogatus
dum petit aut thermas aut Phoebi balnea, dicat
nutricem Anchisae, nomen patriamque novercae
Anchemoli, dicat quot Acestes vixerit annis,   235
quot Siculi Phrygibus vini donaverit urnas.
exigite ut mores teneros ceu pollice ducat,
ut si quis cera voltum facit; exigite ut sit
et pater ipsius coetus, ne turpia ludant.
'haec' inquit 'curas, et cum se verterit annus,   242
accipe, victori populus quod postulat, aurum.'

SATVRA VIII

Stemmata quid faciunt? quid prodest, Pontice, longo
sanguine censeri, pictos ostendere vultus
maiorum et stantis in curribus Aemilianos
et Curios iam dimidios umerosque minorem
Corvinum et Galbam auriculis nasoque carentem?   5
quis fructus generis tabula iactare capaci
Corvinum, posthac multa contingere virga
fumosos equitum cum dictatore magistros,
si coram Lepidis male vivitur? effigies quo
tot bellatorum, si luditur alea pernox   10
ante Numantinos, si dormire incipis ortu
luciferi quo signa duces et castra movebant?
cur Allobrogicis et magna gaudeat ara
natus in Herculeo Fabius lare, si cupidus, si
vanus et Euganea quamtumvis mollior agna,   15
si tenerum attritus Catinensi pumice lumbum

VIII 7 posthac] posse ac *Withof*   8 fumosos *P*: famosos *p*

squalentis traducit avos emptorque veneni
frangenda miseram funestat imagine gentem?
tota licet veteres exornent undique cerae
atria, nobilitas sola est atque unica virtus.   20
Paulus vel Cossus vel Drusus moribus esto,
hos ante effigies maiorum pone tuorum,
praecedant ipsas illi te consule virgas.
prima mihi debes animi bona. sanctus haberi
iustitiaeque tenax factis dictisque mereris?   25
agnosco procerem; salve Gaetulice, seu tu
Silanus, quocumque alio de sanguine, rarus
civis et egregius patriae contingis ovanti;
exclamare libet, populus quod clamat Osiri
invento. quis enim generosum dixerit hunc qui   30
indignus genere et praeclaro nomine tantum
insignis? nanum cuiusdam Atlanta vocamus,
Aethiopem Cycnum, pravam extortamque puellam
Europen; canibus pigris scabieque vetusta
levibus et siccae lambentibus ora lucernae   35
nomen erit pardus tigris leo, si quid adhuc est
quod fremat in terris violentius; ergo cavebis
et metues ne tu sic Creticus aut Camerinus.

    his ego quem monui? tecum est mihi sermo, Rubelli
Blande. tumes alto Drusorum stemmate, tamquam   40
feceris ipse aliquid propter quod nobilis esses,
ut te conciperet quae sanguine fulget Iuli,
non quae ventoso conducta sub aggere texit.
'vos humiles' inquis 'volgi pars ultima nostri,
quorum nemo queat patriam monstrare parentis,   45
ast ego Cecropides.' vivas et originis huius
gaudia longa feras. tamen ima plebe Quiritem
facundum invenies, solet hic defendere causas
nobilis indocti; veniet de plebe togata
qui iuris nodos et legum aenigmata solvat;   50

38 sic *Iunius*: si *P*: sis ω

hic petit Euphraten iuvenis domitique Batavi
custodes aquilas armis industrius. at tu
nil nisi Cecropides, truncoque simillimus Hermae.
nullo quippe alio vincis discrimine quam quod
illi marmoreum caput est, tua vivit imago.   55
dic mihi, Teucrorum proles, animalia muta
quis generosa putet nisi fortia. nempe volucrem
sic laudamus equum, facili cui plurima palma
fervet et exultat rauco victoria circo;
nobilis hic, quocumque venit de gramine, cuius   60
clara fuga ante alios et primus in aequore pulvis.
sed venale pecus Coryphaei posteritas et
Hirpini, si rara iugo victoria sedit;
nil ibi maiorum respectus, gratia nulla
umbrarum; dominos pretiis mutare iubentur   65
exiguis, trito ducunt epiraedia collo
segnipedes dignique molam versare nepotes.
ergo ut miremur te, non tua, privum aliquid da,
quod possim titulis incidere praeter honores
quos illis damus ac dedimus, quibus omnia debes.   70
   haec satis ad iuvenem quem nobis fama superbum
tradit et inflatum plenumque Nerone propinquo;
rarus enim ferme sensus communis in illa
fortuna. sed te censeri laude tuorum,
Pontice, noluerim sic ut nihil ipse futurae   75
laudis agas. miserum est aliorum incumbere famae,
ne conlapsa ruant subductis tecta columnis,
stratus humi palmes viduas desideret ulmos.
esto bonus miles, tutor bonus, arbiter idem
integer; ambiguae si quando citabere testis   80
incertaeque rei, Phalaris licet imperet ut sis

51 hic] hinc *Weidner*
62 Coryphaei] Cory*e* *P*: Corythe *p*   67 nepotes *P*:
Nepotis *p*   68 privum *Salmasius*: primum *Pω*   78 desi-
derat *P*: discinderet *in margin of P*

falsus et admoto dictet periuria tauro,
summum crede nefas, animam praeferre pudori
et propter vitam vivendi perdere causas.
dignus morte perit, cenet licet ostrea centum   85
Gaurana et Cosmi toto mergatur aeno.
expectata diu tandem provincia cum te
rectorem accipiet, pone irae frena modumque,
pone et avaritiae, miserere inopum sociorum—
ossa vides rerum vacuis exucta medullis—   90
respice quid moneant leges, quid curia mandet,
praemia quanta bonos maneant, quam fulmine iusto
et Capito et Numitor ruerint damnante senatu,
piratae Cilicum. sed quid damnatio confert?
praeconem, Chaerippe, tuis circumspice pannis,   95
cum Pansa eripiat quidquid tibi Natta reliquit,
iamque tace; furor est post omnia perdere naulum.
non idem gemitus olim neque vulnus erat par
damnorum sociis florentibus et modo victis.
plena domus tunc omnis, et ingens stabat acervus   100
nummorum, Spartana chlamys, conchylia Coa,
et cum Parrhasii tabulis signisque Myronis
Phidiacum vivebat ebur, nec non Polycliti
multus ubique labor, rarae sine Mentore mensae.
inde Dolabella †atque hinc† Antonius, inde   105
sacrilegus Verres referebant navibus altis
occulta spolia et plures de pace triumphos.
nunc sociis iuga pauca boum, grex parvus equarum,
et pater armenti capto eripietur agello,
ipsi deinde Lares, si quod spectabile signum,   110
si quis in aedicula deus unicus; haec etenim sunt
pro summis, nam sunt haec maxima⟩despicias tu
forsitan inbellis Rhodios unctamque Corinthon,

---

88 accipiet ω: accipiat P   90 rerum P: regum p   93 Nu-
mitor P: Tutor ω   105 Dolabella atque hinc Pω: Dolabellae
Ruperti   112 nam sunt] iam sunt *conj. Büch.*

despicias merito; quid resinata iuventus
cruraque totius facient tibi levia gentis?   115
horrida vitanda est Hispania, Gallicus axis
Illyricumque latus; parce et messoribus illis
qui saturant urbem circo scaenaeque vacantem;
quanta autem inde feres tam dirae praemia culpae,
cum tenuis nuper Marius discinxerit Afros?   120
curandum in primis ne magna iniuria fiat
fortibus et miseris. tollas licet omne quod usquam est
auri atque argenti, scutum gladiumque relinques
et iaculum et galeam; spoliatis arma supersunt.
quod modo proposui, non est sententia: verum est,   125
credite me vobis folium recitare Sibyllae.
si tibi sancta cohors comitum, si nemo tribunal
vendit acersecomes, si nullum in coniuge crimen,
nec per conventus et cuncta per oppida curvis
unguibus ire parat nummos raptura Celaeno,   130
tu licet a Pico numeres genus, altaque si te
nomina delectant, omnem Titanida pugnam
inter maiores ipsumque Promethea ponas,
de quocumque voles proavum tibi sumito libro.
quod si praecipitem rapit ambitio atque libido,   135
si frangis virgas sociorum in sanguine, si te
delectant hebetes lasso lictore secures,
incipit ipsorum contra te stare parentum
nobilitas claramque facem praeferre pudendis.
omne animi vitium tanto conspectius in se   140
crimen habet, quanto maior qui peccat habetur.
quo mihi te solitum falsas signare tabellas
in templis quae fecit avus, statuamque parentis
ante triumphalem? quo, si nocturnus adulter
tempora Santonico velas adoperta cucullo?   145
    praeter maiorum cineres atque ossa volucri
carpento rapitur pinguis Lateranus, et ipse,

147 Lateranus *P*: Damasippus ω

ipse rotam adstringit sufflamine mulio consul,
nocte quidem, sed Luna videt, sed sidera testes
intendunt oculos. finitum tempus honoris   150
cum fuerit, clara Lateranus luce flagellum
sumet et occursum numquam trepidabit amici
iam senis ac virga prior annuet atque maniplos
solvet et infundet iumentis hordea lassis.
interea, dum lanatas robumque iuvencum   155
more Numae caedit, Iovis ante altaria iurat
solam Eponam et facies olida ad praesepia pictas.
sed cum pervigiles placet instaurare popinas,
obvius adsiduo Syrophoenix unctus amomo
currit, Idymaeae Syrophoenix incola portae   160
hospitis adfectu dominum regemque salutat,
et cum venali Cyane succincta lagona.
defensor culpae dicet mihi 'fecimus et nos
haec iuvenes.' esto, desisti nempe nec ultra
fovisti errorem. breve sit quod turpiter audes,   165
quaedam cum prima resecentur crimina barba.
indulge veniam pueris: Lateranus ad illos
thermarum calices inscriptaque lintea vadit
maturus bello Armeniae Syriaeque tuendis
amnibus et Rheno atque Histro. praestare Neronem   170
securum valet haec aetas. mitte Ostia, Caesar,
mitte, sed in magna legatum quaere popina;
invenies aliquo cum percussore iacentem,
permixtum nautis et furibus ac fugitivis,
inter carnifices et fabros sandapilarum   175
et resupinati cessantia tympana galli.
aequa ibi libertas, communia pocula, lectus
non alius cuiquam, nec mensa remotior ulli.
quid facias talem sortitus, Pontice, servum?

---

148 sufflamine mulio *Florileg. Sangall.*: 'mulio est qui consul
fertur' *S on l.* 157: multo sufflamine *Pω*   155 robumque *S*
Florileg. Sangall.: ***umque *P*: torvumque *pω*

nempe in Lucanos aut Tusca ergastula mittas.   180
at vos, Troiugenae, vobis ignoscitis, et quae
turpia cerdoni, Volesos Brutumque decebunt.
quid si numquam adeo foedis adeoque pudendis
utimur exemplis, ut non peiora supersint?
consumptis opibus vocem, Damasippe, locasti   185
sipario, clamosum ageres ut Phasma Catulli.
Laureolum velox etiam bene Lentulus egit,
iudice me dignus vera cruce. nec tamen ipsi
ignoscas populo; populi frons durior huius
qui sedet et spectat triscurria patriciorum,   190
planipedes audit Fabios, ridere potest qui
Mamercorum alapas. quanti sua funera vendant,
quid refert? vendunt nullo cogente Nerone,
nec dubitant celsi praetoris vendere ludis.
finge tamen gladios inde atque hinc pulpita poni,   195
quid satius? mortem sic quisquam exhorruit, ut sit
zelotypus Thymeles, stupidi collega Corinthi?
res haut mira tamen citharoedo principe mimus
nobilis. haec ultra quid erit nisi ludus? et illic
dedecus urbis habes, nec murmillonis in armis   200
nec clipeo Gracchum pugnantem aut falce supina—
damnat enim tales habitus, sed damnat et odit,
nec galea faciem abscondit, movet ecce tridentem,
postquam vibrata pendentia retia dextra
nequiquam effudit, nudum ad spectacula voltum   205
erigit et tota fugit agnoscendus harena.
credamus, tunicae de faucibus aurea cum se
porrigat et longo iactetur spira galero.
ergo ignominiam graviorem pertulit omni
vulnere cum Graccho iussus pugnare secutor.   210
    libera si dentur populo suffragia, quis tam
perditus ut dubitet Senecam praeferre Neroni?
cuius supplicio non debuit una parari

simia nec serpens unus nec culleus unus.
par Agamemnonidae crimen, sed causa facit rem   215
dissimilem. quippe ille deis auctoribus ultor
patris erat caesi media inter pocula, sed nec
Electrae iugulo se polluit aut Spartani
sanguine coniugii, nullis aconita propinquis
miscuit, in scaena numquam cantavit Orestes,   220
Troica non scripsit. quid enim Verginius armis
debuit ulcisci magis aut cum Vindice Galba,
quod Nero tam saeva crudaque tyrannide fecit?
haec opera atque hae sunt generosi principis artes,
gaudentis foedo peregrina ad pulpita cantu   225
prostitui Graiaeque apium meruisse coronae.
maiorum effigies habeant insignia vocis,
ante pedes Domiti longum tu pone Thyestae
syrma vel Antigonae personam vel Melanippae,
et de marmoreo citharam suspende colosso.   230
   quid, Catilina, tuis natalibus atque Cethegi
inveniet quisquam sublimius? arma tamen vos
nocturna et flammas domibus templisque paratis,
ut bracatorum pueri Senonumque minores,
ausi quod liceat tunica punire molesta.   235
sed vigilat consul vexillaque vestra coercet;
hic novus Arpinas, ignobilis et modo Romae
municipalis eques, galeatum ponit ubique
praesidium attonitis et in omni monte laborat.
tantum igitur muros intra toga contulit illi   240
nominis ac tituli, quantum †in† Leucade, quantum
Thessaliae campis Octavius abstulit udo
caedibus adsiduis gladio, sed Roma parentem,
Roma patrem patriae Ciceronem libera dixit.
Arpinas alius Volscorum in monte solebat   245

223 quod *Madvig*: quid *Pω*   226 Graiaeque] grataeque *P*
233 paratis *P*: parastis *ω*   239 monte *PS*: gente *p*
241 in Leucade *P*: non Leucade *pω*: vi Leucade *Owen*

poscere mercedes alieno lassus aratro,
nodosam post haec frangebat vertice vitem,
si lentus pigra muniret castra dolabra;
hic tamen et Cimbros et summa pericula rerum
excipit et solus trepidantem protegit urbem.   250
atque ideo, postquam ad Cimbros stragemque volabant
qui numquam attigerant maiora cadavera corvi,
nobilis ornatur lauro collega secunda.
plebeiae Deciorum animae, plebeia fuerunt
nomina; pro totis legionibus hi tamen et pro   255
omnibus auxiliis atque omni pube Latina
sufficiunt dis infernis Terraeque parenti;
pluris enim Decii quam quae servantur ab illis.
ancilla natus trabeam et diadema Quirini
et fasces meruit, regum ultimus ille bonorum;   260
prodita laxabant portarum claustra tyrannis
exulibus iuvenes ipsius consulis et quos
magnum aliquid dubia pro libertate deceret,
quod miraretur cum Coclite Mucius et quae
imperii fines Tiberinum virgo natavit.   265
occulta ad patres produxit crimina servus
matronis lugendus, at illos verbera iustis
adficiunt poenis et legum prima securis.
    malo pater tibi sit Thersites, dummodo tu sis
Aeacidae similis Vulcaniaque arma capessas,   270
quam te Thersitae similem producat Achilles.
et tamen, ut longe repetas longeque revolvas
nomen, ab infami gentem deducis asylo;
maiorum primus, quisquis fuit ille, tuorum
aut pastor fuit aut illud quod dicere nolo.   275

LIBER QVARTVS

The Vanity of Human Wishes

SATVRA X

Omnibus in terris, quae sunt a Gadibus usque
Auroram et Gangen, pauci dinoscere possunt   distinguish
vera bona atque illis multum diversa, remota
erroris nebula. quid enim ratione timemus
aut cupimus? quid tam dextro pede concipis, ut te   5
conatus non paeniteat votique peracti?   accomplished
evertere domos totas optantibus ipsis
di faciles. nocitura toga, nocitura petuntur
militia; torrens dicendi copia multis
et sua mortifera est facundia, viribus ille   10
confisus periit admirandisque lacertis,
sed plures nimia congesta pecunia cura
strangulat et cuncta exsuperans patrimonia census
quanto delphinis ballaena Britannica maior.
temporibus diris igitur iussuque Neronis   15
Longinum et magnos Senecae praedivitis hortos
clausit et egregias Lateranorum obsidet aedes
tota cohors: rarus venit in cenacula miles.
pauca licet portes argenti vascula puri
nocte iter ingressus, gladium contumque timebis   20
et motae ad lunam trepidabis harundinis umbram:
cantabit vacuus coram latrone viator.
prima fere vota et cunctis notissima templis
divitiae, crescant ut opes, ut maxima toto
nostra sit arca foro. sed nulla aconita bibuntur   25
fictilibus: tunc illa time, cum pocula sumes
gemmata et lato Setinum ardebit in auro.
iamne igitur laudas quod de sapientibus alter
ridebat, quotiens de limine moverat unum
protuleratque pedem, flebat contrarius auctor?   30

sed facilis cuivis rigidi censura cachinni:
mirandum est unde ille oculis suffecerit umor.
perpetuo risu pulmonem agitare solebat
Democritus, quamquam non essent urbibus illis
praetextae trabeae fasces lectica tribunal.   35
quid si vidisset praetorem curribus altis
extantem et medii sublimem pulvere circi
in tunica Iovis et pictae Sarrana ferentem
ex umeris aulaea togae magnaeque coronae
tantum orbem, quanto cervix non sufficit ulla?   40
quippe tenet sudans hanc publicus, et sibi consul
ne placeat, curru servus portatur eodem.
da nunc et volucrem, sceptro quae surgit eburno,
illinc cornicines, hinc praecedentia longi
agminis officia et niveos ad frena Quirites,   45
defossa in loculos quos sportula fecit amicos.
tunc quoque materiam risus invenit ad omnis
occursus hominum, cuius prudentia monstrat
summos posse viros et magna exempla daturos
vervecum in patria crassoque sub aere nasci.   50
ridebat curas nec non et gaudia vulgi,
interdum et lacrimas, cum Fortunae ipse minaci
mandaret laqueum mediumque ostenderet unguem.
ergo supervacua aut vel perniciosa putentur,
propter quae fas est genua incerare deorum?   55
quosdam praecipitat subiecta potentia magnae
invidiae, mergit longa atque insignis honorum
pagina. descendunt statuae restemque sequuntur,
ipsas deinde rotas bigarum inpacta securis
caedit et inmeritis franguntur crura caballis,   60
iam strident ignes, iam follibus atque caminis
ardet adoratum populo caput et crepat ingens

35 praetexta etrabeae P: praetexta et trabeae p 46 loculos P:
loculis ω   54 aut vel perniciosa Döderlein: aut perniciosa
Pω putentur Richards: petuntur Pω

Seianus, deinde ex facie toto orbe secunda
fiunt urceoli pelves sartago matellae.
pone domi laurus, duc in Capitolia magnum 65
cretatumque bovem: Seianus ducitur unco
spectandus, gaudent omnes. 'quae labra, quis illi
vultus erat. numquam, si quid mihi credis, amavi
hunc hominem. sed quo cecidit sub crimine? quisnam
delator quibus indicibus, quo teste probavit?' 70
nil horum; verbosa et grandis epistula venit
a Capreis. 'bene habet, nil plus interrogo. sed quid
turba Remi?' sequitur fortunam ut semper, et odit
damnatos. idem populus, si Nortia Tusco
favisset, si oppressa foret secura senectus 75
principis, hac ipsa Seianum diceret hora
Augustum. iam pridem, ex quo suffragia nulli
vendimus, effudit curas; nam qui dabat olim
imperium fasces legiones omnia, nunc se
continet atque duas tantum res anxius optat, 80
panem et circenses. 'perituros audio multos.'
nil dubium, magna est fornacula. 'pallidulus mi
Bruttidius meus ad Martis fuit obvius aram;
quam timeo, victus ne poenas exigat Aiax
ut male defensus. curramus praecipites et 85
dum iacet in ripa, calcemus Caesaris hostem.
sed videant servi, ne quis neget et pavidum in ius
cervice obstricta dominum trahat.' hi sermones
tunc de Seiano, secreta haec murmura vulgi.
visne salutari sicut Seianus, habere 90
tantundem atque illi summas donare curules,
illum exercitibus praeponere, tutor haberi
principis angusta Caprearum in rupe sedentis
cum grege Chaldaeo? vis certe pila cohortes

64 matellae *P*: patellae *pω* 70 indicibus *PT*: indiciis *p*
73 Remi *P*: tremens *ω* 78 effudit *P*: effugit *p*
93 angusta *p*: augusta *P Büch.*

egregios equites et castra domestica. quidni   95
haec cupias? et qui nolunt occidere quemquam,
posse volunt. sed quae praeclara et prospera tanti,
ut rebus laetis par sit mensura malorum?
huius qui trahitur praetextam sumere mavis,
an Fidenarum Gabiorumque esse potestas   100
et de mensura ius dicere, vasa minora
frangere pannosus vacuis aedilis Ulubris?
ergo quid optandum foret, ignorasse fateris
Seianum; nam qui nimios optabat honores
et nimias poscebat opes, numerosa parabat   105
excelsae turris tabulata, unde altior esset
casus et inpulsae praeceps inmane ruinae.
quid Crassos, quid Pompeios evertit et illum,
ad sua qui domitos deduxit flagra Quirites?
summus nempe locus nulla non arte petitus   110
magnaque numinibus vota exaudita malignis.
ad generum Cereris sine caede ac vulnere pauci
descendunt reges et sicca morte tyranni.
    eloquium ac famam Demosthenis aut Ciceronis
incipit optare et totis quinquatribus optat   115
quisquis adhuc uno parcam colit asse Minervam,
quem sequitur custos angustae vernula capsae.
eloquio sed uterque perit orator, utrumque
largus et exundans leto dedit ingenii fons.
ingenio manus est et cervix caesa, nec umquam   120
sanguine causidici maduerunt rostra pusilli.
'o fortunatam natam me consule Romam':
Antoni gladios potuit contemnere, si sic
omnia dixisset. ridenda poemata malo
quam te, conspicuae divina Philippica famae,   125
volveris a prima quae proxima. saevus et illum
exitus eripuit, quem mirabantur Athenae
torrentem et pleni moderantem frena theatri.

114 ac *pω*: aut *P Büch.*   116 parcam *P*: partam *pω*

dis ille adversis genitus fatoque sinistro,
quem pater ardentis massae fuligine lippus 130
a carbone et forcipibus gladiosque paranti
incude et luteo Vulcano ad rhetora misit.
bellorum exuviae, truncis adfixa tropaeis
lorica et fracta de casside buccula pendens
et curtum temone iugum victaeque triremis 135
aplustre et summo tristis captivus in arcu
humanis maiora bonis creduntur. ad hoc se
Romanus Graiusque et barbarus induperator
erexit, causas discriminis atque laboris
inde habuit; tanto maior famae sitis est quam 140
virtutis. quis enim virtutem amplectitur ipsam,
praemia si tollas? patriam tamen obruit olim
gloria paucorum et laudis titulique cupido
haesuri saxis cinerum custodibus, ad quae
discutienda valent sterilis mala robora fici, 145
quandoquidem data sunt ipsis quoque fata sepulcris.
expende Hannibalem: quot libras in duce summo
invenies? hic est, quem non capit Africa Mauro
percussa oceano Niloque admota tepenti
rursus ad Aethiopum populos aliosque elephantos? 150
additur imperiis Hispania, Pyrenaeum
transilit. opposuit natura Alpemque nivemque:
diducit scopulos et montem rumpit aceto.
iam tenet Italiam, tamen ultra pergere tendit.
'actum' inquit 'nihil est, nisi Poeno milite portas 155
frangimus et media vexillum pono Subura.'
o qualis facies et quali digna tabella,
cum Gaetula ducem portaret belua luscum.
exitus ergo quis est? o gloria, vincitur idem
nempe et in exilium praeceps fugit atque ibi magnus 160
mirandusque cliens sedet ad praetoria regis,
donec Bithyno libeat vigilare tyranno.

150 aliosque *p*: altosque *P* (*but with gloss* praeter Indicos)

finem animae quae res humanas miscuit olim,
non gladii, non saxa dabunt nec tela, sed ille
Cannarum vindex et tanti sanguinis ultor 165
anulus. i demens et saevas curre per Alpes,
ut pueris placeas et declamatio fias.
unus Pellaeo iuveni non sufficit orbis,
aestuat infelix angusto limite mundi
ut Gyari clausus scopulis parvaque Seripho; 170
cum tamen a figulis munitam intraverit urbem,
sarcophago contentus erit. mors sola fatetur
quantula sint hominum corpuscula. creditur olim
velificatus Athos et quidquid Graecia mendax
audet in historia, constratum classibus isdem 175
suppositumque rotis solidum mare; credimus altos
defecisse amnes epotaque flumina Medo
prandente et madidis cantat quae Sostratus alis;
ille tamen qualis rediit Salamine relicta,
in corum atque eurum solitus saevire flagellis 180
barbarus Aeolio numquam hoc in carcere passos,
ipsum conpedibus qui vinxerat Ennosigaeum—
mitius id sane, quod non et stigmate dignum
credidit; huic quisquam vellet servire deorum?
sed qualis rediit? nempe una nave, cruentis 185
fluctibus ac tarda per densa cadavera prora.
has totiens optata exegit gloria poenas.

  'da spatium vitae, multos da, Iuppiter, annos'
hoc recto vultu, solum hoc et pallidus optas.
sed quam continuis et quantis longa senectus 190
plena malis. deformem et taetrum ante omnia vultum
dissimilemque sui, deformem pro cute pellem
pendentisque genas et talis aspice rugas
quales, umbriferos ubi pandit Thabraca saltus,
in vetula scalpit iam mater simia bucca. 195
plurima sunt iuvenum discrimina, pulchrior ille

184 credidit *P*: crederet *ω*

hoc atque ille alio, multum hic robustior illo:
una senum facies. cum voce trementia membra
et iam leve caput madidique infantia nasi;
frangendus misero gingiva panis inermi;   200
usque adeo gravis uxori natisque sibique,
ut captatori moveat fastidia Cosso.
non eadem vini atque cibi torpente palato
gaudia.   aspice partis
nunc damnum alterius. nam quae cantante voluptas,   210
sit licet eximius, citharoedo sive Seleuco
et quibus aurata mos est fulgere lacerna?
quid refert, magni sedeat qua parte theatri
qui vix cornicines exaudiet atque tubarum
concentus? clamore opus est, ut sentiat auris   215
quem dicat venisse puer, quot nuntiet horas.
praeterea minimus gelido iam in corpore sanguis
febre calet sola, circumsilit agmine facto
morborum omne genus, quorum si nomina quaeras,
promptius expediam quot amaverit Oppia moechos,   220
quot Themison aegros autumno occiderit uno,
quot Basilus socios, quot circumscripserit Hirrus,
percurram citius quot villas possideat nunc   225
quo tondente gravis iuveni mihi barba sonabat.
ille umero, hic lumbis, hic coxa debilis; ambos
perdidit ille oculos et luscis invidet; huius
pallida labra cibum accipiunt digitis alienis,
ipse ad conspectum cenae diducere rictum   230
suetus hiat tantum ceu pullus hirundinis, ad quem
ore volat pleno mater ieiuna. sed omni
membrorum damno maior dementia, quae nec
nomina servorum nec vultum agnoscit amici
cum quo praeterita cenavit nocte, nec illos   235
quos genuit, quos eduxit. nam codice saevo
heredes vetat esse suos, bona tota feruntur

232 mater ieiuna] materiae luna P

ad Phialen; tantum artificis valet halitus oris,
quod steterat multis in carcere fornicis annis.
ut vigeant sensus animi, ducenda tamen sunt   240
funera natorum, rogus aspiciendus amatae
coniugis et fratris plenaeque sororibus urnae.
haec data poena diu viventibus, ut renovata
semper clade domus multis in luctibus inque
perpetuo maerore et nigra veste senescant.   245
rex Pylius, magno si quidquam credis Homero,
exemplum vitae fuit a cornice secundae.
felix nimirum, qui tot per saecula mortem
distulit atque suos iam dextra conputat annos,
quique novum totiens mustum bibit. oro, parumper   250
attendas quantum de legibus ipse queratur
fatorum et nimio de stamine, cum videt acris
Antilochi barbam ardentem, cum quaerit ab omni
quisquis adest socius, cur haec in tempora duret,
quod facinus dignum tam longo admiserit aevo.   255
haec eadem Peleus raptum cum luget Achillen,
atque alius cui fas Ithacum lugere natantem.
incolumi Troia Priamus venisset ad umbras
Assaraci magnis sollemnibus, Hectore funus
portante ac reliquis fratrum cervicibus inter   260
Iliadum lacrimas, ut primos edere planctus
Cassandra inciperet scissaque Polyxena palla,
si foret extinctus diverso tempore, quo non
coeperat audaces Paris aedificare carinas.
longa dies igitur quid contulit? omnia vidit   265
eversa et flammis Asiam ferroque cadentem.
tunc miles tremulus posita tulit arma tiara
et ruit ante aram summi Iovis ut vetulus bos,
qui domini cultris tenue et miserabile collum
praebet ab ingrato iam fastiditus aratro.   270
exitus ille utcumque hominis, sed torva canino

253 cum quaerit PT: nam quaerit ω

latravit rictu quae post hunc vixerat uxor.
festino ad nostros et regem transeo Ponti
et Croesum, quem vox iusti facunda Solonis
respicere ad longae iussit spatia ultima vitae.
exilium et carcer Minturnarumque paludes
et mendicatus victa Carthagine panis
hinc causas habuere; quid illo cive tulisset
natura in terris, quid Roma beatius umquam,
si circumducto captivorum agmine et omni
bellorum pompa animam exhalasset opimam,
cum de Teutonico vellet descendere curru?
provida Pompeio dederat Campania febres
optandas, sed multae urbes et publica vota
vicerunt, igitur Fortuna ipsius et urbis
servatum victo caput abstulit. hoc cruciatu
Lentulus, hac poena caruit ceciditque Cethegus
integer, et iacuit Catilina cadavere toto.

    formam optat modico pueris, maiore puellis
murmure, cum Veneris fanum videt, anxia mater
usque ad delicias votorum. 'cur tamen' inquit
'corripias? pulchra gaudet Latona Diana.'
sed vetat optari faciem Lucretia qualem
ipsa habuit, cuperet Rutilae Verginia gibbum
accipere atque suam Rutilae dare. filius autem
corporis egregii miseros trepidosque parentes
semper habet; rara est adeo concordia formae
atque pudicitiae. sanctos licet horrida mores
tradiderit domus ac veteres imitata Sabinos,
praeterea castum ingenium vultumque modesto
sanguine ferventem tribuat natura benigna
larga manu—quid enim puero conferre potest plus
custode et cura natura potentior omni?—
non licet esse viro. nam prodiga corruptoris
improbitas ipsos audet temptare parentes;

295 suam *pω*: suum *P Büch.*   304 viro *Jahn*: viros *P*: viris *ω*

tanta in muneribus fiducia.
i nunc et iuvenis specie laetare tui, quem   310
maiora expectant discrimina. fiet adulter
publicus, et poenas metuit quascumque mariti
irati debet, nec erit felicior astro
Martis, ut in laqueos numquam incidat. exigit autem
interdum ille dolor plus quam lex ulla dolori   315
concessit; necat hic ferro, secat ille *flagellis*.
'sed casto quid forma nocet?' quid profuit immo
Hippolyto grave propositum, quid Bellerophonti?   325
erubuit nempe haec ceu fastidita repulsa,
nec Stheneboea minus quam Cressa excanduit, et se
concussere ambae. mulier saevissima tunc est,
cum stimulos odio pudor admovet. elige quidnam
suadendum esse putes cui nubere Caesaris uxor   330
destinat. optimus hic et formosissimus idem
gentis patriciae rapitur miser extinguendus
Messalinae oculis; dudum sedet illa parato
flammeolo Tyriusque palam genialis in hortis
sternitur et ritu decies centena dabuntur   335
antiquo, veniet cum signatoribus auspex.
haec tu secreta et paucis commissa putabas?
non nisi legitime vult nubere. quid placeat dic.
ni parere velis, pereundum erit ante lucernas;
si scelus admittas, dabitur mora parvula, dum res   340
nota urbi et populo contingat principis aurem.
dedecus ille domus sciet ultimus; interea tu
obsequere imperio, si tanti vita dierum
paucorum. quidquid levius meliusve putaris,
praebenda est gladio pulchra haec et candida cervix.   345
    nil ergo optabunt homines? si consilium vis,
permittes ipsis expendere numinibus quid
conveniat nobis rebusque sit utile nostris.
nam pro iucundis aptissima quaeque dabunt di,

326 haec] hac *Haupt*

carior est illis homo quam sibi. nos animorum   350
inpulsu et caeca magnaque cupidine ducti
coniugium petimus partumque uxoris, at illis
notum qui pueri qualisque futura sit uxor.
ut tamen et poscas aliquid voveasque sacellis
exta et candiduli divina tomacula porci,   355
orandum est ut sit mens sana in corpore sano.
fortem posce animum mortis terrore carentem,
qui spatium vitae extremum inter munera ponat
naturae, qui ferre queat quoscumque labores,
nesciat irasci, cupiat nihil et potiores   360
Herculis aerumnas credat saevosque labores
et venere et cenis et pluma Sardanapalli.
monstro quod ipse tibi possis dare, semita certe   *path*
tranquillae per virtutem patet unica vitae.
nullum numen habes si sit prudentia, nos te,   365
nos facimus, Fortuna, deam caeloque locamus.

SATVRA XI

Atticus eximie si cenat, lautus habetur:
si Rutilus, demens. quid enim maiore cachinno
excipitur vulgi quam pauper Apicius? omnis
convictus thermae stationes, omne theatrum
de Rutilo. nam dum valida ac iuvenalia membra   5
sufficiunt galeae dumque ardet sanguine, fertur
non cogente quidem sed nec prohibente tribuno
scripturus leges et regia verba lanistae.
multos porro vides, quos saepe elusus ad ipsum
creditor introitum solet expectare macelli,   10
et quibus in solo vivendi causa palato est.
egregius cenat meliusque miserrimus horum
et cito casurus iam perlucente ruina.

> 365 habes *P*: abest *p*
> XI 5 iuvenalia *PT*: iuvenilia *p*

interea gustus elementa per omnia quaerunt
numquam animo pretiis opstantibus; interius si   15
adtendas, magis illa iuvant quae pluris ementur.
ergo haut difficile est perituram arcessere summam
lancibus oppositis vel matris imagine fracta,
et quadringentis nummis condire gulosum
fictile; sic veniunt ad miscillanea ludi.   20
refert ergo quis haec eadem paret; in Rutilo nam
luxuria est, in Ventidio laudabile nomen
sumit et a censu famam trahit. illum ego iure
despiciam, qui scit quanto sublimior Atlans
omnibus in Libya sit montibus, hic tamen idem   25
ignoret quantum ferrata distet ab arca
sacculus. e caelo descendit γνῶθι σεαυτὸν
figendum et memori tractandum pectore, sive
coniugium quaeras vel sacri in parte senatus
esse velis—neque enim loricam poscit Achillis   30
Thersites, in qua se traducebat Ulixes—
ancipitem seu tu magno discrimine causam
protegere adfectas, te consule, dic tibi qui sis,
orator vehemens an Curtius et Matho buccae.
noscenda est mensura sui spectandaque rebus   35
in summis minimisque, etiam cum piscis emetur,
ne mullum cupias, cum sit tibi gobio tantum
in loculis. quis enim te deficiente crumina
et crescente gula manet exitus, aere paterno
ac rebus mersis in ventrem faenoris atque   40
argenti gravis et pecorum agrorumque capacem?
talibus a dominis post cuncta novissimus exit
anulus, et digito mendicat Pollio nudo.
non praematuri cineres nec funus acerbum
luxuriae, sed morte magis metuenda senectus.   45
hi plerumque gradus: conducta pecunia Romae
et coram dominis consumitur; inde ubi paulum

16 ementur P: emuntur pω

nescio quid superest et pallet faenoris auctor,
qui vertere solum, Baias et ad ostrea currunt.
cedere namque foro iam non est deterius quam   50
Esquilias a ferventi migrare Subura.
ille dolor solus patriam fugientibus, illa
maestitia est, caruisse anno circensibus uno.
sanguinis in facie non haeret gutta, morantur
pauci ridiculum et fugientem ex urbe pudorem.   55
    experiere hodie numquid pulcherrima dictu,
Persice, non praestem vitae tibi moribus et re,
si laudem siliquas occultus ganeo, pultes
coram aliis dictem puero sed in aure placentas.
nam cum sis conviva mihi promissus, habebis   60
Euandrum, venies Tirynthius aut minor illo
hospes, et ipse tamen contingens sanguine caelum,
alter aquis, alter flammis ad sidera missus.
fercula nunc audi nullis ornata macellis.
de Tiburtino veniet pinguissimus agro   65
haedulus et toto grege mollior, inscius herbae
necdum ausus virgas humilis mordere salicti,
qui plus lactis habet quam sanguinis, et montani
asparagi, posito quos legit vilica fuso.
grandia praeterea tortoque calentia faeno   70
ova adsunt ipsis cum matribus, et servatae
parte anni quales fuerant in vitibus uvae,
Signinum Syriumque pirum, de corbibus isdem
aemula Picenis et odoris mala recentis
nec metuenda tibi, siccatum frigore postquam   75
autumnum et crudi posuere pericula suci.
haec olim nostri iam luxuriosa senatus
cena fuit. Curius parvo quae legerat horto
ipse focis brevibus ponebat holuscula, quae nunc
squalidus in magna fastidit compede fossor,   80
qui meminit calidae sapiat quid vulva popinae.

    57 vitae tibi Büch.: vitae. *** P: vita vel ω

sicci terga suis rara pendentia crate
moris erat quondam festis servare diebus,
et natalicium cognatis ponere lardum
accedente nova, si quam dabat hostia, carne.   85
cognatorum aliquis titulo ter consulis atque
castrorum imperiis et dictatoris honore
functus ad has epulas solito maturius ibat
erectum domito referens a monte ligonem.
cum tremerent autem Fabios durumque Catonem   90
et Scauros et Fabricium, rigidique severos
censoris mores etiam collega timeret,
nemo inter curas et seria duxit habendum
qualis in Oceano fluctu testudo nataret,
clarum Troiugenis factura et nobile fulcrum,   95
sed nudo latere et parvis frons aerea lectis
vile coronati caput ostendebat aselli,
ad quod lascivi ludebant ruris alumni.
tales ergo cibi, qualis domus atque supellex.
tunc rudis et Graias mirari nescius artes   100
urbibus eversis praedarum in parte reperta
magnorum artificum frangebat pocula miles,
ut phaleris gauderet ecus caelataque cassis
Romuleae simulacra ferae mansuescere iussae
imperii fato, geminos sub rupe Quirinos,   105
ac nudam effigiem clipeo venientis et hasta
pendentisque dei perituro ostenderet hosti.
ponebant igitur Tusco farrata catino:
argenti quod erat, solis fulgebat in armis.
omnia tunc, quibus invideas si lividulus sis.   110
templorum quoque maiestas praesentior, et vox
nocte fere media mediamque audita per urbem
litore ab Oceani Gallis venientibus et dis
officium vatis peragentibus. his monuit nos,

91 rigidique ω: postremo *P*   93 habendum *p*ω: habendam *P*
Büch.   94 Oceano *P*: Oceani *p*ω   106 venientis *P*: fulgentis *p*ω

hanc rebus Latiis curam praestare solebat   115
fictilis et nullo violatus Iuppiter auro.
illa domi natas nostraque ex arbore mensas
tempora viderunt; hos lignum stabat ad usus,
annosam si forte nucem deiecerat eurus.
at nunc divitibus cenandi nulla voluptas,   120
nil rhombus, nil damma sapit, putere videntur
unguenta atque rosae, latos nisi sustinet orbes
grande ebur et magno sublimis pardus hiatu
dentibus ex illis quos mittit porta Syenes
et Mauri celeres et Mauro obscurior Indus,   125
et quos deposuit Nabataeo belua saltu
iam nimios capitique graves. hinc surgit orexis,
hinc stomacho vires; nam pes argenteus illis,
anulus in digito quod ferreus. ergo superbum
convivam caveo, qui me sibi comparat et res   130
despicit exiguas. adeo nulla uncia nobis
est eboris, nec tessellae nec calculus ex hac
materia, quin ipsa manubria cultellorum
ossea. non tamen his ulla umquam obsonia fiunt
rancidula aut ideo peior gallina secatur.   135
sed nec structor erit cui cedere debeat omnis
pergula, discipulus Trypheri doctoris, aput quem
sumine cum magno lepus atque aper et pygargus
et Scythicae volucres et phoenicopterus ingens
et Gaetulus oryx hebeti lautissima ferro   140
caeditur et tota sonat ulmea cena Subura.
nec frustum capreae subducere nec latus Afrae
novit avis noster, tirunculus ac rudis omni
tempore et exiguae furtis inbutus ofellae.
plebeios calices et paucis assibus emptos   145
porriget incultus puer atque a frigore tutus.
non Phryx aut Lycius, non a mangone petitus
quisquam erit et magno: cum posces, posce latine.

118 hos ω: hoc *P Büch.*   148 et magno *p*: in magno *P Büch.*

idem habitus cunctis, tonsi rectique capilli
atque hodie tantum propter convivia pexi. 150
pastoris duri hic est filius, ille bubulci.
suspirat longo non visam tempore matrem,
et casulam et notos tristis desiderat haedos,
ingenui vultus puer ingenuique pudoris,
quales esse decet quos ardens purpura vestit. 155
hic tibi vina dabit diffusa in montibus illis
a quibus ipse venit, quorum sub vertice lusit; 160
namque una atque eadem est vini patria atque ministri.
forsitan expectes ut Gaditana canoro
incipiant prurire choro, plausuque probatae
ad terram tremulo descendant crure puellae—
spectant hoc nuptae iuxta recubante marito, 165
quod pudeat narrare aliquem praesentibus ipsis—
non capit has nugas humilis domus. ille fruatur
vocibus obscaenis omnique libidinis arte,
qui Lacedaemonium pytismate lubricat orbem; 175
namque ibi fortunae veniam damus. alea turpis
turpe et adulterium mediocribus: haec eadem illi
omnia cum faciunt, hilares nitidique vocantur.
nostra dabunt alios hodie convivia ludos,
conditor Iliados cantabitur atque Maronis 180
altisoni dubiam facientia carmina palmam.
quid refert, tales versus qua voce legantur?
    sed nunc dilatis averte negotia curis
et gratam requiem dona tibi: quando licebat
per totum cessare diem? non faenoris ulla 185
mentio nec prima si luce egressa reverti
nocte solet, tacito bilem tibi contrahat uxor.
protinus ante meum quidquid dolet exue limen, 190
pone domum et servos et quidquid frangitur illis
aut perit, ingratos ante omnia pone sodales.

178 faciunt *some mss.*: faciant *PT Büch.*
184 licebat *P*: licebit *p*ω

interea Megalesiacae spectacula mappae
Idaeum sollemne colunt, similisque triumpho
praeda caballorum praetor sedet, ac mihi pace   195
inmensae nimiaeque licet si dicere plebis,
totam hodie Romam circus capit, et fragor aurem
percutit, eventum viridis quo colligo panni.
nam si deficeret, maestam attonitamque videres
hanc urbem veluti Cannarum in pulvere victis   200
consulibus. spectent iuvenes, quos clamor et audax
sponsio, quos cultae decet adsedisse puellae:
nostra bibat vernum contracta cuticula solem
effugiatque togam. iam nunc in balnea salva
fronte licet vadas, quamquam solida hora supersit   205
ad sextam. facere hoc non possis quinque diebus
continuis, quia sunt talis quoque taedia vitae
magna; voluptates commendat rarior usus.

SATVRA XII

Natali, Corvine, die mihi dulcior haec lux,
qua festus promissa deis animalia caespes
expectat. niveam reginae ducimus agnam,
par vellus dabitur pugnanti Gorgone Maura,
sed procul extensum petulans quatit hostia funem   5
Tarpeio servata Iovi frontemque coruscat,
quippe ferox vitulus templis maturus et arae
spargendusque mero, quem iam pudet ubera matris
ducere, qui vexat nascenti robora cornu.
si res ampla domi similisque adfectibus esset,   10
pinguior Hispulla traheretur taurus et ipsa
mole piger nec finitima nutritus in herba,
laeta sed ostendens Clitumni pascua sanguis
iret et a grandi cervix ferienda ministro,

195 praeda *P*: praedo *p*   199 videres *p*: videret *P*
XII 2 deis] dies *P*: diis *p*

ob reditum trepidantis adhuc horrendaque passi   15
nuper et incolumem sese mirantis amici.
nam praeter pelagi casus et fulminis ictus
evasit. densae caelum abscondere tenebrae
nube una subitusque antemnas inpulit ignis,
cum se quisque illo percussum crederet et mox   20
attonitus nullum conferri posse putaret
naufragium velis ardentibus. omnia fiunt
talia tam graviter, si quando poetica surgit
tempestas. genus ecce aliud discriminis audi
et miserere iterum, quamquam sint cetera sortis   25
eiusdem pars dira quidem sed cognita multis,
et quam votiva testantur fana tabella
plurima; pictores quis nescit ab Iside pasci?
accidit et nostro similis fortuna Catullo.
cum plenus fluctu medius foret alveus et, iam   30
alternum puppis latus evertentibus undis,
arboris incertae, nullam prudentia cani
rectoris cum ferret opem, decidere iactu
coepit cum ventis, imitatus castora qui se
eunuchum ipse facit cupiens evadere damno   35
testiculi; adeo medicatum intellegit inguen.
'fundite quae mea sunt' dicebat 'cuncta' Catullus
praecipitare volens etiam pulcherrima, vestem
purpuream teneris quoque Maecenatibus aptam,
atque alias quarum generosi graminis ipsum   40
infecit natura pecus, sed et egregius fons
viribus occultis et Baeticus adiuvat aer.
ille nec argentum dubitabat mittere, lances
Parthenio factas, urnae cratera capacem
et dignum sitiente Pholo vel coniuge Fusci;   45
adde et bascaudas et mille escaria, multum
caelati, biberat quo callidus emptor Olynthi.

18 evasit *PT*: evasi ω   32 arboris incertae *P*: arbori incertae
Lachmann   47 callidus *p*: pallidus *P*

sed quis nunc alius, qua mundi parte quis audet
argento praeferre caput rebusque salutem?
non propter vitam faciunt patrimonia quidam,   50
sed vitio caeci propter patrimonia vivunt.
iactatur rerum utilium pars maxima, sed nec
damna levant. tunc adversis urguentibus illuc
reccidit ut malum ferro summitteret, ac se
explicat angustum: discriminis ultima, quando   55
praesidia adferimus navem factura minorem.
i nunc et ventis animam committe dolato
confisus ligno, digitis a morte remotus
quattuor aut septem, si sit latissima, taedae;
mox cum reticulis et pane et ventre lagonae   60
aspice sumendas in tempestate secures.
sed postquam iacuit planum mare, tempora postquam
prospera vectoris fatumque valentius euro
et pelago, postquam Parcae meliora benigna
pensa manu ducunt hilares et staminis albi   65
lanificae, modica nec multum fortior aura
ventus adest, inopi miserabilis arte cucurrit
vestibus extentis et, quod superaverat unum,
velo prora suo. iam deficientibus austris
spes vitae cum sole redit. tunc gratus Iulo   70
atque novercali sedes praelata Lavino
conspicitur sublimis apex, cui candida nomen
scrofa dedit, laetis Phrygibus mirabile sumen,
et numquam visis triginta clara mamillis.
tandem intrat positas inclusa per aequora moles   75
Tyrrhenamque pharon porrectaque bracchia rursum
quae pelago occurrunt medio longeque relinquunt
Italiam; non sic igitur mirabere portus
quos natura dedit. sed trunca puppe magister
interiora petit Baianae pervia cumbae   80

59 taedae *P*: taeda *pω*   61 aspice] respice *Jahn*   62 iacuit]
tacuit *P*   73 mirabile *pS*: miserabile *PS Büch*.

tuti stagna sinus. gaudent ibi vertice raso
garrula securi narrare pericula nautae.
　ite igitur, pueri, linguis animisque faventes
sertaque delubris et farra inponite cultris
ac mollis ornate focos glaebamque virentem.   85
iam sequar et sacro, quod praestat, rite peracto
inde domum repetam, graciles ubi parva coronas
accipiunt fragili simulacra nitentia cera.
hic nostrum placabo Iovem Laribusque paternis
tura dabo atque omnis violae iactabo colores.   90
cuncta nitent, longos erexit ianua ramos
et matutinis operatur festa lucernis.
nec suspecta tibi sint haec, Corvine. Catullus
pro cuius reditu tot pono altaria, parvos
tres habet heredes. libet expectare quis aegram   95
et claudentem oculos gallinam inpendat amico
tam sterili, verum haec nimia est inpensa, coturnix
nulla umquam pro patre cadet. sentire calorem
si coepit locuples Gallitta et Pacius, orbi
legitime fixis vestitur tota libellis   100
porticus, existunt qui promittant hecatomben,
quatenus hic non sunt nec venales elephanti,
nec Latio aut usquam sub nostro sidere talis
belua concipitur, sed furva gente petita
arboribus Rutulis et Turni pascitur agro,   105
Caesaris armentum nulli servire paratum
privato, siquidem Tyrio parere solebant
Hannibali et nostris ducibus regique Molosso
horum maiores ac dorso ferre cohortis,
partem aliquam belli et euntem in proelia turrem.   110
nulla igitur mora per Novium, mora nulla per Histrum
Pacuvium, quin illud ebur ducatur ad aras
et cadat ante Lares Gallittae victima sola

　　　　81 tuti stagna sinus *P*: tunc stagnante sinu *p*
　　　110 belli et] bellique et *p*

tantis digna deis et captatoribus horum.
alter enim, si concedas, mactare vovebit   115
de grege servorum magna aut pulcherrima quaeque
corpora, vel pueris et frontibus ancillarum
inponet vittas, et siqua est nubilis illi
Iphigenia domi, dabit hanc altaribus, etsi
non sperat tragicae furtiva piacula cervae.   120
laudo meum civem, nec comparo testamento
mille rates; nam si Libitinam evaserit aeger,
delebit tabulas inclusus carcere nassae
post meritum sane mirandum atque omnia soli
forsan Pacuvio breviter dabit, ille superbus   125
incedet victis rivalibus. ergo vides quam
grande operae pretium faciat iugulata Mycenis.
vivat Pacuvius quaeso vel Nestora totum,
possideat quantum rapuit Nero, montibus aurum
exaequet, nec amet quemquam nec ametur ab ullo.   130

116 aut *some mss.*: ut *P*: et *pω*

LIBER QVINTVS

SATVRA XIII

Exemplo quodcumque malo committitur, ipsi
displicet auctori. prima est haec ultio quod se
iudice nemo nocens absolvitur, improba quamvis
gratia fallaci praetoris vicerit urna,
quid sentire putas omnes, Calvine, recenti 5
de scelere et fidei violatae crimine? sed nec
tam tenuis census tibi contigit, ut mediocris
iacturae te mergat onus, nec rara videmus
quae pateris; casus multis hic cognitus ac iam
tritus et e medio fortunae ductus acervo. 10
ponamus nimios gemitus. flagrantior aequo
non debet dolor esse viri nec vulnere maior.
tu quamvis levium minimam exiguamque malorum
particulam vix ferre potes spumantibus ardens
visceribus, sacrum tibi quod non reddat amicus 15
depositum? stupet haec qui iam post terga reliquit
sexaginta annos Fonteio consule natus?
an nihil in melius tot rerum proficit usu?
magna quidem, sacris quae dat praecepta libellis,
victrix fortunae sapientia, ducimus autem 20
hos quoque felices, qui ferre incommoda vitae
nec iactare iugum vita didicere magistra.
quae tam festa dies, ut cesset prodere furem
perfidiam fraudes atque omni ex crimine lucrum
quaesitum et partos gladio vel pyxide nummos? 25
rari quippe boni: numera, vix sunt totidem quot
Thebarum portae vel divitis ostia Nili.
nunc aetas agitur peioraque saecula ferri

XIII 5 omnes] homines *Ribbeck*   18 proficit *P*: proficis *pω*
26 numera *P*: numero *pω*   28 nunc *P*: nona *pω*

temporibus, quorum sceleri non invenit ipsa
nomen et a nullo posuit natura metallo.   30
nos hominum divumque fidem clamore ciemus,
quanto Faesidium laudat vocalis agentem
sportula? dic, senior bulla dignissime, nescis
quas habeat veneres aliena pecunia? nescis
quem tua simplicitas risum vulgo moveat, cum   35
exigis a quoquam ne peieret et putet ullis
esse aliquod numen templis araeque rubenti?
quondam hoc indigenae vivebant more, priusquam
sumeret agrestem posito diademate falcem
Saturnus fugiens, tunc cum virguncula Iuno   40
et privatus adhuc Idaeis Iuppiter antris,
nulla super nubes convivia caelicolarum
nec puer Iliacus formonsa nec Herculis uxor
ad cyathos, et iam siccato nectare tergens
bracchia Vulcanus Liparaea nigra taberna: 45
prandebat sibi quisque deus, nec turba deorum
talis ut est hodie, contentaque sidera paucis
numinibus miserum urguebant Atlanta minori
pondere: nondum aliquis sortitus triste profundi
imperium aut Sicula torvus cum coniuge Pluton,   50
nec rota nec Furiae nec saxum aut vulturis atri
poena, sed infernis hilares sine regibus umbrae.
inprobitas illo fuit admirabilis aevo,
credebant quo grande nefas et morte piandum
si iuvenis vetulo non adsurrexerat et si   55
barbato cuicumque puer, licet ipse videret
plura domi fraga et maiores glandis acervos;
tam venerabile erat praecedere quattuor annis,
primaque par adeo sacrae lanugo senectae.
nunc si depositum non infitietur amicus,   60
si reddat veterem cum tota aerugine follem,
prodigiosa fides et Tuscis digna libellis

57 fraga *P*: farra *pω*

quaeque coronata lustrari debeat agna.
egregium sanctumque virum si cerno, bimembri
hoc monstrum puero vel miranti sub aratro   65
piscibus inventis et fetae comparo mulae,
sollicitus, tamquam lapides effuderit imber
examenque apium longa consederit uva
culmine delubri, tamquam in mare fluxerit amnis
gurgitibus miris et lactis vertice torrens.   70
        intercepta decem quereris sestertia fraude
sacrilega. quid si bis centum perdidit alter
hoc arcana modo? maiorem tertius illa
summam, quam patulae vix ceperat angulus arcae?
tam facile et pronum est superos contemnere testes,   75
si mortalis idem nemo sciat. aspice quanta
voce neget, quae sit ficti constantia vultus.
per Solis radios Tarpeiaque fulmina iurat
et Martis frameam et Cirrhaei spicula vatis,
per calamos venatricis pharetramque puellae   80
perque tuum, pater Aegaei Neptune, tridentem,
addit et Herculeos arcus hastamque Minervae,
quidquid habent telorum armamentaria caeli.
si vero et pater est, 'comedam' inquit 'flebile nati
sinciput elixi Pharioque madentis aceto.'   85
        sunt in fortunae qui casibus omnia ponant
et nullo credant mundum rectore moveri
natura volvente vices et lucis et anni,
atque ideo intrepidi quaecumque altaria tangunt.
est alius metuens ne crimen poena sequatur,   90
hic putat esse deos et peierat, atque ita secum:
'decernat quodcumque volet de corpore nostro
Isis et irato feriat mea lumina sistro,
dummodo vel caecus teneam quos abnego nummos.
et phthisis et vomicae putres et dimidium crus   95

    65 miranti *p*: mirandis *P*
    70 miris] miniis (*i.e. sanguineis*) *Porson*: diris *conj. Büch.*

sunt tanti. pauper locupletem optare podagram
nec dubitet Ladas, si non eget Anticyra nec
Archigene; quid enim velocis gloria plantae
praestat et esuriens Pisaeae ramus olivae?
ut sit magna tamen, certe lenta ira deorum est;   10
si curant igitur cunctos punire nocentes,
quando ad me venient? sed et exorabile numen
fortasse experiar, solet his ignoscere. multi
committunt eadem diverso crimina fato:
ille crucem sceleris pretium tulit, hic diadema.'   10
sic animum dirae trepidum formidine culpae
confirmat, tunc te sacra ad delubra vocantem
praecedit, trahere immo ultro ac vexare paratus.
nam cum magna malae superest audacia causae,
creditur a multis fiducia. mimum agit ille,   11
urbani qualem fugitivus scurra Catulli:
tu miser exclamas, ut Stentora vincere possis,
vel potius quantum Gradivus Homericus: 'audis,
Iuppiter, haec, nec labra moves, cum mittere vocem
debueris vel marmoreus vel aeneus? aut cur   11
in carbone tuo charta pia tura soluta
ponimus et sectum vituli iecur albaque porci
omenta? ut video, nullum discrimen habendum est
effigies inter vestras statuamque Vagelli.'

    accipe quae contra valeat solacia ferre   12
et qui nec cynicos nec stoica dogmata legit
a cynicis tunica distantia, non Epicurum
suspicit exigui laetum plantaribus horti.
curentur dubii medicis maioribus aegri:
tu venam vel discipulo committe Philippi.   12
si nullum in terris tam detestabile factum
ostendis, taceo, nec pugnis caedere pectus
te veto nec plana faciem contundere palma,
quandoquidem accepto claudenda est ianua damno,

125 venam *p*: veniam *P*

et maiore domus gemitu, maiore tumultu   130
planguntur nummi quam funera; nemo dolorem
fingit in hoc casu, vestem diducere summam
contentus, vexare oculos umore coacto:
ploratur lacrimis amissa pecunia veris.
sed si cuncta vides simili fora plena querella,   135
si decies lectis diversa parte tabellis
vana supervacui dicunt chirographa ligni,
arguit ipsorum quos littera gemmaque princeps
sardonychum, loculis quae custoditur eburnis,
ten—o delicias—extra communia censes   140
ponendum, quia tu gallinae filius albae,
nos viles pulli nati infelicibus ovis?
rem pateris modicam et mediocri bile ferendam,
si flectas oculos maiora ad crimina. confer
conductum latronem, incendia sulpure coepta   145
atque dolo, primos cum ianua colligit ignes;
confer et hos, veteris qui tollunt grandia templi
pocula, adorandae robiginis, et populorum
dona vel antiquo positas a rege coronas;
haec ibi si non sunt, minor exstat sacrilegus qui   150
radat inaurati femur Herculis et faciem ipsam
Neptuni, qui bratteolam de Castore ducat—
an dubitet solitus totum conflare Tonantem?—
confer et artifices mercatoremque veneni
et deducendum corio bovis in mare, cum quo   155
clauditur adversis innoxia simia fatis.
haec quota pars scelerum, quae custos Gallicus urbis
usque a lucifero donec lux occidat audit?
humani generis mores tibi nosse volenti
sufficit una domus; paucos consume dies et   160
dicere te miserum, postquam illinc veneris, aude.
quis tumidum guttur miratur in Alpibus aut quis
in Meroe crasso maiorem infante mamillam?
caerula quis stupuit Germani lumina, flavam

caesariem et madido torquentem cornua cirro?   165
nempe quod haec illis natura est omnibus una.
ad subitas Thracum volucres nubemque sonoram
Pygmaeus parvis currit bellator in armis,
mox inpar hosti raptusque per aera curvis
unguibus a saeva fertur grue. si videas hoc   170
gentibus in nostris, risu quatiare; sed illic,
quamquam eadem adsidue spectentur proelia, ridet
nemo, ubi tota cohors pede non est altior uno.
   'nullane peiuri capitis fraudisque nefandae
poena erit?' abreptum crede hunc graviore catena   175
protinus et nostro—quid plus velit ira?—necari
arbitrio: manet illa tamen iactura nec umquam
depositum tibi sospes erit, sed corpore trunco
invidiosa dabit minimus solacia sanguis.
'at vindicta bonum vita iucundius ipsa.'   180
nempe hoc indocti, quorum praecordia nullis
interdum aut levibus videas flagrantia causis:
quantulacumque adeo est occasio sufficit irae.
Chrysippus non dicet idem nec mite Thaletis
ingenium dulcique senex vicinus Hymetto,   185
qui partem acceptae saeva inter vincla cicutae
accusatori nollet dare. plurima felix
paulatim vitia atque errores exuit omnes,
prima docet rectum sapientia. quippe minuti
semper et infirmi est animi exiguique voluptas   190
ultio. continuo sic collige, quod vindicta
nemo magis gaudet quam femina. cur tamen hos tu
evasisse putes, quos diri conscia facti
mens habet attonitos et surdo verbere caedit,
occultum quatiente animo tortore flagellum?   195
poena autem vehemens ac multo saevior illis
quas et Caedicius gravis invenit et Rhadamanthus,

178 sed] si *Weidner*   188 exuit omnes] exuit, omnes *Büch.*
191 ultio] uitio *P*

nocte dieque suum gestare in pectore testem.
Spartano cuidam respondit Pythia vates
haut inpunitum quondam fore quod dubitaret   200
depositum retinere et fraudem iure tueri
iurando. quaerebat enim quae numinis esset
mens et an hoc illi facinus suaderet Apollo.
reddidit ergo metu, non moribus, et tamen omnem
vocem adyti dignam templo veramque probavit   205
extinctus tota pariter cum prole domoque
et quamvis longa deductis gente propinquis.
has patitur poenas peccandi sola voluntas.
nam scelus intra se tacitum qui cogitat ullum,
facti crimen habet. cedo, si conata peregit.   210
perpetua anxietas, nec mensae tempore cessat
faucibus ut morbo siccis interque molares
difficili crescente cibo, sed vina misellus
expuit, Albani veteris pretiosa senectus
displicet; ostendas melius, densissima ruga   215
cogitur in frontem velut acri ducta Falerno.
nocte brevem si forte indulsit cura soporem
et toto versata toro iam membra quiescunt,
continuo templum et violati numinis aras
et quod praecipuis mentem sudoribus urguet,   220
te videt in somnis; tua sacra et maior imago
humana turbat pavidum cogitque fateri.
hi sunt qui trepidant et ad omnia fulgura pallent,
cum tonat, exanimis, primo quoque murmure caeli,
non quasi fortuitus nec ventorum rabie sed   225
iratus cadat in terras et iudicet ignis.
illa nihil nocuit, cura graviore timetur
proxima tempestas velut hoc dilata sereno.
praeterea lateris vigili cum febre dolorem
si coepere pati, missum ad sua corpora morbum   230

205 probavit] probabit *P*   208 sola voluntas ω: saeva voluptas
P   213 sed vina] Setina *Withof*

infesto credunt a numine, saxa deorum
haec et tela putant. pecudem spondere sacello
balantem et Laribus cristam promittere galli
non audent; quid enim sperare nocentibus aegris
concessum? vel quae non dignior hostia vita? 235
mobilis et varia est ferme natura malorum;
cum scelus admittunt, superest constantia: quod fas
atque nefas, tandem incipiunt sentire peractis
criminibus. tamen ad mores natura recurrit
damnatos fixa et mutari nescia. nam quis 240
peccandi finem posuit sibi? quando recepit
eiectum semel attrita de fronte ruborem?
quisnam hominum est quem tu contentum videris uno
flagitio? dabit in laqueum vestigia noster
perfidus et nigri patietur carceris uncum 245
aut maris Aegaei rupem scopulosque frequentes
exulibus magnis. poena gaudebis amara
nominis invisi tandemque fatebere laetus
nec surdum nec Teresian quemquam esse deorum.

Influence of Parental Example

SATVRA XIV

(1-100)

Plurima sunt, Fuscine, et fama digna sinistra
et nitidis maculam haesuram figentia rebus,
quae monstrant ipsi pueris traduntque parentes.
si damnosa senem iuvat alea, ludit et heres
bullatus parvoque eadem movet arma fritillo. *dice-box* 5
nec melius de se cuiquam sperare propinquo
concedet iuvenis, qui radere tubera terrae, *truffles*
boletum condire et eodem iure natantis
mergere ficedulas didicit nebulone parente *good for O*
small birds
et cana monstrante gula; cum septimus annus 10
transierit puerum, nondum omni dente renato,
barbatos licet admoveas mille inde magistros,

XIV 2 maculam haesuram *PT*; maculam ac rugam ω

on I side

hinc totidem, cupiet lauto cenare paratu
semper et a magna non degenerare culina.
mitem animum et mores modicis erroribus aequos   15
praecipit atque animas servorum et corpora nostra
materia constare putat paribusque elementis,
an saevire docet Rutilus qui gaudet acerbo
plagarum strepitu et nullam Sirena flagellis
conparat, Antiphates trepidi laris ac Polyphemus,   20
tunc felix, quotiens aliquis tortore vocato
uritur ardenti duo propter lintea ferro—
quid suadet iuveni laetus stridore catenae,
quem mire adficiunt inscripti, ergastula, carcer?
rusticus expectas ut non sit adultera Largae   25
filia, quae numquam maternos dicere moechos
tam cito nec tanto poterit contexere cursu,
ut non terdecies respiret? conscia matri
virgo fuit, ceras nunc hac dictante pusillas
implet et ad moechum dat eisdem ferre cinaedis.   30
sic natura iubet: velocius et citius nos
corrumpunt vitiorum exempla domestica, magnis
cum subeunt animos auctoribus. unus et alter
forsitan haec spernant iuvenes, quibus arte benigna
et meliore luto finxit praecordia Titan,   35
sed reliquos fugienda patrum vestigia ducunt
et monstrata diu veteris trahit orbita culpae.
abstineas igitur damnandis. huius enim vel
una potens ratio est, ne crimina nostra sequantur
ex nobis geniti, quoniam dociles imitandis   40
turpibus ac pravis omnes sumus, et Catilinam
quocumque in populo videas, quocumque sub axe,
sed nec Brutus erit Bruti nec avunculus umquam.
nil dictu foedum visuque haec limina tangat,

24 inscripti, ergastula *Richards*: scripta ergastula *P*: inscripta
ergastula *pω*   38 damnandis. huius enim *P*: damnis.
huiusce etenim *pω*   43 umquam *P*: usquam *pω*

intra quae pater est; procul, a procul inde puellae   45
lenonum et cantus pernoctantis parasiti.
maxima debetur puero reverentia, siquid
turpe paras, nec tu pueri contempseris annos,
sed peccaturo obstet tibi filius infans.
nam siquid dignum censoris fecerit ira   50
quandoque et similem tibi se non corpore tantum
nec vultu dederit, morum quoque filius et qui
omnia deterius tua per vestigia peccet,
corripies *nimirum* et castigabis acerbo   *of course*
clamore ac post haec tabulas mutare parabis? *alter your will* 55
unde tibi frontem libertatemque parentis,
cum facias peiora senex vacuumque cerebro   *cupping-glass*
iam pridem caput hoc ventosa cucurbita quaerat?
      hospite venturo cessabit nemo tuorum.
'verre pavimentum, nitidas ostende columnas,   60
arida cum tota descendat aranea tela;
hic leve argentum, vasa aspera tergeat alter'
vox domini furit instantis virgamque tenentis.
ergo miser trepidas, ne stercore foeda canino
atria displiceant oculis venientis amici,   65
ne perfusa luto sit porticus, et tamen uno
semodio scobis haec emendat servulus unus:
illud non agitas, ut sanctam filius omni
aspiciat sine labe domum vitioque carentem?
gratum est quod patriae civem populoque dedisti,   70
si facis ut patria sit idoneus, utilis agris,
utilis et bellorum et pacis rebus agendis.
plurimum enim intererit quibus artibus et quibus
      hunc tu
moribus instituas. serpente ciconia pullos   *stork*
nutrit et inventa per devia rura lacerta: *lizard*   75
illi eadem sumptis quaerunt animalia pinnis.

45 pater *Pω*: puer *some mss.*   52 qui *PT*: cum *p*
63 furit *P*: fremit *or* fremat *ω*   71 patria *P*: patriae *ω*

vultur iumento et canibus crucibusque relictis
ad fetus properat partemque cadaveris adfert:
hic est ergo cibus magni quoque vulturis et se
pascentis, propria cum iam facit arbore nidos.   80
sed leporem aut capream famulae Iovis et generosae
in saltu venantur aves, hinc praeda cubili
ponitur: inde autem cum se matura levavit
progenies, stimulante fame festinat ad illam
quam primum praedam rupto gustaverat ovo.   85

    aedificator erat Cretonius et modo curvo
litore Caietae, summa nunc Tiburis arce,
nunc Praenestinis in montibus alta parabat
culmina villarum Graecis longeque petitis
marmoribus vincens Fortunae atque Herculis aedem,   90
ut spado vincebat Capitolia nostra Posides.
dum sic ergo habitat Cretonius, inminuit rem,
fregit opes, nec parva tamen mensura relictae
partis erat. totam hanc turbavit filius amens,
dum meliore novas attollit marmore villas.   95

    quidam sortiti metuentem sabbata patrem
nil praeter nubes et caeli numen adorant,
nec distare putant humana carne suillam
qua pater abstinuit, mox et praeputia ponunt;
Romanas autem soliti contemnere leges   100
Iudaicum ediscunt et servant ac metuunt ius,
tradidit arcano quodcumque volumine Moyses,
non monstrare vias eadem nisi sacra colenti,
quaesitum ad fontem solos deducere verpos.
sed pater in causa, cui septima quaeque fuit lux   105
ignava et partem vitae non attigit ullam.

    sponte tamen iuvenes imitantur cetera, solam
inviti quoque avaritiam exercere iubentur.
fallit enim vitium specie virtutis et umbra,

83 levavit] levaret P: levavit T: levabit pω   86 Cretonius
P: Cetronius pω   87 Caietae] Caletae P

cum sit triste habitu vultuque et veste severum,   110
nec dubie tamquam frugi laudetur avarus,
tamquam parcus homo et rerum tutela suarum
certa magis quam si fortunas servet easdem
Hesperidum serpens aut Ponticus. adde quod hunc de
quo loquor, egregium populus putat adquirendi   115
artificem; quippe his crescunt patrimonia fabris,
sed crescunt quocumque modo, maioraque fiunt
incude adsidua semperque ardente camino.
et pater ergo animi felices credit avaros,
qui miratur opes, qui nulla exempla beati   120
pauperis esse putat, iuvenes hortatur ut illa
ire via pergant et eidem incumbere sectae. *way of life*
sunt quaedam vitiorum elementa, his protinus illos
inbuit et cogit minimas ediscere sordes;
mox adquirendi docet insatiabile votum.   125
servorum ventres modio castigat iniquo
ipse quoque esuriens, neque enim omnia sustinet
        umquam
Mouldy mucida caerulei panis consumere frusta,
hesternum solitus medio servare minutal *dish of mince*
Septembri nec non differre in tempora cenae   130
beans alterius conchem aestivam cum parte lacerti
signatam vel dimidio putrique siluro, *fish*
locked up filaque sectivi numerata includere porri; *leeks*
invitatus ad haec aliquis de ponte negabit.
sed quo divitias haec per tormenta coactas,   135
cum furor haut dubius, cum sit manifesta phrenesis, *Madne*
ut locuples moriaris, egentis vivere fato?
interea pleno cum turget sacculus ore,
crescit amor nummi, quantum ipsa pecunia crevit,

113 fortunas ω: fortuna P Büch.   120 miratur P:
mirantur pω   122 pergant ω: peragant P Büch.   131 con-
chem aestivam P: concham aestivi pω   134 negabit ω:
negavit P (cf. III. 168)   139 crevit PT: crescit pω

et minus hanc optat qui non habet. ergo paratur 140
altera villa tibi, cum rus non sufficit unum,
et proferre libet fines maiorque videtur
et melior vicina seges, mercaris et hanc et
arbusta et densa montem qui canet oliva.
quorum si pretio dominus non vincitur ullo, 145
nocte boves macri lassoque famelica collo
iumenta ad virides huius mittentur aristas,
nec prius inde domum quam tota novalia saevos
in ventres abeant, ut credas falcibus actum.
dicere vix possis quam multi talia plorent 150
et quot venales iniuria fecerit agros.
sed qui sermones, quam foedae bucina famae.
'quid nocet haec?' inquit 'tunicam mihi malo lupini
quam si me toto laudet vicinia pago
exigui ruris paucissima farra secantem.' 155
scilicet et morbis et debilitate carebis,
et luctum et curam effugies, et tempora vitae
longa tibi posthac fato meliore dabuntur,
si tantum culti solus possederis agri,
quantum sub Tatio populus Romanus arabat. 160
mox etiam fractis aetate ac Punica passis
proelia vel Pyrrhum inmanem gladiosque Molossos
tandem pro multis vix iugera bina dabantur
vulneribus, merces haec sanguinis atque laboris
nullis visa umquam meritis minor aut ingratae 165
curta fides patriae; saturabat glaebula talis
patrem ipsum turbamque casae, qua feta iacebat
uxor et infantes ludebant quattuor, unus
vernula, tres domini, sed magnis fratribus horum
a scrobe vel sulco redeuntibus altera cena 170
amplior et grandes fumabant pultibus ollae.
nunc modus hic agri nostro non sufficit horto.
inde fere scelerum causae, nec plura venena

149 abeant] habeant PT

miscuit aut ferro grassatur saepius ullum
humanae mentis vitium quam saeva cupido   175
inmodici census. nam dives qui fieri vult,
et cito vult fieri; sed quae reverentia legum,
quis metus aut pudor est umquam properantis avari?
'vivite contenti casulis et collibus istis,
o pueri' Marsus dicebat et Hernicus olim   180
Vestinusque senex 'panem quaeramus aratro,
qui satis est mensis; laudant hoc numina ruris,
quorum ope et auxilio gratae post munus aristae
contingunt homini veteris fastidia quercus.
nil vetitum fecisse volet quem non pudet alto   185
per glaciem perone tegi, qui summovet euros
pellibus inversis: peregrina ignotaque nobis
ad scelus atque nefas, quaecumque est, purpura ducit.'
haec illi veteres praecepta minoribus, at nunc
post finem autumni media de nocte supinum   190
clamosus iuvenem pater excitat: 'accipe ceras,
scribe, puer, vigila, causas age, perlege rubras
maiorum leges. aut vitem posce libello,
sed caput intactum buxo naresque pilosas
adnotet et grandes miretur Laelius alas;   195
dirue Maurorum attegias, castella Brigantum,
ut locupletem aquilam tibi sexagesimus annus
adferat. aut longos castrorum ferre labores
si piget et trepidum solvunt tibi cornua ventrem
cum lituis audita, pares quod vendere possis   200
pluris dimidio, nec te fastidia mercis
ullius subeant ablegandae Tiberim ultra,
neu credas ponendum aliquid discriminis inter
unguenta et corium; lucri bonus est odor ex re
qualibet. illa tuo sententia semper in ore   205
versetur dis atque ipso Iove digna poeta:
"unde habeas quaerit nemo, sed oportet habere."'

176 inmodici *P*: indomiti *pω*   199 trepidum *P*: trepido *pω*

hoc monstrant vetulae pueris repentibus assae,
hoc discunt omnes ante alpha et beta puellae.'
talibus instantem monitis quemcumque parentem   210
sic possem adfari: 'dic, o vanissime, quis te
festinare iubet? meliorem praesto magistro
discipulum. securus abi: vinceris ut Aiax
praeteriit Telamonem, ut Pelea vicit Achilles.
parcendum est teneris, nondum implevere medullas   215
maturae mala nequitiae. cum pectere barbam
coeperit et longi mucronem admittere cultri,
falsus erit testis, vendet periuria summa
exigua et Cereris tangens aramque pedemque.
elatam iam crede nurum, si limina vestra   *daughter in-law*
mortifera cum dote subit. quibus illa premetur   220
per somnum digitis. nam quae terraque marique
adquirenda putas, brevior via conferet illi;
nullus enim magni sceleris labor. "haec ego numquam
mandavi" dices olim "nec talia suasi."   225
mentis causa malae tamen est et origo penes te.
nam quisquis magni census praecepit amorem,
et laevo monitu pueros producit avaros,
et qui per fraudes patrimonia conduplicari,
dat libertatem et totas effundit habenas   230
curriculo, quem si revoces, subsistere nescit
et te contempto rapitur metisque relictis.
nemo satis credit, tantum delinquere quantum
permittas; adeo indulgent sibi latius ipsi.
cum dicis iuveni stultum, qui donet amico,   235
qui paupertatem levet attollatque propinqui,
et spoliare doces et circumscribere et omni
crimine divitias adquirere, quarum amor in te

208 repentibus assae] reppentibus assae *PT*: poscentibus
assem ω   215 medullas. *Büch.*   216 maturae *some mss.*:
naturae *P* nequitiae *pω*: nequitia est *P Büch.*
229 conduplicari *Pω*: conduplicare *some mss.*

quantus erat patriae Deciorum in pectore, quantum
dilexit Thebas, si Graecia vera, Menoeceus,   240
in quorum sulcis legiones dentibus anguis
cum clipeis nascuntur et horrida bella capessunt
continuo, tamquam et tubicen surrexerit una.
ergo ignem cuius scintillas ipse dedisti,
flagrantem late et rapientem cuncta videbis.   245
nec tibi parcetur misero, trepidumque magistrum
in cavea magno fremitu leo tollet alumnus.
nota mathematicis genesis tua, sed grave tardas
expectare colus; morieris stamine nondum
abrupto. iam nunc obstas et vota moraris,   250
iam torquet iuvenem longa et cervina senectus.
ocius Archigenem quaere atque eme quod Mithridates
composuit; si vis aliam decerpere ficum
atque alias tractare rosas, medicamen habendum est,
sorbere ante cibum quod debeat et pater et rex.'   255

    monstro voluptatem egregiam, cui nulla theatra,
nulla aequare queas praetoris pulpita lauti,
si spectes quanto capitis discrimine constent
incrementa domus, aerata multus in arca
fiscus et ad vigilem ponendi Castora nummi   260
ex quo Mars Ultor galeam quoque perdidit et res
non potuit servare suas. ergo omnia Florae
et Cereris licet et Cybeles aulaea relinquas;
tanto maiores humana negotia ludi.
an magis oblectant animum iactata petauro   265
corpora quique solet rectum descendere funem,
quam tu, Corycia semper qui puppe moraris
atque habitas coro semper tollendus et austro,
perditus ac vilis sacci mercator olentis,
qui gaudes pingue antiquae de litore Cretae   270
passum et municipes Iovis advexisse lagonas?
hic tamen ancipiti figens vestigia planta

    269 ac vilis *P*: a Siculis ω: ac similis (= *concolor*) *Housman*

victum illa mercede parat, brumamque famemque
illa reste cavet: tu propter mille talenta
et centum villas temerarius. aspice portus   275
et plenum magnis trabibus mare: plus hominum est iam
in pelago. veniet classis, quocumque vocarit
spes lucri, nec Carpathium Gaetulaque tantum
aequora transiliet, sed longe Calpe relicta
audiet Herculeo stridentem gurgite solem.   280
grande operae pretium est, ut tenso folle reverti
inde domum possis tumidaque superbus aluta,
Oceani monstra et iuvenes vidisse marinos.
non unus mentes agitat furor. ille sororis
in manibus vultu Eumenidum terretur et igni,   285
hic bove percusso mugire Agamemnona credit
aut Ithacum: parcat tunicis licet atque lacernis,
curatoris eget qui navem mercibus implet
ad summum latus et tabula distinguitur unda,
cum sit causa mali tanti et discriminis huius   290
concisum argentum in titulos faciesque minutas.
occurrunt nubes et fulgura: 'solvite funem'
frumenti dominus clamat piperisve coempti
'nil color hic caeli, nil fascia nigra minatur;
aestivum tonat.' infelix hac forsitan ipsa   295
nocte cadet fractis trabibus fluctuque premetur
obrutus et zonam laeva morsuque tenebit.
sed cuius votis modo non suffecerat aurum
quod Tagus et rutila volvit Pactolus harena,
frigida sufficient velantis inguina panni   300
exiguusque cibus, mersa rate naufragus assem
dum rogat et picta se tempestate tuetur.
    tantis parta malis cura maiore metuque
servantur, misera est magni custodia census.
dispositis praedives amis vigilare cohortem   305
servorum noctu Licinus iubet, attonitus pro

296 cadet ω: cadit P

electro signisque suis Phrygiaque columna
atque ebore et lata testudine. dolia nudi
non ardent cynici; si fregeris, altera fiet
cras domus, atque eadem plumbo commissa manebit.   31◦
sensit Alexander, testa cum vidit in illa
magnum habitatorem, quanto felicior hic qui
nil cuperet quam qui totum sibi posceret orbem
passurus gestis aequanda pericula rebus.
nullum numen habes si sit prudentia, nos te,   31
nos facimus, Fortuna, deam. mensura tamen quae
sufficiat census, siquis me consulat, edam:
in quantum sitis atque fames et frigora poscunt,
quantum, Epicure, tibi parvis suffecit in hortis,
quantum Socratici ceperunt ante penates;   32
numquam aliud natura, aliud sapientia dicit.
acribus exemplis videor te cludere? misce
ergo aliquid nostris de moribus, effice summam
bis septem ordinibus quam lex dignatur Othonis.
haec quoque si rugam trahit extenditque labellum,   32
sume duos equites, fac tertia quadringenta.
si nondum inplevi gremium, si panditur ultra,
nec Croesi fortuna umquam nec Persica regna
sufficient animo nec divitiae Narcissi,
indulsit Caesar cui Claudius omnia, cuius   33
paruit imperiis uxorem occidere iussus.

SATVRA XV

Quis nescit, Volusi Bithynice, qualia demens
Aegyptos portenta colat? crocodilon adorat
pars haec, illa pavet saturam serpentibus ibin.
effigies sacri nitet aurea cercopitheci,
dimidio magicae resonant ubi Memnone chordae   5
atque vetus Thebe centum iacet obruta portis.

xv 1 Volusi] bolusi *P*: volsi *p̄*

illic aeluros, hic piscem fluminis, illic
oppida tota canem venerantur, nemo Dianam.
porrum et caepe nefas violare et frangere morsu;
o sanctas gentes quibus haec nascuntur in hortis   10
numina. lanatis animalibus abstinet omnis
mensa, nefas illic fetum iugulare capellae:
carnibus humanis vesci licet. attonito cum
tale super cenam facinus narraret Ulixes
Alcinoo, bilem aut risum fortasse quibusdam   15
moverat ut mendax aretalogus. 'in mare nemo
hunc abicit saeva dignum veraque Charybdi,
fingentem inmanes Laestrygonas atque Cyclopas?
nam citius Scyllam vel concurrentia saxa
Cyaneis, plenos et tempestatibus utres   20
crediderim aut tenui percussum verbere Circes
et cum remigibus grunnisse Elpenora porcis.
tam vacui capitis populum Phaeaca putavit?'
sic aliquis merito nondum ebrius et minimum qui
de Corcyraea temetum duxerat urna;   25
solus enim haec Ithacus nullo sub teste canebat.
nos miranda quidem sed nuper consule Iunco
gesta super calidae referemus moenia Copti,
nos vulgi scelus et cunctis graviora cothurnis;
nam scelus, a Pyrrha quamquam omnia syrmata volvas,   30
nullus aput tragicos populus facit. accipe, nostro
dira quod exemplum feritas produxerit aevo.

  inter finitimos vetus atque antiqua simultas,
inmortale odium et numquam sanabile vulnus
ardet adhuc Ombos et Tentyra. summus utrimque   35
inde furor volgo, quod numina vicinorum
odit uterque locus, cum solos credat habendos
esse deos quos ipse colit. sed tempore festo
alterius populi rapienda occasio cunctis

7 illic aeluros *Brodaeus*: illicaeruleos *P*: illic caeruleos ω *Büch.*
21 verbere] ververe *P* (cf. I. 161)   27 Iunco *P*: Iunio ω

visa inimicorum primoribus ac ducibus, ne   40
laetum hilaremque diem, ne magnae gaudia cenae
sentirent positis ad templa et compita mensis
pervigilique toro, quem nocte ac luce iacentem
septimus interdum sol invenit. horrida sane
Aegyptos, sed luxuria, quantum ipse notavi,   45
barbara famoso non cedit turba Canopo.
adde quod et facilis victoria de madidis et
blaesis atque mero titubantibus. inde virorum
saltatus nigro tibicine, qualiacumque
unguenta et flores multaeque in fronte coronae: 50
hinc ieiunum odium. sed iurgia prima sonare
incipiunt, animis ardentibus haec tuba rixae,
dein clamore pari concurritur, et vice teli
saevit nuda manus. paucae sine vulnere malae,
vix cuiquam aut nulli toto certamine nasus   55
integer. aspiceres iam cuncta per agmina vultus
dimidios, alias facies et hiantia ruptis
ossa genis, plenos oculorum sanguine pugnos.
ludere se credunt ipsi tamen et puerilis
exercere acies, quod nulla cadavera calcent.   60
et sane quo tot rixantis milia turbae,
si vivunt omnes? ergo acrior impetus, et iam
saxa inclinatis per humum quaesita lacertis
incipiunt torquere, domestica seditioni
tela, nec hunc lapidem, qualis et Turnus et Aiax,   65
vel quo Tydides percussit pondere coxam
Aeneae, sed quem valeant emittere dextrae
illis dissimiles et nostro tempore natae.
nam genus hoc vivo iam decrescebat Homero,
terra malos homines nunc educat atque pusillos;   70
ergo deus quicumque aspexit, ridet et odit.
a deverticulo repetatur fabula. postquam
subsidiis aucti, pars altera promere ferrum

46 turba *PT*: ripa *pω*

audet et infestis pugnam instaurare sagittis.
terga fugae celeri praestant, instantibus Ombis,   75
qui vicina colunt umbrosae Tentyra palmae.
labitur hic quidam nimia formidine cursum
praecipitans capiturque. ast illum in plurima sectum
frusta et particulas, ut multis mortuus unus
sufficeret, totum corrosis ossibus edit   80
victrix turba, nec ardenti decoxit aeno
aut veribus, longum usque adeo tardumque putavit
expectare focos, contenta cadavere crudo.
hic gaudere libet quod non violaverit ignem,
quem summa caeli raptum de parte Prometheus   85
donavit terris; elemento gratulor, et te
exultare reor. sed qui mordere cadaver
sustinuit, nil umquam hac carne libentius edit;
nam scelere in tanto ne quaeras et dubites an
prima voluptatem gula senserit; ultimus autem   90
qui stetit, absumpto iam toto corpore, ductis
per terram digitis aliquid de sanguine gustat.
Vascones, haec fama est, alimentis talibus usi
produxere animas. sed res diversa, sed illic
fortunae invidia est bellorumque ultima, casus   95
extremi, longae dira obsidionis egestas.
huius enim quod nunc agitur, miserabile debet
exemplum esse cibi, sicut modo dicta mihi gens
post omnis herbas, post cuncta animalia, quidquid
cogebat vacui ventris furor, hostibus ipsis   100
pallorem ac maciem et tenuis miserantibus artus,
membra aliena fame lacerabant, esse parati
et sua. quisnam hominum veniam dare quisve deorum
ventribus abnueret dira atque inmania passis

---

75 fugae celeri *P*: fuga sceleri *p*   praestant instantibus Ombis
Mercer: praestant (*the rest erased*) *P*: praestant instantibus orbes
or omnes ω   77 hic *p*ω: hinc *P*   93 usi ω, *Housman*:
olim *PT*   104 ventribus *Valesius*: urbibus *P*: viribus *p*ω

et quibus illorum poterant ignoscere manes,   10
quorum corporibus vescebantur? melius nos
Zenonis praecepta monent, nec enim omnia quidam
pro vita facienda putant, sed Cantaber unde
stoicus antiqui praesertim aetate Metelli?
nunc totus Graias nostrasque habet orbis Athenas,   110
Gallia causidicos docuit facunda Britannos,
de conducendo loquitur iam rhetore Thyle.
nobilis ille tamen populus quem diximus, et par
virtute atque fide sed maior clade Zacynthos
tale quid excusat: Maeotide saevior ara   115
Aegyptos. quippe illa nefandi Taurica sacri
inventrix homines—ut iam quae carmina tradunt,
digna fide credas—tantum immolat, ulterius nil
aut gravius cultro timet hostia. quis modo casus
inpulit hos? quae tanta fames infestaque vallo   120
arma coegerunt tam detestabile monstrum
audere? anne aliam terra Memphitide sicca
invidiam facerent nolenti surgere Nilo?
qua nec terribiles Cimbri nec Brittones umquam
Sauromataeque truces aut inmanes Agathyrsi,   125
hac saevit rabie inbelle et inutile vulgus,
parvula fictilibus solitum dare vela phaselis
et brevibus pictae remis incumbere testae.
nec poenam sceleri invenies nec digna parabis
supplicia his populis, in quorum mente pares sunt   130
et similes ira atque fames. mollissima corda
humano generi dare se natura fatetur,
quae lacrimas dedit; haec nostri pars optima sensus.
plorare ergo iubet causam dicentis amici
squaloremque rei, pupillum ad iura vocantem   135
circumscriptorem, cuius manantia fletu
ora puellares faciunt incerta capilli.
naturae imperio gemimus, cum funus adultae

107 omnia quidam P: omnia, quaedam pω   114 Zacynthos P:
Zaghuntus T: Saguntus ω   138 adultae] aduitae P

virginis occurrit vel terra clauditur infans
et minor igne rogi. quis enim bonus et face dignus   140
arcana, qualem Cereris vult esse sacerdos,
ulla aliena sibi credit mala? separat hoc nos
a grege mutorum, atque ideo venerabile soli
sortiti ingenium divinorumque capaces
atque exercendis pariendisque artibus apti   145
sensum a caelesti demissum traximus arce,
cuius egent prona et terram spectantia. mundi
principio indulsit communis conditor illis
tantum animas, nobis animum quoque, mutuus ut nos
adfectus petere auxilium et praestare iuberet,   150
dispersos trahere in populum, migrare vetusto
de nemore et proavis habitatas linquere silvas,
aedificare domos, laribus coniungere nostris
tectum aliud, tutos vicino limine somnos
ut collata daret fiducia, protegere armis   155
lapsum aut ingenti nutantem vulnere civem,
communi dare signa tuba, defendier isdem
turribus atque una portarum clave teneri.
sed iam serpentum maior concordia, parcit
cognatis maculis similis fera, quando leoni   160
fortior eripuit vitam leo? quo nemore umquam
expiravit aper maioris dentibus apri?
Indica tigris agit rabida cum tigride pacem
perpetuam, saevis inter se convenit ursis.
ast homini ferrum letale incude nefanda   165
produxisse parum est, cum rastra et sarcula tantum
adsueti coquere et marris ac vomere lassi
nescierint primi gladios extendere fabri.
aspicimus populos quorum non sufficit irae
occidisse aliquem, sed pectora bracchia voltum   170
crediderint genus esse cibi. quid diceret ergo
vel quo non fugeret, si nunc haec monstra videret

145 pariendis *Büch.*: ***iendis *P*: capiendis *pω*
154 limine ω: limite *P*

7   PPJ

Pythagoras, cunctis animalibus abstinuit qui
tamquam homine et ventri indulsit non omne legumen?

not on [margin annotation]

SATVRA XVI

Quis numerare queat felicis praemia, Galli,
militiae? nam si subeuntur prospera castra,
me pavidum excipiat tironem porta secundo
sidere. plus etenim fati valet hora benigni
quam si nos Veneris commendet epistula Marti   5
et Samia genetrix quae delectatur harena.

commoda tractemus primum communia, quorum
haut minimum illud erit, ne te pulsare togatus
audeat, immo etsi pulsetur, dissimulet nec
audeat excussos praetori ostendere dentes   10
et nigram in facie tumidis livoribus offam
atque oculum medico nil promittente relictum.
Bardaicus iudex datur haec punire volenti
calceus et grandes magna ad subsellia surae,
legibus antiquis castrorum et more Camilli   15
servato, miles ne vallum litiget extra
et procul a signis. iustissima centurionum
cognitio est igitur de milite, nec mihi derit
ultio, si iustae defertur causa querellae.
tota cohors tamen est inimica, omnesque manipli   20
consensu magno efficiunt curabilis ut sit
vindicta et gravior quam iniuria. dignum erit ergo
declamatoris mulino corde Vagelli,
cum duo crura habeas, offendere tot caligas, tot
milia clavorum. quis tam procul absit ab urbe   25
praeterea, quis tam Pylades, molem aggeris ultra
ut veniat? lacrimae siccentur protinus, et se
excusaturos non sollicitemus amicos.
'da testem' iudex cum dixerit, audeat ille

XVI 1 Galli *P*: Galle *pω*   12 oculum *P*: oculos *pω*
21 efficiunt curabilis *P*: officiunt curabitis *pω*   23 mulino
P: Mutinensi *pω*   24 caligas, tot *Dempster*: caligatos *Pω*

nescio quis pugnos qui vidit, dicere 'vidi,'   30
et credam dignum barba dignumque capillis
maiorum. citius falsum producere testem
contra paganum possis quam vera loquentem
contra fortunam armati contraque pudorem.

   praemia nunc alia atque alia emolumenta notemus   35
sacramentorum. convallem ruris aviti
improbus aut campum mihi si vicinus ademit
et sacrum effodit medio de limite saxum,
quod mea cum patulo coluit puls annua libo,
debitor aut sumptos pergit non reddere nummos   40
vana supervacui dicens chirographa ligni,
expectandus erit qui lites inchoet annus
totius populi. sed tunc quoque mille ferenda
taedia, mille morae; totiens subsellia tantum
sternuntur, iam facundo ponente lacernas   45
Caedicio et Fusco iam micturiente, parati
digredimur, lentaque fori pugnamus harena.
ast illis quos arma tegunt et balteus ambit,
quod placitum est ipsis praestatur tempus agendi,
nec res atteritur longo sufflamine litis.   50
   solis praeterea testandi militibus ius
vivo patre datur. nam quae sunt parta labore
militiae, placuit non esse in corpore census,
omne tenet cuius regimen pater. ergo Coranum
signorum comitem castrorumque aera merentem   55
quamvis iam tremulus captat pater; hunc favor aequus
provehit et pulchro reddit sua dona labori.
ipsius certe ducis hoc referre videtur
ut qui fortis erit, sit felicissimus idem,
ut laeti phaleris omnes et torquibus, omnes   60

    *   *   *   *   *

39 patulo *P*: vetulo ω   45 lacernas *P*: lucernas *p*ω
49 ipsis *PT*: illis ω   56 favor *Ruperti*: labor *P*ω
60 *is the last l. on the last page of the last quire of P*

NOTES

M. refers to Professor J. E. B. Mayor's edition (1886).

Friedl.   to L. Friedländer's edition (Leipzig 1895).

Büch.   to F. Bücheler's text of 1893.

Weidn.   to A. Weidner's edition (Leipzig 1889).

Lewis   to J. D. Lewis's edition (1873).

L.   to Dr Leeper's translation (1892).

P denotes the *Codex Pithoeanus*.

T a MS. of the tenth century in the Library of Trinity College, Cambridge (library mark O, iv, 10).

SATIRE I

ON THE REASONS FOR WRITING SATIRE

'The first satire is a series of incoherent complaints...A married impotent, an athletic lady, a barber rich enough to challenge the fortunes of all the patricians: the Egyptian Crispinus with his ring, the lawyer Matho in his litter: the infamous will-hunter, the robber of his ward, the plunderer of the provinces: the pander husband, the low-born spendthrift, the forger, the poisoner; all these are hurried together in no intelligible order, and with the same introductory *cum hoc fiat*, and the same conclusion in several variations, *non scribam saturam?* Then at l. 81 the satire seems to open again and promise a description of various vices; but instead of this we have an elaborate complaint, extending over many lines, of the poverty of the nobility, with a description of the hardships of a client...The ill-proportioned piece concludes with a promise to write against the dead' (Nettleship, *J. Philol.* XVI, 62).

1–21. *I have to listen to so much poetry, epic and elegy, comedy and tragedy, recited without mercy by our authors, that I have determined to commence poet too. I have had the same education as they have. And why I choose satire in preference to other styles, I will explain, if I can get a quiet hearing.*

1. **semper...tantum** the verb (*ero*) is omitted, as often in colloquial and idiomatic Latin; hence the omission is common in proverbs, e.g. *sus Minervam*.

numquamne the word, which precedes the enclitic *ně* in a direct question, is always emphatic and generally begins the sentence.

reponam, 'retaliate'; the word is properly used of 'repaying' a debt: cf. Sen. *Epp.* 81. 9 *non dicimus 'reposuit beneficium'; nullum enim nobis placuit quod aeri alieno convenit verbum.* It is commonly used transitively, with several meanings, 'to place as before', 'to place in exchange', etc. Here there is no expressed object, and it seems to combine the meanings of *respondebo* and *ulciscar*; in French *riposter* is thus used.

For the whole line, cf. Hor. *Epp.* I. 19. 39 *nobilium scriptorum auditor et ultor*, where also by revenge is meant reply. For the terror inspired by the *recitatores*, see n. to III. 9.

2. totiens: the *Theseid* of Cordus (cf. the *Aeneid* of Virgil and the *Achilleid* of Statius) is an epic, so long that one recitation does not see the end of it. Cordus is an unknown poet. For the spelling of his name, see Introduction, p. xliv ff.

3. ergo is probably derived from *e rego*, 'in accordance with the direction'; it has certainly no connection with ἔργῳ: the word is scanned as a trochee first by Ovid. The early writers of dactylic verse have final *o* short only in iambic words, as *citŏ*, *modŏ, homŏ*,; the Augustan writers have it short in cretics also, as *potiŏ, Poliŏ*; the silver age poets shorten any final *o* except inflections of the 2nd declension; and Juv. ventures even on *vigilandŏ* III. 232, where see n.

recitaverit, 'shall it go for nothing that one has...'; a normal use of the future-perfect, expressing the result, still in the future, of an action, already in the past: cf. *consumpserit* below, and Virg. *Aen.* IX. 781 *tantas strages impune per urbem | ediderit?*

togatas, sc. *fabulas.* These were comedies of Roman life, in which the actors wore the Roman dress; Afranius (born about 150 B.C.) was the most successful writer of these plays (Quint. X. 1. 100). All the Latin plays we possess are *palliatae*, representing Greek life and manners and played by actors wearing Greek dress, the ἱμάτιον or *pallium.* The sole exception is the *Octavia*, formerly attributed to Seneca; this is a *praetexta*, i.e. a tragedy representing Roman life.

5. Telephus and **Orestes** are the names of tragedies: Euripides wrote plays on both subjects, of which the former is not extant.

summi, etc.: 'an *Orestes* which, when the margin at the end of the roll is already crammed, is written also on the back and even then unfinished'. The *liber* is the ordinary papyrus roll;

when this writer got to the end of the roll (*ad summum librum*
or *ad umbilicum*), he went on to fill up the margin there, and
lastly wrote on the back of the roll. *plena margine* is abl. absol.
margo, here fem., is often masc., e.g. Pliny, *Epp.* VIII. 20, 5
cunctis margo derasus. nondum, οὔπω, 'not yet', is to be distin-
guished from *non iam*, οὐκέτι, 'no longer'; cf. Sen. *Epp.* 78. 14
*circumcidenda duo sunt, et futuri timor et veteris incommodi
memoria: hoc ad me iam non pertinet, illud nondum.*

7–13. The places and persons of mythology, which appear
ad nauseam in our modern poetry, have become as familiar to
me as my own name.

lucus Martis, where the golden fleece was deposited, at
Colchis. The Aeolian islands, so called from some imaginary
connection with the fabulous isle of Aeolus mentioned in the
Odyssey, are a group of seven volcanic islands, lying between the
north of Sicily and Lucania, now called Lipari. They are con-
nected with the same legend, as the Argonauts landed there. It
is possible that the *Argonautica* of Valerius Flaccus, who died
about A.D. 90, is the main object of this attack.

9. quid..., unde..., quantas...: these verbal clauses
should all be translated by nouns: 'the doings of..., the
place from which..., the size of....'

10. alius, 'another', i.e. Jason: cf. X. 257.

11. pelliculae, 'sheep-skin'; a contemptuous synonym for
the ἄφθιτος στρωμνά and κῶας αἰγλᾶεν χρυσέῳ θυσάνῳ of Pindar
(*Pyth.* 4. 230). The use of diminutives is characteristic of
satura.

Monychus (cf. Homer's μώνυχας ἵππους) was a Centaur: Juv.
refers to the battle between the Centaurs and Lapiths, a con-
stant theme of Greek art, the chief example being the metopes
of the Parthenon.

12. Fronto is some rich man who lends his house and
gardens, shaded by plane trees and adorned with statues, for
the purpose of recitations. The energy of the reciters is indi-
cated by *convulsa* and *ruptae*.

13. adsiduo ruptae lectore, 'broken, thanks to the con-
stant readers': a person as well as a thing may be regarded as an
instrument: cf. *percussum puero* l. 54, and VI. 29 *qua Tisiphone
...exagitare?* When this constr. is used, there is generally an
absence of *intention*: the readers do not *intend* to break the
pillars. Cf. Cic. *pro Mil.* 54 *cum paenula inretitus,...uxore paene
constrictus esset*; Pliny *Epp.* III. 19. 6 *haec felicitas terrae im-
becillis cultoribus fatigatur*: there is no *intention* on the part of the

uxor or *cultores*. That the adj. is not, as has been urged, essential to this constr., is shown by the passages quoted.

The 'abl. of the agent without *ab*' is a doubtful constr.: *Vario* may be dative, in spite of *alite*, in Hor. *Carm.* I. 6. I *scriberis Vario fortis et hostium | victor Maeonii carminis alite*; and many of the other instances given, e.g. Juv. XIII. 124 *curentur dubii medicis maioribus aegri*, are undoubtedly datives, as is shown by the personal pronouns which have different forms for dat. and abl. The dat. after any part of the passive verb is common in all Latin poets.

14. **expectes**, 'one must look for'. The hortative use of this person of the pres. subj. is of constant occurrence; in Cicero it usually expresses a general maxim: cf. *de Off.* III. 82, where εὐσεβεῖν χρεών is translated by *pietatem colas*. Livy and later writers use it also as an imperative addressed to an individual; e.g. Livy XXVI. 50. 7 *amicus populo Romano sis*.

15. **et nos ergo**, 'well then, I too....'.

ferulae the *ferula* (νάρθηξ), and *scutica* (ἱμάς), the cane and tawse, were the weapons with which the schoolmaster of antiquity carried on war against ignorance and idleness. Juv. says that he, no less than others, has been caned at school, and gone on to college. Roman boys, when they had learnt the three R's from the *litterator* or *ludi magister*, went next to the *grammaticus*, whose chief business was to teach grammar and to expound the poets: see n. to VI. 450. Such time as the *grammaticus* did not claim might be given to mathematics (*geometria*) and music.

16. The education, thus begun, was completed in the school of the *rhetor* or 'professor of rhetoric'. The exercises mainly practised there were of two kinds: (1) *suasoriae*, (2) *controversiae*. The *suasoria* was practised first: the pupil delivered a declamation which might be put in the mouth of some famous man on a historical occasion, or might be addressed to him (hence the name) as, in the instance here quoted, to Sulla. A large number of stock themes have been preserved in M. Seneca and Quintilian: e.g. Quint. III. 8. 19 *deliberat C. Caesar an perseveret in Germaniam ire, cum milites passim testamenta facerent*. For Hannibal, cf. Juv. VII. 160.

The *controversia* was not a monologue, but a debate which professed to observe legal forms and to represent the proceedings of a law-court, the pupils acting as counsel engaged in the case. But the circumstances supposed were generally quite unlike those of real life: the magician, the pirate, and the pirate's sympathetic daughter are some of the standing characters. The same case was debated an infinite number of times; and the ambition of the pupil was to import some novelty into the well-worn theme, to suggest some *color* (line of defence) more absurd and ingenious than all its predecessors. Thus the *controversia* served the same purpose as the famous suit of *Peebles against Plainstanes* in Scott's *Redgauntlet*, 'a subject upon whilk all the *tyrones* have been trying their whittles for fifteen years'.

The *grammaticus* and *rhetor* are constantly mentioned together by Latin writers, just as we speak of 'school and college'. Ancient education had at least one advantage, that it was directed from the first to one object—the acquisition of rhetorical power.

privatus should be translated as a separate piece of advice: 'to give up his office (of dictator) and sleep sound': cf. Lucan, VII. 28 *unde pares* (adj.) *somnos populis noctemque beatam?*

18. **periturae**, 'that is sure to be spoilt'; by a principle more common in Greek than Latin, *perire* is used as the passive of *perdere*, which hardly occurs except in the participles *perditus* and *perdendus*. See n. to *perit*, VII. 99.

19. **potius**, 'by preference': where the alternatives are more than two, a prose writer would use *potissimum*, but the word is inconvenient metrically.

20. **Auruncae...alumnus**: Lucilius, the representative writer of *satura* (see Introduction, p. xxv), was born at Suessa Aurunca in Latium 180 (?) B.C.: cf. Mart. XII. 94. 7 *audemus saturas: Lucilius esse* (i.e. to cut me out in satire) *laboras*. For the metaphor, cf. Gray's *Ode* III. 2:

> Behold, where Dryden's less presumptuous car
> Wide o'er the fields of Glory bear
> Two Coursers of ethereal race,
> With necks in thunder cloth'd, and long-resounding pace.

21. **vacat**: the pers. constr. *vacatis* is equally good Latin.

22–80. Every form of vice and crime is rampant. The unsexed woman, the upstart millionaire, the informer, the oppressor of the provinces, the spendthrift, the forger, the poisoner—all these are familiar figures in our streets. Indignation will supply the power of writing which nature may have denied me.

22. **Mevia**, a woman of rank, dresses as an Amazon, and takes part in a *venatio* (beast-baiting), in the amphitheatre. Suetonius says of Domitian (4) *nec virorum modo pugnas sed et feminarum* (*edidit*).

23. **venabula**: the plur. is for metrical convenience: it is not meant that she held more than one spear.

24. **patricios** is used for *nobiles* which the verse will not admit: properly, the *patricii* are only a very small class among the *nobiles*: cf. also IV. 102; VIII. 190; X. 332.

The barber is probably one Cinnamus whom Martial often attacks, e.g. VII. 64 *qui tonsor fueras tota notissimus urbe | et post hoc dominae munere factus eques.*

25 is a parody of Virg. *Ecl.* I. 28 *candidior postquam tondenti barba cadebat.* The l. recurs X. 226. For the Roman way of treating the beard, see nn. to VI. 105, VIII. 166. Observe that Juv. speaks of his youth as already past.

26. **Canopus** (Κάνωβος) was a city at the western mouth of the Nile. Crispinus was born there and came to Rome as a seller of salt-fish. He became a knight (*princeps equitum* IV. 32) and a member of Domitian's privy council. Martial speaks more favourably (VII. 99, VIII. 48), as he naturally would, of the emperor's favourite. Juv. has a remarkable dislike of Egypt and Egyptians: see Introduction, p. xxxix.

27. **revocante**, 'hitching up'.

28, 29. It is generally supposed, on the evidence of this passage alone, that lighter rings were worn in summer by men of fashion; but the Latin does not warrant this inference. Juv. only says that Crisp. wears a gold ring as heavy as he can carry, so heavy that in summer it makes his fingers hot. **aestivum** does not imply a special ring, but is only an equivalent for *aestate*, 'in summer'; cf. *matutino...amomo* (IV. 108) where the adj. merely takes the place of *mane*. A good parallel is afforded by Mart. V. 19. 11 *Saturnaliciae ligulam misisse selibrae...*, where the adj. takes the place of *Saturnalibus*, and no one suggests that half-a-pound of plate at the Saturnalia was different from half-a-pound at any other time. An adj. (especially of time) is employed with great freedom by the later Latin poets, to express an adverbial clause. See n. to VI. 5.

29. **maioris...gemmae**, 'a (not 'his') larger ring': *gemmae* is used for the unmetrical *anuli*: so l. 68 and often.

30. **iniquae...urbis:** cf. *saevae urbis* III. 8; either the hard struggle for life in great cities is in the writer's mind; or perhaps *iniquae* is 'unjust'; i.e. Rome rewards and honours criminals.

31. **ferreus**, 'steeled against it'; it means much the same as *patiens*, but is stronger; cf. VII. 150.

33. **plena ipso:** the *lectica* was constructed to carry two; but Matho is fat and needs a whole one for himself.

magni, 'great' in the sense of 'important, highly placed', not

in the sense of 'devoted'; cf. IV. 74, and Eur. *Medea*, 549 μέγας
φίλος with Verrall's note there.

delator: see n. to III. 116. There cannot well be a reference
here to the fate of Barea, there mentioned; for Barea fell under
Nero, while all the informers mentioned here belong to Domi-
tian's time.

amici: cf. Tac. *Hist.* I. 2 *quibus deerat inimicus, per amicos
oppressi.*

34. **rapturus** is joined by the copula to *delator*, the whole
being equivalent to a relative sentence, *qui detulit...et cito
rapiet.* **de** does not follow *rapturus* but *quod superest*; cf. l. 66,
and III. 259.

35. Baebius Massa and Mettius Carus were themselves
infamous as informers under Domitian, but even they do hom-
age to the superior ferocity of their rival. Thymele and Latinus
were actors: cf. Mart. I. 4. 5 *qua Thymelen spectas derisoremque
Latinum, | illa fronte precor carmina nostra legas.* The passage
shows that Latinus was feared as an informer.

45. **siccum iecur,** 'my fevered bosom' L.

46. **hic..., hic,** 'one..., another'; *hic* is the subject;
spoliator = qui spoliavit.

47. **hic** refers, not to Marius, who being in exile cannot be
seen in the streets of Rome, but to some other criminal; then
the instance of Marius follows to show what a farce even
banishment is.

48. **salvis...nummis:** i.e. the criminal, though con-
demned, keeps his money: this was the rule, when the punish-
ment inflicted was *relegatio:* see n. to l. 73.

49. **exul** is emphatic, and opposed to *hic* 'in Rome'; 'even
in exile'.

octava, sc. *hora:* the usual hour to begin dinner was the
ninth: Mart. IV. 8. 6 *imperat exstructos frangere nona toros;* but
Marius takes advantage of his banishment to anticipate and
prolong the pleasures of the table.

Marius: cf. VIII. 120. Marius Priscus was prosecuted by
Tacitus and Pliny for extortion in his province of Africa, and
condemned by the senate to banishment (*relegatio*) from Rome
and Italy. He was sentenced in Jan. A.D. 100; for a full account of
the proceedings, see Pliny *Epp.* II. 11. For Juvenal's chronology,
the mention of Marius here is important; see Introduction, p. xv.

50. **tu:** it should be observed how common apostrophe of
this kind is in the Latin poets. It may often add rhetorical
force: cf. III. 67; VI. 80, 167; VIII. 95, 231; XIV. 316, 319; but in

many instances it is merely a device for employing a case that
is metrically convenient, e.g. VI. 7 *similis tibi, Cynthia*; VI. 124
tuum, generose Britannice, ventrem; VIII. 39 *tecum est mihi sermo,
Rubelli*: *Cynthiae, Britannici, Rubellio* are in this way avoided.

victrix is not an epithet but part of the predicate: it =
quamquam vicisti. vincere causam is a technical phrase of the
law-court.

ploras, 'are left lamenting'. In spite of the sentence against
Marius, the province received no real reparation for its wrongs.

51. **Venusina**: the allusion is to Horace, a satirist of a very
different kind, who was born at Venusia 65 B.C.

lucerna, 'lamp'; cf. Hor. *Epp.* II. 1. 112 *prius orto | sole
vigil calamum et chartas et scrinia posco.* We should rather say
'pen'; but compare our use of 'lucubration'.

52. **agitem** is used here in two senses: (1) am I to attack;
(2) am I to deal with.

Heracleas and **Diomedeas** are names of epics formed on
the analogy of *Odyssēa* from Odysseus. Theseus and the Mino-
taur, Daedalus and Icarus are the stories referred to below.

54. **percussum puero**: see n. to l. 13: *puero* is instrumental
abl., not dat.: if it were dat., it would be = *a puero* which would
give the wrong sense. Icarus was not an agent at all in the matter,
but a passive victim. Of course the phrase adds to the sarcastic
tone of the description.

fabrumque volantem, 'the flying tinman': see n. on
pelliculae l. 11.

55. **leno** is the husband who connives at his wife's dishonour.

si capiendi…: the wife was prevented by law from in-
heriting, not because of her bad character, but because of her
sex: the *lex Voconia* (169 B.C.) originally enforced these dis-
abilities against women. The law could be evaded by bequeath-
ing the estate to a third party in trust for the woman; in the
present case her own husband is the *heres fiduciarius*, who got a
share as such. *capere* = *heres esse*, while *accipere* is properly
used of the provisional possession of the *heres fiduc.*

(But it is doubtful whether the *lex Voconia* was still in force;
and Friedl. supposes that the woman had no children and was
therefore prevented by the *leges Iulia* and *Papia Poppaea* from
inheriting (see n. to III. 116); though her husband could do so,
if he had a child by a former marriage).

58. **curam…cohortis**: the youth, who has spent all his
money in extravagance, hopes to repair his fortunes in the
army. For the army as a profitable career, cf. XIV. 193.

It was customary for both senators and knights to begin public life with some service in the army. In the case of the former, the office was *tribunatus legionis* and was often a mere form; but the *equites* did actually go through three successive steps in the army, known as the *equestres militiae*: these were *praefectura cohortis sociorum, tribunatus legionis*, and *praefectura alae*, usually in this order. They formed a prelude to the lucrative and important posts in the civil service, which the knights filled as *procuratores*. Cf. Pliny, *Epp.* VII. 25. 2 *Terentius Iunior, equestribus militiis atque etiam procuratione Narbonensis provinciae functus*; Tac. *Agric.* 4 *utrumque avum procuratorem Caesarum habuit, quae equestris nobilitas est*: i.e. the knights chosen to be imperial *procuratores* were to other knights what curule magistrates were to other senators.

It is clear from the word *cohortis* that Juv. is here speaking of a youth of equestrian, not of senatorial, rank.

60. dum gives the reason of the previous clause: his ruin was due to his passion for horses. Driving is no less than a vice in Juv.'s eyes, for a man of birth and position: cf. VIII. 146 ff.

61. Flaminiam, *sc. viam*: this was the Great North Road out of Rome, crossing the Tiber by the *pons Mulvius*.

puer Automedon, 'a young Jehu': Automedon is the charioteer of Achilles in the Iliad. Latin poets often apply Homeric names in this way: cf. *Atriden* IV. 65; *dimitte Machaonas* (the doctors) *omnes* Mart. II. 16. 5.

62. lacernatae: the mistress to whom he displays his skill is dressed like a man. M.

63. ceras, 'note-books': a wooden surface, with a raised frame at the edges, like a school-slate of modern times, was overlaid with a thin coating of dark wax on which the writing was traced with a sharp point. They were much used for rough drafts, owing to the facility of erasure. 'It is said that until quite recently sales in the fish-market of Rouen were marked on waxen tablets' (E. M. Thompson's *Palaeography*, p. 23).

64. iam to be taken with *sexta*: he has *already* six slaves to his litter—a hint that he will soon have eight, and be carried in an *octophoron*. **sexta cervice** = *sex cervicibus*.

66. multum...Maecenate, 'recalling much of Maecenas'. **de** follows *multum* (cf. *exiguum de* III. 123), not *referens*: *referre Maecenatem* is the Latin expression, not (in this sense) *referre de Maecenate*.

Maecenas is a proverb for effeminacy in post-Augustan

literature; cf. XII. 39: we naturally hear less of this from Horace and Virgil.

67. **signator:** see n. to III. 82: when called in to attest a friend's will by his signature, he has inserted a forged document (*tabulas*) in his own favour and signed it with his seal, thereby robbing the rightful heirs. *falsum* is the legal term for all fraud, including forgery: it must, if the text is sound, mean here 'a forged document'.

68. **fecerit:** so P: though the other relative clauses in the paragraph have the indic., the subj. is necessary here, as the clause depends on *cum feratur* l. 64.

uda: i.e. the ring is breathed upon before being used for sealing.

70. **viro...sitiente** must be abl. absol.; but the constr. is awkward as the reader naturally connects *viro* with *porrectura*. **rubetam:** cf. VI. 659; aconite is the other poison most frequently mentioned by Juv.

71. **Lucusta:** a Gallic woman and skilful poisoner, who employed her art against Claudius and Britannicus; she also provided Nero, her pupil and employer, with means of death in his extremity: cf. Tac. *Ann.* XII. 66, XIII. 15; Suet. *Nero,* 47. She was executed by Galba.

72. **per famam et populum,** 'in the face of scandal and before the eyes of the people': the constr. can hardly be called hendiadys, as *per populum* alludes to the custom of the funeral procession passing through the forum, on the way to the place of interment outside the walls.

nigros: the colour is due to poison: cf. *ad Herenn.* II. 8 *si... livore decoloratum corpus est mortui, significat eum veneno necatum.*

73. **Gyaris:** cf. X. 170. Gyara (or Gyarus) is an island in the Aegean; the bad water-supply made confinement in Gyara a very severe punishment (Tac. *Ann.* IV. 30). Confinement to an island was a common punishment in imperial times; thus there is ancient precedent for the banishment of Napoleon to St Helena, of Garibaldi to Caprera, and of Arabi to Ceylon. *deportatio in insulam,* the punishment meant here, was much more severe than *relegatio:* it involved loss of civil rights and generally confiscation of property (see n. to l. 48): it was the imperial substitute for the republican *aquae et ignis interdictio.*

carcere suggests execution: cf. XIII. 245.

74. **esse aliquid,** 'to be a person of importance'; cf. Pliny *Epp.* I. 23. 2 *cum tribunus essem, erraverim fortasse qui me aliquid putavi.*

probitas...alget may be called a μέν clause, **criminibus...
caprum** a corresponding δέ clause; the clauses are more
strongly opposed in Latin when there are no particles, though
quidem...verum, and other combinations, are sometimes used
for this purpose.

75. criminibus: *crimen* = 'accusation', always in Cicero;
but in poetry and silver Latin it has the meaning 'crime' as
often as the other.

hortos: *horti*, 'a pleasure-garden' or 'park', is to be distin-
guished from *hortus*, 'a kail-yard' or 'kitchen-garden'.

mensas: cf. l. 137. Pliny (*Nat. Hist.* XIII. 91) in his botanical
section takes occasion to reprove the *mensarum insania quas
feminae viris contra margaritas regerunt*, i.e. when women are
attacked for their extravagance in pearls, they taunt men with
their passion, no less ruinous, for tables. The most expensive
and desirable tables were *orbes*, round sections, of the *citrus*
tree, a kind of cypress which grew in Mauretania. The slab was
mounted on a support (*pes*) of ivory or precious metal; cf. XI.
122 ff.

76. stantem...caprum: the goat is embossed in high
relief on the cup: cf. Mart. VIII. 51. 9 (of a cup) *stat caper...
vellere Phrixi | cultus.*

78. praetextatus = in his teens: the *praetexta*, or toga with
a border of purple, was worn not only by magistrates, but by
boys of free birth until they put on the *toga virilis*, which was
pure white; this they usually did before their sixteenth birthday.

79 facit indignatio versum: cf. Quint. VI. 2. 26 *quid est
causae ut...ira nonnunquam indoctis quoque eloquentiam faciat?*

80. Cluvienus is otherwise unknown.

81–116. *My book is a medley, which describes all the occupations
and all the passions of men since the time of the Flood. There was
never a better field for satire, never such reckless gambling as
now, and such selfish extravagance. The rich man consumes an
immense dinner himself and distributes a little dole to his de-
pendants and friends on the door-step. There you may see the
nobles and the magistrates giving place to the wealthy freedman;
for money, though it has no temple yet, is the deity which our age
most devoutly worships.*

81. ex quo is to be taken after *agunt* in l. 85.

82. montem: according to most accounts, Parnassus was
the mountain where Deucalion anchored: the language ('took
ship to climb a mountain') is intended to suggest incredulity.

sortes: the word suggests the old Italian method of divination by strips of wood which were shuffled and drawn; but *sortes* came to be used for any kind of oracular response.

83. **mollia** is not epithet but part of the predicate; 'grew soft and warm with life'.

86. **discursus,** 'running to and fro'; both the noun and verb (*discurrere*) are common in silver Latin, to express the aimless activity of idle persons.

farrago, 'hotch-potch', is meant to suggest *satura*; see Introduction, p. xxiv.

est: the verb regularly takes the number of the predicate, when the predicate is a noun and precedes the verb: e.g. Ovid *Ars*, III. 222 *quas geritis vestes sordida lana fuit.*

87. This is an additional reason for writing satire. **et** expresses indignation: cf. Virg. *Aen.* I. 48 *et quisquam numen Iunonis adorat?*

88. **sinus,** 'pocket' or 'purse'; lit. the fold of the toga.

89. **hos animos,** 'such a zest as now'. For the omission of the verb (here *sumpsit* as VI. 285), see n. to l. 1.

90. **casum,** 'hazard': lit. throw. Our 'chance' is derived from *cadentia.* M.

posita may mean (1) 'by their side:' cf. Mart. II. 1. 9 *ante,* | *incipiat positus quam tepuisse calix:* or (2) 'staked on the game'; cf. Plaut. *Curc.* 355 *provocat me in aleam ut ego ludam; pono pallium.* The first meaning seems the better contrast to *comitantibus.* Juv. uses *positus* thrice elsewhere meaning 'laid aside' (II. 74; VII. 26; XI. 69).

91. **proelia:** for the imagery, cf. XIV. 5, and Mart. XIII. 1. 5 *non mea magnanimo depugnat tessera telo.* Thackeray's Barry Lyndon often compliments his profession of card-sharper with the same metaphor.

92. **simplexne furor,** 'madness and nothing more'; so *mors simplex* is used for execution without torture (Suet. *Iul.* 74).

93. **et,** 'and yet...': this action is not merely added to the preceding but contrasted with it. This use of *et* is peculiar to silver-age Latin: Cicero uses it only when the first of the two clauses is negatived.

reddere, 'to give' as due, not 'to give back'; thus *reddere alicui epistulam* is to 'deliver' a letter, not to 'answer' it.

95. **nunc,** 'now-a-days'. This is closely connected with what goes before: the paltry dole is contrasted with the immense dinner consumed by the rich man himself. But it must be noticed that the mention of the *sportula* tempts Juv. to give a

detailed description of the institution, which is quite irrelevant to the topic in hand; he only returns to his subject, the selfish gluttony of the rich, at l. 135.

sportula, 'dole'; lit. 'little basket'. This dole, in reality a sort of daily wages paid by the rich to their dependants, was a regular feature of Roman life under the Empire. When clients became too numerous to be entertained in the ordinary way, a custom arose of distributing food, which was carried away in baskets (hence the name *sportula*); but it was soon found more convenient to distribute money instead, though an edict of Domitian was for a short time enforced which required that a regular meal (*recta cena*) should be supplied instead of money. The regular sum was *centum quadrantes* (= 25 *asses* or 6¼ sesterces), or about 1s. 3d. The name is used of any sum of money distributed unofficially to a number of people: Trajan uses it (Pliny, *ad Trai.* 117) of a dole given in Bithynian towns to guests on special occasions, the amount being one or two *denarii*. Throughout Juv., the dole seems always to be money, not eatables: cf. l. 118.

Two discrepancies have been noticed between Juv.'s account of the *sportula* and Martial's: (1) that the former speaks of the distribution in the morning, not the evening; (2) that he includes nobles, magistrates, and women among the recipients, whereas Martial does not. As to the first difficulty, it seems natural and probable, that those who saw no more of their patron after the *salutatio*, received their *sportula* then, in the early morning; whereas the *anteambulones*, who attended their patron throughout the day, received theirs at the end of their labours, often in the bath itself (cf. Mart. III. 7. 3; X. 70. 13). The other difficulty may fairly be disposed of by supposing that, after Martial had turned his face westwards, it became the custom to distribute the *sportula* to all, without distinction, who attended the *salutatio*. That one great man attended the *salutatio* of another great man, is perfectly established by many passages from both authors.

96. **togatae**: the recipients of the dole and all *salutatores* had to wear the regular dress of the Roman citizen, the *toga* and *calcei*.

97. **ille** is the giver of the dole.

100. **Troiugenas**, a common name in Juv. for the old Roman nobility; cf. VIII. 181; XI. 95. Ancient legend, glorified by Virgil, connected the origin of Rome with Aeneas and so with Troy. So our own old chronicles generally begin our history with the arrival of Brut the Trojan in England.

vexant, 'infest'. L.

101. The praetor is served first, as the superior magistrate. The aspirant for curule office passed through the *cursus honorum* in the following order: *quaestor, tribunus plebis* or *curulis aedilis, praetor, consul,* with an interval, prescribed by law, of at least one year between any two of these magistracies.

102. **libertinus:** the prominence of freedmen in the first century A.D. was largely due to their position at court, as they filled important administrative posts which were afterwards occupied by Roman knights. Their influence and wealth were enormous, especially under Claudius. Juv. himself is stated to have been the son of a rich freedman: see Introduction, p. xi.

104. **fenestrae,** 'holes': ear-rings, or holes for ear-rings, were a sign of oriental birth in the eyes of Greeks and Romans alike; thus a soldier of Xenophon's Ten Thousand, in spite of his Boeotian dialect, was convicted at once as an Oriental, because his ears were pierced, and was driven out of the ranks: Xen. *Anab.* III. I. 31 ἀλλὰ τούτῳ γε οὔτε τῆς Βοιωτίας προσήκει οὐδὲν οὔτε τῆς Ἑλλάδος παντάπασιν, ἐπεὶ ἐγὼ αὐτὸν εἶδον ὥσπερ Λυδὸν τὰ ὦτα τετρυπημένον.

105. **arguerint,** 'might prove': the subj. is potential rather than conditional; the aorist and present are both used potentially, but the aor. is commoner with the first person (e.g. *dixerim*), the pres. with the others. There is no difference in meaning between the tenses, any more than between λέγοιμ' ἄν and εἴποιμ' ἄν.

106. **quadringenta,** sc. *sestertia,* 400,000 sesterces, about £4,000, was the *census* of a Roman knight; here the amount seems to be income, not capital. Even so, the speaker would not be nearly as rich as Pallas (cf. Tac. *Ann.* XII. 53), than whom he claims to be richer.

purpura maior, the badge of the senators, a broad vertical stripe of purple down the front of the *tunica* (called *t. laticlavia*); the *angustus clavus* was a privilege of the knights.

107. From **Laurenti** to **oves** is a μέν clause, from **ego** to **Licinis** a δέ clause; see n. to l. 74; in English one clause should be subordinated to the other.

108. **Corvinus:** the name of a very ancient family of the *gens Valeria,* now so reduced that one representative accepted a yearly pension from Nero; Tac. *Ann.* XIII. 34.

109. Pallas was a freedman of Claudius and the brother of Felix, procurator of Judaea. Licinus was originally a German captive who served first Caesar and then Augustus in various

capacities, and amassed an immense fortune: cf. XIV. 306. (The plur. *Licini* is used generically for 'people like Licinus'.)

110. **sacro...honori:** Livy (II. 33) speaks of a *lex sacrata* by which the persons of the tribunes were made inviolable; but there is some doubt whether this was a compact between the patricians and plebeians, or merely a solemn engagement of the latter to avenge any wrong done to their representatives. In either case, this magistracy was styled *sacrosanctus* even when it had become a nullity under the empire; cf. Pliny *Epp.* I. 23. 1.

111. **pedibus...albis:** slaves, who had just arrived from abroad for sale in the Roman market, were placed on the dealer's block (*in catasta mangonis*) with their feet whitened with chalk (*cretatis* or *gypsatis pedibus*), to distinguish them from *vernae*: cf. VII. 16.

115. Though we have raised temples to many abstract ideas, we have not yet built one to Wealth.

116. 'And Concord, who clatters, when we hail her nest.' The temple of Concord was close under the Capitoline, just behind where the arch of Severus now stands; built originally by Camillus, it was restored by Tiberius. It appears from this passage that storks nested on the roof; and, when a passer-by hailed the temple, the stork clattering with her bill seemed to give an answer: cf. Ovid *Met.* VI. 97 *crepitante ciconia rostro.* For the Roman custom of offering prayer or worship when passing a temple, cf. X. 290: Ovid *Trist.* II. 291 *adoranti Iunonia templa.*

salutato nido is generally explained 'when she hails her nest': but the sarcastic identification of the temple with the nest seems more characteristic of Juv.

117–26. *The highest magistrates and even women may be seen taking part in the scramble for the dole; the poor suffer by this competition.*

117. **summus honor,** the consul. *honor* is used as in l. 110.

118. **rationibus,** 'income': in VI. 511 it means 'expenditure'; being simply 'accounts', it may stand either for the debit or the credit side of the ledger.

119. **comites** are the poor clients, who accompany their patron during the day.

toga, calceus: without these, they could not venture to present themselves, to claim the dole: see n. to l. 96.

120. **fumus,** 'firing'. **domi** contrasts their wants at home with the clothes they must wear abroad (*foris*).

centum quadrantes: the regular amount of the *sportula*: see n. to l. 95.

122. **circumducitur:** i.e. they go to several houses in succession.

123. **absenti,** sc. *uxori*. **nota** means that he is an old hand at the trick, not that the trick is generally seen through.

125, 126 are spoken by the artful husband and addressed partly to the distributor of the dole, partly to his wife.

126. **quiescet,** 'she will be sleeping', i.e. 'probably she's asleep': cf. *sic erit*, 'it is so, you will find', in Comedy. So in Greek the potential optative with ἄν may express 'what may hereafter prove to be true' (Goodwin, §238).

127–46. *The dole once distributed, there is an adjournment to the law-courts. At the end of the day the clients have to abandon all hope of dining with their patron; in spite of his many marvellous tables, one only is used, and there is no diner there but himself. Yet selfish gluttony brings its own penalty in the shape of a sudden and unlamented death.*

127. The 'glorious succession of events' is ironical.

128. Cf. Martial's account of a Roman day (IV. 8), beginning *prima salutantes atque altera conterit hora: | exercet raucos tertia causidicos.* When the early call (*salutatio*) was paid, business began in the law-courts about 8 a.m. For the time of distribution of the *sportula*, see n. to l. 95.

Apollo: a prominent ornament of the forum of Augustus was an ivory statue of Apollo, who is called 'learned in the law' because of the legal proceedings which were constantly going on in his presence. The chief building of the forum was the temple of Mars Ultor, with a double colonnade in front, something like the present porticoes of St Peter's at Rome; in these colonnades Augustus set up statues (*triumphales*) of all the great Roman generals in triumphal robes. This forum, the third in Rome, was completed 12 B.C.; by A.D. 98 there were four adjacent forums (*fora iuncta quater* Mart. X. 51, 12): *f. Romanum, f. Caesaris, f. Augusti, f. Nervae* or *transitorium*.

130. **titulos:** the inscription on the base of a Roman statue followed a regular order: first came the *praenomen, nomen, praenomen* of father and perhaps grandfather, tribe, and *cognomen*: e.g. *M. Tullius M. f. M. n. Cor(nelia tribu) Cicero*; then followed a complete *cursus honorum*, comprising all the magistracies and public offices in order which had been held by the person in question: this sometimes ran from highest to

lowest, sometimes in the reverse order. The whole of this formula was known as *tituli*: cf. *honorum pagina* x. 57.

Arabarches, 'Mogul': one of Cicero's nicknames for Pompey (*ad Att.* II. 17. 3). It is almost certain that this 'Egyptian Mogul' is Tiberius Julius Alexander, a Jew by birth, who became a Roman knight and was successively procurator of Judaea (A.D. 46) and praefect of Egypt (A.D. 67–70): the fact that he had been an early and zealous supporter of the Flavian dynasty, accounts for this extraordinary distinction. See Furneaux on Tac. *Ann.* xv. 28.

132. Juv. makes a sudden jump from the law-courts to the dinner-hour, though he had hinted in l. 127 that he would treat the whole day in detail; there is a similar instance of carelessness in VI. 474 ff. See Introduction, p. xxxiii.

134. **emendus** is emphatic: they must part with their *centum quadrantes* to *pay* for the food and warmth they had hoped to get for nothing in the rich man's dining-room.

136. **rex**, 'patron', often in this sense as early as Terence.

137. **de**, 'although they have...'; lit. 'out of', not 'off'. It is important to realise that the satire is here directed not against greed but against selfishness: though they have many fine tables, they only use one, and that for themselves. The phrase is quite unlike *cenare de fictilibus* etc.

orbibus: see n. to l. 75.

138. **mensa** is often 'a course', but not here; see above.

139. The parasite, finding his occupation gone, will soon be a thing of the past.

140. **luxuriae sordes**, 'niggardly profusion', is an *oxymoron*. Pliny (*Epp.* II. 6. 7) calls the same conduct on the part of the rich *luxuriae et sordium societatem*.

142. The natural and ordinary time for bathing was just before the *cena*, but the gluttons of this time had discovered that digestion was temporarily promoted by the unhealthy practice of bathing in very hot water immediately after the meal.

143. **crudus**: so P: *crudum* of the inferior MSS. has generally been read; but the reading of P seems adequate with a pause after *turgidus*: cf. Hor. *Epp.* 1. 6. 61 *crudi tumidique lavemur*.

crudus, of persons, means 'over-eaten', 'stuffed', while of food it generally means 'uncooked'; Martial plays on the two senses (III. 13. 3) *accusas rumpisque cocum tanquam omnia cruda | attulerit: nunquam sic ego crudus ero*. However, digestion is a kind of cooking, as is shown by the use of *concoctus*; and Celsus, in his medical treatise, uses *crudum* in the sense of *non concoc-*

tum, e.g. IV. 18 *si adhuc subcruda sunt quae vomuntur.* Hence, if
P read *crudum,* it should be kept.

144. The result of this bathing is sudden death, which
prevents the *senex orbus* from making a will; consequently his
money goes to his natural heirs, and his friends (as Juv. ironic-
ally calls them) are indignant at getting no legacies. The verb
sunt is understood and has three subjects, *mortes, senectus,
fabula.* The old are specially mentioned as the special object of
fortune-hunters' attentions.

146. **plaudendum,** not, as in ordinary cases, *plagendum.*

147–71. *The age is as bad as it can be, and I mean to avail
myself of the opportunity for satire. My friends warn me of the
danger of attacking the great criminals of our own day and
advise me to treat of ancient history and mythology. I shall
direct my attack against the past generation.*

147. **moribus,** 'vices'.

148. **minores,** 'posterity', less common than *maiores,*
'ancestors'.

149. **omne...stetit,** 'vice always stands above a sheer
descent', and therefore soon reaches its extreme point; hence
there is no lower depth left for our posterity. The old interpre-
tation ('every vice has settled at its zenith') is untenable,
because *in praecipiti* regularly suggests the danger of falling,
not height alone: cf. Sen. *Epp.* 23. 6 *in praecipiti voluptas stat;
ad dolorem vergit nisi modum tenuit* (Madv.). A sick man, in a
critical condition, is said *in praecipiti esse.* (This explanation
I owe to Mr H. Richards: see *Classical Review,* VI, 125.) But see
Housman, *Classical Review,* XVII, 466.

stetit is thus a gnomic aorist.

utere velis: addressed by the poet to himself; the metaphor
is continued in the following verse.

150. **totos pande sinus:** a common metaphor: cf. Pliny
Epp. VIII. 4. 5 (to a friend intending to write a history of Tra-
jan's campaigns in Dacia) *immitte rudentes, pande vela, ac, si
quando alias, toto ingenio vehere.*

dices: so P: for the fut., cf. XII. 125, XIV. 295. *forsitan* ought
to take the subj. as much as *haud scio an,* and does so in correct
prose; when it takes the ind., it is regarded as a simple adverb.

Recent edd. read *dicas* with the inferior MSS.: and the ind.
of the 2nd person is unusual, but the fact that the verb comes
first is important.

There follows a dialogue between the poet and an imaginary

friend, in the ancient fashion of *satura* (Introduction, p. xxxii).

151. **materiae:** the last syllable is not elided. This and similar instances of hiatus are common in Juv.; see n. to X. 281.

153, 154. The name Mucius shows that Juvenal is thinking of his predecessor Lucilius, who attacked P. Mucius Scaevola: cf. Pers. I. 114 *secuit Lucilius urbem, | te Lupe, te Muci.* It has been suggested that *cuius...nomen* is a quotation from Lucil.; if so, it is not a verbal quotation, as *audeo* could not scan as a dactyl in a writer of that time: see n. on *ergŏ* l. 3.

cuius...nomen is taken by some editors as a relative clause, not as a question, the antecedent of *cuius* being *simplicitas*; in that case, the words belong not to Juv., but to his interlocutor. But the clause seems to have more force as a question.

155. **pone Tigellinum:** in all languages, an imperat. sometimes takes the place of the protasis in a conditional sentence; see n. to VII. 175. Sofonius Tigellinus was a man of infamous character, who encouraged Nero in all his excesses; he was forced to commit suicide by Otho A.D. 69: on Juv.'s choice of an example here, see Introduction, p. xxxv. The speaker means that it is unsafe to satirise the vices of the great.

taeda...harena: 'you will give light in that coat of pitch, dressed in which men stand with breast fast to the stake and send out fire and smoke; and your body traces out a broad furrow in midst of the sand': i.e. you will be burnt alive in the *tunica molesta* (see n. to VIII. 235), and your remains will then be dragged by the *uncus* through the amphitheatre. The main difficulty of this interpretation is the change of tense, with no apparent reason, from *lucebis* to *deducis*. Also, if a man were burnt to death, there would be no remains to be dragged by the *uncus*, much less to trace a 'broad furrow'.

Some editors keep *deducit* (read by P, as also *lucebit* in 155, which seems a certain error) and supply *quae*, out of *qua*, as nom. to the verb: in this case it is not the body of a single victim, dragged through the arena, that forms the *sulcus*, but, a number of men being burnt in one long row, their flaming bodies are said to mark out a furrow, or series of depressions, in the sand. But *sulcus* surely means a depression of the soil, first the mark of a ploughshare, and then a ditch or trench; whereas, at such a scene as Juv. has in his mind, the row of victims would surely strike the eye as something raised above the soil, not sunk beneath it.

None of the many emendations of this difficult passage has

[handwritten marginal note: the hook with which the executioner dragged away the bodies of dead malefactors to throw them in the Tiber]

any probability: *dant lucis* for *deducis*, which is often attributed to Dobree, is only quoted by him as the suggestion of *adolescens quidam*.

I believe that l. 156 as it stands admits of a quite different interpretation, which accounts for the chief difficulty, the change of tense in *deducis*. The metaphor of 'ploughing the sand' is exceedingly common in Latin in the sense of taking trouble with no result. Cf. VII. 48 *nos tamen hoc agimus tenuique in pulvere sulcos | ducimus et litus sterili versamus aratro*; Ovid, *Heroid.* 5. 115 *quid harenae semina mandas? | non profecturis litora bubus aras*: ibid. 16. 139 *quid bibulum curvo litus proscindere aratro | spemque sequi coner quam locus ipse negat?* id. *ex Pont.* IV. 2. 16 *siccum sterili vomere litus aro*; idem, *Trist.* V. 4. 48 *nec sinet ille tuos litus arare boves*; Virg. *Aen.* IV. 212 *cui litus arandum | ...dedimus*; Sen. *de Ben.* IV. 9. 2 *quia ne agricolae quidem semina harenis committant*.

It seems then that Juv. is using a common proverb, and means: 'if you attack the vices of the great who are still living, you will be severely punished and your task is a fruitless one'. For just the same reasons Pliny (*Epp.* v. 8. 12) declines to write a history of his own times: *graves offensae, levis gratia*, 'I shall give serious offence and get little thanks'. If the epithets *latum* and *media* are objected to, the answer is that Juv. intends by them to denote the parade and circumstance with which the fruitless task is set about.

(I had arrived at this explanation some time before I found that it had been already suggested by Professor Palmer in his edition of Ovid's *Heroides*, 1874, p. 44.)

158. **dedit:** *dare*, διδόναι, are the technical words for administering medicine; hence our 'dose' from δόσις. M.

vehatur: deliberative subj.

159. **plumis,** down pillows, a sign of great luxury: cf. VI. 88; x. 362.

161. "'Tis defamation but to say, "that's he!"'" The antecedent of *qui* is the subj. of *erit*. To point out the criminal in the street is as dangerous for yourself as to accuse him formally in court.

Some edd. supply *ei* as antecedent to *qui*: thus the sentence is equiv. to *reus erit qui dixerit*. This is equally good Latin but less forcible.

verbum, 'a single word': the omission of *unum* is characteristic of the best and most idiomatic Latin: cf. Ter. *Andr.* 860 *verbum si addideris*; Cic. *pro Quinct.* 9; idem, *pro Rosc. Am.* 2; idem, *in Verr.* II. 46.

For *verbum*, P has *vervum*: the substitution of *v* for *b* is one of its commonest errors: thus it gives *ververe* for *verbere* (XV. 21), and *vivat* for *bibat* (XI. 203). The corrector of P corrupted *vervum* to *verum*.

162. The 'haughty Rutulian' is Turnus who is 'matched' with Aeneas in Virgil's epic.

163. **nulli...Achilles**, 'nobody resents the slaughter of Achilles'; *nulli* is used for *nemini* for metrical reasons.

164. **Hylas**, the favourite page of Heracles who fell into a well when drawing water, in the Argonautic expedition: cf. Theocr. 13.

165. Here begins the δέ clause, in opposition to ll. 162 ff.

167. **criminibus**: see n. to l. 75.

tacita...culpa is not co-ordinate with *rubet auditor* but with the clause, *cui...criminibus*: the asyndeton is common even where there is no opposition between the clauses.

168. **inde...lacrimae**: cf. Ter. *Andr.* 126 *hinc illae lacrumae*; a common proverb, which perhaps Terence gave form to.

169. **galeatum**: the *galea* was carried by the soldier and not put on till the enemy were sighted.

170. Thus Juv. gives us plain warning that we must not expect to find in him satire upon living persons. See Introduction, p. xxxiv.

171. **Flaminia**: see n. to l. 61. **Latina**: this was a road branching off from the Via Appia, the great road to the South from Rome.

The Scholiast explains *viae in quibus nobiles sepeliebantur*: i.e. the aristocracy is to be the object of satire. The custom of burying the dead in tombs by the side of roads (cf. VIII. 146) accounts for the fashion, common in Latin epitaphs, of appealing to passers-by; such a phrase as *siste, viator*, though quite natural and appropriate on an ancient tomb, is out of place in a modern churchyard, through which no road passes.

SATIRE III

HOW THE POOR LIVE AT ROME

Some remarks on this Satire, in some respects Juvenal's masterpiece, will be found in the Introduction, p. xl ff. Of many translations or imitations the most famous is Samuel Johnson's 'London, a Poem in Imitation of the Third Satire of Juvenal', published in May 1738, which at once brought fame, but little profit, to its author.

1–20. *Though sorry to lose him, I approve of my friend's purpose to leave Rome and settle at Cumae, a pleasant sea-side place. Any desert would be preferable to Rome with all its dangers of fires, falling houses, and reciting poets. Umbricius stopped at the city gate, while his furniture was being put up on the cart; and then we went down together into the valley of Egeria, thronged with begging Jews, where vulgar magnificence has spoilt the simplicity of the old cavern and spring of water issuing from it.*

1. **quamvis** may mean here 'however much' or 'although': in classical prose it is in general restricted to the former meaning; so *quamvis aeger sit* is correct, but *quamvis mortuus sit* is not, because a man cannot be more or less dead.

confusus, 'upset'.

2. Cumae, on the Campanian coast, was the oldest Greek colony in Italy, founded by Chalcidians from Euboea and Aeolians from Cyme. It had at this time few inhabitants and little importance.

3. The Sibyl, who sold her prophetic books on such strange terms to Tarquinius Superbus, had her cavern at Cumae; and those who live there are said to be her citizens. There is an odd legend about her in Petronius (c. 48): Trimalchio, himself a resident at Cumae, says he once saw her hanging up in a bottle and heard her say to the children who asked her what she wanted 'I want to die'. Dr James (*Classical Review*, VI, 74) has shown that this is a Tithonus-myth: as Tithonus shrunk to a grasshopper, so the Sibyl has shrunk through age to the proportions of an insect and is anxious, like him, to give up her immortality. It is some confirmation of this that Statius (*Silv.* IV. 3. 151) mentions Tithonus and the Sibyl together.

4. Baiae, the Roman Brighton, stood on the peninsula of which Cape Misenum forms the apex; Cumae is called the gate of Baiae, because it stood at the base of the peninsula.

5. **secessus:** gen. of definition, like *vox voluptatis*, 'the word pleasure'. This should not be explained as a gen. of quality; for a noun, to which the latter is attached, seldom takes an adjective, whereas *littus* here has *gratum*: see n. to l. 48. **Prochyta,** now Procida, a small island off Misenum, serves as a type of desolation; the **Subura** was one of the chief streets of Rome, leading eastwards from the forum to the Esquiline gate.

7. **incendia:** fires were of constant occurrence in ancient

Rome. The houses were high, the streets were narrow and
blocked up with booths and shops, mostly made of wood. The
ground-floor had usually no rooms opening on the street, and
no windows; so that these encroachments on the public way,
however inconvenient for traffic and likely to spread fire, were
attached to many of the dwelling-houses. A special corps of
firemen (*vigiles*) was instituted A.D. 6 by Augustus; they were
7,000 in number, commanded by a *praefectus*, and distributed
among the fourteen *regiones* of the city. Rich individuals seem
to have had private fire-brigades: cf. XIV. 305.

lapsus tectorum: cf. ll. 190–6. The jerry-builder saw his
opportunity in ancient Rome, as he always will, where land is
dear and population constantly increasing. On the other hand,
the public buildings of that age still astonish us by their
stability.

9. **Augusto...mense:** cf. II. 70 *sed Iulius ardet*. The month
Quintilis was called Iulius after the reform of the Roman calen-
dar by Caesar 46 B.C., his birthday falling in that month. The
ignorance of the pontifices, who inserted an extra day every
third instead of every fourth year, made a further correction
necessary; and Augustus 8 B.C. gave directions that for the next
twelve years there should be no leap-year. On this occasion he
named Sextilis after himself, preferring it, as the date of his
first consulate and chief victories, to September in which he
was born: cf. Suet. *Aug.* 31.

The terrors of the capital reach their climax in the recitations
of poets. All the literature of the time, esp. Pliny's letters and
Martial's epigrams, shows what a nuisance the verse-writer had
become, and what immense demands he made on the time and
patience of his friends whom he invited (*corrogabat*) to hear
him read his own verses.

10. Umbricius' furniture is not bulky and goes into one cart:
it is carried to the gate by his slaves, carts not being allowed in
the streets by day: see n. to l. 236.

11. **arcus madidamque Capenam:** the *porta Capena* was
a southern gate in the Servian wall, by which the *via Appia* led
to Capua. Over the gate passed an aqueduct, a branch of the
aqua Marcia, a chief source of the Roman water-supply. Drops
of water oozed through the channel of the aqueduct down into
the road below: cf. Mart. III. 47. 1 *Capena grandi porta qua
pluit gutta*.

12. **hic ubi** is used here, as often, especially by Ovid, for
ubi, just as *id quod* is used for *quod*: e.g. Ovid *Fast.* II. 193 *idibus*

agrestis fumant altaria Fauni, | hic ubi discretas insula rumpit aquas. The antecedent here is *illuc* understood, which is resumed at *in vallem Egeriae* l. 17. The whole of the four following lines are in a parenthesis.

This arrangement, which is followed by Friedl., seems the simplest for this difficult passage. Munro explained *hic* 'hereupon' as in l. 21, in which case *ubi* must be separated by a comma from *hic*.

nocturnae, 'by night': time is often expressed in Latin and Greek by an adj., where English prefers an adverbial phrase: e.g. *vespertinus pete tectum*, δωδεκαταῖος ἀνεβίω: see n. to l. 28.

constituebat amicae, 'used to hold assignations with his mistress'; purposely contemptuous; see n. to l. 11: Ovid (*Fast.* IV. 669) and Livy (I. 21. 3) call Egeria the *coniunx* of Numa. For *constituere* in this sense (but without dat.), cf. VI. 487. The legend, which Juv. is ridiculing, told that Numa learned his laws by divine communication from the nymph Egeria by night: cf. Livy I. 19. 4, who suggests that Numa invented the story to give sanction to his enactments.

14. Cf. VI. 542. Friedl., following a Scholium on this passage, explains that the 'basket of hay' was indispensable to every Jewish household, because it served to keep their food warm for the Sabbath, on which day the use of fire was forbidden. A hay-box could do this. Yet, as the Jews were beggars, they might be content with a cold meal one day in the week. So perhaps the hay was used merely for bedding, and the basket for scraps. (The Schol. may be based upon Juv. XI. 70 *tortoque calentia faeno | ova*.)

After the destruction of the temple at Jerusalem A.D. 70, the number of Jews in Rome must have largely increased. Most of them were utterly destitute and lived by begging and fortune-telling. The passage shows that for some small payment they were allowed to settle in woods close to the walls.

15. **mercedem**, 'rent'.

16. Outside the *porta Capena*, on the left, there was a wooded valley with a perpetual spring in the middle, gushing out from a dark cavern. The grove was sacred to the Camenae, of whom Egeria was the most famous. Juv. complains that the ancient Latin deities have been evicted to make room for the beggar women of Palestine: *Musis* would not have the same force.

18. **dissimiles veris**, sc. *speluncis*: 'unlike nature', 'travestied'. This 'ancient monument', though surrounded by

disreputable tenants, had been 'restored' with a magnificence that displeased Juv.'s taste.

20. **ingenuum...tofum,** 'native limestone', is contrasted with *marmora*; cf. XIV. 89 *Graecis longeque petitis | marmoribus.* The quarries of Luna (Carrara) in north Italy were worked in ancient times; but Roman taste preferred the variegated marbles of the Peloponnese, Africa, and Asia Minor.

violarent, 'insulted'; cf. XI. 116.

21–57. *Umbricius then began:—I am leaving Rome, because it is no longer a place for an honest man to live in. Let those live there who have no inconvenient scruples, who can carry on any degrading occupation, and then make a fraudulent bankruptcy; such pursuits are impossible for me. I neither can nor will earn the friendship of the great by sharing in their guilty secrets; all the gold in the world is not worth having, if you must pay for it with your peace of mind.*

21. **hic,** 'hereupon'.

23. **here:** *heri* is the older form, obsolete in Quintilian's time: cf. 1. 7. 22 *here nunc E littera terminamus, at veterum comicorum adhuc libris invenio 'heri ad me venit'.*

eadem...aliquid, 'and, again to-morrow, will rub away (i.e. lose) something from the little left'; *res* is the subject, *eadem* is epithet of *res, cras* an adverb. Cf. XVI. 50.

25. **Daedalus:** cf. Virg. *Aen.* VI. 14 *Daedalus, ut fama est, fugiens Minoia regna, | ...insuetum per iter gelidas enavit ad Arctos, | Chalcidicaque levis tandem superadstitit arce,* i.e. at Cumae. The roundabout description of Cumae is to be noticed: Juv. has a great liking for describing places and persons by a periphrasis giving some historical or mythological details about them. Cf. the periphrasis for Tarsus l. 117, for Aeneas V. 45, for Palestine VI. 159, for Pluto X. 112, for Alba XII. 70, and many others.

27. **dum...torqueat,** i.e. while I have some life left: when the Fate has spun all the wool, the man's life is over: cf. Mart. I. 88. 9 *cum mihi supremos Lachesis perneverit annos* (= *cum mortuus ero*).

29. Of Artorius and Catulus nothing is known.

31. This probably refers to the *curatores riparum et alvei Tiberis et cloacarum urbis:* the *curatores* were senators but would employ contractors who might cheat the government.

32. The cleansing of sewers was generally done by convicts; it is described by Trajan (*ad Plin.* 32. 2) as one of the *ministeria*

quae non longe a poena sunt; and we see that discredit fell also on the contractors (*mancipes, redemptores*) who had this work done.

A Roman funeral was generally performed by contract, in many cases by the *Libitinarii*, a guild of undertakers who had their headquarters at the temple of Libitina. This calling also was discreditable, which seems less reasonable.

33. 'And to be sold up beneath the spear of ownership'; this applies to all the contracts above-mentioned: they accept each and all of them, embezzle the money, and make a fraudulent bankruptcy: cf. *cedere foro* XI. 50, and Plaut. *Persa*, 435 (of the bankers of the day) *ubi quid credideris, citius extemplo foro | fugiunt, quam ex sporta* (Palmer, *porta* MSS.) *ludis quom emissust lepus.*

domina…hasta: a spear, as the symbol of conquest and ownership, was stuck up at state-auctions: in Italian *asta* means 'an auction', not 'a spear': we speak of 'the hammer'. An auction of this kind, i.e. a bankrupt sale where the state sold the bankrupt's property, was called *sectio*; and at it, the *hasta* took the place of the *aes et libra* of an ordinary sale. The *hasta* is called *domina*, because the purchasers at a *sectio* acquired legal ownership (*dominium*), just as if they had bought the property *per aes et libram*: cf. Sen. *Controv.* II. 1. 1 *limina… sub domino sectore venalia.*

The 'spear' was also stuck up in the chief civil court at Rome, the *centumviri*, which for this reason is often called *hasta*, e.g. Stat. *Silv.* IV. 4. 43 *centeni moderatrix iudicis hasta*; and in the praetor's court it was represented by the *festuca* used in manumissions.

Others explain: 'and to sell slaves by auction': the trade of *praeco* (cf. l. 157) or *mango*, though despised, comes easy to these adventurers. But *praebere* (cf. v. 172) implies personal disgrace, and cannot be used for *vendere*. Also, this seems to have less point, as a climax is required, and the business of a *praeco*, though not distinguished, was not in itself discreditable.

34. **cornicines:** gladiators are often represented in works of art as fighting to the music of horns.

35. **comites:** the word shows that these men were attached to a *travelling* company of gladiators, that went round the *municipia*.

notaeque…buccae, 'whose swollen cheeks were a common sight throughout the provinces': their cheeks are swollen, when they blow the horn. For *buccae*, applied to persons, cf. XI. 34.

36. **munus edere** (or *dare*) and *dare gladiatores* are the two phrases used by good writers in this sense; the word *munus* suggests the original purpose of gladiatorial shows, to do honour to the dead at funeral games. The same disgust at these shows given by upstarts is expressed by Mart. III. 59 *sutor cerdo dedit tibi, culta Bononia, munus;* | *fullo dedit Mutinae; nunc ubi caupo dabit?*

verso pollice: when a gladiator was wounded and at his antagonist's mercy, the spectators gave the sign for the death-stroke by this gesture; whereas *premere pollicem* was a sign of good-will. It is generally believed that the former gesture was to turn the thumb up towards the breast in imitation of the fatal weapon; and that *premere pollicem* was to turn the thumb downwards towards the ground, a suggestion that the weapon should be thrown down. For this view, see Wilkins on Hor. *Epp.* I. 18. 66. Friedl. supposes that the thumb was turned down as a sentence of death, relying on the evidence of an ancient relief, where the spectators' thumbs are turned up and the inscription shows that the gladiators were spared. When the people wished a brave man to be spared (*missionem flagitabant*), they also waved their handkerchiefs; cf. Mart. XII. 29. 7 *nuper cum Myrino peteretur missio laeso,* | *subduxit mappas quattuor Hermogenes.*

38. **cur non omnia?** sc. *faciant*, 'why should they refrain from anything?' Fortune, for her own amusement, has exalted nobodies; why then should the nobodies regard any employment as degrading for them?

41. **faciam,** 'am *I* to do': the emphatic pronoun is required in English, where the persons are not distinguished by inflections of the verb.

mentiri nescio: cf. Petron. 116 *sin autem...sustinetis semper mentiri, recta ad lucrum curritis.*

42. **motus astrorum:** I am no astrologer, says U., and cannot, as they do, gladden the expectant heir by an assurance of his father's speedy death.

44. **ranarum viscera** refers either to poisoning with the *rubeta* or to divination by opening the bodies of animals (*extispicium*). This latter purpose of the frog is not mentioned elsewhere; but cf. VI. 551.

45. **quae mittit,** 'the presents', **quae mandat,** 'the messages': the strength of English lies in its nouns, of Latin in its verbs.

The use of *mittere* even without a dat., 'to give away', is peculiar to this period: cf. Mart. quoted on 1. 28, idem, VI. 75. 4 (to a lady who sends him dainties which he suspects of being poisoned) *has ego non mittam, Pontia, sed nec edam*; idem, VII. 78. 3 *sumen, aprum, leporem, boletos, ostrea, mullos | mittis*.

47. As I refuse to wink at the thefts of a governor in his province, therefore no governor takes me out as a member of his household (*comes*). For the meaning of **fur**, see n. to l. 53.

48. The subject (*homo* understood) has two epithets, **mancus** and **extinctae dextrae** (= *debilis*); the gen. of quality is used as a round-about adj., cf. XI. 96; **corpus non utile** is in apposition with the subject.

Strictly used, the gen. of quality should be preceded by a class-noun such as *homo, animal*, etc.: but even Horace has such phrases as *multi Damalis meri, magni formica laboris*, where *femina* and *animal* are omitted as too prosaic. Silver-age Latin dispenses with a noun altogether, as here: cf. Pliny *Epp.* III. 5. 8 *erat somni paratissimi*, with Mayor's note.

The honest man is powerless because of his isolation, ὥσπερ ἄζυξ ἐν πεττοῖς. Cf. Aristot. *Pol.* 1253 a 7.

49. **conscius,** 'the sharer in a guilty secret': cf. Mart. VI. 50. 5 *vis fieri dives, Bithynice? conscius esto*.

cui is probably a pyrrhic (two short syllables); again VII. 211; Martial scans it thus four times (I. 104. 22, VIII. 52. 3, XI. 72. 2, XII. 49. 3), always in the second foot of a hendecasyllable.

51, 52. The gifts of the rich to their friends are in reality black-mail.

52. *secreti:* a noun; cf. l. 113; VI. 190.

53. **Verres,** infamous for his extortions as propraetor in Sicily, is used as the type of a robber; cf. II. 24 *quis tulerit Gracchos de seditione querentes? | quis caelum terris non misceat et mare caelo, | si fur displiceat Verri* (i.e. *furi*), *homicida Miloni?*

54. **tanti...non sit...harena Tagi...ut somno careas,** 'do not regard the sand of Tagus as so great a prize, that you should be willing to lose your sleep for it'. The constr. *tanti est ut*...goes through a peculiar development. In its original form, which we have here, the subject of the *tanti est* clause is the *prize we wish to gain*, while the *ut* clause expresses the *price we have to pay*; here the prize is wealth, and the price paid for it is peace of mind. Cf. X. 97 *quae praeclara et prospera tanti, | ut rebus laetis par sit mensura malorum?* It must be observed that in this form of the constr., the *ut* clause is elliptical, and in English we must supply the notion of 'being willing': *ut somno careas* is elliptical for *ut somno carere velis*; and *ut rebus laetis* etc.

is elliptical for *ut propter ea parem quis esse velit mensuram malorum*. Cf. also Ovid *Met.* x. 618 *tanti...putat conubia nostra | ut pereat*, 'he thinks a marriage with us so great a prize that he *is willing* to pay for it with his life', where *ut pereat* is elliptical for *ut perire velit*.

In the second form of the constr., the meaning of the two clauses is exactly reversed; for the subject of the *tanti est* clause represents the price to be paid, while the *ut* clause expresses the prize to be gained. The change of meaning is probably due to the use of *tanti est* without any expressed subject, and without an *ut* clause: in such cases the true meaning of the phrase became obscured. The second form, or a part of it, occurs once in Juv. XIII. 95 *et phthisis et vomicae putres et dimidium crus | sunt tanti*, sc. *ut teneam nummos*, 'disease and deformity are a price worth paying for the prize of keeping the money unlawfully'. The *ut* is often represented by *dum modo*, and the *ut* clause is often suppressed altogether.

The distinction will be clearly seen by an epigram from Martial (VIII. 69):

> *miraris veteres, Vacerra, solos,*
> *nec laudas nisi mortuos poetas.*
> *ignoscas petimus, Vacerra; tanti*
> *non est, ut placeam tibi, perire.*

This is an instance of the reversed constr.: 'death is too high a price to pay for the prize of your applause'. But Martial might have used the original form with precisely the same meaning, the clauses being reversed:

> *tanti*
> *non est, ut peream, tibi placere.*

where *ut peream* would be elliptical for *ut perire velim*: 'your applause is not so high a prize, that for it I am willing to die' (see Madv. *Opusc.* II. 187–95).

Both constructions occur in Cicero: for the first, cf. *de Off.* III. 82, for the second, *in Cat.* II. 15. Both are very common in Ovid and all later writers, particularly Seneca: 'I almost think it was Seneca who crystallized the everlasting *tanti* of the Silver Age' (Heitland, *Pharsalia*, p. cxxix). Horace seems to have avoided it; or, for *pretium aetas altera sordet* (*Epp.* I. 18. 18), he would surely have used *non aetas altera tanti est*.

opaci, 'shaded by trees': Martial, himself a Spaniard, speaks of the shady banks of the Tagus, I. 49. 15 *aestus...aureo*

franges (= *mitigabis*) *Tago* | *obscurus umbris arborum*; cf. also
Stat. *Silv.* II. 3. 1 *opacat* | *arbor aquas*; Virg. *Aen.* VI. 195
opacat | *ramus humum. opaci* could not by itself mean 'thick
with gold', though Sen. *Herc. F.* 1332 has *Tagusve Hibera
turbidus gaza fluens.*

55. **harena** and **aurum** may be translated as a hendiadys:
'all the golden sand that wooded Tagus rolls down into the
sea'. The deposits of gold in the bed of the Tagus were pro-
verbial: cf. XIV. 299: *aureus, aurifer, dives* are some of Martial's
epithets for the river.

56. Sleep is the consequence and reward of a good con-
science. **ponenda,** 'which you must soon resign': the gerun-
dive here serves as a fut. pass. participle; it serves as a pres. pass.
participle in such phrases as *studium evertendae reipublicae.*

57. **magno...amico:** see n. to I. 33.

58–125. *One of my chief reasons for leaving Rome is the crowds of
Greeks who infest it. And yet the Greeks are not the worst: all
the refuse of Asia flows together here. The simple Roman apes
but awkwardly the dress and manners of the versatile and
unscrupulous Greek. It is insulting to the true-born Roman to
give place to them on every occasion. Their most barefaced
flattery carries conviction; in private life they are consummate
actors who outdo the heroes of the stage. Even their philosophers
turn informers and betray their friends and disciples to ruin.
The native Roman, however long and faithful his service to his
patron, is speedily outsted by the intrigues of a Greek.*

58. Greeks were numerous at Rome, where many of them
filled higher positions than that of flatterer and parasite. In the
arts and sciences, particularly in medicine, they distanced all
rivals; but the genuine Roman looked with no friendly eye even
on their medical skill: Cato the censor, in a treatise addressed to
his son, quoted by Pliny *Nat. Hist.* XXIX. 14, says of their doc-
tors: *iurarunt inter se barbaros necare omnes medicina.* Squire
Western might have talked thus of the French; but it is strange
to hear such a monstrous fiction quoted with approval by Pliny.

60. **Quirites** is specially appropriate as the domestic name
of the Romans, applied by themselves to one another; to foreig-
ners they were *Romani.* The Scholiast here quotes *porro,
Quirites!* the street-cry of porters, corresponding to our 'by
your leave!'

61. **Graecam urbem,** 'a Greek Rome': *urbem* would be
τὴν πόλιν in Greek.

quamvis, 'and yet': in classical prose *quamquam*, not *quamvis*, serves, like καίτοι, to begin a correction of a previous statement.

Achaei is subject to *sunt* understood, *quota portio* is the predicate. For a possible explanation of *Achaeae*, the reading of P, see Introduction, p. xlviii. Achaia was the name given to the Roman province of Greece, including all the Peloponnese with most of Hellas proper and the adjacent islands; hence *Achaei* stands for Greeks generally.

62. By the mixing of the rivers the mixing of the populations is meant. Orontes is the chief river of Syria.

63. **chordas obliquas:** the *sambuca*, a kind of harp, is the instrument meant: for a figure, see Rich's *Companion*.

66. **ite,** sc. *ad Circum.*

67, 68. The Greek words and Greek customs, current in Roman society, are ridiculed.

The force of the passage will be lost if the Greek words are translated into English: 'his ceromatic neck' may sound odd, but that is exactly the effect Juv. wishes to produce.

ille: the pronoun throws the mind back to the beginning of Roman history, when no foreign influences had made themselves felt. **trechedipna** are probably a sort of shoe.

68. **ceromatico:** κήρωμα was a mixture of oil, earth, and wax which wrestlers rubbed on their limbs. The sports and rewards of the *palaestra* were always despised by the Romans: cf. Trajan *ad Plin.* 40. 2 *gymnasiis indulgent Graeculi*, where the diminutive suggests that the Greeks are poor creatures for doing so; and see n. to l. 115.

69, 70. The Greeks swarm from all parts of their country to Rome: Sicyon, an important city of Argolis; Amydon, a town in Macedonia; Andros and Samos, islands of the Aegean; Tralles and Alabanda, cities of Caria in Asia Minor.

70. **Samo:** for the hiatus, see n. to x. 281.

71. **Esquilias:** there were many houses of the great on the Esquiline; the *collis Viminalis*, also in the east of the city, was the site of the house of C. Aquilius, the most splendid in Rome about 100 B.C., eclipsing even the palaces of Crassus the orator and Q. Catulus on the Palatine (Pliny *Nat. Hist.* XVII. 2). The periphrasis, in order to avoid the unmetrical Vīmĭnālis, should be noticed.

72. **viscera,** 'the cherished inmates': the word, when used of persons, is usually applied by parents to children.

domini, 'owners', is nom. plur.

73. **perdita**, 'desperate'; more often applied to persons than things: so v. 130.

74. **Isaeo** = *sermone Isaei*; this elliptical form of comparison is common both in Greek and Latin; cf. Cic. *de Or.* I. 197 *si cum...Lycurgo et Dracone et Solone nostras leges conferre volueritis*, where *cum Lycurgo* stands for *cum legibus Lycurgi*.

Isaeus is not the Attic orator of that name, but an Assyrian rhetorician who came to Rome about A.D. 97 and made a great sensation by his eloquence; cf. Pliny *Epp.* II. 3.

ede, 'say'; cf. I. 21. **illum** is any Greek.

75. **secum**, 'in his own person'.

76. **grammaticus, rhetor:** see nn. to I. 15 and 16, VI. 450, VII. 230.

geometres is the Greek γεωμέτρης: the two first vowels probably coalesce here to make one long syllable.

77. **schoenobates:** there is a good Latin word of the same meaning, *funambulus*; but Juv. is purposely using Greek names wherever he can.

medicus: see n. to l. 58.

78. **iusseris** is the protasis of a conditional sentence; cf. VI. 526; *si* is understood, as it can be when the protasis consists of only one or two words; cf. Virg. *Aen.* VI. 31 *partem, sineret dolor, Icare, haberes*. For another case where *si* may be left unexpressed, see n. to l. 100.

This line suggests the exploit of Daedalus which follows in the next.

79. **in summa**, 'to put it shortly': i.e. a single example will save more talking; a favourite phrase with the younger Pliny. Cf. the use of *post* at the end of a long description, Plaut. *Miles*, 648 *post, Ephesi sum natus, noenum in Apulis, noenum Aminulae.*

80. In the legend Athens was the birth-place of Daedalus.

82. **signabit:** a Roman invited his friends to append their seals to various legal documents: (1) to wills: there is extant a parody of a formal Roman will, purporting to be the last testament of a pig, M. Grunnius ('grunter') Corocotta (Bücheler's Petronius, 1882, p. 241); this is attested by seven friends with suitable names such as *Lucanicus*, 'sausage'; each name is followed by *signavit*. (2) to marriage-contracts: cf. X. 336. (3) to manumission of slaves: cf. Mart. IX. 87. 3 *affers nescio quas mihi tabellas | et dicis 'modo liberum esse iussi | Nastam— servolus est mihi paternus— | signa'.* In all these cases the most important persons would affix their seal first.

fultusque toro meliore, 'reclining on a better couch'. For the arrangement of the *triclinium* with its three *lecti* with three places on each, see Hor. *Sat.* II. 8. 20 and Palmer's note there. By the end of the first century this arrangement was often modified: the table was round (see n. to *mensas* I. 75) instead of square, and the guests commonly reclined on a couch in the shape of a horse-shoe called *sigma* (from the capital C) or *stibadium*; cf. Mart. XIV. 87 *stibadia.*

> *accipe lunata scriptum testudine sigma;*
>   *octo capit; veniat quisquis amicus erit.*

idem, x. 48. 6 *septem sigma capit.* This was no doubt more convenient, as the guests had no longer to be a multiple of three. The places of honour on the *sigma* were the two ends (*cornua*). One side of the table, whether square or round, was always unoccupied for the convenience of service; and the habit, which seems to us uncomfortable, of reclining on the left elbow was still kept up.

83. **pruna** were largely imported from Damascus and were called *Damascena* (our 'damsons'); cf. Mart. XIII. 29 *vas damascenorum.*

> *pruna peregrinae carie rugosa senectae*
>   *sume: solent duri solvere ventris onus.*

cottona were a small kind of fig: cf. Mart. XIII. 28 *vas cottonorum.*

> *haec tibi quae torta venerunt condita meta* (drum)
>   *si maiora forent cottona, ficus erant.*

Juv. means that the orientals are *municipes* (cf. IV. 33) of these imported groceries.

84. **usque...est,** 'is it such a mere nothing'; see n. to XIII. 183. **nostra** is not equivalent to *mea*, as Juv. was not a Roman born.

85. **hausit,** 'drew in'; cf. Quint. VI, prooem. 12 *auram communem haurire amplius potui?*

The **baca Sabina** is the olive. To us the olive is a mere relish: to the ancients, as still, in a great measure, to the nations of southern Europe, the oil from this berry was one of the necessaries of life, being used wherever we use butter or animal fats. Butter was used only for medicinal purposes by the ancients, and generally applied externally; it was so little used that there was no Latin name for it, the Greek *butyrum* being found sufficient. We should consider the lemon and orange to be more characteristic of Italy than the olive; but the lemon was hardly known in Italy in the first century, and the orange was not naturalised there till the time of the crusades.

86. quid quod..., 'besides', 'moreover'; lit. 'what of the fact that?' again l. 147; the constr. is found in Horace, *Carm.* II. 18. 23, common in Ovid (e.g. *Met.* V. 528, IX. 595, X. 616) and in all silver-age writers.

adulandi...prudentissima, 'past masters in flattery'.

88. cervicibus, 'the brawny neck'; the word suggests power of endurance. The sing. *cervix* is not used in correct prose, but often in poetry: cf. X. 345.

89. Antaeus, an African giant, was invincible so long as he remained in contact with his mother earth; but Hercules held him off the ground and then strangled him. Juv. may often have seen representations of this feat in sculpture or painting.

90. He admires the squeaky voice of a friend, though no more euphonious than the crowing of a cock.

nec, 'not even', used for *ne...quidem*; cf. II. 149.

> *esse aliquos manes et subterranea regna...*
> *nec pueri credunt nisi qui nondum aere lavantur.*

'not even boys believe, except those who are too young to pay at the baths'. For another sense of *nec = ne quidem*, see n. to XI. 7. *ne...quidem* is obviously inconvenient in poetry; but even prose-writers after the Augustan period constantly use *nec* for it in both senses; perhaps from the influence of Greek analogy. See Madvig's third Excursus to Cic. *de Fin.*

91. quo...marito, 'the husband by whom the hen is pecked', a roundabout phrase for *gallus. quo* is instrumental: see n. to I. 13: the lower animals are seldom regarded as agents in Latin. **mordetur** refers to the cock's way of holding the hen by the crest. *marito* is out of place in the relative clause; the sentence re-arranged and fully expressed would run *qua deterius sonat ne mariti quidem vox illius, quo mordetur gallina.*

92. 'We Romans also may praise these same defects'; but we don't do it so naturally.

93–100. Actors may be so good that they can act the part even of a woman to the life; but among Greeks nothing would be thought of them, as all are equally good actors there. Madvig was the first to explain this passage correctly (*Opusc.* I. 50 ff.).

93. an melior, sc. *comoedus est Graecis?* i.e. any Greek can

beat an actor in his own business. For the position of the main
subject in the dependent clause, cf. *fortuna* l. 40, *potestas* IV. 71.
Women's parts in the *palliatae* are of three kinds, which are all
mentioned here: (1) *meretrices*; of these *Thais* was a notorious
representative; she supplied the title for the master-piece of
Menander; (2) *matronae*; (3) *ancillae*; these are represented by
Doris, a natural name for a Greek slave. She has work to do and
therefore discards the *pallium* or upper-garment, just as the
lower class of Romans (*tunicatus popellus* of Horace) discarded
the *toga* when hard at work.

95. **palliolo:** the diminutive is used for metrical conven-
ience; cf. Mart. IX. 32. 1 *hanc volo quae facilis, quae palliolata
vagatur*; and see n. to *flammeolo* Juv. X. 334.

nempe, 'it is true that...'.

Women's parts were played by male actors in Greece:
Gellius (VI. 5) tells how Polus, a famous actor, when playing the
part of Electra in Sophocles' play, brought on the stage an urn
holding the ashes of his own son. At Rome the *mimus*, a short
character-sketch from common life, was the only theatrical
performance in which women took part. In England, women's
parts were played by men until the Restoration; so that Shake-
speare's Viola and Rosalind were boys playing the part of girls
disguised as boys.

98. Antiochus, Haemus, Stratocles, and Demetrius were all
actors of *palliatae*; the two last had a warm admirer in Quintilian,
who gives a detailed criticism of their acting (XI. 3. 178);
Demetrius, he says, was excellent in married women and strict
old ladies as well as in other parts.

illic, 'in Greece': the reason follows in the next clause.

100. **rides:** what might have been stated as the protasis of a
conditional sentence, is stated as a separate fact: cf. XIII. 227.
Similar clauses in Greek are often introduced by καὶ δή: Eur.
Med. 386 καὶ δὴ τεθνᾶσι· τίς με δέξεται πόλις;

102. **nec dolet,** 'and yet he feels no pain': cf. XIII. 114 *nec
labra moves*. The use of *nec = nec tamen* is peculiar to silver-
age Latin; see n. to I. 93.

poscas: the potential subj. is used in the 2nd person where
we might say, 'if one asks'; the mood is due entirely to this
generic notion: if *amicus* were the subject, the verb would be
poscit. See Munro on Lucr. I. 327.

103. **endromidem:** a heavy woollen garment; see n. to VI. 246.

104. Cf. Mart. II. 18. 2 *iam sumus ergo pares*.

melior, 'he is my superior': cf. l. 93.

105. **aliena...facie:** so Statius says of a devoted atten-
dant, *Silv.* II. 6. 52 *tecum tristisque hilarisque nec unquam | ille
suus, vultumque tuo sumebat ab ore*: contrast with this what
Martial says to a friend whom he is inviting to dine with him
(v. 78. 24) *et vultu placidus tuo recumbes,* 'you will sit at ease
with your natural expression'.

106. **iactare manus,** 'to throw up the hands', as a sign of
admiration; cf. Mart. X. 10. 10 *geminas tendis in ora manus* (of a
flatterer listening to a recitation).

107. This is imitated by Amm. Marcell. XXVII. 3. 5 *homo in-
dignanter admodum sustinens, si, etiam cum spueret, non laudaretur.*

108. 'if the golden cup, when the bottom is turned upwards,
has given forth a smacking sound'; i.e. if the rich man has
drunk off his glass with no heel-taps. Cf. *Ham.* I. 4:

> And, as he drains his draughts of Rhenish down,
> The kettle-drum and trumpet thus bray out
> The triumph of his pledge.

crepitus is the gurgling sound with which the last drops leave
the cup. Another explanation, which makes **trulla** = *lasanum*,
suits the context better and gives a better climax; but *trulla* is
unexampled in this sense and regularly means a vessel for
drinking out of; cf. Hor. *Sat.* II. 3. 144; Mart. IX. 96. 1; Cic.
in Verr. II. 4. 62. Others think that a kind of κότταβος is meant, a
game of which the object was to throw a small quantity of wine
at a mark; but (1) κότταβος is never mentioned as a Roman game,
(2) *fundus* alone cannot stand for 'cup', nor *trulla* for the πλάστιγξ
or saucer into which the wine was thrown.

113. **inde,** 'in consequence'; cf. VI. 139.

114. **coepit** is a perf. of which *incipio* supplies the present.
In Cicero it is always followed by the inf., which must be active
(except in the case of *fieri* which is regarded as middle); but in
later prose and in poetry the constr. without the inf. is common;
cf. Lucr. v. 1416 *sic odium coepit glandis.*

transi, 'pass by, say nothing of': *transire* has only this mean-
ing in Juv.; cf. VI. 602; VII. 190; X. 273.

115. **gymnasia,** 'the [vices of the] wrestling-schools': the
παλαίστρα was an institution characteristic of Greece and never
popular with Romans, who thought this form of exercise un-
practical, indecent (the combatants being naked), and sure to
produce an indolent type of character: see n. to l. 68. Cf. Tac.
Ann. XIV. 20 *ut...degeneret studiis externis iuventus, gymnasia
et otia et turpes amores exercendo.*

maioris abollae, 'of a greater personage'; the phrase seems
proverbial, like 'big-wig' or *gros bonnet*. Others explain 'of
stouter texture', i.e. a robuster crime. Others take *abollae* as =
philosophi; but that dress was not confined to philosophers: it is
worn by Pegasus (IV. 76), and by such a fop and voluptuary as
Crispinus (Mart. VIII. 48).

116. P. Egnatius Celer, born at Berytus in Phoenicia and
perhaps educated at Tarsus (but see n. to *nutritus* l. 117),
professed the Stoic philosophy, and in that capacity, as guide,
philosopher, and friend, was an inmate in the house of Barea
Soranus, a Roman aristocrat. Under Nero A.D. 66 Barea was
accused with his daughter Servilia and condemned under the
law of *maiestas*; the evidence against him was supplied by
Egnatius. The universal detestation felt for the traitor was
satisfied A.D. 70 when he was put to death under Vespasian:
cf. Tac. *Ann.* XVI. 30–2, *Hist.* IV. 10 and 40.

delator, 'informer'; and often prosecutor too; cf. I. 33;
IV. 48; X. 70. The power of these men and the terror they in-
spired were among the worst features of the imperial rule. The
term does not occur in Augustan literature; but the practice
began under Augustus, who by the *lex Iulia* and the *lex Papia
Poppaea*, restricted the power of inheritance of unmarried
persons, and granted rewards to informers (Tac. *Ann.* III. 28;
Suet. *Nero*, 10) *qui nomen alicuius* deferebant *ad aerarium*; the
treasury then claimed the inheritance. This mattered little; but
at some unknown period, probably when Tiberius revived the
law of treason (Tac. *Ann.* I. 72), similar provisions were added
to the *lex Iulia de maiestate*; so that the informer, who brought
a successful charge of treason against any subject, was rewarded
with a fourth part of his victim's property (Tac. *Ann.* IV. 20).
The natural result followed: informers swarmed, and the most
trifling act or word, which could be construed as disrespect for
the monarch, passed for treason. All emperors, however, did
not treat them alike: while Tiberius gave them needless en-
couragement and called them the guardians of the laws (Tac.
Ann. IV. 30), Titus had them flogged in the amphitheatre and
then exiled or sold as slaves (Suet. *Tit.* 8); and both Nerva and
Trajan did much to suppress them (Pliny, *Paneg.* 34 and 35).

117. **discipulum:** Ritter proposed *discipulam*, i.e. Servilia
the young daughter of Barea, because of the narrative of Tacitus
(*Ann.* XVI. 30–2), where Barea is called *grandis aevo*. But *disci-
pulam* spoils the rhetorical climax; while *discipulum* is not really
inconsistent with Tacitus, as this word and *senex* only serve to

mark the relation between the pupil and the grave philosophic teacher.

ripa...caballi, i.e. at Tarsus on the river Cydnus in Cilicia (for the roundabout description, see n. to l. 25), a famous seat of learning, which was supposed to owe its name to a ταρσός (feather or hoof) of Pegasus which fell there. This is the Scholiast's explanation; but the natural meaning of **nutritus** is 'born', not 'educated': and this can be explained. One legend tells that Perseus alighted with Andromeda on the coast at Iope (now Jaffa), south of Berytus. Thus there may be no allusion to Tarsus here, *ripa* being used for *littore*, and *pinna caballi* for *pinnatus caballus*.

118. **Gorgonei...caballi:** Pegasus was said to have sprung from Medusa's blood when her head was cut off by Perseus. *caballi* is a purposely disrespectful term for the winged steed; cf. Pers. *prol.* 1 *fonte...caballino*, and see n. to *pelliculae* 1. 11.

120. These are all Greek names, but it does not appear there is any reason for choosing them in particular.

121. If a Greek makes a friend, he cannot be happy until he has poisoned the friend's mind against everyone else: he can bear no rival near the throne. This seems to be an imitation of Curio's address to Caesar: Lucan, 1. 290 *partiri non potes orbem, | solus habere potes.*

122. **solus habet,** '*but* monopolises him': the asyndeton is not tolerable in English.

124' **perierunt,** 'has gone for nothing', 'is wasted': the slave has been a slave for nothing (*perdidit operam*): see n. to *pereat* IV. 56.

125. At Rome, more than elsewhere, no one thinks twice of a dependant.

126–89. *Apart from the Greeks, the poor at Rome are sadly handicapped, in the race for the favour of the great, by the competition of the great themselves. Wealth takes rank of birth; wealth is preferred to character. Worse than that, poverty makes men ridiculous. Poor men must give place in the theatre to the rich, however obscure or infamous their birth. For poor men Rome is the worst place in the world: they ought long ago to have left it in a body for the provinces, where expenses are fewer and life is simpler. There, even on a holiday, the chief magistrates do not don the toga, but are content with a clean tunic. At Rome everything costs dear: you pay for a nod of recognition from a great man; you are even taxed for the benefit of his slaves.*

126. **porro,** 'further': a common particle of transition; Juv. has done with Greeks. **ne nobis** (i.e. *Romanis*) **blandiar,** i.e. not to lay all the blame on the Greeks.

127. **hic** = *Romae*, as repeatedly throughout the satire: cf. ll. 160, 180, 182, 232, but not l. 226. **si,** 'even supposing'.

nocte, 'before dawn'; there are constant complaints in Martial of the early rising required by the *salutatio.*

togatus: see n. to I. 96, and cf. Mart. X. 82. 2 *mane vel a media nocte togatus ero.*

128. **cum,** 'seeing that', not 'when'.

129. **orbis** is feminine as appears by the next line: Albina and Modia are unknown. Men, as well as women, if they were rich and childless, were courted by fortune-hunters (*captatores*) with a shamelessness and persistency which are constantly attacked by the satirists and indeed by all the writers of the age: see n. to IV. 19. Where there is a system of domestic slavery, marriage is always infrequent; and where marriage is infrequent, legacy-hunting will always be found; but it can seldom have grown to such a height as in the first century A.D. The stories told by Pliny of M. Aquilius Regulus, a man of senatorial rank (Pliny *Epp.* II. 20; IV. 2. 2) would pass belief, if the whole literature of the time did not bear witness to their truth. The practice was common in Horace's time: cf. *Epp.* I. I. 77 *sunt qui | frustis et pomis viduas venentur avaras, | excipiantque senes quos in vivaria mittant.*

130. **salutet:** the first two hours of the day at Rome were devoted to this social duty; the callers had to wait in the *vestibulum* until the *atrium* was thrown open; but they were often disappointed of seeing the great man, who might be disinclined to rise or might have gone off to pay calls himself. The explanation of this custom is to be found partly in the unwillingness of the humbler Romans under the empire to gain a small but honest living by trade. Though the old names of 'patron' and 'client' were retained, the relationship was entirely changed: and both sides tried to give as little and get as much as they could; the patron got some social importance from a swarm of retainers who received in return a small sum as daily wages.

131. **servo,** the reading of P, is obviously more forcible than *servi* and gives a better contrast to *ingenuorum filius*; cf. Hor. *Sat.* II. 5. 18 *utne tegam spurco Damae latus?* Suet. *Claud.* 24 *revertenti latus texit*; it will be noticed that the dat. (as *servo* here) is used in both instances.

cludit latus (= *tegit latus*) is explained by the Scholiast *in* *sinistra ambulat*; so we see that the inferior walked, as a mark of respect, on the left hand of his superior—a survival no doubt of the times when the great warrior did execution with his right arm, while his humbler friends protected him from assault on the shield-side; cf. the use of *tegumen* for 'shield'.

132. **alter,** 'the other', i.e. the slave's son. Cf. the French *l'autre*. The stricter use of *alter* is seen l. 149 as 'one of a pair'.

The pay of a *tribunus militum* at this time is not known, but was evidently considerable; as Pliny (*Nat. Hist.* XXXIV. 11) protests against buying a candelabrum for such a sum, and just below mentions a bronze candelabrum, with a very ugly slave thrown in, sold for 50,000 sesterces; in the 3rd cent. A.D. their pay was 25,000 sesterces (about £250).

133. His mistress is a lady of rank. Calvina may be identical with the Junia Calvina of Tac. *Ann.* XII. 4.

137. **da,** 'produce', = *si dabis*: the apodosis begins at *protinus*, l. 140.

hospes numinis Idaei: cf. XI. 194: when the image of Cybele, the *Magna Mater*, was brought from Pessinus in Phrygia to Rome 204 B.C., P. Cornelius Scipio Nasica was chosen by the senate, on account of his stainless virtue, to convey the image from the ship to the matrons who were to guard it: Livy, XXIX. 11.

139. When the temple of Vesta was burnt 241 B.C., L. Caecilius Metellus rescued the Palladium, or image of Minerva, from the fire. He lost his sight in doing so, and afterward added to his name the cognomen Caecus; cf. VI. 265.

140. *ad censum*, sc. **itur**: we go straight to the point, i.e. his income.

141. **pascit** is used because slaves received little more than their food: Seneca (*Epp.* LXXX. 7) mentions 5 *denarii* (about 3s. 4d.) and 5 *modii* of wheat as a normal monthly wage; cf. l. 167 and XIV. 126 *servorum ventres*.

142. **paropsis,** originally 'a side-dish', is used here and elsewhere simply as 'a dish'.

144. **tantum...fidei,** lit. 'so much trustworthiness also has he'; i.e. the belief in any man's word is in exact proportion to the money he keeps in his strong-box. *fides* is not used in its commercial sense of 'credit'; there would be nothing unreasonable in a man's credit, in this sense, corresponding to his fortune.

Samothracum: with this and with *nostrorum*, sc. *deorum*. Samothrace is an island in the nothern Aegean, where mysteries, second in fame only to those of Eleusis, were performed in honour of the Cabeiri, mystic divinities of whose worship little is known; but it is said that they were avengers of perjury.

145. fulmina: lightning is regarded as the punishment of Juppiter against perjurers; cf. XIII. 223 ff.; Aristoph. *Clouds* 397 τοῦτον [τὸν κεραυνὸν] γὰρ δὴ φανερῶς ὁ Ζεὺς ἵησ᾽ ἐπὶ τοὺς ἐπιόρκους.

147. quid quod, 'besides': see n. to l. 86.

148. hic idem is the poor man.

149. si toga sordidula est: the toga, the ceremonial dress of any Roman citizen, was not only hot and heavy; but, being white, required constant cleaning, for which the *fullo* had to be paid.

sordidula may mean 'somewhat dirty', but the use of diminutives is characteristic of *satura*: cf. *pallidulus* x. 82.

150. consuto vulnere, 'where a rent has been sewn up': abl. absol.

151. non una, 'more than one': cf. VI. 218; VIII. 213, 214.

152. nil...durius in se, 'no more bitter ingredient'.

153. inquit, 'says some one': the use of *inquit* with no subject expressed is common; cf. VII. 242.

154. si pudor est, 'for very shame': cf. Mart. II. 37. 10 *ullus si pudor est, repone cenam*; Ovid *Am.* III. 2. 23 *tua contrahe crura, | si pudor est.*

In the Roman theatre, the fourteen rows of seats immediately behind the *orchestra* (see n. to l. 178) were reserved for the knights. This privilege was secured by the *lex Roscia theatralis* passed 67 B.C. by the tribune L. Roscius Otho (cf. l. 159). The law, after falling into desuetude, was revived by Domitian when censor; and the writers of the time, especially Martial in his fifth book, are full of allusions to it and the awkward or comical scenes in the theatre to which it gave rise. The seats are generally called *quattuordecim ordines*, or *equestria* in prose; in Martial *bis septena subsellia* and once (V. 41. 7) *equitum scamni*. *Pulvinus equester* seems to be used here only.

155. cuius res legi non sufficit: the knights, originally a military order, had become under the Empire, the lower division of the Roman aristocracy. They had the right to sit in the *quattuordecim ordines*, to wear the gold ring (*anulus*) and the narrow stripe of purple (*tunica angusticlavia*). But to qualify as knights, they had to possess a fortune of at least 400,000 sesterces (*quadringenta*), about £4,000: those who no longer possessed this sum, at once lost the privileges of their rank. But the order was not a mere timocracy: it was necessary that a knight should be of free birth himself and that his father and grandfather should have been so also. Thus, in the present case, the part of the law, which bears hardly on the poor, is insisted on, while no attention is paid to the qualification of birth, which cannot have been satisfied by the persons mentioned in ll. 156, 157, however rich they may have been: see n. to VII. 14.

156. **quocumque** is used in the sense of *quovis*, as often in the silver age. In Cicero *quicunque* is a relative, not an indefinite, pronoun; but the abl. is used by him indefinitely in a certain number of phrases such as *quacunque ratione, condicione, quocunque modo, loco*; in all these cases there is an ellipse of some part of *posse*, so that they are not real exceptions. See Madv. 6th excursus to Cic. *de Fin*. Cf. *utcumque*, x. 271.

157. **praeconis:** the large fortunes of auctioneers are often mentioned; cf. Mart. v. 56. 8 *artes discere vult pecuniosas,* | *...praeconem facias vel architectum* (advice to a father about a profession for his son).

158. **pinnirapi,** lit. 'crest-snatcher', a kind of gladiator.

iuvenes = *filios*, which would be used if metre allowed.

lanistae: gladiators were kept together in schools (*ludi*) under strict training and discipline; the officer in charge of the school was called *lanista*; but the name is often applied also to a dealer in gladiators. Their ill-repute and their large gains are illustrated by Mart. XI. 66 *et delator es et calumniator,* | *et fraudator es et lanista. miror* | *quare non habeas, Vacerra, nummos.*

159. **qui nos distinxit,** 'who gave us separate seats'; see n. to l. 154.

160. **gener,** 'as a son-in-law', is predicate not subject.

161. **sarcinulae,** lit. 'baggage', is often used as a sort of slang word for a dowry; cf. our use of 'paraphernalia', which properly means 'dowry' but is used for 'belongings' generally. *collige sarcinulas* is addressed to a wife who is being turned out of doors (VI. 146). The diminutive is perhaps a metrical device. Martial uses *sarcinae* repeatedly in iambic metres, but never has the diminutive; with him the word means 'property' or 'baggage' generally, not a wife's especially. Yet the diminutive belongs also to colloquial language; for Pliny (*Epp*. IV. 1. 2) uses *sarcinulas alligamus* for 'we are packing our traps'.

162. **in consilio est aedilibus,** 'sit as assessor to the aediles': various magistrates, praefects, praetors, and aediles, asked friends to sit with them on the bench and assist their deliberations; the context here seems to indicate that these assessors were paid.

The aediles became, under Augustus, police magistrates with power to supervise the streets, markets, and eating-houses.

agmine facto, 'in a body': again X. 218.

163. **debuerant** is used for *debuerunt*: it is convenient metrically, and also an idiomatic usage, like *tempus erat*: cf. *fuerat*, V. 76.

olim, 'long ago': this use of *olim* for *iam dudum*, πάλαι, is characteristic of silver Latin; Juv. has it IV. 96, VI. 90, 281, 346; IX. 17 *olimque domestica febris*.

tenues, 'needy': cf. *tenuis Afros* VIII. 120.

migrasse, 'to have left Rome': cf. the epigram written when Nero's Golden House was building, Suet. *Nero*, 39 *Roma domus fiet; Veios migrate, Quirites.* The word often means 'to flit', to change one's dwelling in Rome itself, μετοικίσασθαι, *déménager*.

165. **res angusta domi:** see n. to VI. 357.

166. **magno,** sc. *constat*.

167. **frugi** is the dat., used predicatively, of **frux* or **frugis*: its only use is as an indeclinable adj.

170. **contentus...cucullo** is a dependent clause parallel with *translatus...Sabellam*, and both are conditional; but the meaning will be better brought out, if the second clause is translated as a separate statement parallel with *negabis*.

cucullus is the hood of a *lacerna*. Büch. reads *Veneto*, which would mean that this article of dress was connected with Venetia in north Italy. The conjecture *culullo* (bowl) may be more suited to what goes before, but does not lead up so well to the following line.

171. The toga, worn so constantly in Rome (see n. to XI. 204), is only put on by the people of the country-towns for the last ceremony of their lives, their funeral; cf. Mart. IX. 57. 8 *pallens toga mortui tribulis*. Magistrates were dressed after death in the insignia of their office, and citizens in the dress distinctive of their status.

172. **ipsa** may be transl. 'even'.

174. **maiestas,** 'the grand occasion of a holiday'. The little town has no permanent theatre (Rome had none until 55 B.C. when Pompey's was erected) and therefore the play is acted in the open air, on a stage of piled up sods.

(Friedl. explains **herboso** differently: that grass is growing between the stone seats of the open-air theatre. Theatres of stone were common in the Italian cities in the 1st cent.; thus Pompeii, a little town, had two, of which the smaller was roofed. This explanation suits the language and the facts better.)

tandem, διὰ χρόνου, 'after a long interval'.

redit: the contracted perf. has the last syllable long: cf. *abit*, VI. 128; *perit*, VI. 295; *obit*, VI. 559; *perit*, X. 118: in all these cases the last syllable is long though followed by a word beginning with a vowel. The uncontracted perf. also is often lengthened: so *periit*, X. 11.

Lachmann, in a note to Lucr. III. 1042, tried to prove (1) that the contracted perf. of *peto* and *eo* and its compounds is only used before a vowel, by the most careful poets, including Juvenal; (2) that the uncontracted perf. of these verbs also has the last syllable always long, having been originally written with a diphthong. There are, however, instances which do not square with these canons: thus Juv. (VI. 563) has *perit cui*; see also Munro on Lucr. III. 1042. As to the second law that *exiit*, for example, cannot be a dactyl, see excursus to Virgil, *Georgic*, II, in Conington's second edition.

175. **exodium,** 'farce'; the word must originally have meant a piece which came at the conclusion of a performance, as a Satyric drama at the end of three tragedies; but in the simple theatre here spoken of, it furnishes the whole entertainment.

personae: all Roman actors wore masks, except in the *mimus*; their extreme ugliness and wide-open mouths (*hiatus*) may be seen in Baumeister's *Denkmäler*, p. 823. They were generally made of clay (*pallentis*), though coloured to suit the particular character.

176. Cf. Mart. XIV. 176 *persona*.

> sum figuli lusus russi persona Batavi.
> quae tu derides haec timet ora puer.

Juv. is perhaps thinking of the famous scene where Astyanax shrinks from Hector's helmet and plume, *Iliad*, VI. 467 ff.

> ἂψ δ' ὁ πάϊς πρὸς κόλπον ἐϋζώνοιο τιθήνης
> ἐκλίνθη ἰάχων, πατρὸς φίλου ὄψιν ἀτυχθεὶς κ.τ.λ.

177. **habitus,** 'dress'; this meaning is not found in Augustan writers.

178. **orchestram:** there being no chorus in Roman plays, the orchestra had ceased to serve any dramatic purpose; it formed a semi-circle in front of the stage and was reserved at Rome for senators. In the provincial towns the corresponding place would be occupied by the *decuriones*, the provincial senate. Thus *orchestram* here = *decuriones*. The people who sit

behind wear just the same dress as the magnates who sit in front. **honoris,** 'office'; the phrase is playful.

179. **tunicae:** though at Rome the humblest citizen had to wear the *toga* in the theatre; see n. to XI. 204.

summis aedilibus, 'the high and mighty aediles': the style of the chief magistrates, generally two in number like the consuls at Rome, differed in different provincial towns: in most they were called *duumviri*; but in Cumae, for instance, they were *praetores*; in others, as at Arpinum (Cic. *Epp.* XIII. 11. 3), the local magnates were *aediles*; see n. to X. 102.

albae is emphatic; they always wear the *tunica*, but they have a clean one in honour of the holiday.

180. **ultra vires,** 'beyond our means': cf. Hor. *Epp.* I. 18. 22 *gloria quem supra vires et vestit et unguit.*

181. **sumitur** refers to borrowing rather than stealing.

182. **ambitiosa,** 'pretentious'.

183. **quid te moror?** like *quid multa?* 'in a word'.

184. **cum pretio,** 'costs money', the opp. of *sine pretio*. The servants of the rich man have to be bribed in order that you may pay your respects after long waiting (*aliquando*), when you call in the morning; even then he may not speak to you. Cossus (VIII. 21) and Veiento (VI. 113) stand for representative nobles.

185. **clauso...labello,** 'without opening his lips'. How civility from the great was appreciated, may be seen from Petron. 44 *et quam benignus resalutare, nomina omnium reddere, tanquam unus de nobis.* For Veiento, see n. to IV. 113.

186. **barbam,** as well as *crinem*, is to be taken with *amati*: Juv. is pointing out how slaves are exalted at the expense of free-men. The first cutting of the beard was an important occasion: see n. to VIII. 166.

crinem...amati: in great houses there were many *capillati*, young slaves with long hair, chosen for their beauty (cf. V. 56 and 61) to wait at table and pour wine; when they grew to manhood, their hair was cut with some ceremony.

187. **libis:** the cakes are to be offered to the Lares, or the *genius* of the hero of the day. Cf. Ovid *Am.* I. 8. 93 (a lady is instructed how to pillage her lover) *cum te deficient poscendi munera causae, | natalem libo testificare tuum.*

venalibus is emphatic, 'which you must pay for'. If the clients did not actually pay for their *libum*, they had to make some present in return for it.

accipe,...tibi habe, 'take your money and keep your cake', lit. leaven. This is said by the poor client to the pampered slave; he has to pay something for the cake but is too angry to take it. *tibi habe* is generally a rude form of refusal, meaning 'I don't want': cf. v. 118 *tibi habe frumentum*; Sen. *de Ben.* VI. 23. 8 '*nolo. sibi habeat. quis illum rogat?*' *et omnes alias impudentis animi voces his adstrue* (which shows that *sibi habeat = nolo*); Cic. *ad Att.* VII. 11. 1 *sibi habeat suam fortunam*, 'I don't want Caesar's greatness'; and Munro on Lucr. III. 135. In Livy, XXVI. 50. 12 *aurum tollere ac sibi habere iussit*, Scipio's refusal to accept the money is polite, but this is exceptional. Plautus uses the words without any sense of refusal, e.g. *Miles*, 23 *me sibi habeto: egomet me ei mancipio dabo*. Lastly, when *habe* precedes *tibi* (e.g. Catull. 1. 8), the meaning is different.

The common explanation, that the words are addressed to the client and that *libum* is understood as object of *accipe*, is open to two objections: (1) that the meaning of *tibi habe* is ignored, (2) that *fermentum* has to mean 'cause of anger', which is unexampled and here ambiguous. Nor does the pronoun (**istud**) favour this interpretation: *illud* (read only by the worse MSS.) would be required.

189. **cultis**, 'well-dressed', 'smart'.

peculia: in the eye of the law slaves could possess no property, but in practice they were permitted and encouraged to make any savings they could. They could often make something by selling part of their allowance of food: cf. Sen. *Epp.* 80. 4 *peculium suum, quod conparaverunt ventre fraudato, pro capite* (freedom) *numerant*: see n. to l. 141.

190–231. *In a country town no one runs any risk from falling houses or from fires. At Rome half the houses are on the brink of falling; and the dangers of fire are worst for the poor who live up in the attics and only get the alarm when it is too late. When the poor man loses his all by fire, no one will contribute to his relief-fund; but the rich and childless, after a similar misfortune, meet with such generous aid, that they are better off than they were before. An excellent house in the country costs less than the rent of an ill-lighted room in the capital; besides you can have a garden which will give you wholesome exercise and plenty of vegetables.*

190. Cf. l. 7.
Praeneste (now Palestrina), here fem., is generally neut. as Hor. *Carm.* III. 4. 22 *frigidum | Praeneste*. Praeneste, Tibur (Tivoli), and Gabii are in Latium, Volsinii in Etruria.

192. **proni Tiburis arce**: cf. xiv. 87. Horace *l.l.* calls Tibur *supinum*: a city on the side of a hill may be considered to lean either forward (*prona*) or backward (*supina*).

193. 'But we live in a city that rests, a great part of it, upon a slender prop': i.e. many of the houses in Rome are just prevented from falling and nothing more.

tibicine: cf. Ovid *Fast.* iv. 695 *stantem tibicine villam*: the force of the word is well shown by a phrase of Virgil quoted in his life by Donatus, 23 *ac ne quid impetum moraretur, quaedam imperfecta transmisit, alia levissimis verbis veluti fulsit, quae per iocum pro tibicinibus interponi aiebat ad sustinendum opus, donec solidae columnae advenirent.* Festus says this meaning is derived from the support given by the instrument (*tibia*) to the voice; the derivation seems dubious.

194. **magna parte sui**: the phrase sounds unusual, because metre requires an inversion of the regular order, *magna sui parte*: cf. Tac. *Ann.* iii. 43. 2 *quinta sui parte*; Pliny, *Paneg.* 52 *magna sui parte*; idem, *Epp.* v. 6. 7 *summa sui parte*; ibid. 15 *magna sui parte.*

sic, i.e. *tenui tibicine.*

labentibus may be neuter, 'the falling building', or masc. like *securos* below, *nobis* being understood.

195. **vilicus**, 'house-agent', who looks after the lodging-houses (*insulae*) on the part of the landlord; properly the 'bailiff' of a farm.

196. **pendente ruina**, 'when the crash is all but come'.

198. **poscit aquam**, 'is crying, fire!', in the English idiom.

frivola, 'odds and ends'; again v. 59.

199. **Ucalegon** is borrowed from Homer through Virg. *Aen.* ii. 311 *iam proximus ardet | Ucalegon*, which Horace also imitates *Epp.* i. 18. 84 *nam tua res agitur paries cum proximus ardet.* Thus it means here 'your next neighbour'.

tabulata...fumant, 'your third storey is already smoking'. The house is three storeys high, and in the highest, just below the roof, lives the poor man in his garret. Martial, speaking as if he was particularly inaccessible, says (i. 117. 7) *et scalis habito tribus sed* (= καὶ ταῦτα) *altis.* The height of Roman houses, though marvellous to the ancient world, was trifling compared with the old houses in Edinburgh or the modern flats in London.

200. **gradibus...imis**, 'the alarm is raised at the foot of the stairs': *ab* denotes the source whence the alarm comes.

201. **ardebit**: possibly *ardebis* should be read: P has *lucebit* wrongly for *lucebis*, i. 155, and *audebit* for *audebis*, ii. 82.

tegula is used for *tegulae* for metrical reasons, or may be used collectively like κέραμος, 'tiling'.

202. **reddunt**, 'lay': the word, rare in this sense, is commonly used of *living* offspring: cf. Ovid *Fast.* II. 429 *maritae | reddebant uteri pignora rara sui*; ibid. IV. 771 *conceptaque semina coniunx* (the ewe) | *reddat*. Perhaps it is a metaphor from agriculture where the word is constantly used, e.g. Mart. II. 38 *quid mihi reddat ager, quaeris*.

203-7. Codrus is a poor man but a lover of Greek art and literature.

203. **Procula minor**, 'too small for Procula': Procula is probably the name of a dwarf; so *Corbulo* stands for a giant (l. 251), *Hispulla* for a fat lady (XII. 11). For the constr., see n. to IV. 66 *privatis maiora focis*.

204. **ornamentum abaci**, 'to adorn his sideboard': the *abacus* was a square table, in this case a marble slab, used for the display of plate; it had a support, called τραπεζοφόρον, which was often a piece of sculpture in marble, bronze, or silver. Codrus' sideboard rests on the recumbent figure of a Centaur; he has no plate to display on it but only six earthen pipkins. Yet it seems as if he might have placed his beaker (*cantharus*) above the board and not below it.

It must be allowed that a *recumbent* figure would not make a good support; hence some suppose that **Chiro** is the name of Codrus' dog.

205. **marmore** = *abaco*.

206. **iam** goes closely with *vetus*; when joined, like ἤδη, with participles and adjectives, **iam** may be transl. 'quite': cf. Hor. *Sat.* I. 1. 5 *iam fractus*, 'quite broken-down'.

cista: this serves for a *scrinium* which Codrus has not got; that its proper function was to hold clothes, appears from a verse quoted by Quintilian, VIII. 3. 19 *praetextam in cista mures rosere Camilli*.

207. **et** may be transl. 'where', as the *Graeci libelli* are identical with the *divina carmina*: both may well refer to Homer.

opici...mures, 'barbarous Roman mice': they do not spare the Greek poems for their beauty. *opici* = ignorant of Greek: cf. VI. 455: Ὀπικία is the old Greek name for Latium and Campania; and ὀπικοί was applied to the Romans in the same disparaging sense as βάρβαροι to all non-Hellinic peoples. That it was resented, we see from Cato in Pliny, *Nat. Hist.* XXIX. 14 *nos quoque dictitant barbaros [Graeci,] et spurcius nos* *quam alios opicon appellatione foedant.* For such an epithet applied to mice, cf. Hor. *Epp.* I. 20. 12 *tineas inertes.*

210. P reads **aerumnae est;** but *est* is expunged by the corrector. That the sentence is more vigorous without it, proves nothing; but it may be observed: (1) that the contraction *ẽ* may have arisen from a repetition of the last letter of *aerumnae*; (2) that in any MS. provided with glosses, *est* would certainly be supplied over the line and may have found its way thence into the text.

211. **hospitio tectoque,** 'with the shelter of a roof'.

212. **Asturici,** a noble. The matrons and aristocrats put on mourning as if for a public calamity; there is a *iustitium* in the law-courts. For *mater*, cf. VIII. 267; for *pullati*, x. 245.

213. **differt...praetor,** 'the magistrate postpones the bail-bonds', i.e. adjourns the business and remands the cases before him.

214. **odimus,** 'express our hatred of', almost 'curse': see n. to VII. 35.

215. **ardet,** sc. *domus.* Asturicus can hardly be the subject; *ultimus ardebit* (l. 201) is not analogous.

et serves as a kind of temporal particle; cf. the use of καί in Greek. The parataxis is quite possible, though less common, in English.

marmora, 'marble', to build a new house; to translate 'statues' anticipates what is coming.

216. **conferat inpensas** is usually explained 'to contribute money'; but this is unsatisfactory, as *impensae* is 'expenditure' not 'money': *impensis* would be needed to give a suitable meaning. Perhaps *impensae* is used in the sense of 'materials'; in Frontinus, a contemporary of Juv., *de Aquis*, 124 *impensa* evidently means 'material' for the repair of aqueducts and is explained (c. 125) by *terra, limus, lapides, testa* (bricks), *harena, ligna ceteraque quibus ad eam rem opus esset*; cf. also Amm. Marcell. XXVII. 3. 10 *aedificia erigere exoriens nova...non ex titulis solitis parari iubebat impensas, sed, si ferrum quaerebatur aut plumbum aut aes aut quicquam simile, apparitores inmittebantur, qui velut ementes diversa raperent*, where *impensae* is explained by *plumbum, aes, ferrum*, etc. This meaning suits the passage well, coming just after *marmora donet.*

217. Euphranor was a statuary and painter at Athens about 336 B.C.; Polycleitus, a still more famous sculptor, flourished a century earlier; his chief work was the ivory and gold statue of Hera at Argos.

218. **haec,** 'another, a lady'. P and the Scholiast agree in this reading; *phaecasiatorum,* 'slippered', is the ingenious conjecture of Roth, based on *fecasianorum* of the inferior MSS.; but there is no evidence that gods were represented with this shoe. The *ornamenta deorum* are the statues which used to adorn the temples in Asia: cf. Livy, XXVI. 30. 9 *dis ipsis ornamentisque eorum ablatis, nihil relictum Syracusis esse.*

219. **mediam,** 'to put in the middle'; some translate, 'a bust of Minerva', but it seems doubtful whether Juv. could use *media* in the sense of *dimidia*. Minerva, as the goddess of learning, is appropriately placed over a book-case.

220. **modium** suggests that there was so much plate that it was measured, not weighed. **argenti** is not 'money' but 'silver plate'. Yet, as each article of plate had its exact weight stamped upon it, plate had a kind of currency. A piece of plate of a certain weight, e.g. *libra,* was a common form of present; cf. esp. Mart. VIII. 71.

reponit: the prefix means 'in place of what he had before'.

221. Persicus seems to have lived in a house, which, from some previous owner, was called *domus Asturici* (l. 212); or else he is a second instance of profitable losses.

iam, 'actually'.

222. **suspectus tamquam...incenderit,** 'suspected of having burnt'. *tamquam* and *quasi* are constantly used by silver-age writers to express any *opinion* or *report*, without conveying any notion that the opinion or report is not true; in such cases Cicero would use the infinitive. Cf. Mart. quoted on I. 143; Tac. *Ann.* XIV. 22 *de quo vulgi opinio est tanquam...portendat;* Suet. *Tib.* 11 *increbrescente rumore quasi...commoraretur.*

Juv. seems to refer to the same incident as Mart. III. 52:

> *empta domus fuerat tibi, Tongiliane, ducentis:*
> *abstulit hanc nimium casus in urbe frequens.*
> *collatum est decies; rogo, non potes ipse videri*
> *incendisse tuam, Tongiliane, domum?*

223. **avelli circensibus,** 'tear yourself from the races'; cf. XI. 53. The *circenses,* the most popular of all spectacles at Rome, could be seen nowhere else. The provincial towns had theatres for plays, and amphitheatres for gladiatorial shows; but they had no *Circus*. At Rome this occupied the whole valley between the Palatine and Aventine hills; the ascending slopes were covered with immense tiers of seats in wood and marble accommodating, at this time, 250,000 spectators. After the Colosseum

(*amphitheatrum Flavianum*) was opened by Titus A.D. 80, the *Circus* was used almost exclusively for chariot-races.

The number of days devoted to the *Circus* at this time is not certainly known: it was about 21 under Tiberius, and 64 in the middle of the 4th cent. But it must be remembered that the regular holidays of the calendar were largely added to on extra-ordinary occasions. Thus Titus celebrated the opening of the Colosseum by a festival of 100 days, and the shows of Trajan on his second Dacian triumph (A.D. 106) lasted 123 days; on each of these occasions races would be part of the amusement provided.

There were generally 24 races (*missus*) in the course of each day; and in each race the chariots went round the course seven times (*septem spatia*), the distance covered being about five miles. No other form of public amusement caused such passionate excitement: the wishes of the people, says Juv. (x. 81), are confined to two objects, *panem et circenses*. For the *pompa circensis*, see n. to x. 36; for *factiones* and *panni*, see nn. to XI. 193–202.

Sora, Fabrateria, and Frusino were all country-towns in Latium.

224. paratur, 'you can buy outright', is emphatic; this meaning is preserved in the Italian *comprare*, 'to buy'.

225. tenebras, 'a black hole', i.e. an ill-lighted room in a lodging-house (*insula*): cf. Mart. II. 14. 12 *Grylli tenebras* (of a dark bath). The rent (*pensio*) is as much as the price of a good house elsewhere.

226. hortulus: see n. to I. 75.

hic, in the country-towns. Because the well is shallow (*brevis*) no rope is needed to work it, the bucket being dipped in by the hand.

227. tenuis, 'tender'.

229. The Pythagoreans ate no animal food but only vegetables; and, even among vegetables, they abstained from beans, either as having souls or as flatulent food: cf. xv. 174.

230, 231. 'One of the company asked him the meaning of the expression in Juvenal, *unius lacertae*. JOHNSON. "I think it clear enough; as much ground as one may have a chance to find a lizard upon"' (Boswell's *Johnson*, 1874, II, 175). The lizard is so common in Italy that the smallest spot of ground would be sure to harbour at least one. So Martial (XI. 18. 11) describes a very small piece of ground as a place in which *non serpens habitare tota possit.*

232–67. Men fall sick at Rome and sick men die for want of sleep. The rich man makes way with ease and comfort in his litter through the crowded streets; while the poor foot-passenger meets with every inconvenience and danger. Accidents often happen through the sudden collapse of a loaded dray; the bodies of the victims are crushed and annihilated beneath the mass of marble; and they never return to the household which is making preparations for their reception.

232. vigilandŏ: for the quantity, see n. to *ergo,* I. 3.: Juv. does not elsewhere have this *o* short: Seneca, in his tragedies, begins iambic lines with *petendŏ, solvendŏ,* etc.

sed ipsum…, 'yes, and the sickness itself was produced…'. One step in the argument is omitted, i.e. that good digestion depends on sleep.

233. inperfectus is unusual in this sense: *concoctus* or *confectus* are the usual words for 'digested': but cf. Celsus, IV. 23 *quidquid assumptum est* (has been eaten), *imperfectum protinus reddunt* (*intestina*).

235. admittunt, 'make sleep possible', not 'let sleep in'; cf. *non admittentia morsum,* v. 69.

dormitur: for the impers., cf. Mart. XII. 68. 6 *otia me somnusque iuvant, quae magna negavit | Roma mihi: redeo si vigilatur et hic.* In this epigram, which was written in Spain, Mart. is not complaining of the noises in Rome but of the necessity of rising early to perform the *salutatio:* so X. 74: but elsewhere he often speaks of the street-cries and noises which made sleep impossible: cf. esp. XII. 57.

236. raedarum transitus: the *raeda* was a heavy four-wheeled carriage used by travellers. Their noise was troublesome by night, because no vehicles except litters were allowed to pass through the streets of Rome for ten hours after sunrise; an exception was made in the case of materials for public buildings; see ll. 254 ff. **arto…inflexu:** when Rome was rebuilt after the great fire of A.D. 64, the streets were made less narrow and winding than they had been before; but, owing to the hills on which it is built, the city always has been and still is inferior in convenience to other European capitals.

237. stantis, 'blocked'. **convicia** must be understood of the men who accompany the *mandra:* 'the omnibus, as Miss La Creevy protested, swore so dreadfully that it was quite awful to hear it' (*Nicholas Nickleby*).

238. Druso is supposed to mean the emperor Claudius,

whose full name is Ti. Claudius Drusus Caesar. Suetonius (*Claud.* 8 and 33) says that he constantly fell asleep after dinner and sometimes on the judgment-seat. But it is strange that the name Drusus, which was borne by so many of the imperial family, should be used to designate Claudius; also the fut. (*eripient*) is inappropriate. Hence there is much probability in Speyer's conjecture *surdo*. If *Druso* is kept, it seems better to suppose, with the Scholiast, a reference to a contemporary of Juv., unknown to us: cf. *Procula*, l. 203.

vitulisque marinis: cf. Pliny, *Nat. Hist.* IX. 42 (of this animal) *nullum animal graviore somno premitur*.

239. **officium,** 'a social duty', e.g. a call: cf. v. 13, VI. 203, x. 45: that the word is generally restricted to this narrow meaning, is characteristic of life at Rome under the Empire.

240. **super ora:** cf. Sen. *Epp.* 80. 8 *idem de istis licet dicas quos supra capita hominum supraque turbam delicatos lectica suspendit.*

241. So the elder Pliny always read or wrote while being carried about Rome in his litter, and rebuked his nephew for walking instead of following his example (Pliny, *Epp.* III. 5. 16).

obiter again VI. 481.

242. **fenestra:** cf. *specularibus*, IV. 21. Discoveries at Pompeii and elsewhere show that window-glass was already known and used in the 1st cent.; but it is probable that talc (*lapis specularis*) is the material meant in both these passages; it was cheaper for a long time and served to exclude the sun as well as to admit light.

243. **tamen,** although he has taken no trouble.

nobis, i.e. *pauperibus*: in English, 'for' must be supplied at the beginning of this clause.

244. **unda:** cf. Virg. *Georg.* II. 461 *ingentem...salutantum ...undam.* **prior** is opposed to *qui sequitur.*

245. **assere:** probably the pole of a litter.

248. **clavus...militis:** the soldier's shoe (*caliga*) had large nails in the sole: cf. l. 322 and XVI. 24. The emperor Caligula received this nickname when, as a child in the camp, he was dressed like a soldier of the legion.

249. **quanto...fumo,** 'the great cloud of smoke where people throng to a picnic'.

It seems that **sportula** has here a different sense from I. 95, and is used for a 'picnic', i.e. a club-dinner to which each guest brings his share of the eatables. This is generally called *cena collaticia* (ἔρανος); but there is a passage in Suetonius where *sportula* is explained exactly in this sense: *Claud.* 21 *exhibit...*

quod appellare coepit sportulam, quia primum daturus edixerat,
'velut ad subitam condictamque cenulam invitare se populum'.
Such a meal is also called δεῖπνον ἀπὸ σπυρίδος and thus
explained by Athenaeus 365A: ὅταν τις αὐτὸς αὑτῷ σκευάσας
(having cooked) δεῖπνον καὶ συνθεὶς εἰς σπυρίδα παρά τινα
δειπνήσων ἴῃ. In this case the σπυρίς or *sportula* is replaced by
the more elaborate contrivance of a portable stove, carried
together with the food to the *rendez-vous* by a slave. (So Wis-
sowa.)

The common interpretation is that the ordinary 'dole' is here
meant; the clients throng to the spot, each followed by a slave
carrying a stove to keep hot the eatables which they mean to
buy on the spot with their *centum quadrantes*. To this view
there are two objections: (1) the dole is so small that it is absurd
for each slave to carry a whole *batterie de cuisine* to cook the food
purchased; (2) in an enumeration of the hardships of the poor,
it is out of place to count one inconvenience which is due to
themselves.

251. Cn. Domitius Corbulo is probably meant, an eminent
general in the reigns of Claudius and Nero, who is described by
Tacitus (*Ann.* XIII. 8) as *corpore ingens*.

252. **recto,** 'unbent'.

254. **scinduntur tunicae:** cf. Pliny, *Epp.* IV. 16. 2 *adulescens*
scissis tunicis, ut in frequentia solet fieri, sola velatus toga perstitit:
the toga, though more exposed than the tunic, was a much
stouter garment.

255. The timber and marble, carried in waggons along the
crowded streets, are intended for some public building and are
therefore exempt from the usual restrictions: see n. to l. 236.
For similar obstacles in the streets, cf. Mart. v. 22. 7 *vixque*
datur longas mulorum rumpere mandras, | *quaeque trahi multo*
marmora fune vides.

altera: the plur. is due merely to metrical convenience.

257. **saxa Ligustica,** marble from Luna in Etruria, near the
modern Carrara: the place was formerly included in Liguria.

259. **superest de corp.:** cf. I. 34.

260. **perit omne,** 'disappears utterly'; cf. *omne peractum*
est v. 93.

261. **more animae,** 'just like their souls': the disappearance
of the soul, at death, is normal; but in this case the body
vanishes too: cf. Lucan, IX. 788 (to a kind of serpent) *eripiunt*
omnes animam, tu sola cadaver.

262. The slaves at home are making preparations meanwhile

for their master's return for his bath and dinner; but he will
never enjoy them. Juv. may be thinking of the preparations
going on for Hector's return at the very time of his death (*Il.*
XXII. 442).

264. **at ille:** these words are often placed thus, at the end of
the verse, by the Latin poets, apparently for pathetic effect; e.g.
Virg. *Georg.* IV. 513 *at illa | flet noctem* (of the nightingale
robbed of her young).

265. **in ripa:** the river may be any of the four infernal
streams, Acheron, Cocytus, Pyriphlegethon, or Styx; cf.
Homer, *Od.* X. 513.

novicius: adjectives in *-icius* have the first *i* long when they
are formed from participles (as *missicius* from *missus*); the *i* is
short, when the adj. is formed from a noun (so *tribunicius,
latericius, patricius*), or from an adj. (as *natalicius* from *natalis,
aedilicius* from *aedilis*).

Thus *novīcius* appears to be an exception; and hence it has
been argued that it is not formed on the same analogy. (The
quantity of these words is generally wrongly given in Lewis and
Short.)

266. **porthmea:** the *a*, though long in Greek, is generally
shortened by the Latin poets: so *Tereă,* VII. 12; *Prometheă,*
VIII. 133; *Peleă,* XIV. 214.

The 'grim ferryman that poets write of' is Charon, a per-
sonage who does not appear in Homer, although till very recent
times he still figured in the popular beliefs of Greece as the
conductor of the dead. Virgil (*Aen.* VI. 298) calls him *portitor
horrendus.*

Charon would not convey across the river any who had not
been buried with due rites: cf. Virg. *ibid.* 327.

267. Charon expected to receive a fee for his services: cf.
Aristoph. *Frogs,* 270 ΧΑ. ἔκβαιν', ἀπόδος τὸν ναῦλον. ΔΙ. ἔχε δὴ
τὠβολώ. For this purpose it was the custom in Greece to place a
coin in the mouth of the dead; hence **porrigat ore.** A single
obol is the usual sum; cf. Lucian *de Luctu* 10 ἐπειδάν τις
ἀποθάνῃ τῶν οἰκείων, πρῶτα μὲν φέροντες ὀβολὸν εἰς τὸ στόμα κατέθηκαν
αὐτῷ, μισθὸν τῷ πορθμεῖ τῆς ναυτιλίας γενησόμενον. The custom
became common among the Romans under the empire: skeletons
have been found in various parts of Europe with coins of the
early emperors between their teeth.

268-314. *And there are other dangers to be faced in the streets at
night: you may suffer from broken crockery or at least dirty*

water thrown from the upper windows; or you may meet with some drunken roisterer who picks a quarrel to justify him in beating you. The burglar and footpad also ply their trade freely in Rome; our age is rife in crime.

268. Juv. resumes where he left off at 261.

diversus is used here in the sense of *varius*; in classical Latin, it means 'in opposite directions' and is distinct from *varius*; hence the old title of Cicero's letters to his friends, *epistulae ad diversos*, is barbarous and cannot proceed from their author.

269. **respice** is followed by three indirect questions, *quod spatium* sc. *sit, quotiens...cadant*, and *quanto...signent*. For the height of houses in Rome, see n. to l. 199.

270. **testa** seems to be used here for the more common *tegula*, 'a tile'; hence **tectis** means 'roofs' not 'houses'. It is commonly transl. here 'a potsherd'; but this anticipates the next clause.

fenestris: in Rome, as still in the East, the lower storeys of houses had no windows facing the street.

271. **quanto...silicem,** 'with what weight they score and injure the pavement where they strike'.

272. **laedant:** cf. Ovid, *Heroid.* IX. 87 *aper...Erymantho | incubat, et vasto pondere laedit humum.*

silicem: '*silices* denote the hard blocks of volcanic basalt with which the Romans paved their streets and roads' Munro on Lucr. I. 571. *silice sternere* = 'to pave'. **ignavus,** 'unbusinesslike'.

274. **adeo** (lit. 'so true is it that') may be transl. 'for indeed', when it begins a clause which accounts in an emphatic way for a statement immediately preceding, and when it applies, as here, to the whole clause and not to any one word in it; see n. to XIII. 183. *ita* (very common in Plautus), *usque eo, is, tam,* and *tantum* are used in the same way at the beginning of a sentence.

This absolute use of *adeo* is peculiar to silver-age Latin and occurs constantly in Livy and Tacitus. In classical Latin it has two uses: (1) as an enclitic, following pronouns etc.; (2) to mark a gradation, when it is followed by *ut*.

276. **optes,** 'you must pray'; a constant use of the 2nd pers. of the pres. subj.; see n. on *expectes* I. 14. **tecum** = *tacitus*, 'silently'.

277. **sint,** sc. *fenestrae*. The emphasis falls on *defundere*. Juv.

alludes to the unpleasant custom of upsetting slops into the
street from upper windows. Edinburgh was notorious for this
practice in the last century; when the cry '*gardy-loo:*' (i.e.
gardez l'eau) was heard, the passer-by made haste to stand
from below.

278. Milton, *P.L.* Bk. I, speaks of ruffians of this kind whom
he may have met himself:

> And when night
> Darkens the streets, then wander forth the sons
> Of Belial, flown with insolence and wine.

In the beginning of the 18th cent. London was infested by
street ruffians, known by the name of Mohocks (i.e. savages):
'Did I tell you of a race of rakes called the Mohocks, that play
the devil about this town every night, slit people's noses, and
bid them etc.?' (Swift's *Journal to Stella*, 8 March 1711–12).

petulans, 'violent', 'brutal'. The word is only a little less
strong than *furiosus* (l. 291), as may be seen from Cic. *Brutus*,
241 *fervido quodam et petulanti et furiosos genere dicendi*.

279. The reference is to Homer, *Il.* xxiv. 10 where the grief
of Achilles for Patroclus is described: ἄλλοτ' ἐπὶ πλευρὰς
κατακείμενος, ἄλλοτε δ' αὖτε | ὕπτιος, ἄλλοτε δὲ πρηνής: cf. Sen. *Dial.*
IX. 2. 12 *qualis ille Homericus Achilles est, modo pronus, modo
supinus, in varios habitus se ipse componens*.

280. **in faciem** = *pronus*. **mox deinde**, 'a moment after'.

281. **ergō** is a spondee here and in IX. 82; elsewhere in Juv.
it is, or at least may be, a trochee.

aliter = *nisi occiso aliquo*.

282. **somnum...facit**: cf. l. 242.

285. **multum flammarum**, 'a quantity of lights': these
may be torches: cf. *multum caelati*, XII. 46. **aenea lampas**, a
lamp, fed with oil, of Corinthian bronze, and carried by a
servus praelucens. A poor man had to content himself with a
candle of wax or tallow; candles were used in Italy long before
oil-lamps. There was never any system of street-lighting at
Rome.

286. **deducere**, 'to escort'; *reducere* would be more strictly
accurate here.

289. **pulsare** has for a passive **vapulare**, just as τύπτειν
often has πληγὰς λαμβάνειν.

290. **stari**: impersonal; cf. *dormitur* l. 235.

292. **acetum**, often called *posca*, is vinegar mixed with
water, the ordinary drink of the Roman soldier.

293. **sectile porrum:** leeks were eaten at two stages of their growth, when they were just above the soil, or when they had grown to a head; the former was called *porrum sectile* or *sectivum*, the latter *p. capitatum*; cf. Mart. III. 47. 8 *utrumque porrum* (in a list of vegetables).

The aggressor pretends that his victim smells of this vegetable: cf. Mart. XIII. 18 *porri sectivi.*

> *fila Tarentini graviter redolentia porri*
> *edisti quotiens, oscula clusa dato.*

296. **consistas,** 'you have your stand', i.e. as a beggar. **quaero,** 'am I to seek?': cf. *conciditur,* IV. 130. The indic. is used in a deliberative question, when the question is rhetorical and no answer is expected: here the speaker assumes as a matter of course that his victim will be found in such a place. Thus in Catullus, I. 1 (*cui dono lepidum novum libellum?*), the answer (*Corneli, tibi*) follows as a matter of course. **proseucha,** the name applied to a Jewish praying-house, suggests the contempt of the orthodox for the dissenter; we might say 'Little Bethel'.

298. **pariter,** 'in either case'.

299. **haec** is attracted into the gender of its predicate *libertas* as a demonstr. or relative pron. is generally; cf. Virg. *Aen.* VI. 129 *hoc opus, hic labor est.* Tacitus offers most exceptions to the rule.

300. **pugnis concisus:** cf. Cic. *in Verr.* II. 3. 56 *cum pugnis et calcibus concisus esset,...mille promisit* [*medimnos*].

301. **paucis,** 'a few at least'; he sees the necessity of sacrificing some.

302. **metuas,** 'must you fear'; see n. to I. 14.

303. **clausis...tabernae,** 'when houses are shut, and after all the shops everywhere are made fast with bolts, and the shutters are closed and silent'. **compago** is the surface presented by the closed shutters or folding-doors of the shops; when the *compago* was not fastened, the whole breadth of the shop was open to the street, as may be seen at Pompeii.

307. The Pontine Marshes occupied a space of about 30 miles in length by 7 or 8 in breadth, in the south of Latium, and extended from Cisterna to the sea at Terracina. One of the improvements which Caesar had projected at the time of his assassination was to drain these marshes (Suet. *Iul.* 44).

The Gallinarian forest was on the west of Campania, between the Vulturnus and Cumae. It was a *Pineta,* like that at Ravenna.

Both marshes and forest were thinly inhabited and therefore a favourite place of resort for *grassatores*.

308. **sic...tamquam** is not a common combination; and Büch. proposed *sicae*, i.e. *sicarii*; but cf. VI. 341 and Mart. V. 6. 17 *sic tanquam nihil offeras agasque*. **huc,** i.e. to Rome. **omnes,** sc. *grassatores*.

vivaria, 'their feeding-ground' L.: lit. 'preserves' where game or fish (cf. IV. 51) was preserved and fattened for the table.

309. **non** must be supplied before *fornace*.

310. Cf. Seneca, *Hercules F.* 934 *ferrum omne teneat ruris innocui labor,* | *ensesque lateant*.

timeas, 'one may fear'; the mood would be the same without the preceding *ut*.

312. The Latin names, in order of ascent, may be seen in Plaut. *Pers.* 57 *pater, avos, proavos, abavos, atavos, tritavos*. Juv. means 'our remote ancestors'.

313. **tribunis** does not refer to the plebeian *auxilium* but to the military tribunes with consular power first appointed in 445 B.C.; the former could not be said to have been at any time supreme in the state.

314. **uno...carcere:** this was the *carcer Mamertinus* built by Ancus Martius; cf. Livy, I. 33. 8 *carcer ad terrorem increscentis audaciae media urbe imminens foro aedificatur*. The Tullianum was a dungeon added to this prison by Servius Tullius.

Juv. clearly implies that there were other prisons at Rome in his time. Where these were is not certainly known; but it is probable that the barracks of the praetorian and urban cohorts were used for this purpose; cf. VI. 561 *castrorum in carcere*. Cf. also Trajan, *ad Plin.* 57. 2 *vinctus mitti ad praefectos praetorii mei debet*, which suggests that the prefects had prisons at their disposal: though it will be remembered that Paul, who was sent thus to Rome (A.D. 62), was allowed 'to abide by himself with the soldier that guarded him' (Acts XXVIII. 16).

315–22. *But I must be going. Whenever you visit Aquinum, I will come from Cumae to see you and to hear your satires.*

315. **causas,** sc. *migrandi*.

317. **muliŏ:** for the quantity, see n. to *ergo*, I. 3.

virga adnuit: cf. VIII. 153.

319. **tuo:** Juv. was a native of Aquinum in Latium. It appears from the text that the chief deities worshipped there were Diana and Ceres Helvina: for the inscription to Ceres at Aquinum, generally referred to Juv., see Introduction, p. xix.

320. **vestram** = *Aquinatium*; here, as always, *vester* must be distinguished from *tuus*.

321. **ni pudet illas:** it is not suggested that the satires are afraid of criticism but that they might desire a more distinguished critic.

322. **auditor:** Büch. has now (1893) gone back to *adiutor* which he removed in 1886 on Beer's report of P's reading; but his note admits that the first hand of P may have written *auditor*, which is found as a correction of *adiutor* in several MSS. This being so, the latter, which is better in point of meaning, should be kept. It is hard to see how Umbr. could help the satires except by listening to them; cf. Mart. XII. pref. *si quid est enim quod in libellis meis placeat, dictavit auditor*; Ovid *ex Pont.* IV. 2. 35 *excitat auditor studium*.

Although reciting poets are one of the chief terrors of Rome (cf. l. 9), Umbr. is willing to listen to his friend's satires.

caligatus, 'with thick boots'; there is no metaphor here from military service; Umbr. hints that his plain rough dress may not find favour with Juv.'s Muse.

SATIRE IV

A CABINET COUNCIL

It has often been observed that this satire consists of two parts, which have little connection with each other. The first begins by introducing Crispinus as the object of attack, and his vices are satirised (1–33); then follow three lines (34–6) as introduction to the main episode, in which Crispinus has only a passing mention (l. 108), while the real subject is Domitian and his manner of behaviour to the high officials who formed his Privy Council. The poem is certainly ill-constructed; and Weidner concludes that we have here two satires of which the first was left unfinished. This may be so; but we have no reason for supposing that anyone other than the author originally brought them together to form one poem. See Introduction to Sat. VII.

1–33. *Crispinus is once more my theme, that vicious voluptuary. At present I deal rather with his follies than his vices. This Egyptian upstart paid an enormous price for a large mullet; but this was only in humble imitation of his imperial master.*

1. **Crispinus:** see n. to I. 26. Juv. speaks as if he had already attacked Crispinus; but he can hardly refer to Sat. I,

as it was probably written after Sat. IV: see Introduction,
p. xv.

2. **ad partes,** 'to play his part' on the stage of satire. This
promise is not fulfilled: Crispinus is not mentioned in any of
the later satires.

4. **deliciae,** 'a voluptuary'; in a slightly different sense in
XIII. 140. In Mart. VIII. 48. 6 *nec nisi deliciis convenit iste color*
(the subject is the loss of a purple cloak belonging to this same
Crispinus), *deliciae* is explained by edd. in this sense; but a
comparison with Mart. I. 59. 2 will show that it means 'luxury'
in both passages.

viduas tantum = *eas quae viduae* (unmarried) *tantum sunt*:
the notion is that the seduction of a married woman is a greater
crime and therefore more attractive to C.

5. **quantis...porticibus,** 'the size of the colonnades in
which...'. *porticus* is a covered walk in which the rich took
their drives, sheltered from sun or rain; cf. VII. 178 ff. There
were three such walks in Nero's Golden House, each a thou-
sand feet long, and this in the heart of Rome (Suet. *Nero*, 31).

7. Land was naturally most valuable when nearest to the
commercial and political centre of the city. The ground on which
Caesar's forum was laid out cost him over £1,000,000 (Suet.
Iulius, 26). **aedes,** sc. *vicinas foro*.

9. **incestus** refers to the profanation of religion involved in
the act.

cum quo, unusual for *quocum*: again l. 87.

vittata is not a needless epithet; it is emphasised by separa-
tion from the noun and is meant to force on the reader that the
Vestal, even in her dress, was distinguished from other women
and set apart for a sacred calling. Cf. Stat. *Theb.* VII. 758
conatusque toris vittatam attingere Manto | Lampus; Ovid, *Fast.*
VI. 457 *nullaque dicetur vittas temerasse sacerdos | hoc duce, nec
viva defodietur humo*.

10. **terram subitura,** 'doomed to be buried alive': cf.
Ovid above. A Vestal who had broken her vows was buried in an
underground vault in the Campus Sceleratus, a light and some
food being left with her; and this punishment, after long disuse,
was inflicted by Domitian as censor upon Cornelia, one of the
Vestals, A.D. 91; cf. Pliny *Epp.* IV. 11, 6: Crispinus is not men-
tioned there as her accomplice.

If Juv. does not allude here to the fate of Cornelia, then the
participle means, 'who might have been buried alive': this con-
ditional sense of the fut. participle is common in silver-age

Latin: see n. to VI. 277. For it is not probable that any other Vestal actually suffered this punishment during this period, or Pliny would mention it.

11. **sed nunc,** sc. *agendum est.*

12. **fecisset,** 'had been guilty of', is a technical word: cf. *feci*, 'I am guilty', VI. 638, and Mart. IX. 15 *inscripsit tumulis septem scelerata virorum | 'se fecisse' Chloe. quid pote simplicius?*

caderet, 'would have been convicted': also a technical word; cf. *cadere causa.*

iudice morum, the censor: Domitian assumed this title A.D. 84.

13. **Titio Seioque:** the names are used in the sense of 'any ordinary men'; cf. Mart. V. 14. 5 *post Gaiumque Luciumque consedit* where the names stand not for particular people but for any two knights; so Natta and Pansa VIII. 96. In the jurists the standing names for plaintiff and defendant were Gaius Seius and Lucius Titius: cf. Plutarch, *Quaest. Rom.* 30 ὥσπερ οἱ νομικοὶ Γάϊον Σήϊον καὶ Λούκιον Τίτιον παραλαμβάνουσι. So Gaius and Gaia were used as equivalents of 'husband' and 'wife' in some declarations which formed part of the Roman ceremony of marriage.

14. *quid agas* = 'one is helpless'; cf. III. 291. **omni crimine,** 'than any charge' you can bring.

15. **persona,** 'the individual'; cf. Mart. I. pref. *cum salva infimarum quoque personarum reverentia ludant*; idem, X. 33. 10 *parcere personis, dicere de vitiis.* In classical Latin the word does not bear this meaning, but that of III. 175 and others nearly akin to it.

mullum: one of the reasons of Tiberius' sumptuary laws was the sale of three mullets for 30,000 sesterces (Suet. *Tib.* 34). A mullet weighing 4½ lb., presented to Tiberius, was sent by him to the market for sale; Apicius and Octavius bid against each other for it, as the emperor had predicted; Octavius secured it for 5,000 sesterces (Seneca, *Epp.* 95, 42). In each case the price is about 1,000 sesterces (£10) for each lb. in weight.

16. **sane,** 'it is true', 'I grant you'.

19. **praecipuam...ceram,** 'the first place in the will', wills being written on wax tablets; cf. Hor. *Sat.* II. 5. 53 *quid prima secundo | cera velit versu*; Mart. IV. 70 *nihil Ammiano praeter aridam restem | moriens reliquit ultimis pater ceris.*

For *captatio* (legacy-hunting) see n. to III. 129; also V. 98, VI. 39, X. 202, XII. 93–130. A vivid representation of the practice is given in Petronius, 116 (the place is Crotona and the

9

time about A.D. 57) *quoscunque homines in hac urbe videritis,
scitote in duas partes esse divisos. nam aut captantur aut captant.
in hac urbe nemo liberos tollit, quia quisquis suos heredes habet,
non ad cenas, non ad spectacula admittitur sed omnibus prohibetur
commodis, inter ignominiosos latitat. qui vero nec uxores unquam
duxerunt nec proximas necessitudines habent, ad summos honores
perveniunt...adibitis, inquit, oppidum tanquam in pestilentia
campos, in quibus nihil aliud est nisi cadavera quae lacerantur,
aut corvi qui lacerant.*

20. **ratio ulterior,** 'a secondary motive'.

magnae, 'noble'; see n. to I. 33.

misit, 'has given it away': see n. to III. 45.

21. **specularibus:** see n. to III. 242. **antro** is used for
lectica to suggest coolness and space.

23. This refers no doubt to the incident related by Seneca;
see n. to l. 15. M. Gavius Apicius lived in the reigns of Augustus
and Tiberius: his love of gastronomy has made his name pro-
verbial. When his fortune was reduced to 10,000,000 sesterces
(about £100,000), he poisoned himself, thinking the sum in-
sufficient for the tastes of such an epicure (Mart. III. 22).

miser, 'miserly'; cf. Mart. I. 99. 9 *abisti | in tantam miser
esuritionem.*

hoc tu, sc. *fecisti?*

24. **Crispinus** was an Egyptian; and it was in the Nile that
the papyrus chiefly grew. Pliny (*Nat. Hist.* XIII. 72) says that
the inner bark of the plant was used by the natives to make
sails, mats, and clothes. **succinctus** is meant to suggest the
menial offices performed by C. in his native country; cf. VIII.
162.

25. **squamae,** sc. *emptae sunt?* The word is used contemp-
tuously for *piscis*: 'did you pay that price for scales?' For the
omission of the verbs here, see n. to I. 1. **minoris:** gen. of
price.

26. **provincia,** 'the provinces', as opposed to Italy: cf. V.
97.

27. **sed** may be transl. 'and indeed'; it serves rather to
emphasise than to contradict the previous statement; cf. V. 147
and Mart. I. 117. 7 *et scalis habito tribus, sed altis* (and high ones
too). It seems that the pasture-lands of Apulia sold cheap,
owing to the decay of agriculture in Italy; cf. Sen. *Epp.* 87. 7
*tantum suburbani agri possidet quantum invidiose in desertis
Apuliae possideret.* Even in Cicero's time Apulia was thinly
populated: cf. Cic. *ad Att.* VIII. 3. 4 *Apulia delecta est, inanissima*

pars Italiae. With the second **vendit,** *tanti* must be supplied again.

28. **putamus,** 'must we suppose'; cf. *conciditur* l. 130.

29. **induperatorem:** again x. 138. Some edd. see sarcasm in the archaic form; but, if so, Juv. has made a virtue of necessity, as metre prevented him from using any other; *imperator* is one of the many common words which no poet could ever use in dactylic verse.

30. **de margine:** the mullet was a mere side-dish, not, as you might expect, the *caput cenae* or *pièce de résistance.*

31. **purpureus:** cf. 1. 27 and Mart. VIII. 48. 1 *nescit cui dederit Tyriam Crispinus abollam.*

scurra Palati is contemptuous for *amicus principis.*

32. **princeps equitum** (cf. Pliny, *Epp.* I. 14. 5 *equestris ordinis princeps*) is not an official title: but it probably means that Crispinus was *praefectus praetorio,* one of the two commanders of the praetorian guard, Fuscus being the other; this was the highest position which any man of the equestrian order could occupy. There were other important and lucrative posts filled also by *equites*: the chief of these were the *praefectura vigilum, praefectura annonae, praefectura Aegypti.* These offices might be, and often were, held one after the other; but the first was regarded as the crown of the equestrian career.

33. **municipes:** i.e. the sprats also came from Egypt; cf. XIV. 271 and Mart. X. 87. 10 *Cadmi municipes* (i.e. *Tyrias*) *lacernas.*

fracta de merce is explained 'from a stock of damaged wares'; C. bought salt fish, which had gone bad, from the wholesale dealers and hawked it through the streets.

But the sense of *fracta* (which was read by the Scholiast) is very unusual; and, as T agrees with P in reading *facta,* it seems likely that the true reading is *Pharia,* a common equivalent for *Aegyptiaca*: cf. XIII. 85; Stat. *Silv.* II. 1. 73 *mixtus Phariis venalis mercibus infans. Pharia* would be likely to appear as *faria* in a Latin MS.: see Hosius on Lucan, VIII. 546, X. 171: then *faria* became *farta* and finally *facta,* of which *fracta* is a correction. *Pharia* was proposed by Muretus.

The statement of Crispinus' birth and occupations is probably an exaggeration.

34. **considere:** poetry was generally recited standing, but this is prose.

35. **res vera:** this is not impossible, considering the story told by Dio Cassius (LXVII. 9) of how Domitian entertained the senate.

36. **puellas:** the jest, a poor one, seems to be that the Pierides, after their centuries of service at Parnassus, must be old women; but the poet is willing to earn their gratitude by understating their age.

37–149. *A huge turbot was once brought to Alba as a present to Domitian from a fisherman who knew that it would not be safe to keep or sell such a prize. There was no dish large enough to hold the monster; and in this dilemma the emperor summoned his council. Pegasus was present, Crispus, Acilius and his ill-fated son, Rubrius, Montanus, Crispinus, Pompeius, the strategist Fuscus, Fabricius Veiento and the blind informer, Catullus. Montanus, a past master in gastronomy, solved the difficulty, and the council broke up.*

37. **Flavius ultimus:** Domitian was the younger of Vespasian's two sons and the last of his race who sat on the throne. The Scholiast on this line cites as the end of an epigram by Martial: *Flavia gens, quantum tibi tertius abstulit heres! | paene fuit tanti non habuisse duos,* 'how much glory the Flavian house lost by its third representative! it would almost have been better not to have had the other two' i.e. than to have had all three, including Domitian. The epigram is not found in our MSS. of Martial, but may very well be his.

38. **calvo...Neroni:** i.e. Domitian was as cruel as Nero but had not his youthful beauty. He was bald, and wrote a book *de Cura Capillorum*: cf. Suet. *Dom.* 18 *calvitio quoque deformis ...calvitio ita offendebatur, ut in contumeliam suam traheret si cui alii ioco vel iurgio obiectaretur*: this is difficult to reconcile with Martial's frequent jests at the expense of bald men.

39. '**There turned up in the Adriatic a wonderful monster of a turbot**': *spatium rhombi* is like μέγα ὑὸς χρῆμα (Herod. 1. 36).

40. **Ancon** (ἀγκών 'the elbow'), a city of Picenum on the Adriatic, was founded about 380 B.C. by exiles from Syracuse; hence it is called *Dorica*. Venus was its tutelary deity and had a temple there.

41. **sinus,** sc. *retis,* 'the folds of the net'.

42. **Maeotica:** *palus Maeotica* is the ancient name of the sea of Azof. 'It is usually frozen every year from November to the beginning of March. There is perhaps no equal extent of water on the whole surface of the globe which abounds in fish so much as this sea' (*Nat. Encycl.*). Juv. supposes the fish to grow fat and lazy during their confinement beneath the ice.

43. **effundit,** 'sends in shoals'. Cf. Tac. *Ann.* XII. 63 *vis*

piscium immensa, Pontum erumpens...hos ad portus (i.e. Byzantium) *defertur.*

torrentis, not *torpentis,* is now (1893) found to be the reading of P as well as of the Scholiast; it is certainly right: the steady current in the Black Sea is often mentioned by the ancients; cf. Lucr. v. 507 *Pontos, mare certo quod fluit aestu;* Sen. *Nat. Quaest.* IV. 2. 29 *ob hoc Pontus in infernum mare assidue fluit rapidus...in unam partem semper pronus et torrens;* Lucan, III. 277 *quaque fretum torrens Maeotidos egerit undas | Pontus.*

46. **summo** for *maximo,* for metrical convenience. Augustus, and all his successors, assumed the title of *pontifex maximus.*

47. **et litora:** elsewhere they were a matter of course; here they were less to be looked for.

48. **delatore:** see n. to III. 116. The variety here mentioned is less dangerous and corresponds more closely to the συκοφάνται in Aristophanes.

dispersi...nudo, 'the inspectors of sea-weed, scattered everywhere, would at once bring an action against the naked boatman'. **algae** suggests that the most worthless trifle could not escape their prying eyes.

nudo, 'in his shirt-sleeves'; cf. John XXI. 3 'Simon Peter saith unto them, I go a fishing...7 So when Simon Peter heard it was the Lord, he girt his coat about him (for he was *naked*).'

51. **vivaria:** see n. to III. 308. Mart. (IV. 30) speaks of the fish in the Lucrine lake as the property of Domitian and sacred; he actually says that they recognise their master and lick his mighty hand!

53. **Palfurius Sura** and **Armillatus, two jurists,** are mentioned as zealous maintainers of the imperial claims: of the latter nothing is known: Suet. (*Dom.* 13) tells us that the former was once expelled from the senate and was a distinguished orator.

55. **res fisci est,** 'is the property of the imperial treasury'. There were at this time three treasuries at Rome: (i) the *aerarium Saturni,* in the temple of Saturn; it received the taxes and dues levied in Rome and Italy and the revenues of the senatorial provinces, and defrayed the expenses of administration in these parts of the empire; (ii) the *aerarium militare,* which was instituted by Augustus A.D. 6 in order to provide pensions for his soldiers: its main source of income was the *vicesima hereditatum,* a 5 per cent succession duty; (iii) the *fiscus,* or imperial treasury, probably instituted by Claudius, at least not by Augustus, as he never speaks of it in the *Monumentum Ancyranum* but

always of *mea pecunia*. The *fiscus* was managed by an army of
imperial *procuratores*; and from it were provided all sums re-
quired for the defence of the empire and for the administration
of the imperial provinces. The same provinces originally sup-
plied its revenue, but it gradually extended its claims (as the
present instance may show), until it entirely superseded the
old *aerarium* or 'senatorial treasury'.

56. **ne pereat,** 'that he may get some good by it', lit. 'that
it may not go for nothing': the same sense would be conveyed
by *ut imputare possit*. Juv. does not mean that the fish would be
spoilt but that it would be forcibly taken from its captor, who
would get no thanks.

iam letifero cet.: the language is mock-heroic; *letifero*,
because autumn is the season of fevers and agues; cf. X. 221.

57. **quartanam:** this is a fever between the attacks of which
there was an interval of two (not three) days: it was never fatal:
cf. Celsus III. 15 *quartana neminem iugulat*: hence it is welcomed
by invalids.

58. **recentem** is predicate.

59. **hic,** i.e. *piscator*. **auster:** this wind (the Scirocco) is
spoken of as particularly unfavourable to fresh meat or fish;
cf. Hor. *Sat.* II. 2. 40 *at vos, | praesentes austri, coquite horum
obsonia*.

60. **lacus:** there are two lakes below where the fisherman
now found himself, *lacus Albanus* and *l. Nemorensis*, now *lago
Albano* and *lago di Nemi*.

quamquam: Cicero and his contemporaries never use
quamquam with an adj., participle (as here), or adverb, but
always with a finite verb following: they might so use *quamvis*,
etsi, or *etiamsi*, but only where one verb (*servat* here) is under-
stood a second time. But the use is common in silver-age Latin;
cf. l. 79 and Tac. *Ann.* VI. 30 *haec, mira quamquam, fidem ex eo
trahebant* cet. In both the latter instances the same verb is not
understood; and Cicero must have written *quamquam dira erant
tempora, quamquam mira erant*. (See Madv. on Cic. *de Fin.* V. 68.)

61. **Alba Longa,** according to the legend, was founded by
Ascanius, son of Aeneas, who made it his seat of government
instead of Lavinium; cf. XII. 70 ff. It was destroyed (*diruta*) by
the Romans under Tullus Hostilius; but it appears that it still
retained a sacred fire which purported to have been brought
from Troy; and the celebration of the *feriae Latinae* on the
Alban Mount indicates the historical fact that Alba was once
the head of a Latin confederacy.

Domitian constantly resided at his *arx Albana*; see n. to l. 145.
minorem, sc. *Vestā Romanā.*

62. **miratrix:** cf. *turba salutatrix,* V. 21; *victrix turba,* XV. 81;
in Cicero the verbal nouns thus formed denote a permanent
quality; but, already in Livy, they are used as here to describe
a single action or a passing emotion.

64. **admissa,** 'with the right of entrée': the word is tech-
nically used of admission at court: cf. Pliny, *Paneg.* 48 (of
Domitian's court) *obversabantur foribus horror et minae et par
metus admissis et exclusis.*

65. **Atriden,** i.e. the emperor; see n. to *Automedon,* I. 61.

66. **privatis...focis,** 'too great for a subject's kitchen';
cf. *Procula minor,* III. 203. Classical Latin expresses this by the
positive and *pro*; cf. Caes. *Bell. Gall.* I. 2 *pro multitudine homi-
num...angustos se fines habere arbitrabantur.* **genialis...dies,**
'spend today in enjoyment'; compare such phrases as *genio
indulgere,* 'to treat yourself well'.

67. **iste** is used where classical usage requires *hic*; this
confusion is common in poets of the silver age; cf. *istos colles,*
VI. 295. **stomachum...sagina,** lit. 'to distend your stomach
with cramming'. The phrase is remarkable, esp. from a fisher-
man to an emperor: does Juv. mean to parody the common
animum laxare, 'to unbend the mind'? Cf. Mart. IV. 8. 9 (of
Dom.) *laxatur nectare Caesar.* There is no foundation for the
old explanation that Dom. is asked to prepare for the pleasures
of the table by taking an emetic.

68. **tua...saecula** = your happy reign. The plur. is due
to metre: Pliny, writing to (and of) Trajan, often uses *tuum
saeculum* in this sense but never the plural.

69. **ipse capi voluit:** cf. Mart. IX. 31. 5 (of a goose sacrificed
in honour of Domitian) *ipse suas anser properavit laetus ad aras.*

quid apertius? i.e. no flattery could be more barefaced; cf.
Cic. *Lael.* 99 *aperte adulantem nemo non videt, nisi qui admodum
est excors.*

70. **cristae:** so we call a man 'crest-fallen': we may trans-
late, 'he began to plume himself'.

71. **dis aequa:** cf. Pliny, *Paneg.* 4 (of the Roman emperors)
principem quem aequata dis immortalibus potestas deceret.
Domitian himself began a public circular, professing to come
from his treasury officials, with the words *dominus et deus noster
hoc fieri iubet* (Suet. *Dom.* 13); Martial constantly uses such
language to him. For the elliptic comparison (*dis = deorum
potestati*), see n. to III. 74.

72. **patinae mensura,** 'a dish of the right size'; cf. Thuc.
III. 20 τὴν ξυμμέτρησιν τῶν κλιμάκων, 'the right length for the ladders'.

73. **proceres** are not the same as *patres* in l. 64; the latter
are casual senators waiting for Domitian's levée; but the *pro-
ceres* (as appears from *amicitiae* l. 75, *comes* l. 84, *amici* l. 88) are
the personages technically styled *amici Caesaris* or 'Privy
Councillors'. This Cabinet of advisers originated by Augustus
was more formally constituted by Tiberius and continued by
his successors: cf. Suet. *Tib.* 55 *viginti sibi e numero principum
civitatis depoposcerat veluti consiliarios in negotiis publicis*;
idem, *Tit.* 7 *amicos elegit, quibus etiam post eum principes ut et
sibi et rei p. necessariis adquieverunt praecipueque sunt usi.* The
Cabinet generally included not only the chief senators, such as
the consuls, *praefectus urbi* and other consulars, but also those
knights who occupied the three or four highest posts accessible
to their order; see note to l. 32. Eleven are mentioned as
present on this occasion.

Thus those who were called the Emperor's 'friends' might
be, as in this case, those to whom he was least friendly.

74. **magnae:** 'with Majesty'; see n. to I. 33.

75. **Liburno:** the name of this Dalmatian people is com-
monly used for slaves of different kinds, for an usher here and
Mart. I. 49. 33, for a sedan-bearer Juv. VI. 477. Cf. the use of
Suisse in French for a beadle or ornamental porter.

76. **abolla:** see n. to III. 115.

77. Pegasus, a learned jurist (according to the Schol. known
as *liber, non homo*) had risen to the dignity of *praefectus urbi.*
There is an interesting account of this office in Tacitus (*Ann.*
VI. 10 and 11). In Republican times the title was given as a mere
compliment to any young man of rank; he had no real powers
but was supposed to be the formal representative of the consuls
during their compulsory absence from Rome for the cere-
monies of the *feriae Latinae* on the Alban Mount. Augustus
conferred the title upon the holder of a really great office: the
praefectus urbi was regularly a consular, and the post was the
crown of the consular career. It was held for a number of years,
sometimes for life. This magistrate had represented Augustus
during his absences from Rome, and, during the latter years of
Tiberius' reign had become the greatest power in the state
next to the emperor.

attonitae does not express the surprise of the city at P.'s
appointment but the reign of terror caused by Domitian's
ferocity; cf. XI. 199.

positus would be *praepositus* in prose.

vilicus: where all were slaves (cf. l. 38), the chief magistrate could only be called a slave-driver and a slave himself.

78. **aliud,** sc. *erant quam vilici?*

quorum: the plur. does not necessarily mean that there was more than one *praefectus urbi* at the same time. There is an isolated statement that Domitian created a number of *praefecti urbi*: but it occurs in a very late writer whose unsupported evidence is worth little. Juv. may be comparing Pegasus with his predecessors in office, or may be thinking of the equestrian *praefecti* as well: see n. to l. 32.

79. **quamquam:** see n. to l. 60; the clause introduced by it here is adverbial.

80. **inermi,** 'without her sword': this is a criticism on P.; though good and honest, he was too weak to fill his place properly in such troublous times.

81. Vibius Crispus, *vir ingenii iucundi et elegantis* says Quintilian, who gives an instance of his wit (v. 13. 48). Tacitus (*Hist.* II. 10) calls him 'great rather than good'. When Domitian had just succeeded his brother, Crispus was asked by a courtier whether anyone was then in presence of the emperor; he replied, *ne musca quidem*, alluding to the imperial passion for killing flies (Suet. *Dom.* 3). With *Crispi senectus*, comp. *spatium rhombi* in l. 39.

82. **mite ingenium,** 'a gentle spirit', is in apposition with *Crispi senectus.*

84. **comes Caesaris** is used technically like *amicus Caesaris*; when used strictly, the terms are not identical, the *comites* being specially appointed for each progress or campaign; thus every *comes* would be an *amicus* but not every *amicus* a *comes*. Our 'count' is derived from this sense of *comes*.

clades and **pestis** are names for Domitian. Pliny, with less restraint, calls him *immanissima belua* (*Paneg.* 48), *optimi cuiusque spoliator et carnifex* (*ibid.* 90), and *avidissimus praedo* (*ibid.* 94).

85. **honestum,** 'good', 'morally right', not 'honest' in our sense of the word.

liceret, not *licuisset*, because, if the condition were changed to an affirmation, the verb would be *non licebat*, not *licuit*; cf. Cic. *Orat.* 29 *Pericles si tenui genere uteretur* (the affirmation would be *non utebatur*), *nunquam fulgere, tonare, permiscere Graeciam dictus esset.*

violentius, 'more dangerous'.

87. **cum quo** (see n. to l. 9) is governed by *locuturi*, which

is equivalent to *si locuturus erat*: the meaning is that a man's life was not safe in talking to the emperor, however innocent or trivial the subject of conversation might be; cf. Tac. *Ann.* VI. 7 (of the victims of the *delatores*) *perinde in foro, in convivio, quaqua de re locuti incusabantur*.

aut nimboso: the spondaic hexameters in Juv. (proper names and Greek words being left out of account) are as follows: III. 17, 273; IV. 87; VI. 429; IX. 111; X. 88, 304, 332; XI. 68, 71, 133; XII. 117, 121; XIII. 191; XIV. 115, 165, 326; XV. 36. He also gives variety to his verse by ending it often with a monosyllable and occasionally with a word of four or five syllables.

88. **amici** is emphatic and ironical; see n. to l. 73.

90. **civis,** not merely 'citizen', but 'good citizen' or 'patriot': cf. the use of *civilis*. **posset:** subj., because of the consecutive force contained in *qui*.

91. **vitam inpendere vero:** cf. Tac. *Ann.* XII. 65 *sed ita de se meritum Caesarem ut vitam usui eius inpenderet*, 'Claudius has treated me so well, that I am willing to sacrifice my life to his interests'.

93. **solstitia,** 'summers'; cf. Virg. *Georg.* I. 100 *umida solstitia atque hiemes orate serenas*; the sing. would be used but for the metrical inconvenience of *octogesimum*.

armis is correctly used for 'defensive armour' not 'weapons'. Cf. Sen. *Dial.* IV. 33, 2 *notissima vox est eius qui in cultu regum consenuerat. cum illum quidam interrogaret, quomodo rarissimam rem in aula consecutus esset, senectutem: 'iniurias' inquit 'accipiendo et gratias agendo'.*

94. **proximus,** sc. *Crispo. Acilius Glabrio*, an *aequalis* of Crispus, and his son came next; the latter was consul together with Trajan, the future emperor, in A.D. 91; he was exiled and afterwards put to death by Domitian in spite of the artifices mentioned here.

95. **iuvene** = *filio*: see n. to *iuvenes* III. 158.

96. **olim...est,** 'has long been'; see n. to III. 163.

97. **in nobilitate,** 'in the case of an aristocrat'.

98. **malim,** 'I prefer to be', as I am, a nobody.

fraterculus gigantis: the little brother of a giant is a giant himself; and all the giants were γηγενεῖς, 'born of earth'; but the Latin equivalent, *terrae filius*, was used to express 'the son of nobody in particular'; cf. Pers. VI. 57 and 59 *progenies terrae, terrae filius*. Juv. uses the ambiguity to confer a comic distinction upon those who have no ancestors (cf. VI. 13).

100. **Albana…harena:** i.e. in the amphitheatre attached to Domitian's villa at Alba.

101. **venator,** 'as a beast-fighter'; the word is technically used for those who fought wild beasts in the arena; this was a regular part of a show, coming at the beginning, and was called *venatio*; the spear which the men carried was called *venabulum*; cf. I. 23. The combatant would naturally be lightly dressed (*nudus*). Glabrio tried in vain to escape Domitian's jealousy by simulating these low tastes.

iam, 'by this time'.

103. L. Junius Brutus simulated idiocy to escape the jealousy of his uncle Tarquin: cf. Livy, I. 56. 7 *neque in animo suo quicquam regi timendum neque in fortuna concupiscendum relinquere statuit contemptuque tutus esse.* Hence he got the name Brutus.

barbato inponere regi, 'to outwit a prince with much more beard than brains' (Gifford). Cf. XVI. 31: εὐήθεια is supposed to be characteristic of the good old times. That the razor is a very ancient invention may be seen from the Homeric phrase ἐπὶ ξυροῦ ἵσταται ἀκμῆς: the Romans however let their beards grow, till Africanus Minor set the fashion of shaving daily; cf. Pliny, *Nat. Hist.* VII. 211 *primus omnium radi cotidie instituit Africanus sequens.* The custom of shaving, at least after middle life (see n. to VI. 105), was universal until the emperor Hadrian brought the beard back into fashion and was followed by his successors.

104. **melior vultu,** 'more cheerful'; cf. Mart. IV. 1. 4 *semper et hoc vultu vel meliore nite*; Juv. IX. 12 *vultus gravis.*

105. Rubrius Gallus is mentioned elsewhere as a general under Otho and Vespasian: the 'unspeakable crime in the past', with which he was charged, is said to have been an intrigue with Julia, the daughter of Titus.

106. 'And nevertheless more impudent than a profligate turned satirist': for the restricted meaning of the word *satura*, as used by Juv., see Introduction, pp. xxxi–xxxiv.

107. **Montani…venter:** cf. *spatium rhombi*, l. 39. It is not certain whether the Montanus, who distinguished himself so greatly on this occasion, is mentioned elsewhere: he was evidently an old man at this time; cf. ll. 136 ff.

108. Crispinus only comes in for this passing and comparatively lenient mention, in spite of the beginning of the satire.

To use scent in the morning is a sign of an idle and dissolute life: the custom was to use perfumes, together with chaplets of flowers, only at the *comissatio*, which followed the *cena*; cf. IX.

128 *dum bibimus, dum serta, unguenta, puellas | poscimus, obrepit non intellecta senectus*; also XI. 122; XV. 50.

109. **funera:** Roman poetry is full of allusions to the custom of pouring perfumes and scented oils over the pyres of the dead; the body also was perfumed: cf. Propert. V. 7. 32 *cur nardo flammae non oluere meae?* Pers. III. 104 *compositus lecto crassisque lutatus* (daubed) *amomis.*

110. **Pompeius,** a *delator,* is not mentioned elsewhere.

aperire ('to slit') is governed by *saevior*: cf. Sen. *Dial.* III. 2. 2 *ira iussit...alium servili manu regalem aperire iugulum* (surely a quotation from tragedy though not so marked by Haase).

susurrus has here, like ψιθυρισμός, the notion of slandering.

111–12. Cornelius Fuscus, then praefect of the praetorian guard, was defeated and killed as chief commander in the Dacian war A.D. 86–88; there is an epitaph for his grave in Mart. VI. 76, and some account of his earlier years in Tac. *Hist.* II. 86. Now, says Juv., he was studying the art of war in the lap of luxury.

113. Fabricius Veiento and L. Valerius Catullus Messalinus were two of the most infamous *delatores* under Domitian. The former continued to be about the court under Nerva; cf. Pliny, *Epp.* IV. 22 for a story in which both he and Catullus come in. Catullus was consul A.D. 73; he was living in 93 (cf. Tac. *Agric.* 45 *intra Albanam arcem sententia Messalini strepebat*) but appears, from Pliny, *l.l.,* not to have survived Domitian (A.D. 96).

114. **numquam visae:** Catullus was blind; cf. Pliny, *l.l. luminibus orbatus ingenio saevo mala caecitatis addiderat.*

115. Many as were the scandals of the age, Catullus surpassed them all; cf. II. 143 *vicit et hoc monstrum tunicati fuscina Gracchi,* 'even this scandal was surpassed by the tunic and trident of a Gracchus'.

116. **a ponte satelles:** 'a beggar-courtier': *a ponte = mendicus*: the phrase is similar to *meretrix Augusta,* VI. 118, and *mulio consul,* VIII. 148: in each the juxta-position of the two nouns is a kind of *oxymoron.* Catullus is a *satelles,* 'a tyrant's minion', these *satellites* (δορυφόροι) being, like *arx* (see n. to l. 145), a regular attribute of the tyrant: so Pliny (*Paneg.* 23) says of Trajan, contrasting him with Domitian, that he entered Rome, not *stipatus satellitum manu*; so Pothinus, Pompey's murderer, is constantly called the *satelles* of King Ptolemy: e.g. Lucan, VIII. 675 *Pharius...satelles.* Such persons live in kings' houses and are as far removed as possible from beggars on the bridge. But the blindness and servile manners of Catullus make

Juv. call him a beggar, just as he calls the consul, who holds the reins himself, *mulio*.

For *a ponte*, cf. XIV. 134 *aliquis de ponte*, 'a beggar': bridges were favourite stands (*stationes*) of beggars: cf. v. 8; Mart. x. 5. 3 *erret per urbem pontis exul et clivi* (ie.. driven even from the beggars' stands).

The epithet **dirus**, 'formidable', itself indicates that *satelles* means a powerful and wicked man, not, as it is generally explained, a beggar; applied to the latter it can only mean 'ill-omened', and so M.; but the first meaning would at once occur to the reader's mind in connection with such a notorious *delator* as Catullus.

117. **Aricinos...axes:** Aricia was in a hollow on the Appian Way about 20 miles from Rome: *Aricini* may be said of carriages coming to, or leaving, the village: see n. to VI. 5. For Aricia as a resort of beggars, cf. Mart. II. 19. 3 *debet Aricino recumbere clivo,* | *quem tua felicem, Zoile, cena facit.*

118. **devexae**, 'descending the slope': the beggar is demonstrative in his gratitude, when the carriage leaves him behind; cf. Mart. I. 3. 7 (of a reciter thanking an audience for their applause) *dum basia iactas.*

119. **plurima dixit**, i.e. he was loudest of all in its praise: cf. Pliny, *Epp.* II. 14. 8 (of an audience at a speech) *nam plerique non audiunt* (listen), *nec ulli magis laudant.*

120. **at** may be transl. 'unfortunately': it seems to have sometimes an ironical force which *sed*, for instance, has not: cf. Cic. *ad Att.* VII. 21. 2 *consul ei rescripsit, ut prius ipse in Picenum* (sc. *iret*); *at illud totum erat amissum*; ibid. VIII. 8. 2 *fulsisse mihi videbatur* τὸ καλὸν *ad oculos eius...: at ille tibi* πολλὰ χαίρειν τῷ καλῷ *dicens pergit Brundisium.*

121. **sic**, i.e. with equal knowledge of the facts.

pugnas is perhaps 'feints'; cf. the Comic *pugnae aliquid dare*, 'to play a trick'; this avoids the tautology. **Cilix** is a gladiator, perhaps named after the country of his birth.

122. **pegma** was a scaffold of wood, erected in the amphitheatre or at other shows, to produce certain stage effects; it could be heightened or lowered by machinery: cf. Sen. *Epp.* 88. 22 *machinatores, qui pegmata per se surgentia excogitant.* Our 'pageant' is derived from it. The *pueri* here mentioned may be actors (so of acrobats Mart. v. 12. 4 *septem...pueros levat vel octo*); or possibly Cupids who were raised to the roof by help of the machine at some point of the spectacle. **velaria** seems to be used first in this passage for *vela*.

123. **fanaticus:** the priests of Bellona were so called, because of the frantic excitement displayed in their ritual: cf. Lucan, I. 565 *quos sectis Bellona lacertis | saeva movet*; Juv. applies the word also to the Galli or priests of Cybele (II. 112); Martial applies *entheus, entheatus* to both (XII. 57. 11; XI. 84. 4).

oestro, 'frenzy': so *thyrsus* is used for Bacchic inspiration, and then metaphorically as in Lucr. I. 922 *acri | percussit thyrso laudis spes magna meum cor.*

126. **temone Britanno:** this is a reference to the *esseda* or scythed chariots in which the Britons used to fight: they ran out upon the pole in battle: cf. Caes. *Bell. Gall.* IV. 33 *per temonem percurrere consuerint.*

127. **excidet,** 'will be hurled'. The name Arviragus, used by Shakespeare in *Cymbeline* for a British prince, must be taken, indirectly, from here.

That the fish, thought caught off Picenum, came from a distant country, is proved by its hostile appearance.

128. **erectas in terga sudes:** a difficult phrase, because of the acc. *terga. in,* followed by the acc., should mean 'against', so that *sudes* are probably the lateral fins. The fins are called *sudes* to give a military sound to the description.

129. **Fabricius** is one of Veiento's names. In a fragment of Statius, quoted by the Schol. on l. 94, he is called Fabius Veiento; but the text is sound: Statius means that in the German war Veiento's strategy (he is called *prudens,* l. 113) was worthy of Q. Fabius Maximus, the hero whose inaction foiled Hannibal.

130. The Emperor, as president of the Council, now asks the different members of it for their *sententia.* **quid censes** 'what do you move?' the word is technical. **conciditur?** 'is it to be cut up?'; see n. to *quaero,* III. 296.

131. **testa alta,** 'a deep dish'.

132. **muro,** 'circuit'; it does not mean that the turbot is to be baked in a pie. **orbem** is the circumference of the fish.

133. 'The dish demands a mighty Prometheus, and without delay'; Prometheus is a nickname for a potter, as Prometheus was the first worker in clay when he made the human race from that material; M. quotes Lucian, *Prom.* 2 οἱ Ἀθηναῖοι τοὺς χυτρέας καὶ ἱπνοποιοὺς καὶ πάντας ὅσοι πηλουργοὶ Προμηθέας ἀπεκάλουν.

134. **properate:** the plur. shows that the order is given to attendants.

135. **castra,** perhaps 'court'; as the monarch was *imperator,* so his residence, whether inside or outside Rome, was called

castra; and *castrensis* is often found in the sense of 'imperial' in inscriptions. There is no earlier instance of this sense in literature.

136. **vicit...sententia:** just so ἐνίκησεν ἡ γνώμη: cf. Pliny, *Paneg.* 76 (of business in the senate) *vicit sententia non prima sed melior.* Even in prose, the verb generally precedes the subject in this common phrase.

137. The most dissolute of his successors failed to rival the prodigality of Nero: his birth made things possible for him which a Flavius could not venture upon; and also he left the state practically bankrupt at his death.

138. **aliamque famem:** *alteram* would be used in prose. There is no reference to the use of emetics but only to the fact that the appetite revives after drinking wine; cf. Mart. v. 78. 17 *post haec omnia, forte si movebit | Bacchus quam solet esuritionem.*

139. **usus edendi,** 'skill in gastronomy'.

140. **tempestas,** though not used in classical prose for *tempus,* is common in the poets and also in Livy and Sallust; cf. Cic. *de Orat.* III. 153.

It appears that the best oysters came from Circei on the Latin coast, the Lucrine lake near Baiae on the coast of Campania, and Rutupiae (now Richborough) in Kent; this connoisseur could distinguish each kind at the first bite.

143. **semel aspecti** = *quem semel aspexerat.*

echini: some sort of sea-urchin which was considered a delicacy; cf. Hor. *Sat.* II. 4. 33 *ostrea Circeis, Miseno oriuntur echini.*

144. **surgitur,** 'the meeting rises'; for the impers., cf. *itur,* l. 65.

145. **Albanam...arcem:** the word *arx* is used here and in Tac. *Agric.* 45 (quoted on l. 113), as an invidious substitute for *villa, arx* being regularly used for the fortified residence of a tyrant; cf. X. 307 *in arce tyrannus;* Senec. *Controv.* 1. 7. 16 *vidi filium in arce* (= τυραννοῦντα εἶδον τὸν υἱόν).

146. **attonitos:** see n. to l. 77.

147. The country of the Chatti was the modern Hesse, the Sycambri lived in what is now Rhenish Prussia; for the former, cf. Mart. II. 2. 5 (published in 85 or 86) *frater Idumaeos meruit cum patre triumphos | (i.e. Vespasian and Titus together triumphed over Syria); quae datur ex Chattis laurea, tota tua est.* If Domitian fought against the Sycambri, we have no mention of it elsewhere.

148. **tamquam et:** so P and T: other MSS. omit *et:* edd. generally read *ex. tamquam et* is harsh for *et tamquam,* but I do not feel sure that emendation is necessary.

diversis partibus, 'a distant quarter': cf. Ovid, *Ars,* I. 685 *iam nurus ad Priamum diverso venerat orbe,* i.e. from Sparta to Troy. L. translates 'from opposite quarters': but the sing. *epistula* makes this less probable: for the meaning of *diversus,* see n. to III. 268.

149. 'Terrified despatches had flown in headlong haste'; the metaphor is from the flight of a terrified bird. The Schol. states that a laurel or a feather was fastened on despatches, one in sign of a victory, the other of a defeat; the latter statement is borne out to some extent by Stat. *Silv.* V. I. 93 *nullaque famosa signatur lancea pinna*; but this explanation is not needed here.

150–4. *Such occupations were undignified enough for an emperor; but he was better so employed than in destroying the noblest Romans.*

151. Suetonius gives a list of the chief victims of Domitian's cruelty (*Dom.* 10): they include Civica Cerealis, Salvidienus Orfitus, Acilius Glabrio (l. 95), Salvius Cocceianus, Mettius Pompusianus (one of whose crimes was that he carried about an atlas!), Junius Rusticus, and Aelius Lamia (l. 154). For his *saevitia,* cf. also Pliny *Paneg.* 48 *domo quam nuper illa immanissima belua plurimo terrore munierat, cum...nunc propinquorum sanguinem lamberet, nunc se ad clarissimorum civium strages caedesque proferret.*

152. **animas,** 'lives'.

153. **cerdonibus:** see n. to VIII. 182. The conspirators, who put an end to Domitian, were all of the meanest rank.

154. **coeperat:** the aorist, not imperf. or pluperf., would be used here after *postquam* in classical Latin.

nocuit, 'proved his ruin'; *nocere* is often more than merely 'to injure'; cf. x. 8.

Lamiarum: Aelius Lamia is taken as a type, because of his especially ancient ancestry; cf. VI. 385, and Hor. *Carm.* III. 17. 1 *Aeli, vetusto nobilis ab Lamo.*

SATIRE V

A DINNER-PARTY; GREAT MEN AND THEIR CLIENTS

A good illustration of this satire is supplied by Pliny, *Epp.* II. 6 *longum est altius repetere...ut homo minime familiaris cenarem apud quendam, ut sibi videbatur, lautum et diligentem, ut mihi, sordidum simul et sumptuosum. nam sibi et paucis opima quaedam,*

ceteris vilia et minuta ponebat. vinum etiam...in tria genera
descripserat, non ut potestas eligendi, sed ne ius esset recusandi,
aliud sibi et nobis, aliud minoribus amicis...aliud suis nostrisque
libertis. Pliny expresses his disgust at these arrangements, and
ends his letter with a piece of advice to his young friend and
correspondent: *igitur memento nihil magis esse vitandum quam*
istam luxuriae et sordium novam societatem. Martial attacks the
same meanness; see esp. III. 60. Lucian (*de Merc. Cond.* 26)
gives a similar account of the treatment of learned Greeks at
the tables of their great Roman patrons, in his time. Julius
Caesar on the other hand sent his baker to prison, because he
had dared to put before him a finer bread than he had given to
his guests (Suet. *Iul.* 48).

Some may be reminded of Macaulay's account of the social
standing of an English domestic chaplain, two centuries ago: 'he
was permitted to dine with the family; but he was expected to
content himself with the plainest fare. He might fill himself with
the corned beef and the carrots: but, as soon as the tarts and
cheesecakes made their appearance, he quitted his seat, and
stood aloof till he was summoned to return thanks for the re-
past, from a great part of which he had been excluded' (*History*
of England, chap. 3). But even this was less insulting than the
treatment described by Juvenal and Pliny.

1–11. *A man's character is worthless when, for the sake of food,*
he can swallow such insults as you submit to, my friend Trebius;
it would be better to beg your bread in the streets.

1. **te:** the whole satire is addressed to a perhaps imaginary
Trebius.

propositi, 'line of life': cf. IX. 20 *flexisse videris | propositum et*
vitae contrarius ire priori.

2. **ut bona summa putes:** the clause may be consecutive or
may merely define *mens.* The plur. *bona* is used for metrical
reasons instead of the *summum bonum* of prose.

aliena...quadra, 'on the crumbs from another man's
table': *quadra* is one of the quarters into which the round loaf of
bread (*placenta*) could be broken up. Our hot-cross buns have
probably been evolved from this ancient form of loaf. Martial
uses the word four times, thrice of the section of a *placenta*
(III. 77. 3; VI. 75, 1; IX. 90. 18), and once of a hunch of cheese
(XII. 32. 18).

3. **illa,** 'the treatment'. **nec** here may be either οὔτε or οὐδέ,
'neither' or 'not even', the Latin of this age making no distinc-

tion between these; see n. to XI. 7. For Sarmentus, who was a
table-wit of Augustus, cf. Hor. *Sat.* I. 5. 52 and Palmer's note
there.

iniquas, 'ill-sorted' M.: at the table of Caesar (i.e. Augustus)
distinctions were made between the guests: there was not the
aequa libertas which also Juv. attacks VIII. 177.

4. Gabba is twice mentioned by Martial as a jester at the
court of Augustus (I. 41. 16; X. 101).

5. **quamvis iurato,** 'even on oath'; *iuratus*, though passive
in form, is active here in meaning; it can have a passive meaning
as in *iurata superis unda* i.e. the Styx. One of the main syntactical
differences between Latin and Greek is the lack of participles
in the former: no Latin verbs have a perf. or aor. participle
active, except (i) deponents, e.g. *sequor* has *secutus*, (ii) a limited
number of verbs of which *iuro* is one; other similar participles
are *adultus* = *qui adolevit, nupta* = *quae nupsit, tempus prae-
teritum* = *quod praeteriit.* Some adjectives, such as *cautus,
quietus, tacitus,* are really participles which belong to this class.

6–9. Suppose you cannot get otherwise what your belly de-
mands—and it does not need much—why not start as a beggar?

6. **hoc...ipsum,** αὐτὸ τοῦτο, 'even the amount'. The clause
introduced by *puta* is substituted for the protasis of a con-
ditional sentence, the apodosis beginning at *nulla crepido* cet.
For the constr., cf. II. 153 *sed tu vera puta: Curius quid sentit?*
'but, if you suppose it true, what does Curius feel?'

8. By **crepido** the steps of a great house or public building
are meant; these, as well as the bridges (see n. to IV. 116), were
a natural place for beggars to take their stand (*consistere*).

tegetis pars, 'a scrap of a mat'; the *teges* is often mentioned
as a badge of poverty: slaves and beggars used it to sleep on; it
was, sometimes at least, made of rushes: cf. Mart. XI. 32. 2
de bibula sarta palude teges.

9. **dimidia:** abl. of amount of difference.

tantine iniuria cenae 'is the insult of a dinner worth so
much', i.e. that you should prefer it to a beggar's fare: for
tanti, see n. to III. 54. *cenae* is gen. of definition: the *cena* itself is
an *iniuria*; hence the phrase = *iniuriosa cena.*

10. **ieiuna,** 'importunate'. The subject of **possit** is *fames*
understood.

illic, i.e. *in crepidine vel ponte.*

11. **canini,** 'thrown to dogs'; cf. Mart. X. 5. 3 *erret per
urbem pontis exul* (banished from) *et clivi,* | *interque raucos
ultimus rogatores* | *oret caninas panis improbi buccas.*

12–23. *An invitation to dinner seldom comes your way; when it does, it is considered to repay in full all the services you have rendered to your patron at great inconvenience to yourself.*

12. **fige,** 'consider'; cf. XI. 28. **discumbere,** 'to fill a place', is regularly used by silver-age writers for reclining at a meal, either of a number or of one person; cf. VI. 434; Suet. *Aug.* 74 *cum convivae cenare inciperent prius quam ille discumberet.* Cicero sometimes uses it in the sense of 'to go to bed': e.g. *de Invent.* II. 14 *cenati discubuerunt ibidem.*

13. **officiorum** = what is called elsewhere *opera togata*, 'attendance as a client': see n. to III. 239.

14. **magnae,** 'with the great'; see n. to I. 33.

inputat, 'claims credit for it, makes a favour of it'; this word is not used by the Augustan writers, except twice by Ovid, but is constantly found in both prose and poetry of the silver age. It is a metaphor taken from book-keeping, and means 'to set down to' an account with someone, and so often takes a dat. of the person as well as an acc. of the thing. In its metaphorical use it has two distinct senses: (i) to claim credit from someone for something, (ii) to throw the blame for something on someone. For the former, which is much commoner, cf. VI. 179, Mart. V. 80. 1 *non totam mihi, si vacabis, horam | dones et licet imputes;* Suet. *Tib.* 53 *imputavit* (he regarded it as a favour) *quod non laqueo strangulatam in Gemonias abiecerit.* For the second sense, cf. Juv. II. 16 *hunc ego fatis | imputo* (I throw the blame for him on fate); Tac. *Hist.* II. 31 *illi initium belli nemo imputabat.* Cicero often uses *assignare* in the latter sense.

rex is constantly used in the sense of 'great man, patron': see n. to I. 136. Virro, the great man of this satire, is generally referred to as *ipse* or *dominus*, again as *rex* l. 130.

17. **tertia..culcita:** a common custom was to have three *lecti*, each with three places, round a square table, the fourth side being left unoccupied for the convenience of service; each guest had a *culcita* (often called *pulvillus*) on which he rested his left elbow. For a different arrangement of the table, see n. to III. 82.

Trebius is to have the worst place at table: he is to be *imus in imo lecto.*

vacuo is part of the predicate.

cessaret: the historic tense where the principal verb (**ait**) is primary, should be noted: **libuit** is a perfect of repeated action ('whenever the fancy has taken him'): a perfect of this kind may be followed (1) by historic tenses, e.g. Cic. *de Orat.* III. 196

in his si (if ever) *paulum modo offensum est, ut aut contractione brevius fieret aut productione longius, theatra tota reclamant*; (2) by primary tenses, e.g. Cic. *pro Flacco*, 11 *Graecus testis, cum* (whenever) *ea voluntate processit ut laedat, non iurisiurandi sed laedendi verba meditatur.* The former is probably more usual. For consecution of tenses after an ordinary perfect, cf. *Classical Review*, III. 6 ff.

18. **una simus,** 'let us have your company'.

votorum summa, 'the height of your ambition' is attained; *est* is understood; cf. Pliny, *Paneg.* 44 *erat summa votorum melior pessimo princeps.*

19. Trebius, the dependant, whose adventures are described throughout the satire, can require nothing further and has sufficient motive for all his exertions. One of the principal duties of a client was to rise early in order to call on (*salutare, mane domi videre*) his patron. Martial is full of complaints on this head: after he had quitted Rome for Spain, he writes to Juv. (XII. 18. 13 about A.D. 101) *ingenti fruor improboque somno | ...et totum mihi nunc repono quidquid | ter denos vigilaveram per annos; | ignota est toga*; it was necessary for *salutatores* to wear this garment; see n. to *togatae*, I. 96.

20. **ligulas dimittere,** 'to leave his shoe-strings untied'; *demittere* would suggest that the shoe-strings are trailing on the ground; *dimittere* means that they fly in all directions as he walks: cf. Mart. III. 36. 3 *horridus* (i.e. in disorder) *ut primo te semper mane salutem.*

21. **orbem,** 'their round' of visits.

22, 23. The client goes his round of calls before dawn, or at midnight. The untimeliness of the hour is of course exaggerated: cf. VII. 222.

dubiis: i.e. the stars are waning before the coming dawn.

23. The constellation of the Bear was also known to the ancients as *Plaustrum*, 'the Waggon' (here **serraca**); it was supposed to be driven by the 'Herdsman', **Bootes**; it is, or was, called 'Charles' Wain' in some parts of England. Cf. Hom. *Od.* v. 272 ὀψὲ δύοντα (*pigrum*) Βοώτην | ἄρκτον θ' ἣν καὶ ἅμαξαν ἐπίκλησιν καλέουσιν. It is called **frigida** as a northern constellation.

24–79. *And what a dinner after all! Your host drinks the best wine from jewelled cups handed him by the fairest of Greek pages; your wine is poison, your cup coarse and broken, your attendant an African negro who looks like a highwayman. The bread, the very water, supplied to you, is of a different and inferior quality. Was it worth while, you ask yourself, going through so much to get so little?*

24. **vinum,** sc. *tibi ponitur*: 'wine that greasy wool would refuse to put up with'. **sucida lana,** wool with the natural oil in it, is distinguished from *lota lana*: it was an important item in the medicine-chest of antiquity, being used, like a sponge, to apply fomentations of wine and vinegar: cf. Celsus, II. 33 *lana sucida ex* (previously steeped in) *aceto vel vino*. Good wine would naturally not be wasted on external applications.

25. The wine is so bad that it turns those who drink it into madmen; the Corybantes are the frenzied priests of Cybele; see n. to *fanaticus*, IV. 123.

26. **iurgia proludunt,** 'wrangling begins the fray'; cf. xv. 51 and Quint. v. 10. 71 *habent omnia initium, incrementum, summam, ut iurgium deinde caedes et strages. prolusio* is regularly opposed to *pugna* as 'skirmish' to 'battle'; cf. Cic. *in Caecil.* 47 *sin... in hac prolusione nihil fueris, quem te in ipsa pugna...fore putemus?* The phrase *prolusiones academicae* is barbarous and absurd. **et,** 'also', i.e. 'you go further and...'.

27. **saucius,** 'drunk': cf. Mart. IV. 66. 12 *incaluit quotiens saucia vena mero.* The word goes with the previous l.; for the pause after *saucius*, cf. I. 143. The other meaning of the word, 'wounded', is flat here and involves a tautology.

mappa: each guest brought a napkin of his own, which was used, not only for wiping the hands (hence χειρόμακτρον), but also for carrying away articles of food or trifling presents (ἀπο-φόρητα); cf. Petron. 66 *ego tamen duo [mala] sustuli et ecce in mappa alligata habeo; nam si aliquid muneris meo vernulae non tulero, habebo convicium* (he will scold me).

28. **vos** are the *clientes* of whom Trebius is one: Virro's freedmen form the opposing regiment.

29. **lagona** is an earthenware wine-jug, with a narrow neck and a handle; they were largely manufactured at Saguntum in Spain; cf. Mart. XIV. 108 *quae non sollicitus teneat servetque minister,* | *sume Saguntino pocula facta luto.* The abl. (*lagona*) is instrumental, the jugs being the actual weapons used; cf. Mart. VIII. 6. 7 *hoc cratere ferox commisit praelia Rhoetus.*

30. **ipse** here, and often in the satire, stands for the host to whom Juv. gives the name of Virro. 'The great man drinks wine that was bottled when the consul of the year wore long hair'; for **capillato,** see n. to *barbato*, IV. 103. Wine was placed, after

making, in *dolia*, and afterwards transferred (**diffusum**) to
smaller vessels known as *amphorae*; each *amphora* had a ticket or
inscription (*nota*) giving the year (i.e. consul) in which it was
filled and the place from which it came; cf. l. 34 and Petron.
34 *statim allatae sunt amphorae vitreae diligenter gypsatae,
quarum in cervicibus pittacia erant affixa cum hoc titulo: Falernum
Opimianum annorum centum.*

31. The Social Wars lasted from 91–88 B.C., so that this
wine would be nearly 200 years old; some Roman vintages
actually reached this age; thus Pliny (*Nat. Hist.* XIV. 55) says
of *Opimianum*, which was made in 121 B.C., the year when L.
Opimius was consul and C. Gracchus fell, *durant adhuc vina
ea CC fere annis, iam in speciem redacta mellis asperi.*

32. **cardiaco,** 'dyspeptic'; so καρδία in Thuc. II. 49 is 'the
stomach' and is wrongly translated *cor* by Lucr. VI. 1151; see
Munro, *ad loc.* Celsus uses **cardiacus** in this sense (where,
however, it is plain that acute disease of this kind is meant) and
recommends as a remedy *sorbere vini cyathum interpositaque
hora sumere alterum* (III. 19). The **cyathus** was a sort of spoon,
like a toddy-ladle, holding $\frac{1}{12}$ of a *sextarius*; it was used to
transfer wine from the *crater* (mixing-bowl) to the drinking-
cup.

numquam missurus, 'he never would give away': the fut.
partic. is used as the apodosis of a conditional sentence, as ἄν
διδούς might be: see n. to VI. 277: here the protasis is contained
in *cardiaco* (= *si c. sit*). For this sense of *mittere* see n. to III. 45.

33. Pliny (*Nat. Hist.* XIV. 59 ff.) arranges the chief Italian
wines in the following order, though admitting that tastes differ:
(i) Setine, the favourite of Augustus and many of his successors;
(ii) Falernian; (iii) Alban and Surrentine.

34. **patriam titulumque:** the *titulus* or *nota* included the
patria, i.e. the vineyard from which the wine came: see n. to
l. 30.

35. **fuligine:** in order that wine might mature more rapidly,
it was often placed after bottling in an *apotheca*, an upper room
to which smoke had free access: cf. Hor. *Carm.* III. 8. 11
amphorae fumum bibere institutae | consule Tullo. **testa** is a
common equivalent for *amphora*, and for any vessel made of
clay: cf. XV. 311.

36. Thrasea Paetus and his son-in-law Helvidius Priscus
both lost their lives from their love of liberty, the former under
Nero A.D. 66, the latter under Vespasian; so it was natural they
should keep the birthdays of the great *liberatores*, Brutus and

Cassius. It appears from this passage that, notwithstanding their Stoic principles, they appreciated good wine.

37. **Brutorum:** plur., because both Decimus and Marcus Brutus took part in the conspiracy against Caesar.

38. **Heliadum crustas,** 'cups overlaid with amber'; the Heliades are the daughters of Helios and sisters of Phaethon who, after his death, were changed into poplars and their tears into amber (*electrum* or *sucinum*): cf. Ovid *Met.* II. 340–66.

inaequales: cf. XIV. 62 *vasa aspera*. This is the only l. in Juv. which ends with three spondees.

39. **aurum,** 'a gold cup'; the context shows that such a cup is supposed also to have jewels on it.

41. It is feared that Trebius may use his nails to prise off one of the precious stones: cf. Mart. VIII. 33. 5 *an magis astuti derasa est ungue ministri | brattea* (a thin plate of metal), *de fulcro quam reor esse tuo*; Suet. *Claud.* 32 *convivae, qui pridie scyphum aureum subripuisse existimabatur, revocato in diem posterum calicem fictilem apposuit*.

For **acutos** there is a v. l. *amicos,* which is found in many MSS. and is written over *acutos* by the first hand in P. I suggest that the real text is *observet. amico | da veniam*. The corruption began by putting the pause at the end of the line: *amico* then became *amicos,* which being obviously absurd, was 'corrected'. The sarcastic use of *amicus* pervades the whole satire: cf. l. 173. If Juv. wrote *amico* here, *illi* in l. 42 is decidedly less harsh.

42. **illi,** 'belonging to him', i.e. Virro. He must not be blamed; the jasper on the cup so carefully watched is a famous jewel. The reading *illic,* 'on the cup', has little authority.

43. For the custom of transferring jewels from rings to cups, cf. Mart. XIV. 109 *calices gemmati. gemmatum Scythicis ut luceat ignibus* (i.e. emeralds) *aurum, | aspice. quot digitos exuit iste calix!*

45. 'The youth preferred to Iarbas' is Aeneas, whom Dido preferred to her African suitors; cf. Virg. *Aen.* IV. 36 *despectus Iarbas.* For the roundabout way of mentioning Aeneas, which is characteristic of Juv., see n. to III. 25. Virgil speaks of jasper on Aeneas' scabbard, *Aen.* IV. 261 *illi stellatus iaspide fulva | ensis erat.*

46. These cups were called *calices Vatinii,* but the metre makes a periphrasis necessary; Martial (XIV. 96) describes them as *vilia sutoris…monumenta Vatini.* Vatinius, who is to be distinguished from Caesar's instrument of the same name, had originally been a shoemaker at Beneventum and became power-

ful at Nero's court. Martial shows that the cups were named
after him because their shape recalled his long nose: they had
four spouts or nozzles and were evidently not valuable.

48. **rupto...vitro**, 'demanding sulphur with its glass
broken': there are two possible explanations: (1) the cup is
said to be asking for sulphur because it was customary to
barter broken glass for sticks tipped with sulphur, which served
as matches: cf. Mart. I. 41. 4 *qui pallentia sulfurata fractis |
permutat vitreis*; Stat. *Silv.* I. 6. 73 *quique comminutis | permutant
vitreis gregale sulfur*; (2) sulphur was used as cement: so the
Schol. If the second explanation be adopted, *vitro* is probably dat.

49. Cf. IV. 138.

50. **decocta** was water treated according to an invention of
Nero: it was first boiled (hence the name) to purify it and then
cooled with snow. Martial (XIV. 116) speaks of *decoctae nobile
frigus aquae.*

52. 'The water you clients drink is different.' And then a
fresh subject of complaint begins, the difference in attendance.
tibi is opposed to *ante ipsum* l. 56.

cursor: a groom is good enough to wait upon the client,
though his ordinary duty is to run in front of his master's
carriage; African slaves were commonly used as coachmen and
grooms.

54. **et** connects the whole line, which is an adjectival sen-
tence, with the adj. *Mauri*; it need not be translated.

nolis does not refer specially to Trebius, who would not be
likely to drive about much, but means 'whom one would not
care to meet...'. The meaning is either that he is so ugly as to
be alarming; or, as the darkness would make his ugliness less
striking, that he looks like a dangerous ruffian.

55. **veheris:** journeys were often performed at night; cf. X.
20 and n. to III. 236. **per** should strictly be *praeter*; cf. VIII. 147.
For **Latinae**, see n. to I. 171.

56. **flos Asiae** is a Greek page (*capillatus*, see n. to III. 186),
young and beautiful: Martial addresses such a slave as *decus
mensae* (VIII. 51. 19): the most valued came from Miletus. The
verb understood is *stat, stare* often meaning 'to wait at table':
see n. to l. 65.

paratus, 'who cost'; both *parare* and *comparare* constantly
have this sense; it may have been the colloquial equivalent of
emere, as *comprare* in Italian has no other meaning.

57. **census**, 'the revenues'; we might say 'Civil List'.
'Fighting Tullus' (who is to be distinguished from Servius

Tullius) and Ancus are perhaps mentioned with a reference to Hor. *Carm.* IV. 7. 15 *dives Tullus et Ancus.*

58. **ne te teneam,** 'to be short'.

59. **frivola,** 'trumpery', 'odds and ends'; cf. III. 198. Ganymede is used as a synonym for *pincerna* because of his functions at the divine banquets; youth and beauty are suggested by his name, whereas the real cup-bearer is black and ugly; cf. *a ponte satelles,* IV. 116.

60. **tot milibus emptus:** immense prices were given for slaves of this kind (*delicati* or *capillati*); Martial often speaks of 100,000 sesterces (about £1,000) and even twice as much being given for one of them.

61. Both **forma** and **aetas** are expressed by ὥρα, which would be the Greek equivalent.

62. **digna supercilio,** 'excuse his insolence'; the ordinary use of *dignus* (e.g. *dignus est propter aetatem qui superbiat*) is slightly modified.

ille is the *Gaetulus Ganymedes:* even he, who has less excuse, is inattentive to your wants.

63. With **calidae** and **gelidae,** *aquae* is understood.

64. **quippe,** 'for'; the only meaning of the word in Juv.: see n. to XIII. 26.

indignatur, 'he thinks it beneath him'.

65. **poscas:** subj. because it expresses the thought of the servant, 'because (thinks he) you ask . . .'.

se stante, 'while he waits'; cf. Mart. IV. 66. 9 *nec tener argolica missus de gente minister,* | *sed stetit inculti rustica turba foci;* Suet. *Iul.* 49 *Memmius etiam ad cyathum et vinum Nicomedi stetisse obicit (Caesari).*

66. **servis:** the abl. had become the normal case after *plenus* by Quintilian's time; cf. Quint. IX. 3. 1; Cicero and Caesar almost always use the gen.

67. **murmure:** the slave who hands you bread is sulky also; bread was handed round at table in baskets (*canistra,* l. 74).

68. **fractum** may = *qui fractus est,* and then means that the bread, being too stale and tough to cut, has been broken into lumps by the slave; but it also may stand for *qui frangitur* 'which costs you an effort to break'. In the latter case the time of *fractum* is not anterior to *porrexit;* cf. Livy, II. 36. 1 *servum sub furca caesum* (τυπτόμενον) *medio egerat circo;* these constructions are due to the lack of Latin participles. There is no tautology: Trebius has to break his bread before putting it into his mouth.

iam mucida: for *iam* with adjectives, see n. to III. 206.

69. 'To give your grinders exercise, and impossible to bite'; for **admittentia**, cf. III. 235: **agitent** is subj. because of the consecutive force of *quae*.

70. fictus, 'kneaded', though more commonly used of pottery, is perfectly appropriate here, and, as the reading of P, should be preferred to *factus* of other MSS.; cf. Suet. *Aug.* 4 *materna tibi farina est ex crudissimo Ariciae pistrino; hanc finxit ...mensarius*; Ovid, *Fasti*, III, 670 *fingebat tremula rustica liba manu*; Sen. *Epp.* 90, 23 *farinam aquā sparsit et adsidua tractatione perdomuit finxitque panem*. Perhaps the word here refers to the fancy shapes in which Roman loaves were often baked. **siligine:** cf. Sen. *Epp.* 119, 3 *utrum hic panis plebeius sit an siligineus, ad naturam nihil pertinet*.

72. artoptae, 'the bread-pan'; bread was sometimes served hot in the pans in which it had been baked.

finge...inprobulum: this clause, just like that introduced by *puta*, l. 7, is equivalent to the protasis of a conditional sentence; Browning often uses a clause like this, beginning 'put case that' etc.

73. inprobulum, 'somewhat venturesome'; for the diminutive, see n. to *lividulus*, XI. 110.

superest illic, 'there is someone there set over you', i.e. to watch you: *superest* is apparently used in the sense of ἐφέστηκε.

ponere, 'to lay down' the bread; cf. *ponenda*, III. 56.

74. vis tu is regularly used to express a peremptory command, the pronoun being a necessary part of the phrase: *vin tu* is merely interrogative; cf. Hor. *Sat.* II. 92 *vis tu homines urbemque feris praeponere silvis?* (where Bentley calls the constr. *elegantissimus Idiotismus*). *visne* is very seldom used thus, but occurs in Cic. *ad Fam.* IV. 5. 4 (where Servius Sulpicius tells how, when he saw the ruined cities of Greece, he reproved himself for resenting the short life of individuals) *visne tu te, Servi, cohibere et meminisse hominem te esse natum?* *vis tu* is common in Seneca. We may transl. here, 'I beg that you will...'.

75. colorem, 'colour': this seems the natural rendering, comparing *niveus* of l. 70 with Mart. IX. 2. 4 *convivam pascit nigra farina tuum*. But the word can also mean 'quality' in silver-age Latin, and may mean it here: cf. IX. 28 *pingues... lacernas,* | *munimenta togae, duri crassique coloris*, where the epithets show that *coloris = fili*.

76–9 is what Trebius is supposed to say to himself, when rebuked for his presumption: 'this is the reward of all my exertions in attending my patron's *levée*'.

77. **per montem** cet., 'I breasted the hill and hastened up the chilly Esquiline'; cf. Mart. v. 22 *mane domi nisi te volui meruique videre,* | *sint mihi, Paulle, tuae longius Esquiliae.* l. 5 *alta Suburani vincenda est semita clivi.* In Juv. also, *mons* may refer to the steep ascent from the Subura, or to the Esquiline itself. This hill, in the east of the city, had been made a popular site for great houses by Maecenas: see edd. on Hor. *Sat.* I. 8. 14. All the three poets are obliged by their metre to avoid the word *Esquilinus*; as Juv. had to express *Viminalis* by a periphrasis III. 71.

79. **Iuppiter** is here, and often elsewhere, used for 'sky', 'air', 'weather'; so *sub Iove*, 'in the open air'. **paenula** (Φαι-νόλης) was a thick sleeveless over-coat, of frieze or leather, worn by men and women, as a protection against bad weather.

80–106. *The same difference is kept up throughout the courses. Your host has the finest and rarest fish that Sicily or Corsica can supply and the best of oil to pour over it; you must be content with an eel from the Tiber and stinking oil for sauce.*

80. **distinguat**, 'marks out', 'reserves' for the host. Some translate, 'parts in two', which may be right; but the rendering 'adorns' is inadmissible: for all Latin writers use *distinguere* in this sense ('to vary', cf. ποικίλλειν), only where there is *variety* or *number*; thus the sky is *stellis distinctum* but not *sole distinctum*; a cup is *gemmis distinctum*; and a number of oysters might be said *lancem distinguere*, but not a single lobster.

81. **squilla** is probably a lobster. **saepta**, 'walled-in', is used in much the same sense as *constrictus* below.

82. **asparagis**: Mart. (XIII. 21) has a couplet on asparagus, which he calls also **spina** (cf. spinach): the best came from Ravenna. **qua...cauda**, 'how its tail looks down upon the party'; the lobster, being carried high up by a tall servant, seems to look down in scorn on the clients.

84. **dimidio...ovo**, 'garnished' (lit. 'hemmed in') 'by half an egg'; cf. Mart. v. 78. 5 *divisis cybium latebit ovis.* That the *cammarus* was a poor man's fare, appears from Mart. II. 43. 12 *concolor in nostra, cammare, lance rubes*, whereas the rich man eats a huge *mullus*.

85. **feralis cena**: nine days (hence the name) after interment, a meal called *novemdialis cena*, including eggs, lentils, and salt, was placed upon the grave: eggs have often been found in tombs.

86. **Venafrano**, sc. *oleo*: the oil from Venafrum in Samnium was considered the best in the world.

87. **pallidus**, 'colourless': cabbage was boiled in soda, *nitrum*, to give it a brighter green: cp. Mart. XIII. 17 *ne tibi pallentes moveant fastidia caules,* | *nitrata viridis brassica fiat aqua.*

88. **lanternam:** bad oil from Africa, only fit to use in a lamp, is what you are expected to swallow.

alveolis, 'platters'; again VII. 73.

89. **canna** is gen. explained as 'a boat of reeds', *c.* being = *harundo*; a Schol. says it is a *gandeia*, an African kind of ship; a boat of reeds could not have crossed the Mediterranean.

Micipsarum, 'of African kings': the name is used generically, one Micipsa at least having been king of Numidia. **subvexit**, 'imported'; *sub* means 'up the Tiber'; the true *baca Sabina* (see n. to III. 85) came *down* the Tiber. Of good wine, the reverse was the case: see n. to *devectum* VII. 121.

90. Boccar is an African who uses the oil of his native country at the baths, to the annoyance of other people; cf. Hor. *Sat.* I. 6. 123 *unguor olivo,* | *non quo fraudatis immundus Natta lucernis* (cf. *lanternam* above).

91. This line and VI. 126 are the only two in Juv. which P does not include in its text. These therefore may be considered doubtful; but of the many others which edd. used to bracket, there is not one, says Bücheler, which has been proved not to be written by the satirist himself. No doubt there are weak and otiose lines, the removal of which adds force to the meaning; but it is uncritical to assume that Juv. could never have written such.

92. **mullus:** the bearded mullet was especially esteemed by connoisseurs; see n. to IV. 15.

93. Tauromenium is the ancient Naxos on the east coast of Sicily. In the Phalaris controversy, one of Bentley's arguments was, that since the Tauromenites are mentioned in the letters, and their city was not founded before 396 B.C., the letters cannot be of the age they pretend to. When his opponents took refuge in the desperate suggestion that the Tauromenites might have got their name from a river Tauromenius, before there was a city Tauromenium, Bentley replies: 'Now if the Tauromenites were a sort of fish, this argument drawn from the river would be of great force.'

94. **nostrum mare** (= *proxima*, l. 96) is the Tyrrhene or Tuscan Sea.

dum gula saevit, 'because of the rage of gluttony'; *dum* gives the reason of the preceding statement as in I. 60.

95. scrutante, 'ransacking'; cf. XI. 14 and Sen. *Epp.* 89. 22 *vos...quorum profunda et insatiabilis gula hinc maria scrutatur, hinc terras.*

96. crescere, 'to grow big'; i.e. they are all caught before they are fit to kill.

97. provincia: see n. to IV. 26.

98. 'Dainties for the will-hunter L. to buy and for Aurelia to—sell'. L. is a *captator* or *heredipeta* (see nn. to III. 129; IV. 19); A. is a rich *orba* whose favour he tries to win by presents of dainty dishes: cf. Mart. IX. 88 *cum me captares, mittebas munera nobis: | postquam cepisti, das mihi, Rufe, nihil. | ut captum teneas, capto quoque munera mitte, | de cavea fugiat ne male pastus aper.* But A. is avaricious as well as rich and, instead of eating, sells the good things. Thus *vendat* is παρὰ προσδοκίαν for *edat* or *accipiat* and throws some fresh light on the lady's character. In Mart. VII. 20 the last word *vendit* is παρὰ προσδοκίαν in just the same way.

Perhaps the name *Aurelia* is chosen with reference to the story told by Pliny *Epp.* II. 20. 10.

100. gurges Siculus, 'the flood of Sicily', is Charybdis: the Homeric name was applied to a dangerous local current, still active, in the straits of Messina between Rhegium and Messana. That the current was most dangerous when the south wind was blowing, appears from Sen. *Epp.* XIV. 8 *temerarius gubernator contempsit austri minas, (ille est enim qui Siculum pelagus exasperet)...petit littus...quo propior Charybdis maria convolvit*; Ovid, *Met.* VIII. 121 *austroque agitata Charybdis.*

For the meaning of *gurges* (not 'whirl-pool' but 'flood'), cf. Henry's *Aeneidea,* I. 368 ff.

101. carcere: cf. *Aeolio carcere,* X. 181; the wind flies on wings; when these are wet by flying over the stormy sea, he sits at home to dry them.

102. contemnunt, 'make light of', i.e. venture to fish there; cf. Seneca quoted above.

lina, 'nets', lit. 'flax' from which the net is made; 'line' is *linea.*

104. Tiberinus is the name of a fish bred in the Tiber; of what kind it was, is not known; Juv. asserts that the spots on the fish are due to the ice on the river, but he was probably no naturalist: it is clear that the fish was not valued.

et ipse means 'the fish, like you the client, was bred by the banks of the Tiber'; *et* does not, I think, connect the *Tiberinus* with the *anguilla* (so Weidn.).

106. **crypta** is an underground drain, at the far end of the
cloaca maxima, which runs from the Subura under the Forum
and thence by the Vicus Tuscus into the Tiber; this sewer,
which is of immense antiquity, still serves its original purpose.
The Subura was the most populous part of the city.

107–13. *I should like to say one word to Virro and beg him—
not indeed to be generous—but to treat his guests with decent
politeness.*

107. **velim,** sc. *dicere*.

108. 'No one expects such presents as Seneca used to give to
his poorer friends.' **modicus** is used in the sense of *tenuis*, 'of
moderate fortune'; cf. *minoribus amicis* in Pliny's letter quoted
at the beginning of this satire. In this sense Tacitus speaks of
modici equites (*Ann.* I. 73) and *modici senatores* (*ibid.* XI. 7). For
the sense of *mittere*, see n. to *mittit*, III. 45.

109. For these patrons, cf. Mart. XII. 36. 8 (to Labullus,
who thinks himself generous) *Pisones Senecasque Memmiosque |
et Crispos mihi redde, sed priores: | fies protinus ultimus bonorum,*
where *bonus*, as here, means 'beneficent'. Gaius Calpurnius
Piso, the most prominent subject of his time, lost his life after
an unsuccessful conspiracy against Nero A.D. 65; the whole
Seneca family, including Lucan, were involved in it and perished
at the same time. Cotta is probably a son of M. Valerius Messala
who was adopted into the *gens Aurelia*; he was a patron of Ovid.

110. **titulis** = *honoribus*: see n. to I. 130.

112. **civiliter,** 'as an equal with equals' M. *civilis* before the
Empire means 'belonging to citizens', but is used afterwards in
a special sense: an emperor is called *civilis* when he is condes-
cending and does not use his position to insult his subjects;
thus it was *civilitas* to go about without lictors (Suet. *Tib.* 11,
idem, *Calig.* 3). The word was applied also, by an extension of
the same meaning, to the great and rich; see n. to VIII. 73.

hoc face et esto is a veiled conditional sentence: 'if you do
this, then you may be', etc.

113. **ut...multi:** cf. l. 43.

114–48. *Virro has a goose, a capon, and a boar set before him, and
truffles afterwards; all the carving is done with due ceremony.
Meanwhile, you get nothing and do not venture to open your
mouth; Virro does not even offer to drink wine with you. If you
suddenly became rich, how differently he would treat you! And
if you were childless as well as rich, then you would be the great
man's master; but, being poor, you can afford to have children.*

It must be noticed that the clients have no share at all in this course; this is shown partly by no mention of any dish set before them, partly by l. 121 (where see n.) and ll. 166–9.

114. **anseris...iecur,** 'a goose with swollen liver'; cf. *Montani venter*, IV. 107. Then, as now, geese were crammed in such a way as to produce by disease of the liver a dainty dish for epicures, *foie gras*: cf. Mart. XIII. 58 *iecur anserinum. aspice, quam tumeat magno iecur ansere maius.*

par, 'as large as'.

115. **altilis** is a fowl kept in the dark and crammed with sweetened meal; cf. Mart. XIII. 62 *pascitur et dulci facilis gallina farina, | pascitur et tenebris. ingeniosa gula est.*

flavi is ξανθὸς Μελέαγρος of Hom. *Il.* II. 642; he slew the Calydonian boar; Roman poets, when speaking of a large *aper*, can seldom resist a comparison with the *fera Aetola*. **ferro** is for *venabulo*.

116. **spumat,** as if the animal were still alive; as the word is regularly used of the living boar (cf. Ovid, *Met.* VIII. 288; Mart. XI. 69. 9, XIV. 70), Juv. uses it to show that the boar's head appeared at table. *fumat*, the reading of the inferior MSS., is apparently the correction of someone whose zeal for accuracy was stronger than his imagination.

tubera, 'truffles', called *tubera terrae*, XIV. 7, and Mart. XIII. 50; their growth was thought to be stimulated by thunder: cf. Pliny, *Nat. Hist.* XIX. 37 *de tuberibus haec traduntur... : cum fuerint imbres autumnales ac tonitrua crebra, tunc nasci... tenerrima autem verno esse*: he mentions a kind of *tuber* called in Greece κεραύνιον, which indicates the same belief.

117. **facient...cenas maiores,** 'shall add a dish to the dinner': cf. XII. 56.

118. **tibi habe...,** 'keep your corn to yourself'; I don't want it; in this phrase the pronoun, **tibi, sibi,** etc., regularly precedes the verb: see n. to III. 188.

Alledius, an unknown gourmand, is represented as not only gluttonous but selfish; he considers the corn-supply from Africa of no importance, if the truffles do not fail. Africa, Sicily, and Egypt, were the chief sources from which the huge idle population of Rome was fed; see n. to *messoribus*, VIII. 117.

119. **disiunge boves** = 'stop producing corn'.

dum, 'provided that'.

120. The **structor** is properly the arranger of the table; here and XI. 136 he is also the carver, who is elsewhere called *scissor* or *carptor* (Petron. 36; Juv. IX. 110).

ne...desit, 'that nothing may be wanting to make you angry'; **indignatio** is 'cause of anger'.

121. **spectes** is not a mere synonym of *aspicias*: it means that Trebius is a mere looker-on and gets nothing to eat, which is the just cause of his anger; cf. Mart. I. 43 *bis tibi triceni fuimus, Mancine, vocati | et positum est nobis nil here praeter aprum:...et nihil inde datum est; tantum spectavimus omnes*; Lucian *de Merc. Cond.* 26 (of a Greek philosopher dining in a great Roman house) εἰς τὴν ἀτιμοτάτην γωνίαν ἐξωσθεὶς κατάκεισαι μάρτυς μόνον τῶν παραφερομένων.

chironomunta, 'gesticulating': owing to the importance of the arms in ancient dancing, χειρονομεῖν or *bracchia movere* is often synonymous with ὀρχεῖσθαι or *saltare*: thus Herodotus can say in the famous story of Hippocleides (VI. 129) τοῖσι σκέλεσι ἐχειρονόμησε where the first part of the word has clearly lost its meaning. Similarly *saltantem* here need not imply much, if any, motion of the carver's legs.

volanti, 'in air'.

122. **dictata magistri,** 'his teacher's lessons'; the art of carving was carefully taught by means of wooden models; cf. XI. 137 ff.

123. **omnia,** 'leaving nothing out': the position of the word and its place in the sentence must be taken into account. **sane,** 'I allow'; cf. IV. 16.

124. **secetur:** cf. XI. 135; the common word for 'to carve' is *carpere*: hence Trimalchio's carver was called Carpus, so that when the master called out *Carpe*, it might be understood either as the vocative of his name, or as the imperative of the verb (Petron. 36).

125. The reference is to Virg. *Aen.* VIII. 264 *pedibusque informe cadaver | protrahitur.*

127. **hiscere,** 'to open your mouth'; cf. Livy, XLV. 26. 7 *nemo adversus praepotentes viros hiscere* ('to say a single word') *audebat*; Pliny, *Paneg.* 76 (of Domitian's senate) *quis antea loqui, quis hiscere audebat praeter miseros illos qui primi interrogabantur?*

tria nomina habere = *liber esse*; cf. Quint. VII. 3. 27 *propria liberi quae nemo habet nisi liber, praenomen, nomen, cognomen, tribus*; e.g. *Marcus Tullius Cicero Cor.* (i.e. *Cornelia tribu*). Women had two names, slaves only one. Trebius is freeborn, but he cannot practise the freedom of speech which is his birthright.

prŏpinat: the first syll. is often long elsewhere. Cf. Mart. II. 15 *quod nulli calicem tuum propinas, | humane facis, Horme, non* *superbe*: from these and other passages it is clear that the proposer of a health drank first himself and then sent the cup round to the friend whose health he was drinking: so *propinare alicui* comes to mean 'to drink from the same cup as someone'.

128.   **contacta**, 'polluted'; cf. *contagio*.

130.   **perditus**, 'lost to shame'. **bibe**, 'a glass of wine with you!'

131.   **pertusa...laena**, 'with a coat in holes'; the verb seems to get its meaning more from the preposition than from the simple verb; thus the pitcher of the Danaides is *pertusum vas*, to bore the ears for ear-rings is *pertundere aures*; it is exactly synonymous with *perforare*, of which most inflections are impossible in dactylic verse.

132.   **quadringenta** (sc. *sestertia*) is the *census equester*; see n. to I. 106; in both places it stands for 'a fortune'. **similis dis:** the notion is, that the gods would give the fortune, if the fates would let them.

133.   **homuncio**, 'mere man'; cf. Sen. *Epp.* 116. 7 *nos homunciones sumus* and therefore cannot be without passions like the Stoic *sapiens*. **quantus**, 'how great a man'; the constr. is like *e paupere dives*, τυφλὸς ἐκ δεδορκότος.

134.   **quantus...amicus**, 'how valued a friend'; the sense is different from *magnus amicus*, I. 33, where see n.

135.   This is how Virro would speak then, first to the servants, then to Trebius himself. **frater** is a complimentary term used between equals in rank and age; Horace recommends it to electioneerers (*Epp.* I. 6. 54). With **vis**, supply *tibi detur*.

136.   **ilibus:** this was thought the choice bit of the whole boar; cf. Mart. x. 45. 3 *costam rodere mavis | ilia Laurentis cum tibi demus apri*. Bücheler's explanation, that *ab ipsis ilibus* goes with *frater* in the sense of αὐτάδελφος, is surely impossible.

139.   **luserit:** the subj. is used, by the regular rule, for the imperat., when the prohibition is in the 3rd person; the tense is aorist, not perfect, and is exactly like the present (*ludat*) in meaning; cf. Livy, IX. 11. 13 *mŏratus sit nemo quominus...abeant*; see n. to VII. 93.

Aeneas = *filius*: the allusion is to Dido's words *Aen.* IV. 328 *si quis mihi parvulus aula | luderat Aeneas,...non equidem omnino capta ac deserta viderer.* Juv. means 'you must have no children, you must be *orbus*, so that Virro may hope to inherit your money'.

illo = *quam filius*: the use of *ille* without emphasis is characteristic of silver Latin: see n. to VI. 274.

(Mr Lendrum, however, explains *illo* as = *quam Virro*: 'you must have no son nor daughter to cut Virro out'.)

140. **iucundum** and **carum** are predicate, not epithets.

141. **nunc,** 'as it is', i.e. as you have no money to leave. Mycale stands for the wife of Trebius: cf. Mart. XI. 55. 5 *dicat praegnantem tua se Cosconia tantum,* | *pallidior fiet iam pariente Lupus* (L. is a *captator*).

This explanation, which is given by M., I believe to be right; but it is not even mentioned by Friedl., who adopts the following view: the supposition that Trebius is rich, holds good to the end of the paragraph; Mycale is not a wife, but a *concubina*; hence his children by her have no legal claim to his money, and Virro may hope to inherit. The grounds on which this view is put forward seem to be: (1) that the contrast between Virro's treatment of the father and the children is unaccountable; (2) that Mycale is an impossible name for a Roman wife.

It seems to me certain that *nunc* is used, as constantly, to deny a supposition, which is, in this case (l. 132) that Trebius is rich: such a *nunc* need not begin the sentence. On the new view, *nunc* is, to say the least of it, useless. Again, it is distinctly stated that Trebius has a wife (l. 77): is he in a position to keep up two establishments? Neither law nor public opinion allowed the *uxor* and *concubina* to live under the same roof.

Nor do the objections to the old interpretation convince me: (1) Virro might well be ungracious to Trebius and yet be willing to spend a few pence on the children; nor (2) was it illegal, or at all improbable, that such a person as Trebius should marry a *libertina*, who bore, after the *nomen* of her patron, some name like Mycale (perhaps *Megale*, the exact spelling of the word is quite uncertain). The Index to Wilmanns's *Inscriptions* gives hosts of similar female names, such as *Cedne, Dorcas, Sabbathis* and *Samne*; each of these four have *Claudia* prefixed; but Juv. does not wish to use a complimentary title here.

The digression in ll. 141–5 is strikingly irrelevant; but such irrelevance is common in Juv. The words *sterilis uxor* suggested a new thought, that the poor man is an exception to the rule laid down; and five verses are at once written to express this, in which he makes Virro behave much better to the children than to their father.

142. **semel:** so Büch. with P: but this would surely mean that M. could have no more children: so perhaps Hermann is right in preferring *simul* of the inferior MSS. See Bentley on Hor. *Sat.* II. 8. 24, where there is the same discrepancy between MSS.

143. **nido,** 'your young ones'. The 'green waistcoat' is perhaps a miniature of the green tunic (*viridis panni*, XI. 198) worn by one of the chariot-drivers at the races; boys and men alike took a passionate interest in these contests: cf. Suet. *Nero,* 22 *quondam tractum prasinum agitatorem* (the 'green' driver) *inter condiscipulos querens, obiurgante paedagogo, de Hectore se loqui ementitus est*; Tac. *Dial.* 29 *paene in utero matris concipi mihi videntur histrionalis favor e gladiatorum equorumque studia.* The garment is for the children to play with; so are the nuts, which were to the children of antiquity much what marbles are to our own, with the advantage that they were good to eat. Sometimes a benevolent man provided by will *sparsio nucum inter pueros.* Virro is too economical to provide any but the smallest kind.

145. The 'infant parasite' is Trebius' son.

146. **ancipites fungi,** 'doubtful funguses': i.e. there is considerable risk in eating them: cf. Lucan, VI. 112 *letumque minantes | vellere* (to pluck) *ab ignotis dubias radicibus herbas*; Mart. III. 60. 5 *sunt tibi boleti, fungos ego sumo suillos.*

147. **sed,** 'yes, and...', has the sense of *et quidem*, καὶ ταῦτα: see n. to IV. 27. Juv. seems to borrow here from Mart. I. 20. 4 *boletum, qualem Claudius edit, edas.* Report said that Claudius was poisoned in a *boletus* by his wife Agrippina: cf. VI. 620; Suet. *Claud.* 44.

A Roman dinner at this date was generally divided into three parts: (1) *gustus* or *promulsis* (so called because *mulsum*, wine and honey, was often drunk then); this consisted mainly of vegetables, eggs, and dried fish; (2) *fercula*, the actual dinner, with different kinds of fish and meat; (3) *mensae secundae*, a dessert of fruit. Virro's dinner does not follow this order exactly. Juv. begins with the wine and bread (24–75), goes on with fresh fish, which would naturally be a *ferculum* (80–106), then to the *fercula* (114–45). Now come the *boleti*, which would naturally form part of the *gustus*; and lastly fruit (149–55) in its proper place as *mensae secundae*. I have begun a new paragraph at 149, to show that *boleti* are not a part of the dessert.

149–55. *The same distinction is observed in the dessert supplied to rich and poor.*

149. The 'rest of the Virros' are not the host's relations but those guests whom he treats as equals. For **reliquis,** see n. to X. 260.

150. **poma,** the regular close of a Roman dinner; hence the

proverb *ab ovo | usque ad mala* (Hor. *Sat.* I. 3. 6). The word is used of other fruit, as well as apples: e.g. Ovid *Fast.* II. 253 *stabat adhuc duris ficus densissima pomis*.

pascaris, 'one might feast'; see n. to *nolis*, l. 54. Cf. Mart. quoted on l. 162.

151. The reference is to Hom. *Od.* VII. 114–21, where the never-failing orchard of Alcinous is described. From l. 117 τάων οὔ ποτε καρπὸς ἀπόλλυται οὐδ' ἀπολείπει | χείματος οὐδὲ θέρευς, ἐπετήσιος, it seems that **autumnus** means 'fruit-trees' or 'fruit-time'; cf. Sen. *Thyest.* 168 (of the fruit-trees eluding Tantalus) *totus in arduum | autumnus rapitur silvaque mobilis*.

152. Hercules carried off the golden apples of the Hesperides which were guarded by a dragon.

153. **scabie mali,** 'a rotten apple'. **quod** should strictly be *quale*. **aggere,** 'the embankment'; see n. to VIII. 43.

154, 155 are a roundabout description of a performing monkey, which is taught to shoot, riding on a goat. **flagelli:** cf. *metuens virgae*, VII. 210.

156–73. *This treatment is due, not to Virro's meanness, but to his pleasure in your pain. A free man ought not to submit to such insults, merely in the hope of getting good things to eat; and if you put up with it, you deserve all you suffer.*

157. **hoc agit ut doleas,** not 'he does this to give you pain', but 'his deliberate purpose is to give you pain'; the former would be *hoc facit*: cf. *hoc agite*, VII. 20.

comoedia has special point from the fact that the Romans had performances of some kind going on during their dinners, however modest these were: cf. XI. 179 ff.; Pliny, *Epp*; I. 15. 2 (to a friend who had failed him after accepting an invitation to dinner) *audisses comoedos vel lectorem vel lyristen vel, quae mea liberalitas, omnes*; ibid. III. 1. 9 *frequenter comoedis cena distinguitur*. These were the amusements of respectable people; dancing-girls from Gades (cf. XI. 162) and mimes were preferred at what Martial calls **convivia nequiora.**

158. **plorante gula,** 'than a greedy angry man'; Trebius is angry but too greedy to get up and go.

159. **si nescis** = *ne ignores*, 'I must tell you'. Cf. Ovid *Trist.* IV. 9. 11 *omnia, si nescis, Caesar mihi iura reliquit*; idem, *ex Pont.* III. 3. 28 *quae sunt, si nescis, invidiosa tibi*.

160. **diu** seems to belong to *stridere*; 'to go on grinding your tight-clenched teeth': *dolor* has the epithet *frendens* Sen. *Herc. Fur.* 696.

161. The line, which is co-ordinate with the next, may be subordinated in English: '*though* you think yourself...he thinks...'.

162. **nidore culinae**: cf. Mart. I. 92. 9 *pasceris et nigrae solo nidore culinae.*

164. **Etruscum...aurum:** all free-born children (*ingenui*) wore an amulet round the neck, to avert the evil eye; boys wore it until they put on the *toga virilis*, girls until their marriage. The amulet was carried in a round or heart-shaped case (*bulla*), which was of gold in the case of the children of *ingenui*, while the children of *libertini* had to be content with leather. The *bulla* is called *nobile pectoris aurum* by Statius (*Silv.* V. 3. 120). It is here called Etruscan, because the custom, like most other peculiarities of Roman religion or superstition, came from Etruria.

165. 'Or merely the distinction of a knot made of poor man's leather.' **signum** means *signum ingenuitatis*; for **loro** see preceding n.

166. **vos,** Trebius and others like him. **ecce...altilis** is what they are supposed to say to themselves.

168. **minor altilis,** 'the remains of a fat hen'; cf. VIII. 4 *umerosque minorem*; Mart. II. 37. 4 *mullum dimidium.* Or perhaps 'too small' for Virro.

inde = *ab eo,* 'in consequence of this'.

169. **stricto** would naturally be followed by *ense*; the expected feast is the fight for which the clients are waiting 'with bread ready for action'.

iacetis, the reading of P, which gives a good sense, should be preferred to *tacetis* of the inferior MSS., though the reading cannot be considered certain, owing to the common confusion of *t* and *i* in P: see Introduction, p. xlvi.

170. **omnia...debes,** 'if you are willing to put up with any treatment, then you deserve it also (*et*)', i.e. 'serve you right'.

171. **vertice raso:** professional buffoons (γελωτοποιοί, *moriones*) had their heads shaved; it was their business, like that of our clowns, to look as ridiculous as possible and receive all the kicks and blows (*alapae*). Further than this, degradation cannot go, says Juv.

172. **quandoque,** 'sooner or later', used like *aliquando*; again XIV. 51; the use is found in Cicero's letters (*ad Fam.* VI. 19. 2 *ego me Asturae arbitror commoraturum quoad ille quandoque veniat*) and is common in Livy and later writers, especially Suetonius. It is also used as a conjunction (= *quandocunque*)

but not in classical prose; cf. Hor. *Ars*, 359 *indignor quandoque
bonus dormitat Homerus*; Suet. *Claud.* 1 *pristinum se rei p.
statum, quandoque posset, restituturum.*

SATIRE VI

WOMEN; ADVICE TO ONE ABOUT TO MARRY

This 'Legend of Bad Women' (Mackail) is Juvenal's longest
and most elaborate effort. It displays both his strength and his
weakness in the strongest light, his genius for descriptive epi-
gram and his indifference to order and arrangement.

'A very brief analysis of this celebrated piece will discover the
badness of its composition. 1–59 Do not think of marriage, few
women being both chaste and fair: 60–113 do not look for a wife
in the theatre: all ladies prefer actors and gladiators: 135–160
no men love their wives, but only their wives' fortune or beauty:
161–183 a perfect wife would be intolerable: 184–199 it is very
bad in a lady to talk Greek; 200–224 a wife is always a tyrant:
225–230 she will marry as often as she likes:—and so on, and
so on.

In fact, with all its brilliancy of execution in detail, the piece,
as far as composition is concerned, is a mere chamber of horrors.
The main theme, that it is madness to marry because a good wife
cannot be found, is not so much worked out as illustrated by a
series of pictures quite unconnected, and arguments sometimes
inconsistent...

The inconsistencies might be defended in a humourist: he
would be in his right in saying that a licentious wife or an over-
virtuous wife are equally objectionable. But this ground is not
open to the moralist, who is bound to defend virtue against all
cavil' (Nettleship, *J. Philol.* XVI, 63).

1–32. *It is possible that women were still modest in the golden age,
when life was simple and hard. But soon after Jupiter came to
rule, Modesty migrated from earth to heaven. Yet you, Postumus,
are going to marry in our days; which proves you must be mad.
For who would incur the slavery of marriage, when there are so
many means of putting an end to one's existence?*

1. **Saturno rege:** the reign of Saturn was the golden age of
innocence; cf. XIII. 38 ff.; Virg. *Aen.* VIII. 319 *primus ab aetherio
venit Saturnus Olympo.* | ...*aurea quae perhibent illo sub rege
fuere* | *saecula.*

3. **spelunca:** houses being yet unknown, primitive man lived in caves and holes in the ground; cf. Lucr. v. 955 *nemora atque cavos montes silvasque colebant.* Juv. has this part of Lucretius in mind throughout his description of the world's youth.

ignemque Laremque are governed by *praeberet,* not by *clauderet.*

4. **dominos,** sc. *pecoris.* So Polyphemus, in the Odyssey, lived in the same cave with his sheep.

5. **silvestrem montana** = *in silvis et montibus:* cf. Lucr. quoted above; Cic. *de Orat.* I. 36 *initio genus hominum in montibus ac silvis dissipatum;* Quint. IX. 4. 4 (*nec*) *urbibus montes ac silvas mutari oportuit.* The juxtaposition of the adjectives at the beginning of the sentence shows that they are used instead of an adverbial clause of place: so *Aricinos* = *Ariciae* (at Aricia), IV. 117. For an adj. taking the place of an adverbial clause expressing time, see n. to I. 28.

6. **frondibus:** cf. Lucr. v. 970 *silvestria membra | nuda dabant terrae nocturno tempore capti, | circum se foliis ac frondibus involventes;* and Hom. *Od.* v. 482 (where the shipwrecked Odysseus makes himself a bed of leaves) ἄφαρ δ' εὐνὴν ἐπαμήσατο χερσὶ φίλῃσιν | εὐρεῖαν· φύλλων γὰρ ἔην χύσις ἤλιθα πολλή.

vicinarum: i.e. the wild beasts have not yet been driven back by civilisation.

7. **tibi, Cynthia:** for this device for avoiding the unmetrical dative, see n. to I. 50. Cynthia was the name which Propertius gave his mistress, Hostia, in his poems; his first book begins with the word *Cynthia,* and bore this name; cf. Mart. XIV. 189 *Cynthia, facundi carmen iuvenale Properti, | accepit famam nec minus ipsa dedit.* Her lover often speaks of her grace and accomplishments.

8. The other lady, 'whose pretty eyes were clouded by a sparrow's death', is Lesbia, the mistress of Catullus, who says in his elegy on the bird (III. 17) *tua nunc opera meae puellae | flendo turgiduli rubent ocelli;* Lesbia was probably a pseudonym for the notorious Clodia. Juv. means: 'the women of old were very unlike the modern fashionable ladies: they had less sensibility and more morals'.

10. **horridior,** 'more uncouth': cf. Mart. I. 62 *casta nec antiquis cedens Laevina Sabinis | et quamvis tetrico tristior ipsa viro.* **glandem:** this was the food of men before corn was discovered, and the food of pigs afterwards; cf. XIII. 57 and Lucr. v. 939 *glandiferas inter curabant corpora quercus.*

11. **quippe**, 'for'; see n. to XIII. 26. The line is borrowed from Lucr. V. 907 *tellure nova caeloque recenti*.

12. One legend said that the first men were born from oaks or rocks; another that Prometheus made them by mixing earth with water. For the first, cf. Virg. *Aen.* VIII. 314 *haec nemora indigenae Fauni nymphaeque tenebant*, | *gensque virum truncis et duro robore nata*; Hom. *Od.* XIX. 162 ἀλλὰ καὶ ὣς μοι εἰπὲ τεὸν γένος, ὁππόθεν ἐσσί· | οὐ γὰρ ἀπὸ δρυός ἐσσι παλαιφάτου οὐδ' ἀπὸ πέτρης (i.e. you must have had some parents). For the other legend, cf. XIV. 35 and Ovid *Met.* I. 82 *quam* (i.e. *tellurem*) *satus Iapeto mixtam fluvialibus undis* | *finxit in effigiem moderantum cuncta deorum*.

The conj. *rupe et robore* (Scholte) is very attractive; cf. Homer quoted above, and Stat. *Theb.* III. 559 *at non prior aureus ille* | *sanguis avum scopulisque satae vel robore gentes* | *mentibus his usae*; ibid. IV. 340 *saxis nimirum et robore nati*.

13. In either case, the *indigenae* had no human parents and were *progenies terrae*; see n. to *fraterculus gigantis*, IV. 98.

15. **aliqua**, 'at least some'. **et sub Iove**, 'even when Jupiter was king', which he became by expelling his father Saturn. But by the time his beard had grown, things were different and worse; cf. XIII. 40.

17. **alterius** is παρὰ προσδοκίαν: : it was common to swear by one's head; but the cunning Greek saves himself by swearing by someone else's who will bear the consequences of the perjury; see n. to XIII. 84.

18. **caulibus**: the dat. after *timere* is good and classical, e.g. Virg. *Aen.* II. 729 *comitique onerique timentem*; but the addition of the acc. (*furem*) makes the constr. a rare one; for an instance, cf. Quint. IV. I. 9 *iudex libentissime patronum audit quem iustitiae suae minime timet*. In Mart. I. 82. 7 *domino nihil timebat*, the acc. is adverbial.

aperto, 'unwalled'. **viveret**: *quisque* is to be supplied from *nemo*, as subject: so Hor. *Sat.* I. I. I where see Palmer's note.

19. Astraea, daughter of Zeus, was the last of the immortals to leave the earth when it became too wicked for her: cf. Ovid *Met.* I. 149 *caede madentes*, | *ultima caelestum, terras Astraea reliquit*.

Hence 'Astraea Redux' (= the return of the Golden Age), the heading of a chapter in Carlyle's *French Revolution*.

20. **hac**, i.e. *Pudicitia*. **duae sorores**, '*the* two sisters'.

21. **anticum et vetus**, 'of ancient date and of long standing'; there is no tautology, as the first adj. indicates the point of

time when the practice began, and the second, the period of time during which it has been kept up; cf. xv. 33; Plaut. *Miles,* 751 *quin tu istanc orationem hinc veterem atque antiquam amoves?*

genium contemnere fulcri, 'to disregard the deity of the marriage-bed': the *fulcrum* corresponds to the head of a modern sofa; it is the end of the frame-work on which the pillows of a couch were placed; some specimens are extant, and these usually terminate in an ass's head of bronze: cf. XI. 96 ff.: lower down on the *fulcrum,* there is generally a round boss of metal carrying a bust of a Genius or some god, who is supposed by Juv. to protect the inviolability of wedlock.

This passage was first explained by Professor Anderson (*Class. Rev.* III, 322, where a good illustration will be found). *fulcrum* was before supposed to mean the foot of the couch, that on which the couch rests; it really means the top of the couch, that on which a sleeper rests his head.

23. **crimen:** see n. to I. 75. **mox,** 'afterwards'; the regular meaning of the word in silver Latin.

24. '*But* the silver age...' This age began with the expulsion of Saturn; it was followed by the iron age, which was even more wicked, though in this one respect it had nothing to learn. The inappropriate plural *saecula* is due to metrical convenience; see n. to IV. 68. For ages named after metals, see n. to XIII. 28.

25. **conventum...sponsalia,** 'a contract and agreement and betrothal'; *conventus* is used *metri causa* for *conventio*; this latter and *pactum* being the general legal designations of any contract. The *sponsalia* are so called from the promise of the woman's hand given by her legal representative; cf. Plaut. *Aul.* 255 A. *quid nunc? etiam mihi despondes filiam?* B. *illis legibus,* | *cum illa dote quam tibi dixi.* A. *sponden ergo?* B. *spondeo.* In imperial times a written contract (*tabulae sponsales*) was drawn up, and friends invited to the ceremony.

26. **tempestate:** see n. to IV. 140. **a tonsore** cet., 'you are having your hair dressed by a master-barber', in preparation for the ceremony: cf. XI. 150.

27. **digito pignus,** 'a pledge for her finger', i.e. a ring: cf. Hor. *Carm.* I. 9. 23 *pignusque dereptum lacertis* | *aut digito.* This was given by the man, as a pledge that he would fulfil the contract, and was worn by the woman on her fourth finger; cf. Pliny, *Nat. Hist.* XXXIII. 12 *etiamnunc sponsae muneris vice ferreus anulus mittitur isque sine gemma,* i.e. the simplicity of former times is still kept up in this custom.

28. **eras,** 'you used to be'. Postumus Ursidius is the man to

whom the whole satire is addressed, though his intentions only give a handle for the general attack on women, and he soon passes out of sight.

29. Tisiphone is one of the three Furies, Allecto and Megaera being the others; and because *Tis.* = *furia*, the abl. may be used without *ab*: see n. to I. 13; and cf. Sen. *Dial.* IV. 10. 8 *multi furiis ambitionis agitati*. **colubri:** snakes which they twined in their hair and held in their hands; the effect of the snakes was to produce madness in those they touched; cf. Virgil's account of Allecto's visit to Turnus (*Aen.* VII. 445 ff.).

30. **salvis tot restibus,** 'when there are so many ropes in existence', i.e. when it is so easy to hang yourself: cf. Stobaeus *Floril.* 59. 6 πλεῖς τὴν θάλατταν σχοινίον πωλουμένων; i.e. why go to sea when you can die an easier death? For *restibus*, cf. Mart. IV. 70. 1 *nihil Ammiano praeter aridam restem | moriens reliquit ultimis pater ceris.* **ullam,** 'at all'; the word is emphatic by position.

31. **caligantes,** 'at a dizzy height'; *caligare* (like ἰλιγγιᾶν) is properly said of the eyes of a person looking down from a height, and then of the person himself; cf. Sen. *Epp.* 57. 4 *caligabit, si vastam altitudinem in crepidine eius constitutus despexerit*; Stat. *Theb.* IV. 539 *caligantem longis Ixiona gyris*; but it is further applied as here to the place which causes dizziness.

32. The Aemilian bridge is handy for drowning yourself from. This bridge, later known as the *Ponte Rotto*, was built in 179 B.C. It was the first stone bridge across the Tiber, and led from the *forum Boarium* to the Janiculan Hill.

38–49. *You mean to be the father of a family; well, to secure the joys of fatherhood, you must give up the pleasures of your palate. Strange, that you, once the dread of husbands, should become a husband yourself. And, besides, you actually expect a wife with the old-fashioned strictness of morals; if you get one, you may thank heaven for your extraordinary fortune.*

38. **lex Iulia:** in order to stimulate marriage and repress celibacy, Augustus passed repeated enactments, the chief of which were the *lex Iulia de maritandis ordinibus* and the *lex Papia Poppaea* (A.D. 9); see n. to I. 55: the latter conferred the *ius trium liberorum* and restricted the power of childless and unmarried persons to inherit property. The object was not attained; and for the connection of *delatio* with these laws, see n. to III. 116.

39. **cogitat,** 'he purposes'; so in the elliptical constr. used

by Cic. in his letters, e.g. *ad Att.* V. 15. 3 *inde ad Taurum cogita-bam*, sc. *ire*.

cariturus, 'though he must lose'; the fut. partic. is used, where Cic. would say *quamquam carebit*. If Ursid. has children, the *captatores* will not send him good things to eat; see nn. to III. 129, IV. 19. That a turtle was a dainty appears from Mart. III. 60. 7 *aureus immodicis turtur te clunibus implet;* | *ponitur in cavea mortua pica mihi*.

40. **iubis:** apparently this stands for the mullet's beard: cf. Pliny *Nat. Hist.* IX. 64 *(mulli) barbâ gemina insigniuntur*; Mart. VII. 20. 7 calls the beards of oysters *cirri*. **macellus** is used for the dainties bought there; they themselves are said to hunt for legacies, as they are given away with that object; so in Mart. X. 96. 9 *conturbator macellus*, 'the market which brings men to ruin'.

42. **olim,** i.e. *qui olim est*, 'who has long been...'.

44. 'Who has so often hidden in the chest that held Latinus in danger of his life'; Latinus (see n. to I. 35) played the part of a lover in a farce, who had to save himself thus from a jealous husband; Ursid. has often had to do the same. For *perituri*, see n. to l. 277.

45. **quid quod,** 'moreover'; see n. to III. 86.

antiquis de moribus, the preposition is unusual in this phrase: it seems to be partitive as in *una de multis, nemo de iis*; and then *a. m.* is used for *mulieribus a. m.* For *antiquis*, cf. Mart. quoted on l. 10; Livy, XXXIX. 11. 5 *probam et antiqui moris feminam*.

illi, 'by him': see n. to I. 13.

46. **nimiam...venam,** 'prick a vein, he has too much blood'; *sanguinis missio* was a staple remedy of ancient medicine in fever or insanity; cf. Celsus, III. 6 *in hoc genere morborum* (i.e. *febribus) sanguinem misisse...prodest*. Ursid. has too much blood; he is too *sanguine*, we should say. For **pertundite**, see n. to V. 131.

47. **delicias hominis,** 'how nice the fellow is!' he is *delicatus*, i.e. gives himself airs and is difficult to please; cf. X. 291, XIII. 140. The exclamatory acc. is generally preceded by an interjection; cf. l. 531.

Tarpeium = *Iovis Capitolini*; cf. XII. 6; XIII. 78. Jupiter and *Iuno pronuba* were the chief deities who presided over marriage.

48. **pronus,** 'down on your face and worship...'; for the position, cf. Mart. I. 92. 10 *bibis immundam cum cane pronus aquam*.

auratam: cf. Livy, xxv. 12. 13 where a *bos auratus* is sacri-
ficed to Apollo: the horns of the victim especially were gilded;
cf. Hom. *Od.* III. 384 τήν (i.e. βοῦν) τοι ἐγὼ ῥέξω, χρυσὸν κέρασιν
περιχεύας.

49. **capitis...pudici:** transl. 'a chaste wife'.

60–113. *Do you expect to find such a lady in places of public
resort, or in the theatres? Women go there because they have a
passion for actors and still more for gladiators. One great lady,
a senator's wife, went to Egypt as a gladiator's paramour. She
found it no hardship then to face the sea and live on board ship;
if she had gone with her husband, it would have been a different
story. It was only the profession that attracted her: all women
delight in a man of arms.*

60. **porticibus,** 'arcades': see n. to IV. 5; the word should not
be transl. 'porch' or 'portico'; there were many public *porticus*
at Rome—*p. Argonautarum, p. Europae, p. Pompeii*, etc.; Mart.
sums up public places under *convivia, porticus, theatra* (VII. 76. 2).

61. **spectacula** includes the theatres, amphitheatre, and
circus: in all these the rows of seats were divided by passages
into wedge-shaped blocks (*cunei*); cf. Mart. XII. 29. 15 *quamvis
non modico caleant spectacula sole, | vela reducuntur cum venit
Hermogenes.* **totis** (cf. French *tout*) is used in a sense hardly
distinguishable from *omnibus*: cf. VIII. 255.

62. 'An object for you to love without misgiving and to pick
out thence'; perhaps a *hysteron-proteron*.

82. Juv. is referring to some notorious scandal of the day:
the name of Eppia's husband is not certainly known; but see n.
to l. 113.

For a similar scandal in high life, cf. Pliny, *Epp.* VI. 31. 4
*nupta haec tribuno militum honores petituro (= viro senatorio) et
suam et mariti dignitatem centurionis amore maculaverat.*

ludum, 'a company of gladiators', of whom Sergius was one.
Some inferior MSS. have *ludium*, which, even if it could scan
as a spondee (cf. VII. 185), offers a fresh difficulty, as *ludius* always
means 'an actor', which the context shows that Sergius was not.

83. Pharos is the little island, famous for its light-house, a
mile off Alexandria; see n. to XII. 76; hence *Pharius* is used for
Aegyptiacus.

famosaque moenia Lagi, 'the notorious city of Lagus', i.e.
Alexandria; Lagus was the father of Ptolemy the founder of the
Egyptian empire; the city was notorious for the laxness of its
morals.

84. 'And Canopus cried out against the monstrous corruption of Rome': Eppia's profligacy was too much even for the easy morals of Egypt. For Canopus, see n. to I. 26 and XV. 46; for *urbis = Romae*, see n. to III. 61.

85. **sororis:** this detail seems to show that the story is true.

86. **nil patriae indulsit,** 'she set no store by her country'; cf. Hor. *Sat.* II. 2. 94 *das aliquid famae?* 'do you care at all for reputation?' A personal dat. is commoner in this constr., e.g. *hoc tibi do*, 'I do this for your sake'.

87. **utque...stupeas,** 'and, more surprising still'; for this use of the 2nd pers. pres. subj., cf. Tac. *Germ.* 24 *aleam, quod mirere, sobrii inter seria exercent*.

ludi include races in the Circus and theatrical entertainments: gladiatorial shows and *venationes* were termed *munera*. For the sarcasm, cf. XI. 53. **Paridem:** see n. to VII. 87.

89. **segmentatis,** 'purfled': cf. II. 124 (of a man dressing as a woman) *segmenta et longos habitus et flammea sumit; segmenta* are pieces of cloth, of any shape, generally of purple embroidered with gold, which were sewn, like patches, on the stuff of a garment or coverlet; the coverlet of Eppia's cradle had been so adorned. Though she had been used to luxury from her infancy, still she did not hesitate. A senator called M. Eppius is mentioned by Cicero (*ad Att.* VIII. 11B. 1). **dormisset:** for this mood after *quamquam*, see n. to VII. 14.

90. **contempsit,** 'she made light of'; cf. V. 102. **famam,** 'reputation'.

91. **minima est iactura:** cf. III. 125. **molles cathedras =** *mulieres*, as they are the occupants of 'cushioned chairs'; cf. Virg. *Aen.* VIII. 666 *pilentis matres in mollibus*; Mart. III. 63. 7 *inter femineas tota qui luce cathedras | desidet*.

92, 93. The mention of the Tyrrhene sea shows that Eppia did not leave Italy at Brundisium but on the west coast; when she had passed the straits of Messina, she would be in the Ionian sea. The masc. **Ionius** is unusual for the neut.; hence Bentley (on Hor. *Epod.* 10. 19) proposed to read *sonorum* for *sonantem*, or that *fluctum* should be supplied from *fluctus* above.

94. The meaning is that she had to sail from sea to sea, before reaching Egypt.

95. **ratio pericli,** 'reason for facing danger'. **honesta** is the opposite of *turpis* (l. 97). The subj. to *timent* is the sex generally.

97. '*But* they bring courage etc....'

quas turpiter audent: cf. VIII. 165; the emphasis falls on the adv.

98. For the form of condition, cf. l. 470. The sentence is a μέν clause, the δέ clause beginning l. 100.

99. **tunc,** 'in that case'. **summus vertitur aer,** 'the sky goes round and round', i.e. the lady feels dizzy and sick; cf. l. 304.

100. **illa** refers to l. 98; **haec** is the lady who follows a paramour.

101. **prandet:** not being sick, she is able to eat, and she is not so exclusive as to object to the presence of the sailors, but even takes a share in their duties.

103. 'But where was the beauty with which Eppia fell in love, or the youth that attracted her?' The gladiator she followed had neither.

104. **vidit,** 'did she see in him'. **ludia,** 'a gladiator's wife'; cf. l. 266 and Mart. v. 24. 10 *Hermes, cura laborque ludiarum.*

105. **sustinuit,** 'she stooped'. **nam:** there was no such inducement of youth or beauty; *for* etc.

Sergiolus, 'the beautiful boy, Sergius', a ὑποκόρισμα such as she might use; cf. Catull. 45. 13 where Acme gives her lover Septimius the name *Septimillus*; Tac. *Ann.* vi. 5. 1 where a senator is accused of *majestas* for saying *me tuebitur Tiberiolus meus.* Cf. also Cicero's *pulchellus puer* for Clodius Pulcher, and the Doric diminutives Ἀσώπιχος, Ἀμύντιχος for Ἀσωπός, Ἀμύντας (Pind. *Ol.* 14. 17; Theocr. 7. 132).

radere guttur, 'to shave'; it is shown by other evidence, and especially by coins, that the Romans did not regularly shave the whole face after the *barbae depositio* (see n. to III. 186), but wore a close-clipped beard until their 40th year, when it was removed. Thus the meaning here is that Sergius was over 40: see n. to l. 215.

106. **secto...lacerto:** he was mutilated as well by a cut he had received on the arm: nor was his face calculated to charm.

107. **sicut,** 'as for example': cf. VII. 204; elsewhere in Juv. it has the commoner meaning 'in the same way as'. Büch. suggested *ficus*, 'a swelling', as *sicut* is easily dispensed with, whereas some noun seems to be wanted, other than *gibbus*, for *attritus* to agree with: see the following n.

108, 109. There is some doubt as to the meaning of **gibbus** here: it means 'a hump' x. 294, and generally as an adj. means 'convex' as opposed to 'concave': if it means 'a polypus' or swelling inside the nose, it can hardly be rubbed by the helmet: hence some take **attritus** as a noun, 'a sore place caused by the helmet'. But *ingens* seems in favour of understanding *gibbus* as an excrescence on the nose with which the closed helmet would come in contact. A figure of a *Thraex* (who also is *secto lacerto*) with his helmet closed, is given in Baumeister's *Denkmäler*, p. 2098.

et...ocelli, 'and the malignant trouble of a constant discharge from one eye'; Sergius was blear-eyed (*lippus*); cf. Celsus, VII. 7. 7 *in angulo* (*oculorum*) *qui naribus propior est, ex aliquo vitio quasi parvula fistula aperitur, per quam pituita assidue distillat*; §15 he speaks of cases *quibus non multa sed acris pituita est.*

110. hoc, 'their profession'. Hyacinthus, the favourite of Apollo, is used as a type of youthful beauty: 'Adonis' is our equivalent: both are used in Greek: cf. Lucian *Merc. Cond.* 35 εἰσὶ δ' οἳ καὶ ἐπὶ κάλλει θαυμάζεσθαι ἐθέλουσι, καὶ δεῖ Ἀδώνιδας αὐτοὺς καὶ Ὑακίνθους ἀκούειν.

112. amant, sc. *mulieres.*

113. accepta rude, 'if he had got his discharge from the amphitheatre'; such a gladiator was called *rudiarius*; phrases with *rudis* are often used figuratively: cf. VII. 171; Hor. *Epp.* I. 1. 2 *spectatum satis et donatum iam rude.*

Veiento videri: perh. Fabricius Veiento (III. 185): if Veiento was the senator to whom Eppia was married, the meaning is that Sergius, no longer a gladiator, would be as unattractive to Eppia as her own husband; if Veiento was not her husband, Juv. must mean the name to be typical of repulsive qualities. The former explanation seems to give more point.

136–41. *Some men give the highest character of their wives; but they have some special reason for doing so. Thus Censennia can do no wrong in her husband's eyes because she is rich.*

137. bis quingena dedit, 'she brought him a million sesterces; for that price he gives her the reputation of chastity'; *bis quingena*, usually expressed by *deciens*, is about £10,000. This was the amount of the senatorial *census* and is often mentioned as a rich woman's dowry: cf. X. 335 (where see n.) and Mart. XI. 23. 3 *deciens mini dotis in auro | sponsa dabis.* For this use of *dedit*, cf. Mart. II. 65. 4 *illa, illa dives mortua est Secundilla, | centena deciens quae tibi dedit dotis?* For **tanti,** see n. to III. 54; this differs slightly from the instances there quoted, *tanti* being = *hac mercede, hoc accepto*, and no *ut* clause following.

138, 139. He is not really in love with her at all: the passion which he pretends, is all due to her money. Cupid is generally represented as carrying bow and arrows and a torch (*lampas*, *faces*) with which to attack his victims; for illustrations from monuments, see Baumeister, *Denkmäler*, p. 499.

140. **libertas emitur,** 'she buys the right to do as she pleases'. **coram** cet., 'in her husband's presence she may make love by signs and answer lovers' letters'. **innuat:** cf. Mart. IX. 37. 5 though your hair and teeth are false, Galla, yet *innuis illo*, | *quod tibi prolatum est mane, supercilio*.

141. **rescribat:** cf. XIV. 29 and Mart. II. 9 *scripsi; rescripsit nil Naevia; non dabit* (= χαριεῖται) *ergo*. **vidua,** 'unmarried'; cf. IV. 4; *vidua* is used as the feminine of *coelebs* and thus is often applied to the Amazons and their queen, e.g. in Seneca's plays.

142–60. Nor is Sertorius an exception to the rule. He is in love with Bibula, but only with her face. Whenever her beauty goes, she will be turned out of doors, double-quick. Meanwhile she makes the most of her time and uses her influence to ruin her husband by boundless extravagance.

142. This l., like l. 136, is put in the mouth of someone objecting to Juv.'s cynicism. **desiderio,** 'love'; the word is sometimes used, as here, in the sense of *amor*: cf. Hor. *Epod.* 17. 80 *desiderique temperare pocula* (i.e. love-potions); Ovid, *Remed.* 646 *dum desideriis effluat illa tuis*. The plur. is used by silver-age writers merely for 'wishes', 'wants': e.g. Pliny, *ad Trai.* 94 *peto a te, cuius in omnibus desideriis meis indulgentiam experior*.

143. **si verum excutias,** 'if you elicit the facts': the metaphor is taken from shaking out any receptacle, e.g. the *sinus* of the *toga*, to get what is inside; the phrase is elliptical for *si rem excutias et verum invenias*.

144. **subeant, laxet, fiant** are hypothetical subjunctives: 'suppose there appear', etc.; so in the phrase *velis, nolis*.

arida must be part of the predicate: 'suppose her skin becomes loose and dry': the wrinkles are caused by the skin becoming loose, while *arida* denotes a parchment-like complexion; cf. X. 192 ff.

146. **collige sarcinulas,** 'pack up your baggage'; see n. to III. 161; there is a reference to the formula which was used by both parties to a divorce, *res tuas tibi habeto*; cf. Mart. X. 41. 1 *Proculeia, maritum* | *deseris atque iubes res sibi habere suas*.

The freedman undertakes this unpleasant job for his master, as Palaestrio does for Pyrgopolinices in the *Miles* of Plautus.

147. **iam gravis,** 'a perfect nuisance': for *iam* with an adj., see n. to III. 206. **emungeris** is middle; cf. X. 199.

149. **calet et regnat,** 'she is in high favour and a queen'; I know no precise parallel to *calere* in this sense, but *frigere* often means 'to be distasteful'; Lucan, VII. 734 *dum fortuna calet* is hardly similar. Can *calet* mean, 'she is a novelty'? For *regnare*, cf. Ovid, *Am.* II. 19. 33 *si qua volet regnare diu, deludat amantem.*

150. **pastores,** sc. *Canusinos*. Dark wool from Canusium in Apulia seems to have been used specially for making liveries for *lecticarii*; cf. Mart. IX. 22. 9 *ut canusinatus nostro Syrus assere sudet.*

ulmosque Falernas: Italian vines were trained upon trees, and the elm, because of its scanty foliage, was found to answer the purpose best; *maritare ulmos* is 'to train vines on the elm', whereas the *platanus*, a thick-leaved tree, is *coelebs*. Thus *ulmi* comes to mean 'a vineyard'.

151. **quantulum in hoc,** 'how little there is in this', i.e. 'how little this amounts to', compared with what follows. **pueri** are slaves for personal attendance; **ergastula** (see n. to VIII. 180) is properly the prison in which chained labourers are confined, but is also used for the labourers themselves: cf. Pliny, *Nat. Hist.* XVIII. 36 *coli rura ab ergastulis pessumum est et quidquid agitur a desperantibus.*

152. **quod** must be supplied a second time with **habet,** and in a different case.

ematur, 'must be bought'; an old complaint against women, that they do not like to be outdone by their neighbours in display.

153, 154. **mense quidem brumae,** 'at least in the month of mid-winter'; cf. Mart. XIII. 1. 4 *ebria bruma*. The Saturnalia, an exceptional opportunity for extravagance, were celebrated as a public holiday from 17–19 Dec.; and a fair, called *Sigillaria* from the statuettes in clay (*sigilla*) which were a main article on sale there, went on for four days after 17 Dec. For the purpose of this fair, canvas booths (*casa candida*) were erected near the Saepta, in the Campus Martius; the effect of these booths was to cover up the walls of the *porticus Agrippae*, and perhaps other buildngs. This *porticus* was adorned with frescoes representing the voyage of the Argonauts; consequently it was often called *porticus Argonautarum*, as the *porticus Pollae*, erected by Agrippa's sister, Polla Vipsania, was known for similar reasons as the *porticus Europae* (cf. Mart. III. 20. 12,

XI. 1. 12, II. 14. 15–18). Jason would certainly be a prominent figure in the frescoes: he is called *mercator* sarcastically, because of the purpose of his voyage; the Argonautae are degraded to *nautae*.

The Saturnalia and Sigillaria are coupled in a letter of Tiberius quoted by Suet. *Claud.* 5: when Claudius, then a young man of little promise, pressed his uncle to make him consul, Tiberius *id solum codicillis rescripsit quadraginta aureos in Saturnalia et Sigillaria misisse ei*, a present as uncomplimentary as the tennis-balls sent to Herny V by the Dauphin.

154. armatis: i.e. the heroes on board the Argo carried weapons and were not dressed like ordinary sailors: a ship is often said to have *arma* (or *armamenta*) but sailors had none: cf. Ovid *Fast.* II. 101 *quid tibi cum gladio? dubiam rege, navita, pinum.*

155. tolluntur, sc. *a Bibula*, 'she carries off'. **crystallina** are vases made, not of glass, but of rock-crystal; their excellence consisted in having no flaw (*vitrum*) in the substance; cf. Mart. IX. 59. 13 *turbata brevi questus crystallina vitro*; III. 82. 25 *crystallinisque myrrhinisque propinat.* It was for breaking a vase of this kind that one of his slaves was sentenced by Vedius Pollio to be thrown as food to his pet lampreys; Augustus, who was dining with Pollio at the time, ordered all the *crystallina* to be broken in his presence and the fish-pond to be filled up (Sen. *Dial.* v. 40. 2).

156. myrrhina or *murrea* are constantly mentioned in ancient writers together with *crystallina* as vases of great value. But no specimen is now extant, and there has been much controversy as to what substance is meant by *murra*. One fact is clear from the ancient authorities (cf. esp. Pliny, *Nat. Hist.* XXXVII. 204) that *murra* was a mineral dug out of the earth, not an artificially manufactured porcelain. Some good authorities hold that *murra* was 'a variety of agate, containing shades of red and purple'. The mineral was not transparent and was very fragile.

Berenices: this Jewish princess played a part, somewhat like Cleopatra's, in Roman history. As a young widow she lived long with her brother, King Agrippa II, and it was believed she had incestuous relations with him (cf. *incestae*, l. 158); Paul defended himself before them at Caesarea A.D. 62 (Acts xxv and xxvi). Titus, when commanding in Syria, fell in love with her but was prevented from actually marrying her by the loathing the Romans felt for such an alliance. She spent a considerable time at Rome, and it is quite possible that a Roman jeweller may have had for sale a famous diamond belonging to her. For works of art with a pedigree, cf. XII. 44–7.

157. **pretiosior:** cf. Mart. VII. 17. 8 (of a copy of his own works corrected by him) *haec illis pretium facit litura.*

159, 160. A description, in Juv.'s manner, of Palestine: cf. XV. 5 and 6. **mero** = *nudo*, which is actually read in P, a remarkable instance of a gloss ousting the original word: see Introduction, p. xlvii.

160. Cf. XIV. 98; Tac. *Hist.* v. 4. 3 *sue abstinent memoria cladis, quod ipsos scabies quondam turpaverat cui id animal obnoxium.* **vetus,** 'long-established'. **senibus** is part of the predicate, 'spares them to grow old'.

161–83. *Even a wife, who has every virtue and every perfection, is apt to be a great nuisance to her husband, from her just pride and self-satisfaction. In this very way Niobe, a model wife and mother, caused the death of her husband and her whole family. There is more bitter than sweet in a faultily faultless wife.*

161 is a further objection put in the mouth of some critic. **tanti greges** refers to the female population of Rome.

162. **sit** is hypothetical subjunctive: 'suppose a woman is...'. **decens,** 'graceful'; *indecens,* 'awkward', inadmissible in Juv.'s. verse, occurs often in Martial's iambic metres.

163. **porticibus disponat avos:** either this refers to the *imagines,* portrait-masks of ancestors, which ornamented the *atrium* in the houses of nobles: see n. to VIII. 1: or, as a *porticus* was not the usual place to display these, the reference may be to triumphal statues of ancestors which would naturally be placed in a *porticus* or *vestibulum:* cf. VII. 126.

omni, 'than any...'.

164. The reference is to Livy, I. 13. 1 *tum Sabinae mulieres ...crinibus passis scissaque veste...ausae se inter tela volantia inferre...dirimere infestas acies, dirimere iras.*

Sabina: the Sabine women had a high reputation for chastity: cf. X. 299; and Martial quoted on l. 10.

165. **rara avis:** cf. Sen. *de Matr.* 56 (Haase, III, 430) *si bona fuerit et suavis uxor, quae tamen rara avis est, cum parturiente gemimus.* **niger cycnus,** like *albus corvus* (VII. 202), is a type of something uncommon, a freak of nature.

166. **feret uxorem:** cf. V. 164. **cui constant omnia,** lit. 'in whose case all things are correct', i.e. who is complete per-

fection; the metaphor is from book-keeping; e.g. *ratio constat*, 'the account tallies, is correct'.

167. **Venustinam** is read by Büch. for *Venusinam* of all MSS. and the Scholiast. The latter would mean 'a wife from a small country-town like Venusia'; but the word is elsewhere scanned *Venŭsinus*, e.g. I. 51. Venustina must be some woman, well known at Rome, of no distinction and no reputation. Martial has a *Vetustina* and a *Vetustilla*, both of whom answer to this description.

CORNELIA MATER GRACCHORUM was inscribed upon the statue erected to this famous woman by the Roman people. She was the younger daughter of the elder Scipio, whose chief exploits are referred to in the following verses: he surprised Syphax, a Numidian prince, burnt his camp, and destroyed his forces in 203; he conquered Hannibal at Zama in 202 B.C. For her betrothal, cf. Livy, XXXVIII. 57.

169. **supercilium,** 'pride'; cf. v. 62. **numeras:** see n. to l. 382. **in dote,** 'as a part of your dowry'.

171. **cum tota C. migra,** 'go elsewhere, Carthage and all': so αὐτῇ τῇ Καρχηδόνι: cf. XIV. 61.

172. Niobe, proud of her seven sons and seven daughters (*greges natorum*), presumed to claim superiority over Latona, who sent her one son and one daughter to avenge her by shooting down all Niobe's children.

dea pone, the emendation of Graevius for *depone* of MSS., seems necessary and certain; **tu** (which has a gloss, *Diana*, in many MSS.) seems meaningless without it; and in the legend Artemis joins with Apollo in inflicting the punishment; cf. Ovid *Met.* VI. 216 ff. where the story is told at length.

173. **faciunt:** *fecerunt* would be more usual in this sense: see n. to IV. 12. There is something intentionally ludicrous in Amphion directing the divine wrath against his wife.

174. Amphion was Niobe's husband; according to Ovid (*l.l.* 271) he stabbed himself when his sons were killed. **contrahit,** 'bends': lit. 'brings the two ends of the bow together'.

175. **extulit,** 'Niobe carried to the grave'; cf. I. 72; *effert*, l. 182, is quite different: the subject is kept for the dependent clause. **greges natorum:** so *populo natorum* Ovid, *l.l.* 198.

176. **dum,** 'because', a regular use of the conjunction; cf. I. 60. **gente** is used in the sense of *genere*, and does not mean Apollo and Artemis; in Ovid (*l.l.* 185) Niobe points out the undistinguished ancestry of Latona.

177. **eădem,** 'also'. **scrofa...alba,** '*the* white sow' known to Roman legend; see n. to XII. 72.

178. Expressed more fully the sentence would run: *nulla gravitas nec forma maritae tanti est ut velis eam tibi ab illa semper imputari*: for the constr. of **tanti** and the ellipse of a word like **velis,** see n. to III. 54; for **imputare,** see n. to v. 14. Transl. 'no character and no beauty is worth so much, that you would be willing to have it always thrown in your teeth'.

179. The *rarum summumque bonum* is a virtuous and beautiful wife, as the next line shows.

181. **habet:** the subject is *uxor* understood.

deditus, 'devoted'; again l. 206.

183. **in diem,** 'every day', is correctly used with a distributive sense after **septenis;** cf. Livy, XXII. 23. 6 *argenti pondo bina et selibras in militem praestaret,* '2½ lb. a man'; just so '7 hours a day', i.e. more than half the day.

horis; the abl. to express duration of time is common in silver Latin: cf. VII. 235, X. 239, XI. 53 and 72.

184–90. *Another tiresome trick of our Roman women is the way they insist on talking Greek on all occasions.*

185. **nam,** 'for example'. **rancidius,** 'in worse taste'; *putidus* is used in the same sense.

187. **Sulmonensi:** Sulmo, the birthplace of Ovid, is a provincial town in the country of the Peligni. Even out of Rome, the women think they have no pretensions to beauty unless they speak Greek, and that pure Attic. Cicero seems to have admired the Latin spoken by his countrywomen very highly: cf. *de Orat.* III. 45 (Crassus is speaking) *equidem cum audio socrum meam Laeliam—facilius enim mulieres incorruptam antiquitatem conservant, quod multorum sermonis expertes ea tenent semper quae prima didicerunt—sed eam sic audio, ut Plautum mihi aut Naevium videar audire.* It appears from Quintilian (I. 1. 12 and 13) that it was a common practice to teach children Greek first; he approves of this, but thinks that Latin should soon follow.

graece, sc. *loquuntur*: but with *latine* below, *loqui* should not be understood, *scire* (or *nescire*) *Latine* being the common idiom: cf. Cic. *Phil.* v. 13 *num Latine scit?* idem, *pro Flacc.* 10 *Graece nesciunt.*

189. **pavent,** 'they express fear', crying ὤμοι and the like.

200–30. *If you are not likely to feel affection for the lady to whom you are betrothed, it seems a pity to marry her and to incur the expenses of the ceremony. If, on the other hand, your heart is*

*hers, you must be prepared for the most absolute slavery. She
will decide all your expenditure for you and choose your friends;
she will name, as your heir, a lover of her own; she will insist on
punishing your servants, even with death. And soon she will
leave you and marry someone else; perhaps she will marry you
again afterwards. Some ladies marry oftener than once a year.*

200. **legitimis...tabellis**, 'by a contract in due form': for
tabellae, see n. to l. 25; an action for breach of this contract
was not possible by Roman law.

201. **ducendi**, sc. *eam*.

202. **quare...donanda**, 'why you should waste money on a
feast, and on cakes which you must present to the guests, after
a good dinner, when the company is dropping off': the *cena*
was a regular part of the ceremony, usually eaten in the bride's
house, before the *deductio* began to her new home: Juv. here
speaks as if the bridegroom provided the entertainment. The
custom of distributing *mustacea* after a wedding is not else-
where mentioned, but still survives in a similar shape.

203. **officio**: the word is used, especially of a marriage, in
the sense of 'ceremony'; cf. Suet. *Calig.* 25 *Liviam C. Pisoni
nubentem, cum ad officium et ipse venisset, ad se deduci imperavit*;
idem, *Claud.* 26 *cuius nuptiarum officium et ipse...celebravit.*
Here it is used in a concrete sense of the company at the cere-
mony; cf. x. 45; Ovid, *ex Pont.* iv. 4. 42 *officium populi vix
capiente domo*; Pliny, *Epp.* iii. 12. 2 *officia antelucana* (= *salu-
tatores matutini*) with Mayor's n.

crudis = *post cenam*; it also suggests they have eaten more
than is good for them already; cf. Mart. xii. 76 (when harvest
has been abundant but prices are low) *ebrius et crudus nil habet
agricola.*

illud, 'the present'; the custom of presenting a sum of
money to the bride the day after marriage, is not mentioned
elsewhere.

204, 205. 'When on the splendid salver the conqueror of
Dacia and of Germany glitters with lettered gold'; i.e. the
present consisted of *aurei* (gold coins worth about £1. 1s.),
issued by Trajan and bearing his *cognomina* and his image.
Both titles are given to Domitian by Martial in dedicating his
8th book (date A.D. 93) to that emperor; but no coin of his is
inscribed *Dacicus*. Trajan (and Nerva also) took the *cognomen* of
Germanicus in Oct. or Nov. A.D. 97 at the end of the Suebian war;
and added *Dacicus* in 103 or the end of 102 after the war in Dacia.

Consequently, this verse cannot well have been written earlier than the latter date.

The **lanx** is itself part of the present; gladiators in the arena were often rewarded with *lances* filled with *aurei*: cf. Mart. *Epig. Lib.* 29. 6 *quod licuit, lances donaque saepe dedit.*

205. **scripto:** cf. Stat. *Silv.* III. 3. 105 *quid Ausoniae scriptum crepet arce monetae.*

206. This is the alternative to the supposition in l. 200. **uni** = *soli.*

207. **summitte caput:** cf. v. 172.

208. **ferre iugum:** cf. *iactare iugum*, XIII. 22.

209. 'Even if she is in love herself, she delights to torment and rob her lover.' **spoliis** is not used metaphorically; cf. l. 232 and Ovid *Am.* I. 10. 29 *sola viri mulier spoliis exultat ademptis.*

210. **igitur:** cf. Quint. I. 5. 39 *an sit 'igitur' initio sermonis positum, dubitari potest, quia maximos auctores in diversa fuisse opinione video:* for Cicero's practice, see Madv. on *de Fin.* I. 61.

A wife is a worse evil to a good and kind man, because she will never fail to take advantage of his good-nature.

214. **dabit affectus,** 'will prescribe your feelings (likes and dislikes) for you'; cf. Lucan, x. 94 (of young Ptolemy) *puer ipse sororem, | sit modo liber, amat: sed habet sub iure Pothini | affectus ensesque suos. affectus* are not necessarily feelings of affection; but the word inclines to that meaning in silver-age Latin: cf. Pliny, *Paneg.* 79 (of Trajan) *praesidebit laetitiae publicae...temptabitque adfectus nostros, ut solet, cohibere, nec poterit.* See n. to XV. 150.

ille excludatur is the lady's imperious command; *ille* is explained in what follows.

215 'Whose beard your door saw', i.e. who often came to your house while his beard was still allowed to grow; see n. to l. 105.

216. **cum,** 'although'. Not even persons of the most discreditable occupations are prevented from leaving their property as they please; but the married man is practically *intestabilis.*

217. **iuris idem:** *idem* is given the constr. of *tantundem.* **harenae** = *gladiatoribus.* For the conjunction of these classes, cf. III. 155 ff.

218. **non unus,** 'more than one'; cf. VIII. 213.

rivalis is the French *amant*; cf. Mart. III. 70 *moechus es Aufidiae, qui vir, Scaevine, fuisti; | rivalis fuerat qui tuus, ille vir est.* **dictabitur,** sc. *ab uxore.*

219. A dialogue follows between the husband and wife.
crucem: this was essentially the *servile supplicium*: cf. Plaut.
Miles, 372 *noli minitari: scio crucem futuram mihi sepulcrum:* |
ibi mei sunt maiores siti, pater, avos, proavos, abavos. The cruci-
fixion of slaves was abolished by the Christian emperors, to
avoid profanation of the sacred emblem.

220. **quis...audi:** 'who informed against him? hear what
he has to say'; cf. x. 69.

221. **de morte hominis,** 'when a man's life is concerned'.
Cf. Amm. Marcell. XXIX. 2. 18 *de vita et spiritu hominis...
laturum sententiam diu multumque cunctari oportet.*

222. **ita servus homo est?** 'is a slave a man, then?' cf. Sen.
Epp. 47. 1 '*servi sunt.*' *immo homines.* '*servi sunt.*' *immo con-
tubernales.* '*servi sunt.*' *immo humiles amici*; Petron. 71 *et servi
homines sunt et aeque unum lactem biberunt.* **nil fecerit, esto,**
'grant that he has done nothing; very well'; the subj. is con-
cessive and is akin to the hypothetical in l. 144; cf. Cic. *pro
Lig.* 18 *fuerint cupidi, fuerint irati, fuerint pertinaces: sceleris vero
crimine, furoris, parricidii liceat...carere.*

223. Cf. Shak. *Jul. C.* II. 2. 'The cause is in my will, I will not
come.'

224. **regna,** 'place of power': cf. *regnat*, l. 149.

225. **permutat domos,** 'she goes from one house to an-
other'; cf. l. 94. **flammea conterit,** 'wears out her wedding
veil', by figuring so often as a bride; the *flammeum* was a red or
yellow veil worn by Roman brides: the plur. cannot be pressed
here, as the poets, for metrical reasons, generally use either the
plur. *flammea* (so Juv. II. 124; Lucan, II. 361; Stat. *Theb.* II. 341,
and Mart. twice) or the dimin. *flammeolum* (so Juv. X. 334);
but in prose *flammeum* is regularly found and so in the glyconics
of Catullus (61. 122).

inde, i.e. from her new home.

226. **vestigia** is 'the imprint' left by herself on the bed she
has forsaken: it is constantly used in this sense: cf. Ovid *Her.*
X. 53 (the deserted Ariadne to Theseus) *et tua, quae possum, pro
et vestigia tango | strataque quae membris intepuere tuis*; Cic. *in
Verr.* II. 3. 79 *cum in lecto...mulieris vestigia viderent recentia*;
Sen. *Epp.* 108, 23 *laudare solebat Attalus culcitam quae resisteret
corpori: tali utor etiam senex, in qua vestigium apparere non possit*;
Aesch. *Ag.* 420 ἰὼ λέχος καὶ στίβοι φιλάνορες. Often *vestigia
tua = tua pars lecti.*

227. This refers to the decorations in honour of her second
marriage; before the boughs have withered, she is off again.

The doors of the bridegroom's house were wreathed with laurel garlands: cf. l. 79 *ornentur postes et grandi ianua lauro*; Lucan, II. 354 (of Cato's marriage with maimed rites owing to civil war) *festa coronato non pendent limine serta*; Stat. *Theb.* II. 248 *fractis obtendunt limina silvis*.

228. **vela** is an awning hung over the *vestibulum* or open space between the street and the house-door.

229. At the end of the Republic, and still more under the Empire, marriage had come to be regarded by many as a temporary connection which might be dissolved for the most trivial reasons: thus Caesar and Antony were each married four times, Pompey five times. The moralists naturally exaggerate the evil: cf. Mart. VI. 7. 3 *aut minus aut certe non plus tricesima lux est,* | *et nubit decimo iam Telesilla viro*; Sen. *de Ben.* III. 16. 2 *numquid iam ulla repudio erubescit, postquam inlustres quaedam ac nobiles feminae non consulum numero sed maritorum annos suos computant?*

230. **autumnos** is used in the sense of *annos*.

titulo...sepulcri, 'a fact worthy to be inscribed on her tombstone'; the irony gains from the fact that it was considered a distinction to have had only one husband and that this was often recorded on tombstones: cf. Prop. v. 11. 36 *in lapide huic uni nupta fuisse legar*; Wilmanns *Inscript.* 2063 *uni nupta viro summa cum concordia ad ultimum diem pervenit* cet. Such women were called *univiriae*.

231–41. *A mother-in-law makes peace impossible in a household during her lifetime. She teaches her daughter how to deceive you and makes her as bad as herself.*

231. **salva** = *superstite* or *vivente*; cf. l. 30; Mart. v. 10. 7 *Ennius est lectus salvo tibi, Roma, Marone* (i.e. in Virgil's life-time).

232. **nudi**, 'ruined', is part of the predicate, as if it were *spoliare et nudare maritum*: cf. l. 210.

233. **corruptore**: cf. IV. 8; the word suggests that the wife is a young bride: cf. Pliny, *Nat. Hist.* XXII. 3 *nec quaerit (Gallia) in profundis murices...ut inveniat per quod facilius matrona adultero placeat, corruptor insidietur nuptae.*

234. **nil...rescribere**, 'to send an artful and ambiguous answer'; *simpliciter rescr.* would be, to say yes or no: for *rescribere*, see n. to l. 141.

235. **custodes**: cf. l. 348: this refers to a peculiar custom by which a male chaperon was put in charge of a wife, to ensure her fidelity: cf. Tac. *Ann.* XI. 35 *Titium Proculum custodem a Silio*

Messalinae datum, with Furneaux's note. Sometimes the husband was put in charge of such a chaperon by a jealous wife: cf. Mart. x. 69 *custodes das, Polla, viro, non accipis ipsa.*

corpore sano, 'though she is quite well'; the subject to **advocat** must be the daughter, not the mother; when the latter has made all preparations, the wife pretends to have a fever and sends for the doctor, on purpose to keep her husband out of the room; meanwhile the lover is hidden somewhere on the premises.

236. Archigenes was a famous physician of the time, a Greek, like most of his profession; see n. to III. 58; he is mentioned again XIII. 98, XIV. 252. **pallia,** 'bed-clothes'.

239. **scilicet expectas** has the same force as *i nunc et expecta.*

240. **quam quos habet:** in English we must supply 'herself'. **utile:** it is natural the daughter should resemble the mother; and further it is to the mother's advantage.

241. **filiolam:** the dimin. is metrical; see n. to *iuvenes,* III. 158.

242–5. *Some women are litigious and never happy if they are not managing their own case as plaintiff or defendant.*

242. 'There is hardly any matter about which a woman will not stir up a law-suit'; though *lis* and *litigare* are often used of angry discussion, the context shows that actual proceedings in court are meant here; she insists on going to law. Of Manilia nothing else is known. **moverit** is a potential aorist, = οὐκ ἂν κινήσειε..

244. **libellos,** 'pleadings': they prepare the documents for their barrister with no professional advice. Quintilian censures pleaders who sanction this practice of suitors, and says that the barrister ought to have long and repeated interviews with his client: XII. 8. 4 *pessimae consuetudinis, libellis esse contentum, quos componit...litigator.*

245. **principium,** 'exordium': cf. Quint. IV. 1. 1 *quod principium Latine vel exordium dicitur, maiore quadam ratione Graeci videntur* προοίμιον *nominasse:* it is the technical word for the beginning of a speech in court. **locos:** the word is used in several senses in rhetorical treatises: it may mean 'passages', any considerable divisions of a speech; or 'beauties', parts of the speech with which especial pains have been taken; or lastly it may be = *loci communes,* 'declamation on general subjects', for which cf. Cic. *de Orat.* III. 106 and 107: all the three senses, and perhaps others, are common in Quintilian.

Celsus is probably A. Cornelius Celsus to whom Quintilian often refers as a writer on rhetoric; a treatise by him on medicine is extant. There were two famous jurists of this century, called Iuventius Celsus; but a teacher of rhetoric is more in point here. These women are prepared to teach eloquence to a master of the art.

246–67. *There are also the ladies who practise swordsmanship; you can hardly expect modesty from them. A happy husband you will be, if there is an auction of your wife's property, and all her fighting gear comes to the hammer. Just see how a woman, who finds the lightest silk too heavy, goes through all the prescribed exercise in fighting order; such a thing as you would never see done by a gladiator's wife.*

246. **endromidas:** cf. III. 103; this was a thick wrap, put on in the intervals of severe exercise, to guard against a chill; these ladies wear it of purple, though it was usually of rough, plain stuff; cf. Mart. IV. 19. 3 *sordida sed gelido non aspernanda Decembri | dona, peregrinam mittimus endromidam, | seu lentum ceroma teris tepidumve trigona.*

femineum ceroma, 'the wrestling of women'; *ceroma* is properly the mixture of oil and wax rubbed by wrestlers on their limbs, as in Mart. quoted above; hence it also = wrestling; cf. III. 53 *luctantur paucae.*

247. **vulnera pali:** the *palus* was a wooden stump on which the gladiator practised his cuts and thrusts with a wooden sword (*rudis*); cf. Mart. VII. 32. 8 *nudi stipitis ictus hebes*, 'blows with a blunt sword at a defenceless stump'.

248. **rudibus** seems the word required: P has *udibus, the other MSS. *sudibus*; cf. Ovid, *Ars*, III. 515 *sic ubi prolusit, rudibus puer ille relictis | spicula de pharetra promit acuta sua*; Tac. *Dial.* 34 *ferro non rudibus dimicantes.* In Livy, XL. 6. 6 (of a sham fight) *multa vulnera rudibus facta, nec praeter ferrum quicquam defuit ad iustam belli speciem, rudibus* is generally read for *sudibus* of MSS.

lacessit, 'challenges', not 'attacks': to do everything according to rule, she carries a shield though there is no adversary.

249. **numeros,** lit. 'rhythm', is a technical word for the prescribed rhythmical movements of the fencing or wrestling school: cf. Quint. XII. 2. 12 *palaestrici doctores illos, quos numeros vocant, non idcirco discentibus tradunt, ut his omnibus…utantur …sed ut sit copia illa ex qua unum aut alterum…efficiant*; idem, X. 1. 4. Thus the phrase is not similar to *omnibus numeris absolutus.*

250. **Florali matrona tuba:** the Floralia (28 Apr.–3 May)
were games at which farces (*mimi*) were played, and custom
sanctioned unusual freedoms on the part of the actresses: cf.
Mart. I. *praef. epigrammata illis scribuntur qui solent spectare
Florales. Floralis tuba = Florales ludi*, as all public shows began
with the blowing of trumpets; cf. X. 214. *matrona* comes here
where one would expect *meretrix*, that class being prominent at
these games: cf. Ovid, *Fast.* V. 349 *turba quidem cur hos celebret
meretricia ludos, | non ex difficili causa petita subest.*

nisi si cet.: there is one shameful excuse for her, that she has
a further purpose in her mind, and is practising for fighting in
real earnest in the amphitheatre: cf. I. 22 and Stat. *Silv.* I. 6. 53
stat sexus rudis insciusque ferri | et pugnas capit improbus viriles.

255. **auctio:** an auction was not always a sign of insolvency
but an ordinary expedient of Romans, who possessed, by in-
heritance or otherwise, property they did not want: so, when
Hortensius, a very rich man, died, Cicero writes of the young
Hortensius (*ad Att.* VII. 3. 9) *aveo scire...quarum rerum
auctionem instituat.* With **quale decus** supply *sint*, to which
balteus etc. are subjects.

256, 257. These are parts of the armour worn by the gladia-
tors known as *Samnites*: **balteus** is a sword-belt; **manicae** are
rings of armour which most gladiators wore round the right
arm, the left being protected by the shield; **tegimen cruris**
is a metal greave (*ocrea*), in shape and size like a cricket-pad,
which the *Samnites* wore only on the left leg: cf. Livy, IX. 40. 3
(of the Samnite army in 310 B.C.) *sinistrum crus ocrea tectum:
galeae cristatae, quae speciem magnitudini corporum adderent.
Ibid.* §17 *Campani, ab superbia et odio Samnitium, gladiatores...
eo ornatu armarunt Samnitiumque nomine compellarunt.* A good
illustration of a *secutor* dressed as a *Samnes* will be found in
Baumeister's *Denkmäler*, p. 2097.

257. **si diversa movebit proelia,** 'if she engages in a dif-
ferent kind of fighting', i.e. as a *Thraex*, who wore greaves on
both legs; cf. VIII. 201 and n. there.

258. **vendente** refers to the auction.

puella, 'your young wife'.

259. **cyclade:** the *cyclas* is an article of female attire, so
named from a fringe of purple or gold embroidery which ran
round it; this passage shows that it was made of some very light
material.

260. 'For whose delicate charms even a little scrap of silk is
too heavy'; silk is first mentioned by the Augustan writers

under three names, *bombycina*, *serica*, and *Coae vestes*; pro-
bably they were spun by different kinds of worms. The best
bombycina came from Assyria.

261. **fremitu**, 'loud noise'. **monstratos ictus:** cf. *numeri*,
l. 249: a trainer stands beside her and tells her how to thrust;
dictata is the common phrase: cf. v. 122; Petron. 45 *Thraex qui
et ipse ad dictata pugnavit*; Suet. *Iul.* 26 *tirones...per equites
Romanos...erudiebat, precibus enitens...ut...ipsi dictata
exercentibus darent*.

262. **quanta** is an epithet of **fascia**, just as **quam denso
libro** is: 'see the size, and thickness of pith, of the bandages
fitting close to her legs': she is wearing *fasciae* made of bark, as
a protection for her legs in default of greaves.

265. **neptes** cet. = great ladies. The Aemilii Lepidi were
one of the most illustrious Roman families; for Metellus, see n.
to III. 139; Fabius Gurges (the glutton) got this *agnomen* from
the dissoluteness of his youth, but became a distinguished man,
and was thrice consul and *princeps senatus*.

266. **ludia:** cf. l. 104.

267. **gemat** refers to the gasping or groaning caused by the
severe exercise: cf. Hor. *Epod.* v. 30 *ligonibus duris humum* | *ex-
hauriebat ingemens laboribus*; Sen. *Epp.* 56. 1 *cum fortiores exer-
centur et manus plumbo graves iactant...gemitus audio*. The con-
text shows that Asylus is a gladiator.

268–85. *Your wife will rob you of your sleep by her curtain-
lectures in which she invents causes for jealousy to hide her own
misdoings. When she cries, which she can do when she likes, you
think it is a sign of her love for you; you would find out your
mistake, if you were to rifle her desk. But however clearly her
guilt is demonstrated, however hopeless her position seems to be,
she never wants arguments to defend herself.*

268. **alterna iurgia**, 'mutual recriminations'.

270. **tunc gravis:** Büch. reads *cum gravis* with P. That Juv.
might use *cum...*, *tunc...* for *cum...*, *tum...*, is shown by
IX. 118 *vivendum est recte cum propter plurima, tunc his* | *prae-
cipue causis*: but here the constr. can only be explained by
supplying *semper*: 'always a nuisance to her husband, she is
worse then than a tigress robbed of her cubs'. As such an ellipse
is surely impossible, it seems better to read *tunc* or *tum* with the
other MSS., while recognising that the text is probably corrupt.
Perhaps *cum* might be retained, if a comma is placed after *illo*
and a stop after *viro*.

orba tigride: cf. Mart. III. 44. 6 *non tigris catulis citata raptis* | *...nec sic scorpios improbus timetur.*

271. **facti,** i.e. her own misconduct.

272. **odit,** 'abuses'; see n. to VII. 35. **paelice** is abl. of cause, 'because of an imaginary rival'.

273–5. A new sentence should begin here in English; the ablatives are absolute: 'she has tears always in abundance and always ready in their place and only waiting to see in what way she may bid them flow'. **statio** suggests a picket of soldiers. **manare** is governed by **iubeat,** not by **expectantibus.** For *iubeat,* cf. Mart. I. 33 *amissum non flet, cum sola est, Gellia patrem:* | *si quis adest, iussae prosiliunt lacrimae.* With **amorem,** sc. *esse.*

276. **tibi...places,** 'are conceited', 'think well of yourself'; cf. X. 41; Ovid, *Remed.* 685 *desinimus tarde quia nos speramus amari;* | *dum sibi quisque placet, credula turba sumus.* **uruca:** a word of uncertain meaning, of which various explanations are offered by the Scholiast: (1) the name given to the *stupidus* (see n. to VIII. 192) in a farce, (2) a kind of animal like a hedgehog, (3) an animal like a weevil that grows on beans. The word occurs in Pliny, *Nat. Hist.* XI. 112 where it means 'a caterpillar'.

Büch. explains 'you worm', suggesting that as *curculio* is the name of a parasite in Plautus, so *uruca* is applied to a creeping, crawling, cringing man.

277. **exsorbes,** 'kiss away'. **lecture:** the voc. is used by attraction for the nom.; cf. Virg. *Aen.* II. 282 *quibus, Hector, ab oris* | *expectate venis? lecture* contains the apodosis of a conditional sentence, *quae scripta legas, si retegantur scrinia?* See n. to *missurus,* V. 32. This use of the fut. partic., unknown to Cicero, is common in all silver-age writers, especially in Seneca and Pliny, and is often introduced as here by the interrog. pronoun: e.g. *Nat. Hist.* VII. 105 (of a distinguished soldier) *quae omnia ex oratione eius apparent habita cum in praetura sacris arceretur a collegis ut debilis, quos hic coronarum acervos constructurus hoste mutato:* i.e. the other praetors tried to prevent his sacrificing because he had lost an arm; but what heaps of decorations he would have amassed, if his campaigns had not been against Hannibal, when they were seldom won.

The **scripta** and **tabellae** are the wife's compromising letters.

278. **tibi,** 'by you'. **zelotypae:** cf. V. 45; the word is post-Augustan though Cicero uses ζηλοτυπία. There is no Latin word which exactly expresses the 'jealousy' of lovers; though

dolor is used for the resentment of an injured wife or husband: see n. to X. 315.

280. **Quintiliane:** see n. to VII. 186.

colorem, 'line of defence'; cf. VII. 155; *color* (χρῶμα) is a term of the rhetorical schools, constantly used by Seneca and Quintilian: it is the favourable light in which a speaker endeavours to place an action which he is defending: cf. Quint. IV. 2. 100 *ne illud quidem ignorare oportet, quaedam esse quae colorem non recipiant sed tantum defendenda sint*; Ovid, *Trist.* I. 9. 63 *ergo ut defendi nullo mea posse colore,* | *sic excusari crimina posse puto*; the passages quoted show that *color* is not a mere synonym of *excusatio*, the latter being used where the former was impossible.

281. **haeremus,** 'I am puzzled'; cf. Trajan, *ad Plin.* 82 *potuisti non haerere*, 'you need not have been puzzled'. This is supposed to be Quint.'s answer. **convenerat,** 'it was settled long ago'; nothing is here expressed by the plpf., which the aorist would not equally convey.

283. **mare caelo confundas:** used proverbially of making a great disturbance; cf. II. 25 *quis caelum terris non misceat et mare caelo?* Livy, IV. 3. 6 *quid tandem est cur caelum ac terras misceant?*

284. **homo sum** has two common meanings: (1) I have the feelings of a man; so Terence, *Haut. Tim.* 77 *homo sum; humani nihil a me alienum puto.* (2) I have the weaknesses of a man; so Petron. 75 *nemo, inquit, nostrum non peccat. homines sumus, non dei.* The latter meaning is required here. But perhaps a third meaning is possible, i.e. I have the rights of a man and will not be treated like a slave (cf. l. 222) or an animal.

nihil or **quid** is generally preferred in such sentences to *nemo* or *quis:* cf. Mart. I. 10. 3 *adeone pulcra est? immo foedius nil est:* no doubt *nihil* is more comprehensive.

285. **deprensis,** 'when they are detected'; said of the crime l. 640. **crimine** is 'guilt', not 'accusation'. Cf. Tac. *Ann.* I. 38 *praesidium ab audacia mutuatur.*

286–309. *This corruption of morals was unknown in the days when life at Rome was hard and simple; peace has brought luxury in its train, a worse enemy than war; and now all the corruptions of foreign nations have found a home here. How can you expect virtue in a woman who is not even sober? It is easy to understand their contempt for the altar of chastity.*

286. **monstra:** cf. l. 645 and *prodigia* l. 84.

288. **contingi**, 'to be polluted'; cf. *contacta*, v. 128. It is difficult to determine whether **tecta** is acc., or the first of a number of subjects to *sinebant*, the object being *eas* understood from *Latinas*.

289. **vellere Tusco**: the women of old spent much of their time in spinning wool, which is mentioned as a virtue in their epitaphs; cf. Wilmanns, *Inscript.* 549 *domum servavit, lanam fecit*.

291. Hannibal marched suddenly under the walls of Rome in 211 B.C.: the consular army encamped between the Colline and Esquiline gates, outside the Agger: Livy (XXVI. 10) gives a vivid picture of the terror in Rome.

292. **longae pacis mala**: cf. Shak. 1 *Hen. IV*, IV. 2 'the cankers of a calm world and a long peace'.

294. Cf. the beginning of Sallust's *Catiline*, esp. c. 10 *qui labores, pericula, dubias atque asperas res facile toleraverant, eis otium divitiae...oneri miseriaeque fuere. igitur primo pecuniae, deinde imperi cupido crevit: ea quasi materies omnium malorum fuere.*

ex quo, 'from the time when', 'ever since': cf. I. 81; X. 77.

295. For the quantity of **perit**, see n. to III. 174.

hinc = *a luxuria*. **istos** is used for *hos*, as often in silver Latin: cf. IX. 131 *salvis his collibus* (= *incolumi Roma*).

296. For the thought and expression, cf. III. 62. **Sybaris** is used for 'the morals of Sybaris'; the Sybarites were, and still remain, proverbial for τρυφή: hence Thackeray's name 'Percy Sibwright' for a self-indulgent young man in *Pendennis*. The other places mentioned were famous in the same way but in a less degree.

297. The line refers to the scene in the theatre of Tarentum, 300 years before, when the Roman ambassador was grossly insulted by a drunken buffoon and said, 'It will take much blood to cleanse this robe'. The people in the theatre wore garlands (*coronatum*), as there was a feast of Dionysus being celebrated; and at such a time they were not likely to be sober (hence *madidum*); cf. Dio Cass. *fragm.* 145. **petulans,** 'insolent', differs from III. 278; so ὕβρις means both 'violence' and 'insolence'.

298. **obscaena**, 'filthy', in reference to the corruption it brought with it: cf. *funesta pecunia*, I. 113.

299. **fregerunt saecula**, 'enervated the age'. It is difficult to say, perhaps Juv. himself could not have said, what age is meant. Tacitus, whose evidence is more to be trusted, was of opinion that luxury reached its height at Rome during the

century between the battle of Actium and the accession of Galba; he attributes the change for the better largely to the character and habits of Vespasian (*Ann.* III. 55). Again, is Juv.'s phrase of the long peace merely rhetorical? The period, to which his satires generally refer, is that between Nero and Trajan; and Nero's death was followed by one civil war after another of which Italy bore the brunt. It is true, of course, that the Roman state had long had no rival, civilised or barbarian; and Juv. must mean this.

300. **quid...curat?** i.e. a drunken woman has no scruples.

302. **mediis iam noctibus:** cf. IV. 137; the Romans in general kept early hours, and a *comissatio* prolonged to such an hour was in itself scandalous. Seneca (*Epp.* 95, 21) says of women: *non minus* (i.e. *quam viri*) *pervigilant, non minus potant; et oleo et mero viros provocant.*

303. **mero:** no water is put in the wine; for the custom of mixing wine and perfumes, cf. Petron. 70 (where the scene is like that described here) *pueri capillati attulerunt unguentum in argentea pelve pedesque recumbentium unxerunt...hinc ex eodem unguento in vinarium atque lucernam aliquantum est infusum.*

304. **concha:** cf. l. 419: it is a broad vessel, shaped like a shell, and generally used (like *pelvis* quoted above) to hold scent: cf. Hor. *Carm.* II. 7. 22 *funde capacibus | unguenta de conchis.* These revellers are not content with drinking from cups, but pour their wine into the great salver that holds *unguenta.*

305. **geminis...lucernis:** the wine makes them see double: cf. Petron. 64 *et sane iam lucernae mihi plures videbantur ardere totumque triclinium esse mutatum.*

306. **i nunc et...:** a formula of derision or remonstrance, in which *et* is sometimes omitted and sometimes inserted between the two imperatives: again X. 310; XII. 57: 'doubt, if you care to', will give the sense here; *nunc* = 'when you have heard of such goings on as these'.

sorbere aera, 'to sniff'.

307. Tullia and two sisters called Maura are supposed to be returning at night from a party such as that described above. They pass through the *forum Boarium*, between the Palatine and the river, and express by gestures and words their contempt for Pudicitia, who had a shrine there.

346–65. *It is no use keeping a strict watch over your wife, as some advise. All women, rich and poor, are alike in their passion*

*for shows and in their extravagance. Men have some notion of
what money is and realise that it can come to an end; a woman,
when she has set her heart on a thing, never thinks of the price.*

346. **quid...olim moneatis,** 'the advice you have long been
giving'; see n. to III. 163.

347. **pone seram:** *oppone* would be used in prose: the *sera*
(μοχλός) is a cross-bar which fitted into the door-posts on both
sides and so prevented the door being opened; cf. Aristoph.
Thesm. 414 εἶτα διὰ τοῦτον ταῖς γυναικωνίτισιν | σφραγῖδας
ἐπιβάλλουσιν ἤδη καὶ μοχλούς.

348. **custodes,** 'the guardians of her honour': see n. to l.
235. **ab illis:** the same constr. as ἄρχεσθαι ἀπό τινος.

350. The poor woman, who walks on foot, is no better than
the great lady in her litter. For **silicem,** see n. to III. 272.
conterit, 'walks on'; *tero* is common in this sense: cf. Mart.
II. 11. 2 *quod ambulator porticum terit seram.*

351. Cf. I. 64; the litter was carried by six or eight slaves,
Liburnians, Cappadocians, or Syrians being preferred: the
poles of the litter (*asseres*) were supported on the shoulder: cf.
Mart. IX. 22. 9 *ut canusinatus nostro Syrus assere sudet.*

352. Ogulnia is so impoverished that she has to hire a dress
and attendants to make a respectable appearance in public; yet
she has still some family plate left, which she gives away to
athletes.

conducit...vestem: so Simaitha (Theocr. 2. 74) borrows
τὰν ξυστίδα τὰν Κλεαρίστας, in order to see a grand procession
in fine clothes.

353. **comites:** see n. to I. 119. **cervical** (προσκεφαλαῖον) is a
cushion or padded back for the *sella,* or sedan-chair.

354. **nutricem:** the presence of an old servant would give
an air of respectability. **flavam:** the colour suggests that the
girl is a captive from some northern nation. (As the Scholium on
flavam is *puellam minorem,* Scaliger suggested that the original
word was *faveam,* which he restored elsewhere, i.e. Plaut.
Miles, 797, where see Tyrrell: *favea* would here be an adj. if
puellam is sound.)

355. **argenti,** 'silver plate', not 'money'.

356. **levibus,** 'beardless', i.e. young: cf. Mart. XIV. 205
sit nobis aetate puer, non pumice, levis. Or the word may mean
'smooth', with reference to the oil used by wrestlers: comp.
Mart. III. 58. 25 *lubricus palaestrita* with idem, IX. 56. 11 *dum
puer es, redeas, dum vultu lubricus.*

357. **res angusta domi:** cf. III. 165: in this phrase *domi* is a device for expressing *familiaris* of prose; cf. Cic. *de Part. Orat.* 112 *angustiae* (nom.) *rei familiaris*; Tac. *Ann.* XII. 52 *ob angustias familiares.*

358. **se metitur:** cf. XI. 35 and Hor. *Epp.* I. 7. 98 *metiri se quemque suo modulo ac pede verumst.* The emphatic **illum** is characteristic of silver Latin; cf. l. 269, v. 139.

359. **tamen,** 'after all', i.e. in spite of extravagance up to a certain point, a man does draw the line somewhere. The contrast is heightened by the asyndeton of the opposing clause (l. 362).

361. **formica...magistra:** moralists of all times have used the example of the ant to enforce industry and prudence: cf. Hor. *Sat.* I. I. 32–7 where she is called *haud ignara ac non incauta futuri.* **expavere** is the aor. of frequency.

362. '*But* an extravagant *woman*...'

363. **velut** is used for *velut si,* as often by the poets: cf. IV. 59: so *tamquam* often, and *perinde ac* less often.

pullulet: the word, here metaphorically used, is properly said of living things, *pulli* being the young of any animal, or the shoots of a plant: cf. Sen. *de Clem.* I. 8. 7 *praecisae arbores plurimis ramis repullulant.*

364. The line may be a reminiscence of Hor. *Sat.* I. I. 51 *at suave est ex magno tollere acervo.*

365. **usquam,** 'in any case'.

380–97. *Some women are musical enthusiasts and delight to hold in their hands the lyre which their favourite virtuoso has honoured by using. One great lady actually enquired by sacrifice and augury whether her favourite harper would be successful in winning the prize at a future contest. The gods seem to have plenty of leisure if they have time to answer such enquiries as this.*

380. **cantu** includes playing as well as singing.

organa, 'musical instruments': cf. Suet. *Nero,* 41 *organa hydraulica.* The only instrument mentioned here is the *cithara,* an instrument with seven strings, which we may call 'lyre'. It is also called *testudo,* from the legend that the first instrument of the kind was made by Hermes from the shell of a tortoise over which he stretched sheep's gut; and *fides* (l. 388), from its strings. The strings were struck partly by the fingers, partly by an instrument of ivory or metal called the *plectrum* or *pecten.*

381. **densi...sardonyches,** 'her rings glitter thick all over

the sounding-board': the rings come close one after the other on her fingers: cf. Quint. XI. 3. 142 (advice to barristers) *manus non impleatur anulis*; Mart. XI. 59 *senos Charinus omnibus digitis gerit | nec nocte ponit anulos*. The sardonyx is mentioned in a ring again VII. 144 (where it is fem. as in Pliny, *Nat. Hist.* XXXVII. 85 ff.), and XIII. 139; it was often used for this purpose, whereas a diamond ring is hardly mentioned in ancient times except Berenice's in l. 156: cf. Mart. XI. 37 *Zoile, quid tota gemmam praecingere libra | te iuvat et miserum perdere sardonycha?* In Mart. *sardonyx* is always masc.

382. **crispo...chordae,** 'the strings are struck in succession (by her) with the vibrating quill'. *crispus*, of a man, means 'with curly hair', of wood, 'with a ripple in the grain' (Mart. XIV. 90); applied to the *pecten*, it expresses the vibration at each blow on the strings.

numerantur: this does not mean that the strings are *counted* with the *pecten*, but that they, being *numerous*, are struck: so l. 169 *numeras triumphos = multos adfers triumphos*: cf. Mart. VIII. 28. 7 *an tua...numeravit lana Timavum*, 'was your fleece washed in Timavus of many mouths?'; *ibid.* 65. 9 *currus numerant elephanta frequentem*, 'the chariot is drawn by a number of elephants': in each case the idea of number belongs not to the verb but to the object. The silver-age use of *perdere* is analogous: *perdere cenam* (l. 202), 'to give a feast to no purpose'; *perdere scelus*, 'to commit a crime to no purpose'; *perdere naulum*, 'to spend your fare to no purpose': thus *frustra dare, frustra admittere, frustra solvere* are all expressed by the same word.

383. Hedymeles, whose name is invented to express his accomplishments, stands for any *virtuoso* of the day. **operas dedit,** 'has been kind enough to oblige'; this looks like a technical phrase for giving a performance; the survival of the sense in the modern *opera* is curious: *operas edere* is used by Seneca of giving lectures: *Epp.* 29. 6 *philosophum Aristonem, qui in gestatione disserebat. hoc enim ad edendas operas tempus ceperat.* With **hunc,** supply *pectinem.*

384. **solatur,** in the absence of the man himself.

indulget, 'lavishes'.

385. **de numero Lamiarum** = *nobilissima*; cf. IV. 154.

nominis Appi, i.e. a lady of the *gens Claudia*; *Appi* is gen. of the adj.; cf. Tac. *Ann.* I. 8 *Livia in familiam Iuliam nomenque Augustum adsumebatur.* This lady's descent on both sides is given.

386. She sacrificed to Janus and Vesta and asked some sign of them whether Pollio would be successful. **Ianum Vestamque** is equiv. to 'all the gods from first to last': cf. Cic. *de Nat. De.* II. 67 (quoted by Lewis) *cumque in omnibus rebus vim haberent maxumam prima et extrema, principem in sacrificando Ianum esse voluerunt...in ea dea* (i.e. Vesta), *quod est rerum custos intumarum, omnis et precatio et sacrificatio extrema est.* Yet the custom of naming Vesta last in a prayer does not seem to have been invariable: cf. Ovid, *Fast.* VI. 303 *inde precando | praefamur Vestam quae loca prima tenet.*

387. **Capitolinam...quercum:** cf. Suet. *Dom.* 4 *instituit et quinquennale certamen Capitolino Iovi triplex, musicum, equestre, gymnicum.* The *agon Capitolinus* (for the name, as well as the thing, was Greek) was instituted by Dom. to celebrate his restoration of the Capitoline temple in A.D. 86. The prizes were wreaths of oak-leaves (*Tarpeiae quercus,* Mart. IV. 54), and to win one was a great object of ambition: Statius repeatedly expresses his disappointment at being defeated in the verse competition (*Silv.* III. 5. 31, V. 3. 231). This *agon* lasted without interruption down to the 5th cent.

Pollio is the name of a *citharoedus* (cf. l. 391) famous at the time: he is mentioned again VII. 176, and by Mart. IV. 61. 9 *here de theatro, Pollione cantante, | cum subito abires* cet.

388. **fidibus promittere,** 'to promise it to his lyre', i.e. to make sure of getting it.

(Friedl. takes *fidibus* as abl. 'to promise a performance on the lyre': cf. *promisit,* VII. 84: but this seems less appropriate here.)

faceret, 'could she have done', in the way of sacrifice and consultation of the gods.

389. **aegrotante** = *si aegrotaret.* **tristibus** cet., 'did not speak hopefully of her little son's case'; cf. XVI. 12. The diminutive here has its proper force and is not a metrical device.

391. **cithara** is contemptuous for *citharoedo.* **velare caput:** the Romans prayed with the head covered, the Greeks with it bare (*aperto capite*): cf. Lucr. V. 1198 *nec pietas ullast velatum saepe videri | vertier ad lapidem.*

dictata verba is the liturgy, the prescribed form of prayer, which was first gone through by the officiating priest and then repeated, word for word, by the person consulting the god.

392. **aperta palluit agna:** her anxiety was so great, that, when the haruspex cut open the victim, to ascertain the god's will by the appearance of the *exta,* her colour went. So Pliny

(*Epp.* VI. 2. 2) tells us that Regulus, a pleader, *semper haruspices consulebat de actionis eventu* (the result of his speech).

393. **antiquissime divum:** cf. Ovid, *Fast.* I. 103 *me Chaos antiqui—nam sum res prisca—vocabant*: Janus there (l. 235) recalls the time of Saturn's first arrival in Italy.

394. **his** is fem., not neut.: 'to such as she'.

395. **quod video,** 'as far as I see': so *ut video,* XIII. 118, in a similar remonstrance with the gods. **aput vos,** 'in heaven'.

397. **varicosus fiet,** i.e. by so much standing. **haruspex,** for which modern English offers no exact equivalent, is constantly rendered 'bowel-prier' by the Elizabethan translators, e.g. Philemon Holland.

398–412. *Even a musical wife is not so bad as an unsexed creature who runs about the town and is not ashamed to give her opinion to experts in questions of war and politics. She knows all the news from abroad and all the scandal of the town; and unburdens herself of her budget to anyone she meets in the open street.*

398. *mulier* is to be supplied as subject to **cantet,** and is followed by two epithets, *audax,* and the clause *quae possit ferre...et...ipsa loqui.* Weidner's *perferre* for *quae ferre* is an arbitrary corruption of the text.

399. **coetus virorum:** men meet by themselves, to discuss business; and there, according to Juv., women are out of place; cf. Stat. *Theb.* IX. 825 *nonne hanc, Gradive, protervam | virginitate vides mediam se ferre virorum | coetibus*; and Inscr. quoted on l. 289. Yet Roman women, after marriage, were always allowed a far freer and less secluded life than the women of enlightened Attica.

400. **paludatis:** the *paludamentum* (the derivation is unknown) was a purple cloak worn by a general when he left Rome in possession of the *imperium*. Cf. Livy, XLIV. 22. 8 (where Aemilius Paullus is objecting to the criticism of civilians) *in omnibus circulis atque etiam, si dis placet* (if you please!), *in conviviis sunt qui exercitus in Macedoniam ducant, ubi castra locanda sint sciant* cet.; these *convivia* would include women.

401. **recta facie,** 'with unflinching face'; the phrase is used and explained by Quint. IX. 3. 101 *orator habet rectam quandam velut faciem, quae ut stupere immobili rigore non debebit, ita saepius in ea, quam natura dedit, specie continenda est.* Cf. *recto vultu,* X. 189.

siccis...mamillis, i.e. like an unsexed creature.

402. **eadem,** 'also'. Martial (IX. 35) describes a man of the same kind: *artibus his semper cenam, Philomuse, mereris,* | *plurima dum fingis sed quasi vera refers.* | *scis quid in Arsacia Pacorus deliberet aula,* | *Rhenanam numeras Sarmaticamque manum;* | *verba ducis Daci chartis mandata resignas,* | *victricem laurum quam venit ante vides;* | *scis quotiens Phario madeat Iove* (i.e. the Nile) *fusca Syene,* | *scis quota de Libyco litore puppis eat,* | *cuius Iulaeae capiti nascantur olivae,* | *destinet aetherius cui sua serta pater* (see n. to l. 387). Such inventors of idle rumours were said, in a phrase first invented at this time, *vendere fumos*: cf. Mart. IV. 5. 7.

403. The Chinese represent the extreme East, the Thracians the North. **secreta...pueri,** 'what passes in secret between the step-mother and step-son', i.e. the cruelty he suffers in secret from her.

404. **diripiatur,** lit. 'is torn to pieces', i.e. is extremely popular, is scrambled for: cf. Stat. *Silv.* V. 3. 130 *Maeoniden...* *urbes* | *diripiunt*; idem, *Theb.* V. 721 *matremque avidis complexibus ambo* | *diripiunt*; Mart. VII. 76 *quod te diripiunt potentiores,* | *...nolito tibi nimium placere*; Persius has *rapere* in the same sense, 2. 37 *puellae* | *hunc rapiant*.

407. **instantem,** 'threatening': comets (*stellae crinitae* in Latin) have always been popularly supposed to presage misfortune, especially to kings: cf. Stat. *Theb.* I. 708 *quae* (neut.) *mutent sceptra cometae*; Suet. *Nero,* 36 *stella crinita, quae summis potestatibus exitium portendere vulgo putatur*; idem, *Claud.* 46 and *Vesp.* 23 *cum...stella crinita in caelo apparuisset...pertinere dicebat ad Parthorum regem qui capillatus esset* (so that his end, and not Vespasian's own, was clearly foreshadowed by a star with long hair). So Milton, of the sun eclipsed, *P.L.* Bk. 1: 'or, from behind the moon, | in dim eclipse disastrous twilight sheds | on half the nations, and with fear of change | perplexes monarchs.'

The natural phenomena mentioned here help to assign a date for the composition of the satire. See Introduction, p. xv.

409. **ad portas,** sc. *urbis.*

Niphatem: the name of a great mountain-range in Armenia; but, in spite of its name, the later Roman poets generally speak of it as a river: so Juv. here, and Lucan, III. 245 *Armeniusque tenens volventem saxa Niphaten*; the language of neither poet suits a mountain, though Hor. *Carm.* II. 9. 20 speaks of *rigidum Niphaten,* showing that he knew what it was.

412. For **quocumque** as an indef. pron. (*cuicumque* in the same line is a relative), see n. to III. 156.

413–28. *Even such a wife is better than one whose selfishness and want of control are a terror to her neighbours. When her sleep is disturbed by barking, woe to the guilty dog and to the dog's owner! She goes late to the bath with great fuss and parade and keeps her guests waiting for their dinner. When she does appear, she whets her appetite by huge potations.*

413. **vitium,** 'failing': it need not mean, as it does in II. 34, 'vicious person'.

414. **concidere,** like *caedere*, 'to beat', 'to belabour': cf. III. 300.

415. **exorata,** 'when appeased': i.e. when she is in a good humour, her neighbours get off with stripes; *for*, when she is really angry, she inflicts the severer punishment with cudgels (*fustes*). This is the Scholiast's explanation.

The reading however is uncertain. P has *exortata*, and Büch. gives *exhortata* which has no tolerable meaning. As the Schol. on l. 417 contains the word *experrecta*, it may be that this is the true reading, which was ousted by a gloss *exorta*, and then 'corrected' to *exorata*. Thus the reading of P is accounted for. Then the sentence beginning *nam si* would merely be an explanation of the preceding sentence, no contrast between the two kinds of punishment being intended. The military punishment with *fustes* meant death; but, in such an instance as the present, nothing so severe need be meant.

417. **illis,** instrumental. **dominum,** 'the dog's owner'.

418. **canem:** watch-dogs were commonly kept in Rome, especially to guard shops; hence the *humiles vicini* were perhaps shopkeepers; cf. Tibull. II. 4. 32 *hinc coepit custos liminis esse canis*.

419. **balnea nocte subit:** cf. XI. 205; the ordinary hour for bathing was the 8th or 9th hour, i.e. 2 or 3 p.m. at the equinox, the *cena* following immediately after: Mart. speaks of the 10th hour as late, X. 70. 13 *balnea post decimam lasso centumque petuntur | quadrantes.* **nocte** need not mean late at night.

conchas: see n. to l. 304: she uses them instead of the usual *gutus.* **castra moveri,** 'her quarters to be shifted': the phrase is used of a single person moving; cf. Mart. v. 14. 3 *bis excitatus terque transtulit castra* (he decamped); Sen. *Epp.* 83. 5 *ille tantus psychrolutes qui...auspicabar in Virginem* (i.e. water from the aqueduct Virgo) *desilire,...ad Tiberim transtuli castra*; Prop. v. 8. 28 *mutato volui castra movere toro.* No doubt, the word suggests here that she has a numerous retinue.

420. magno...tumultu, 'with great fuss and noise'.
sudare refers to the *Laconicum* or Turkish bath, a room immediately over the hot-air furnace in a bathing-house, where the bather sat to promote perspiration, afterwards plunging into the *piscina*, a cold swimming-bath; cf. Mart. VI. 42. 16 *ritus si placeant tibi Laconum,* | *contentus potes arido vapore* | *cruda Virgine Marciave mergi.*

421. Like a man she takes exercise before her bath, choosing for the purpose *halteres*, pieces of stone or metal swung in the hands like dumb-bells: they were often held by persons jumping, whence the name: cf. Mart. VII. 67. 5 (also of a woman) *gravesque draucis* | *halteras facili rotat lacerto.*

ceciderunt, 'have given out'.

424. fameque: for the long *e*, cf. XIV. 84; Ovid, *Met.* XI. 369 *saevit pariter rabieque faméque*; Lucan, III. 352 *ubera sicca famé*; Mart. I. 99. 18 *famé peribis.*

426. oenophorum sitiens: the acc. is cognate, as in *sitire sanguinem*: the *oenoph.* is the same as *vinarium*, a large wine-vessel, which in this case holds 'a full *urna*', i.e. 24 *sextarii* or about 3 gallons. For the use of *plena*, cf. Mart. I. 99. 1 *non plenum modo viciens* (not fully two millions) *habebas.*

427. pedibus: cf. Mart. VII. 20. 19 *mixto lagonam replet ad pedes vino.* **sextarius alter:** she drinks off two pints, before beginning her meal, to whet her appetite. **orexim,** again XI. 127: the metre forbids *esuritionem*, which occurs several times in Martial. Celsus, the medical writer, uses *cibi cupiditas* for 'appetite', and *cibi fastidium* for the want of it.

Seneca reproves this habit, *Epp.* 122. 6 *non videntur tibi contra naturam vivere, qui ieiuni bibunt, qui vinum recipiunt inanibus venis et ad cibum ebrii transeunt?*

434–56. Next there is the literary lady, a worse nuisance still, who will discuss the poets and compare their merits at dinner. While she speaks, no one, not even another woman, can hope to get in a single word. Moderation is desirable even in virtuous pursuits. The lady, whose society you are to enjoy at bed and board, had better not be a mistress of literature and rhetoric; a husband should have leave to make slips in his grammar.

434. discumbere coepit, 'has begun her dinner': see n. to v. 12.

435. 'Has a compliment for Virgil and excuses for Dido at death's door', i.e. thinks her sin atoned for by her death; cf. *Aen.* IV. 610 *morientis Elissae.*

436. **committit vates,** 'pits the poets against each other', i.e. arranges them in order of merit; cf. I. 163: this is a commoner sense of *comparare*, which is properly 'to match a pair (*par*) e.g. of gladiators against one another'; but, to avoid tautology, it seems needful to give **comparat** here the meaning, 'compares': *comparare* of n. to v. 56 is an entirely different word.

Cf. FitzGerald, *Letters*, p. 386: 'Macaulay's Memoirs were less interesting to me...It is wonderful how he, Hallam and Macintosh could roar and bawl at one another over such Questions as Which is the Greatest Poet? Which is the greatest Work of that Greatest Poet? etc. like Boys at some Debating Society.'

437. **alia pars** is the other scale of the balance, answering to **inde**, 'in one scale'; *altera* would be more correct but unmetrical; again VII. 114. For the metaphor, cf. Hor. *Epp.* II. 1. 29 *Romani pensantur eadem | scriptores trutina*: in Aristoph. *Frogs*, 1365–410 the test of a material balance is applied to poetry. For comparisons between Homer and Virgil, see n. to XI. 180.

438. Teachers of literature and professors of rhetoric alike are silenced: see nn. to I. 15 and 16. **rhetorĕs** has its Greek quantity: so Mart. II. 64. 5 *tres uno perierunt rhetores anno.*

439. **loquetur,** 'will get a word in'.

441. That **tot** is not parallel with *tanta* has been indicated by a slight change of punctuation. **pariter,** 'all together'.

442. **iam nemo** cet., 'let no one in future trouble the trumpet or beat on brass; for she unaided...'; superstition believed that the moon's eclipses were caused by witch-craft, which might be frightened away by these noises: cf. Livy, XXVI. 5. 9 *cum aeris crepitu, qualis in defectu lunae silenti nocte cieri solet*; Mart. XII. 57. 15 *numerare pigri damna qui potest somni, | dicet quot aera verberent manus urbis, | cum secta Colcho luna vapulat rhombo*; Stat. *Theb.* VI. 663 *sic cadit...solis opaca soror* (the moon); *procul auxiliantia gentes | aera crepant.* The moon is tortured by the witch's magic wheel till the spell is broken by the sound of beating on copper.

444. 'The philosopher sets a limit even to virtues'; i.e. he lays down that even a virtue may be carried to excess: cf. Hor. *Epp.* I. 6. 15 *insani sapiens nomen ferat, aequus iniqui, | ultra quam satis est virtutem si petat ipsam.* This lady goes too far with her learning.

445. There is an ellipse before *nam*: 'and rightly, for...'.

docta nimis: cf. Mart. II. 90. 9 *sit mihi verna satur, sit non doctissima coniux.*

videri implies that she cannot have more than a pretence to literature and eloquence.

446, 447. She ought to identify herself altogether with the other sex, wearing, instead of the matron's *stola* coming down to the feet, the man's *tunica* which did not reach below the knee. Men only were permitted to sacrifice to Silvanus; and men paid a *quadrans* ($\frac{1}{4}$ *as*) for admission to the baths; it seems from this passage that women paid more; boys below a certain age paid nothing; cf. II. 152.

448. **recumbit,** i.e. at meals; cf. l. 434.

449. **dicendi genus,** 'a style of oratory'; this is a technical phrase of rhetoric, and the matter is discussed by Quintilian, XII. 10, where he recognises three *genera* of Greek oratory, the Attic, Asian, and Rhodian styles.

aut curvum...enthymema, 'nor let her brandish the rounded syllogism in hurtling phrase': the *enthymema* was a regular weapon of the rhetorician, and is explained at length by Quintilian; cf. esp. V. 10: it is an argument in syllogistic form, though not corresponding exactly to the syllogism of dialectic. For the epithet *curvum* 'rounded', complete', 'compact', cf. Quint. XI. 3. 102 *enthymemata sua gestu velut corrotundant*; Plato, *Protag.* 342E ἐνέβαλε ῥῆμα ξυνεστραμμένον ὥσπερ δεινὸς ἀκοντιστής. The whole language is borrowed from the brandishing and throwing of actual weapons; cf. *sagittae*, VII. 156; Sen. *Epp.* 108. 10 *sententia velut lacerto excussa torquetur.*

450. **historias:** this (as all the acquirements mentioned to l. 456) is the province, not of the *rhetor*, but of the *grammaticus*: cf. Quintilian quoted on VII. 230; Sen. *Epp.* 88. 3 *grammaticus circa curam sermonis versatur et, si latius evagari vult, circa historias, iam ut longissime fines suos proferat, circa carmina* (where surely *historias* and *carmina* should change places).

451. 'Let there be some things also in books which she does not understand.' **et,** i.e. as well as the many things she does know: cf. Quint. I. 8. 21 *mihi inter virtutes grammatici habebitur aliqua nescire.*

452. **quae...artem,** 'who consults and turns over the grammar of Palaemon'. **volvit** (*evolvere* is the prose word) refers to the shape of the ancient book, which was a roll: see n. to VII. 23. **ars** (τέχνη) is the regular word for a systematic treatise on any science: such a treatise written by a *grammaticus* would be a grammar: cf. VII. 177. For Palaemon, see n. to VII. 215.

453. **lege et ratione loquendi**, 'the rules and laws of grammar'; cf. *verborum regula*, VII. 230, and Quint. I. 9. I (the *grammaticus* undertakes two parts of education) *ratio loquendi et enarratio auctorum*.

454. **tenet**, 'remembers'; *tenere* is constantly used by the poets for the inconvenient *reminisci*; for *oblitus sum* they often substitute *hoc mihi excidit*.

antiquaria: cf. Tac. *Dial.* 21 *nec quemquam adeo antiquarium puto ut Caelium ex ea parte laudet qua antiquus est*. The reaction in favour of ancient writers, and even obsolete words, of which Fronto (consul A.D. 143) is the chief example extant, was already in progress; but Juv. and Martial (XI. 90) both held entirely aloof from it.

versus are 'lines', either of verse or prose: see n. to VII. 153.

455. **nec curanda viris** = *et non-curanda viris*; 'things men should not trouble themselves about'.

For **opicae**, see n. to III. 207.

456. **soloecismum**, 'a slip in syntax': the word was supposed to have come into use from the bad grammar spoken at Soli in Asia: *barbarismus*, the other fault which the *ratio loquendi* was intended to guard against, is 'a slip in accidence'; cf. Suet. (Roth, p. 319) *soloecismus in sensu fit, barbarismus in voce*; see also n. to VII. 230.

457–73. *Women think that wealth and fine jewellery give them licence to do as they please. They smear their faces with cosmetics and pastes, to the disgust of their husbands; but all these are washed off before they meet their lovers. As many dressings and poultices are applied to a face as if it were an open sore.*

458. **virides gemmas**, 'emeralds', *smaragdi*, which the ancients ranked third of precious stones, next to the diamond and pearl; the best came from Scythia.

459. **extentis**: the ears are 'stretched' by the weight of the jewels. **commisit**, lit. 'has joined'; so *commissura*, 'a joint'.

elenchos are long pear-shaped pearls: cf. Pliny, *Nat. Hist.* IX. 113 *elenchos appellant fastigata* (tapering) *longitudine...in pleniorem orbem desinentes. hos digitis suspendere* (i.e. in rings) *et binos ac ternos auribus feminarum gloria est*. The origin of the name is uncertain.

460. Some edd. bracket the line; but it seems sufficiently apposite, as a woman with such jewels as are described here must be rich.

461. **interea,** 'in preparation', i.e. before she goes abroad in all her charms.

462. **pane,** 'a bread-poultice'; cf. II. 107 (of Otho's effeminacy) *pressum in facie digitis extendere panem.*

pinguia, 'sticky'. Poppaea, Nero's wife, was famous for the use of cosmetics, and may have invented some paste or enamel for the skin which was subsequently known by her name; see n. to l. 469.

463. When the husband wishes to kiss his wife, his lips stick fast in the paste she is wearing. **hinc,** 'by it'.

464. From **quando** to **Indi** (l. 466) is parenthetical.

465. **moechis,** 'for their lovers'. **foliatum** or *nardinum* is the name given to a particular scent, made of seven different ingredients; one of these was *nardum*, of which the best kind came from India: cf. Pliny, *Nat. Hist.* XIII. 15. **parantur:** see n. to III. 224.

466. **mittitis:** the word is often used in the sense of 'exporting': cf. v. 92 and 119; Tibull. II. 2. 3 *urantur odores,* | *quos tener e terra divite mittit Arabs.*

467. **tectoria...reponit,** 'takes off the original enamel': the metaphor is from building; *tectorium opus,* 'stucco', made of pounded marble, was much used by the Romans for wall-surfaces and cornices; many beautiful reliefs have been found modelled in this material. Cf. Pers. v. 25 *pictae tectoria linguae,* 'the plaster of a varnished tongue' (Conington).

468. **agnosci,** 'to be recognisable': for the hiatus, see n. to X. 281.

469. **asellas:** Juv. seems to refer to the story told by Pliny of Poppaea who relied on asses' milk as a cosmetic: *Nat. Hist.* XI. 238 *Poppaea certe Domiti Neronis coniunx quingentas per omnia secum fetas trahens* [*asellas*] *balnearum etiam solio totum corpus illo lacte macerabat.*

471. **inducitur,** 'is coated': cf. Pliny, *Nat. Hist.* XXXV. 102 *huic picturae quater colorem induxit* (he gave this picture four coats of paint).

472. **siliginis:** cf. v. 70 and l. 462 above.

474–507. *Women are unreasonable and cruel to their servants, who have often to suffer for no fault of their own. A lady's dressing-room is often like the torture-chamber of a tyrant. If the mistress is not satisfied with her toilet or her appearance, the maid is cruelly beaten. A sort of cabinet-council goes on in the dressing-room over the important business of the toilet, especially*

*over the arrangement of the hair. This is built up into an im-
mense erection, so as to make the lady, who may be very short,
appear of heroic stature.*

474. penitus, 'thoroughly', is often coupled with *cognoscere*;
cf. Cic. *de Orat.* II. 99 *ut eas* (i.e. *causas*) *diligenter penitusque
cognoscat.*

toto: see n. to *horis* l. 183.

475. agitent is used in the sense of *agant*, 'their business and
occupations'. Juv. seems to forget what he promises; he gives
the episode of the dressing-room but does not go on to the rest
of the day's occupations. See n. to I. 132.

476. libraria occurs only here in literature, and seems to
mean the woman whose business it was to weigh out wool for
the other maids to spin; she is called *lanipendia* elsewhere.

ponunt...tunicas: they 'take off their clothes' in order to
be beaten; cf. Sen. *Dial.* v. 12. 5 *Plato...cum servo suo ira-
sceretur...ponere illum statim tunicam et praebere scapulas ver-
beribus iussit.* A slave, in this plight, is called *despoliatus*: cf.
Petron. 30 *servus nobis despoliatus procubuit ad pedes ac rogare
coepit ut se poenae eriperemus.*

477. Liburnus corresponds to the modern 'coachman'; see
n. to IV. 75.

478. alieni, i.e. the husband's.

479. frangit ferulas, 'has the rod broken on his back'; cf.
VIII. 247. The *ferula* (νάρθηξ) is a cane; *scutica*, also called *lora*,
is a strap; the *flagellum* or *flagrum* is a cat-o'-nine-tails of knotted
cord or wire; that the last weapon was by far the most for-
midable, is proved by Horace, *Sat.* I. 3. 119 *nam, ut scutica dig-
num horribili sectere flagello,* | *ne ferula caedas meritum maiora
subire* | *vulnera, non vereor.*

480. The services of the *tortor* or *carnifex* are in such con-
stant requisition that the lady pays him a yearly salary instead
of by the job; the *tortores* were probably public slaves.

481. verberat: the context shows she does not do the thrash-
ing with her own hands. **obiter:** cf. III. 241. **amicas,** 'her
lady-visitors'.

482. considerat: for this word, used of the careful inspec-
tion of a connoisseur, cf. Suet. *Cal.* 36 *quas...diligenter ac lente
mercantium more considerabat*; Petron. 50 *quam [lancem] cum
Agamemnon propius consideraret.*

483. longi...diurni, 'she reads through the broad-sheet of
the long gazette'; this refers to the Roman substitute for a
newspaper, the *acta diurna* or *publica*, mentioned also II. 136;

VII. 104. This journal was instituted by Caesar in the year of his consulship (59 B.C.); it recorded births, deaths, marriages, divorces (Sen. *de Ben.* III. 16. 2) and other occurrences of general interest, not excluding scandal.

transversa: the ancient book was generally written in columns (*paginae*) on a broad strip of papyrus; but many official documents were written in lines that ran the whole breadth of the strip: cf. Suet. *Iul.* 56 *epistulae quoque eius ad senatum extant, quas primus videtur ad paginas...convertisse, cum antea consules et duces nonnisi transversa charta scriptas mitterent.* This passage shows that the *acta* were written, not *per paginas* 'in columns', but *transversa charta* 'right across the page'.

(Friedl., following the Schol., supposes *diurnum* to be 'an account-book'; but (1) care about her expenditure is quite unlike this lady's character; (2) *transversa* seems not to be used elsewhere in connection with such a book.)

485. **cognitio** is properly applied to any magisterial investigation: cf. Pliny quoted on l. 497.

486. 'Her rule over her household is no milder than that of a Sicilian court'; the Sicilian tyrants, especially Phalaris, were proverbial for cruelty: cf. Hor. *Epp.* I. 2. 58 *invidia Siculi non invenere tyranni | maius tormentum.*

487. **constituit,** 'has made an assignation'; see n. to III. 12; cf. Terence, *Haut. Tim.* 726 *quom venturam dixero et constituero.* **decentius,** 'more becomingly'.

488. **hortis** may be either her own property or a public park; see n. to I. 75.

489. **Isiacae...lenae:** the Iseum, or temple of Isis, in the Campus was notorious as a place of assignations: cf. Ovid, *Ars,* I. 77 *nec fuge linigerae Memphitica templa iuvencae; | multas illa facit quod fuit ipsa Iovi.* The Romans constantly confused Isis with Io: see n. to l. 526.

490. For the cruelty of Roman ladies to their maids, cf. Ovid, *Am.* I. 14. 16 *ornatrix tuto corpore semper erat. | ante meos saepe est oculos ornata, nec unquam | bracchia derepta saucia fecit acu*; Mart. II. 66 *unus de toto peccaverat orbe comarum | anulus, incerta non bene fixus acu. | hoc facinus Lalage speculo quo viderat ulta est, | et cecidit saevis icta Plecusa comis.* We must imagine the mistress, mirror in hand, criticising the growth of the structure on her head and ordering blows to be showered on the maid. Seneca (*Dial.* X. 12. 3) shows that men were no more easily pleased: *illos otiosos vocas, quibus apud tonsorem multae horae transmittuntur...dum de singulis capillis in consilium itur?* (cf.

l. 497)...*quomodo excandescunt, si quid ex iuba sua decisum est, si quid extra ordinem iacuit, nisi omnia in anulos suos reciderunt?*

491. **Psecas:** like Plecusa above, this maid is Greek; the name occurs in Ovid, *Met.* III. 172 as borne by one of Diana's nymphs.

492. **altior,** 'too high'. The **taurea** must have been a kind of *scutica*.

493. **crimen:** see n. to I. 75.

494. **admisit:** see n. to XIII. 1.

hic, 'in this', 'here'.

495. Juv. insinuates that the lady punishes her maid only because she is not satisfied with her own beauty.

With **laevum,** 'on the left side', understand *crinem*.

496. For **orbem,** cf. Mart. quoted on l. 490.

497. The language is taken from the meeting of a body of magistrates or of the Senate. The old woman, who had been maid to the lady's mother, 'sits as assessor', *est in consilio*; see n. to III. 162 and cf. Pliny, *Epp.* VI. 22. 2 *recepta cognitio est; fui in consilio*. She, like the *princeps senatus*, is asked to state her opinion first: she used to stick the hairpins herself, but is now given the work of apportioning wool to the other women; cf. *libraria*, l. 476. **materna,** sc. *ancilla*. **admota lanis:** cf. Sen. *Epp.* 12. 3 *ad ostium admotus*; ibid. 47. 15 *omnes servos admovebo mensae meae?*

498. **emerita,** 'dismissed from service'; the word is properly applied to veteran soldiers. That **acus** is here 'a hairpin', is shown by Mart. quoted on l. 490. **sententia** and **censebunt** are both terms taken from the procedure of the Senate; cf. IV. 130 and 136.

500. **tamquam...animae,** 'as though her reputation were at stake, or her life'.

502. **tot adhuc conpagibus,** 'with so many added erections': this use of *adhuc* is characteristic of silver Latin: in Cicero it is an adverb of time and means 'until now'; poets and prose writers of the imperial age use it as an equivalent of *etiam* or ἔτι: cf. VIII. 36; Sen. *Nat. Quaest.* IV. 8 *unam rem adhuc adiciam*; Mart. V. 22. 9 *illud adhuc gravius*.

503. 'Looked at in front, she will seem to you an Andromache'; as a woman of the heroic age and wife of Hector, Andr. was supposed to be tall. Juv. seems to imitate Mart. XIV. 212 (on a dwarf) *si solum spectes hominis caput, Hectora credas: | si stantem videas, Astyanacta putes*.

504. **aliam,** 'a different person'.

cedo si (again XIII. 210) has the same meaning as the commoner *quid censes, si. . .*; some word like *sententiam* is understood with *cedo*. Juv. means that the effect is especially absurd if the woman is short.

506. **nullis. . .cothurnis,** 'if she have no assistance from high boots': cf. Sen. *Epp.* III. 3 *non exsurgit in plantas nec summis ambulat digitis eorum more qui mendacio staturam adiuvant*; Quint. II. 3. 8 *ut statura breves in digitos eriguntur*.

507. She is so short that she has to stand on tiptoe for a kiss.

508–68. *A woman has no interest in her husband's affairs, but pays the greatest attention to priests of all kinds. Bellona's priest gets food and clothes from her; if Io and Isis require it, she will do the most painful penance and take the longest journeys. No foreign divinity does she disregard; she supports the priests of Osiris and the palsied prophetess of Judaea. Besides, all women are the prey of soothsayers; and the soothsayer they believe in most is the one whose art has got him into trouble, who has been long imprisoned and nearly put to death. They ask of soothsayers how long their husbands and other relatives will live.*

508. **mentiŏ:** for the quantity, see n. to *ergo*, I. 3.

509. **damnorum** means the expense his wife causes him.

511. **rationibus,** 'to his expenses': see n. to I. 118. For the priests of Bellona and their frantic worship, see n. to IV. 123. The passages there quoted show that there was a strong likeness between their ceremonies and those of the priests of Cybele or *Galli*.

515. The 'howling herd' are the other priests and votaries of Cybele. **tympana** (cf. VIII. 176) were always used by these enthusiasts to accompany their cries: cf. Catull. 63. 9 *typanum, tubam Cybelles, tua, mater, initia* (symbols).

cedunt: i.e. they cannot make such a noise: cf. l. 438.

516. 'Whose plebeian cheeks are covered by his Phrygian head-dress': this cap had lappets which fell down the cheek and might be tied under the chin. **Phrygia:** see n. to III. 137.

517. **grande sonat,** 'with loud utterance': many passages show that the fevers of autumn were dreaded at Rome; cf. *letifero autumno*, IV. 56.

septembris, sc. *mensis*; cf. II. 70 *sed Iulius ardet.*

518. **nisi. . .ovis,** 'unless she purify herself with a hundred eggs', eggs being regularly used for this purpose; cf. Ovid, *Ars*, II. 329 *et veniat quae lustret anus lectumque locumque, | praeferat et tremula sulpur et ova manu.*

519. **xerampelinas**, sc. *vestes*: a name to denote the colour of a dead vine-leaf, something between red and purple. The priest is content with cast-off clothes and the cheapest presents. These priests of Cybele were the begging friars of antiquity; hence their Greek name of μητραγύρται (from ἀγείρειν).

521. **in tunicas eat**: the clothes she has given are to serve as a kind of *piaculum* or scape-goat; on them are to fall the perils that might otherwise have befallen their owner: cf. Sen. *Dial.* XII. 18. 6 *quicquid matri dolendum fuit, in me transierit...fuerim tantum nihil amplius doliturae domus piamentum*.

522. The order of words should be kept: 'in winter she will break the ice and etc.'.

525. **regis agrum,** the Campus Martius, which was said to have been occupied by Tarquinius Superbus and to have been restored to the people by Brutus, on the expulsion of the king.

526. **erepet:** the preposition only repeats the sense of *totum* in the l. before: cf. Sen. *Dial.* VII. 26. 8 *cum aliquis genibus per viam repens ululat.*

candida: Io was changed into a white cow. The confusion between Isis and Io, who seem to be identified here, arose from Io's journey to Egypt, and from the fact that both were represented with a cow's horns; cf. Herod. II. 41 τὸ τῆς Ἴσιος ἄγαλμα ἐὸν γυναικήιον βούκερων ἐστι κατά περ Ἕλληνες τὴν Ἰοῦν γράφουσι.

528. **Meroe:** mentioned again XIII. 163: the name of a large island, including a city of the same name, formed by branches of the Nile. It was nearly 900 miles south of Syene (Assouan) which was the southern limit of the Roman Empire.

529. The Iseum was in the Campus Martius, this worship being forbidden within the walls of the city; the site has been well established by discoveries of Egyptian statuary. The *ovile*, so called from its shape, was more generally known as the *Saepta*; it was the place in which the people had been penned to give their votes, in the days of political freedom.

530. **dominae,** 'the goddess', i.e. Isis; it appears from this that votaries of Isis spent the night in the temple in the hope of a personal communication from the goddess; to interpret these visions would be the business of the *Isiaci coniectores* mentioned by Cicero, *de Div.* I. 132. This practice of ἐγκοίμησις (*incubatio*) was constantly observed in the temples of Asclepius; cf. Aristoph. *Plut.* 659–744.

531. 'A likely sort of soul and intellect for the gods to converse with by night!' For this contemptuous use of *en* with the

acc., cf. II. 72 *en habitum quo te...populus modo victor et illud | montanum positis audiret vulgus aratris*; Cic. *in Verr.* II. 1. 93 *haec est istius praeclara tutela. en cui liberos tuos committas* ('a pretty fellow to make your children's trustee!').

532. Osiris, Isis, and Anubis all belong to the same group of deities whose worship was imported from Egypt. The first two are generally represented as husband and wife, while Anubis is the guardian and attendant of Isis. His images have a dog's head; hence Virgil's contemptuous *latrator Anubis* (*Aen.* VIII. 698). Ovid speaks of all the three together in a prayer to Isis, *Am.* II. 13. 11 *per tua sistra precor, per Anubidis ora verendi: | sic tua sacra pius semper Osiris amet.* Hence the use of *ergo* to introduce Anubis here.

533. It was the practice of the *Isiaci* to wear linen and to shave the head; no doubt these customs were for the sake of cleanliness, the vegetable substance being thought purer than wool. Cf. Plut. *de Iside*, 3 οὔτε γὰρ φιλοσόφους πωγωνοτροφίαι... ποιοῦσιν οὔτ' Ἰσιακοὺς αἱ λινοστολίαι καὶ ξυρήσεις: Mart. XII. 29. 19 *linigeri fugiunt calvi sistrataque turba.*

534. **plangentis:** at certain times in the year Isis was supposed to be mourning the loss of Osiris; hence she is called by Martial (II. 14. 8) *maesta iuvenca.* The people mourn in sympathy; but the priest of Isis who wears the dog's head of Anubis in the street processions, laughs at their simplicity.

539. **meditata**, 'studied', is passive; the past participles of many other deponent verbs have a passive, as well as an active, sense, e.g. *confessus, ementitus, imitatus, sortitus*, etc.

praestant, 'secure'.

541. **corruptus**, 'bribed'; the priest receives the eatables for the benefit of Osiris.

542. **dedit...locum**, 'has given place', i.e. has retired from the scene. **cophino faenoque:** see n. to III. 14.

543. **Iudaea tremens**, 'a palsied Jewess'. The Jews in Rome seem to have lived, like gypsies in modern Europe, by begging and fortune-telling; cf. Mart. XII. 57. 13 *a matre doctus...rogare Iudaeus.*

545. **arboris:** in what sense is the Jewess a priestess 'of the tree'? It seems most probable that the reference is to the trees outside the Porta Capena where the Jews were allowed to take up their abode as squatters; cf. *arbor* and *silva*, III. 16. It has been suggested that *arbor* stands here for 'the cross'. That Juv. did not distinguish Jews from Christians is possible; but he does not refer elsewhere to any Christian institutions; and *arbor*,

though often used, in the phrase *infelix arbor*, of hanging, sel-
dom means 'a cross', though there are constant references in all
Latin literature to the *servile supplicium*. The word seems then
to be a sneering allusion to the fact that the Jewish temple was
destroyed, so that the Jews had now to worship under the trees
instead.

Juv. may have in mind the priestesses of Dodona and their
oak-trees; he cannot have known of the Hebrew prophetess,
Deborah, who judged Israel and 'dwelt under the palm tree of
Deborah' (Judg. iv. 5).

caeli, 'the sky': for this object of Jewish worship, see n. to
XIV. 97.

546. 'She too' (as well as the priest of Isis) 'gets her palm
crossed'.

548. **amatorem tenerum,** 'a young lover'; it must be kept
in mind that all the different quacks and impostors are dealing
with a Roman lady.

549, 550. *extispicium* was one of the most orthodox methods
of divination in antiquity; the *haruspex* examined the internal
organs, especially the heart and liver, of an animal freshly killed,
and foretold the future according to the appearances observed.
The art came originally from Etruria; the places here mentioned
do not seem to have had any special reputation for proficiency;
they are meant to suggest that any eastern superstition found
patronage at Rome.

calidae, 'still warm'.

552. **faciet...ipse,** 'he will commit a crime, intending him-
self to turn informer'. If the occasion is very important, the
soothsayer will murder a child, in order to divine by means of
the *exta* (cf. Cic. *in Vat.* 14 *cum puerorum extis deos manis
mactare soleas*); but to save himself, he will turn informer
(*delator*), and get his patroness into trouble: cf. Sen. *Dial.* IV. 7.
3 *alius delator věnit eius criminis, cuius manifestior reus est.*

553. **Chaldaeis,** 'astrologers': this name and *mathematici* (l.
562) are generally reserved for fortune-tellers, while *astrologus*
denotes an orthodox student of the science; but here **astrologus**
is not an astronomer, but an astrologer. The *Chaldaica doctrina*
was first systematised about 250 B.C. by Berosus a priest in
Babylon; and the name **Chaldaei** was applied to all who told
fortunes by the stars. Of the immense influence of these men
in that age there is abundant evidence: Tacitus repeatedly
mentions decrees of the senate expelling all *mathematici* from
Italy: cf. *Ann.* II. 32, XII. 52; *Hist.* II. 62; also *Hist.* I. 22

where they are described as *genus hominum potentibus infidum,
sperantibus fallax, quod in civitate nostra et vetabitur semper et
retinebitur.*

554. **a fonte...Hammonis,** 'comes straight from the
spring of Ammon': cf. Tac. *Ann.* VI. 21 *quae dixerat* [*Thrasyllus*]
oracli vice accipiens. There was an ancient and famous oracle
at Ammonium in North Africa where Zeus Ammon was wor-
shipped; one of the curiosities of the place was a spring of water
which was warm by night and cold by day; cf. Herod. IV. 181;
Lucr. VI. 848 ff.

555. **cessant,** 'is dumb', *oracula* being sing. in meaning. Cf.
Lucan, V. 111 *non ullo saecula dono | nostra carent maiore deum,
quam Delphica sedes | quod siluit, postquam reges timuere futura |
et superos vetuere loqui.* The evidence of Strabo shows that, by
the time of Augustus, the most famous oracles, Delphi, Do-
dona, and Ammon, had fallen into neglect. The reason of this
was that the Romans were indifferent to oracular responses,
being content with their *haruspices* and Sibylline books. Yet
under the Antonines the oracles again became of considerable
importance; and the emperor Julian consulted the oracle of
Delphi about A.D. 360.

556. **damnat,** 'pronounces guilty'.

caligo futuri, 'darkness as to the future', *caligo* taking the
constr. of such a word as *ignoratio.* Cf. Hor. *Carm.* III. 29. 29
prudens futuri temporis exitum | caliginosa nocte premit deus.

558, 559. 'By whose friendship and by whose venal horo-
scope a great citizen died, one whom Otho dreaded': the
emperor Galba is meant here. Otho was urged on in his ambi-
tious projects and crimes by an astrologer whom Tacitus (*Hist.*
I. 22) calls Ptolemaeus, but Suetonius (*Otho*, 4), Seleucus. The
text seems to imply that he had once been intimate with
Galba and had transferred his allegiance to Otho for gain;
nothing is known of this from other sources.

conducenda, lit. 'able to be hired'. **tabella** is a document
containing the calculations by which his high destiny was proved
to Otho. For the quantity of *obit*, see n. to III. 174.

These two lines would be better away, as they limit the
reference, in an awkward way, to one astrologer in particular,
who is hardly likely to have survived until this date. They are
inserted in the margin of P by the second hand; yet they
certainly read like Juv.'s writing.

559. **formidatus Othoni:** it would seem more natural to
say that Otho was dreaded by Galba; but the reference may be

to the pretence that his life was in danger, which Otho had re-
course to, in order to lessen the infamy of his action; cf. Tac.
Hist. I. 21 [*Otho*] *fingebat et metum quo magis concupisceret;...
occidi Othonem posse.*

560. **inde... artis,** 'their confidence in his skill depends on
whether...'. **ferro** = *manicis.*

561. **longo:** the sense required seems to be 'for a long time';
the distance or nearness of the prison is irrelevant. The adj. is
used for the adverb *longum*: cf. VIII. 47.

castrorum in carcere: see n. to III. 314.

562. **genium... habebit,** 'will get credit for genius'; *genius*
usually means the attendant spirit which lives in each man dur-
ing his life, and to which bloodless sacrifices were offered on
birthdays; here its meaning seems to approach that of *ingenium.*
The idea of a 'familiar spirit' as a magician's helper seems for-
eign to Juv.'s age. Cf. Mart. VI. 60. 10 *victurus genium debet
habere liber*; Stat. *Silv.* IV. 6. 19 *nox... memoranda diu geniumque
habitura perennem.*

563. **in Cyclada mitti:** for this punishment (*deportatio in
insulam*), see n. to I. 73. The group of little islands surrounding
Delos in the Aegean were and are still known by the common
name of Cyclades.

564. **contigit** suggests that he was lucky to escape a capital
sentence, **tandem** that he was confined for a term of years in
Seriphos, now Serpho, one of the Cyclades; cf. X. 170.

caruisse = ἀπαλλαγῆναι:: cf. X. 287, XIV. 151: *carere* is also
often used of a desirable object, e.g. money (I. 59), sleep (III. 56),
shows (XI. 53).

565. **lento,** 'long delayed': cf. *tardas colus*, XIV. 248; Ovid,
Met. I. 148 *filius ante diem patrios inquirit in annos.* It was a
capital offence for a slave to enquire into his master's horoscope;
and the *delatores* often accused their victims of making similar
investigations as to the fate of reigning emperors: cf. Tac. *Ann.*
III. 22 *adiciebantur adulteria, venena, quaesitumque per Chaldaeos
in domum Caesaris.*

566. **de te,** i.e. the imaginary husband to whom most of the
satire is addressed. The wife is called Tanaquil, because the
wife of Tarquinius Priscus was skilled in magic and thus dis-
covered the destiny of Servius Tullius; cf. Livy, I. 34. 9
Tanaquil, perita, ut volgo Etrusci, caelestium prodigiorum.

567. **efferat:** see n. to l. 175. **an,** 'whether', not 'or': see n.
to VII. 141.

adulter, 'her lover'.

*569-91. An even worse type is the woman who, instead of con-
sulting experts in astrology, is herself an expert, whose life is
ruled by the precepts of the stars in all her doings, small and
great. Poor women go to the circus and consult a common fortune-
teller there, on their matrimonial affairs; the rich, equally super-
stitious, hire a more expensive practitioner to come to their
houses.*

569. Astrologers always held that the planets Saturn and
Mars had an evil influence, while Venus and Jupiter brought
luck and happiness: cf. Lucan, I. 652 *stella nocens...
Saturni; ibid.* 661 *Venerisque salubre| sidus.*

570. quo...astro: 'in what constellation', i.e. sign of the
Zodiac or part of the sky.

572. illius (ἐκείνης δέ) is strongly opposed to **hae** in l. 569;
the contrast is stronger in Latin without particles. **etiam,**
'even', goes with *occursus.*

573. sucina: there are frequent references, beginning with
Ovid, to this custom of Roman ladies: they carried balls of
amber in their hands for the sake of the agreeable smell it gave
out when warm: cf. Mart. v. 37. 9 *fragravit ore, quod rosarium
Paesti, | quod sucinorum rapta de manu gleba.*

574. ephemeridas, 'almanacs': these ἐφημερίδες or πίνακες
were tables on which the risings and settings of the planets and
constellations were registered, to facilitate calculations. Cf.
Scott's *Guy Mannering,* chap. IV.: 'He accordingly erected his
scheme or figure of heaven, divided into its twelve houses,
placed the planets therein according to the Ephemeris, and
rectified their position to the hour and moment of the nativity.'

iam, 'by this time'; so great is her proficiency.

575. castra...patriamque: two different cases are sup-
posed: the husband has to leave Rome, in the first case for a
foreign command, in the second, for his native place, which
might or might not be in Italy.

It is possible however that the true reading is *patriamve,* as
-ve is apt to become *-que* in the MSS. of Juv.: cf. VI. 13 (critical
n.).

576. pariter goes with *ibit.*

numeris...Thrasylli, 'if she is detained by the calculations
of Thrasyllus': he was an astrologer, famous for his intimacy
with Tiberius; he died A.D. 36: cf. Tac. *Ann.* VI. 21; Suet. *Tib.*
14. His *numeri* are books of astrological calculations written by
him.

577. **ad primum lapidem:** miles on the Roman roads were marked by stones and measured from the *aureum miliarium* or golden milestone in the forum, close to the temple of Saturn; cf. Quint. IV. 5. 22 *facientibus iter multum detrahunt fatigationis notata inscriptis lapidibus spatia*; Mart. X. 79. 1 *ad lapidem Torquatus habet praetoria quartum.* This lady will not take the shortest journey without consulting the stars. So it is said that in India a native gentleman will generally consult his astrologer before taking a railway journey.

578. This has no reference to divining by twitching of the eyes or eyebrows, often mentioned elsewhere: the meaning is that she will not apply the simplest remedies for the smallest ailments without the sanction of the planets.

579. **inspecta genesi,** 'she first examines her horoscope and then...'; the emphasis falls on the abl. absol. *genesis* or *genitura* is the calculation founded upon the exact hour of birth.

581. **aptior,** 'the right hour', i.e. fitter than others: cf. Virg. *Georg.* I. 286 *nona fugae melior.* Petosiris, an Egyptian priest, was said to be one of the founders of astrology; cf. *Anth. Pal.* XI. 164 (of an astrologer whose prediction of his own death had not been fulfilled) αἰσχυνθεὶς Πετόσιριν ἀπήγξατο. Books on the subject, professing to be written by him, were in circulation at this time; Pliny quotes from them.

This passage is imitated by Amm. Marcell. XXVIII. 4. 24 *multi...nec in publicum prodeunt nec prandent nec lavari arbitrantur se cautius posse, antequam ephemeride scrupulose sciscitata didicerint, ubi sit verbi gratia signum Mercurii.*

582–4. 'If she be of middle rank, she will traverse the space on each side of the turning-posts, and will draw lots, and offer brow and hand to the fortune-teller, who asks her to smack her lips again and again.' The Circus, or Race-course, at Rome, in the valley between the Palatine and Aventine hills, was a favourite resort of low fortune-tellers; cf. Cic. *de Div.* I. 132 *non habeo nauci...de circo astrologos*; hence Hor. *Sat.* I. 6. 113 *fallacem circum.* There was a long, low platform of marble called *spina*, which filled the centre of the arena; at each end of the *spina* was a plinth supporting three conical pillars of gilt bronze; these were the *metae*, round which the chariots turned. **utrimque metarum** = at each end of the *spina*. **mediocris:** cf. XI. 177.

583. **metarum:** this gen. after *utrimque* has analogies in Greek: e.g. Thuc. III. 6 ἑκατέρωθεν τῆς πόλεως: but I find no parallel in Latin.

sortes ducet: a common method of Roman divination. The *sortes* were generally wooden counters, with words inscribed on them, which were drawn at random from a vessel (*sitella*). See n. to I. 82.

frontem: divination by the forehead was a regular method, practised by the *metoposcopi* (μετωποσκόποι); the fate of Britannicus was foretold in this way (Suet. *Titus*, 2).

584. **poppysma** (from ποππύζειν) is the sound which results when the closed lips are forcibly parted. The various meanings may be seen in the Greek Lexicon; but the meaning 'applause', which is there suggested for this passage, seems unlikely. It is better to understand that this sound is part of the ceremony which the *vates* obliges his client to perform. He tells her fortune by the *sortes* she draws, by the lines of her hand and face, and meanwhile she is instructed to keep up this noise, to assist the magic. Cf. Pliny, *Nat. Hist.* XXVIII. 25 *fulgetras poppysmis adorare consensus gentium est.* For a smacking noise as a lucky omen, cf. Theocr. 3. 29 οὐδὲ τὸ τηλέφιλον ποτεμάξατο τὸ πλατά-γημα, | ἀλλ' αὔτως ἁπαλῷ ποτὶ πάχεος ἐξεμαράνθη.

585. **Indae:** so Büch. for *inde* of P which is really the same: others read *inde*, i.e. *a Phrygia*; but this is not in accordance with **dabunt**, which, as the reading of P, is to be preferred to *dabit* of other MSS. For the sex, cf. *Iudaea*, l. 543; all the different fortune-tellers for women, mentioned in Plaut. *Miles*, 693, *praecantrix, coniectrix, ariola, haruspica*, are women themselves.

586. **conductus**, 'for hire'; cf. *conducenda*, l. 558. **mundus** is constantly used in the sense of *caelum*.

587. **qui...condit:** when any spot was struck by lightning, it was a *prodigium*, and it became the duty of the *pontifices* to make expiation (*procurare*) for it. In practice, this duty was left to the *haruspices*, who gathered together everything scorched by the lightning and buried it with muttered prayers; cf. Lucan, I. 606 *Arruns dispersos fulminis ignes | colligit et terrae maesto cum murmure condit.* The place was then walled round and known as *bidental* or *puteal*.

publica would in prose be *publice* (δημοσίᾳ), 'for the state'. The rich lady employs, so to speak, no obscure dissenter, but a dignitary of the Established Church.

588. 'But the destinies of the poor depend upon the Circus and the Embankment', i.e. they consult fortune-tellers there. For the **agger,** see n. to VIII. 43.

589. **aurum** can hardly mean the woman's hair (so Büch.): though Browning speaks of 'all the gold, | Used to hang and

brush their bosoms', it seems improbable that Juv. could use *aurum* for *capillus aureus*. Friedl. explains that it was the custom (as now in the East) for women of low station, especially *copae* (cf. l. 591), to carry their wealth upon them in the form of gold ornaments. Madvig, reasonably objecting to *aurum* as a sign of poverty, proposed to read *armum* in the sense of *humerum*; the emendation is not convincing, but neither are the explanations of *aurum* given above.

590. **falae** are probably wooden towers, temporarily erected for sham fights in the Circus; they stood between the *spina* (see n. to 582) and the circumference of the arena. **delphinorum columnas**: these columns were erected on the *spina* itself and served to support seven marble dolphins; for an illustration, see Rich's *Companion* or Smith's *Classical Dictionary*, I. 435. Each race was generally seven laps (*spatia*); and the number of dolphins, displayed at any moment on the columns, indicated the number of laps that had been completed.

591. 'Whether she shall leave' (i.e. divorce) 'the publican and marry the slop-seller': either *condicio* shows the woman's rank in life.

592–609. *Poor women at least consent to become mothers and to bear the dangers and troubles of a family; it is otherwise with the rich. Foundlings are smuggled into great houses; and these are the prime favourites of Fortune.*

592. **hae** are the women last mentioned, the poor.

602. 'I say nothing of spurious children': for **transeo**, see n. to III. 114. **suppositus** (ὑποβολιμαῖος) is found also I. 98.

603. **gaudia**, the joy of the father at the birth of a child.

lacus are the tanks or reservoirs all over the city, in which the water of the aqueducts was stored for use. It appears from this passage that it was the custom to expose infants there; they would be placed there during the night in the hope that someone, coming for water in the morning, would take pity on them and rear them. The *lacus* are **spurci** from being thronged by men and animals: cf. Sen. *Epp.* 36. 2 *quemadmodum ad lacum concurritur, quem qui exhauriunt, et turbant.*

(Friedl., however, explains *lacus* here as in Lucr. IV. 1026).

604. These foundlings are fathered upon the noblest houses and in time hold sacred offices reserved for nobles of the highest birth; the Scauri were a distinguished family of the *gens Aemilia.*

falso corpore, 'as pretenders'; cf. *falso nomine,* I. 98.

605. **inproba,** 'mischievous'. Fortune is supposed to stand by the pool in person at night, smiling on the new-born infants exposed there naked, and treating them like a fond mother.

606. **omni...sinu,** lit. 'with all her lap', i.e. with fondest affection: cf. Mart. I. 15. 9 *haec utraque manu complexuque assere toto*; idem, I. 70. 14 *nulla magis toto ianua poste patet.*

608. **secretum:** Fortune only is in the secret of the mystification and enjoys it: cf. Pliny, *Epp.* IV. 11. 2 *quos tibi, Fortuna, ludos facis!*

his se ingerit, 'she forces herself (i.e. her gifts) on them': cf. Lucan II. 263 *ingeret omnis | se belli fortuna tibi*; Sen. *Epp.* 76. 6 *gratia ac dignitas fortasse ingerentur tibi.*

609. **suos,** 'her own', is emphatic. **producit,** 'brings them forward', 'pushes them on'; cf. Livy, XL. 56. 7 *omni genere honoris producere eum non destitit.* For the idea that foundlings are the favourites of Fortune, cf. Soph. *O.T.* 1080 ἐγὼ δ' ἐμαυτὸν παῖδα τῆς τύχης νέμων | τῆς εὖ διδούσης οὐκ ἀτιμασθήσομαι. | τῆς γὰρ πέφυκα μητρός.

610–26. *Wives buy spells and potions to upset their husbands' reason; the result may be idiocy or, even worse, madness: take the case of Caligula, who drank a philtre administered by his wife Caesonia and in his madness nearly threw the world out of joint. The poison administered by Agrippina to Claudius was a trifling crime in comparison. The latter only put an end to a drivelling dotard; the former cost cruel sufferings to the noblest Romans.*

610. **Thessala...philtra:** from ancient times Thessaly was infamous for its witches and their magic arts: Plato and Aristophanes both speak of their reputed power to draw down the moon by spells, and Lucan (VI. 434 ff.) has a long account of their practices. For the use of charms, cf. *ibid.* 452 *carmine Thessalidum dura in praecordia fluxit | non fatis adductus amor.* The madness of Lucretius is said to have been caused by a love-philtre administered by his wife: cf. also Tac. *Ann.* IV. 22 *Numantina, accusata iniecisse carminibus et veneficiis vaecordiam marito, insons iudicatur.*

612. She chastises her husband in a manner usually reserved for children; cf. Persius, V. 169 *solea, puer, obiurgabere rubra.* So Lucian (*de Hist. Conscrib.* 10) speaks of pictures representing Hercules παιόμενον ὑπὸ τῆς Ὀμφάλης τῷ σανδαλίῳ. **inde** = *a philtris.*

615. **et furere,** 'to go mad as well': the mere loss of your

faculties matters less. The 'uncle of Nero' is the emperor Gaius
or Caligula; he was a son, and Agrippina, Nero's mother, was a
daughter, of Germanicus.

616. 'Into whose cup Caesonia poured the whole forehead of
a staggering (i.e. new-born) foal'; for the fact here alleged, cf.
Suet. *Calig.* 50 *creditur potionatus a Caesonia uxore amatorio
quidem medicamento sed quod in furorem verterit*: it seems certain
that Caligula suffered from insanity, whatever its cause. Foals
were born with a membrane on the forehead, called *hippomanes*,
which the dam would bite off if permitted; it was believed that
this membrane was of sovereign efficacy as a love-charm. The
love of the mare for the foal was thus accounted for. Cf. Pliny,
Nat. Hist. VIII. 165 *equis amoris innasci veneficium, hippomanes
appellatum, in fronte, caricae* (a fig) *magnitudine,...quod statim
edito partu devorat feta.*

618. **cuncta**, 'the world', i.e. the Roman Empire, owing to
the madness of its master. **ruebant** = *in ruinam ibant. incen-
dium* and *ruina* are often used together to convey the idea of
destruction: cf. Livy, III. 52. 6 *si decemviri finem pertinaciae non
faciunt, ruere ac deflagrare omnia passuri estis?*

619. The Roman emperor has as much power for the hap-
piness or misery of men as Jupiter himself has.

621. For the **boletus** of Agrippina, see n. to v. 147.

siquidem...senis, 'for it stopped the heart of one old
dotard'; *siquidem*, a favourite word with Suetonius, is rare in
the Latin poets.

622. **tremulumque...caelum**, 'and bade his palsied head
go down to heaven': for this account of Claudius, cf. Suet.
Claud. 30 *risus indecens, ira turpior spumante rictu, umentibus
naribus...caputque cum semper, tum in quantulocumque actu vel
maxime tremulum*; Sen. *Apocol.* 5 (on Claudius' arrival in heaven)
*nuntiatur Iovi venisse quendam...; nescio quid illum minari,
assidue enim caput movere*; *ibid.* 7 *quae patria, quae gens mobile
eduxit caput?*

descendere in caelum is an oxymoron; the expression is
intended to convey the contempt with which his memory was
covered, in spite of the divine honours awarded him by Nero.
Seneca's satirical account, quoted above, of his death and re-
ception in the other world, is a curious contrast to the philoso-
pher's other works; the title *apocolocyntosis* (for *apotheosis*) hints
that Claudius became a pumpkin instead of a god.

623. The construction is improved by putting a comma, for
a full stop (so edd.), after *saliva*: then *siquidem* introduces a pair

of contrasted clauses; see n. to 1. 107. For a similar pair of clauses after *siquidem*, cf. Livy, XXXVIII. 50. 7.

624. potio is Caligula's philtre which drove him to excesses of cruelty. **torquet** governs *patres*: cf. Sen. *Dial.* v. 18. 3 *Gaius Caesar...et senatores et equites Romanos uno die flagellis cecidit, torsit, non quaestionis sed animi causa.* Seneca and Suetonius abound in instances of his savage ferocity.

625. The senators and knights (*uterque ordo*) together formed the Roman aristocracy, under the emperor and above all the rest of the citizens, who formed the third and lowest class of freemen.

626. 'So great the price paid for a mare's offspring and a single witch'; i.e. all this suffering came about from such a trifling cause, the *hippomanes* used by Caesonia.

627–61. *To kill a stepson is no crime nowadays. Even her own children, if they are heirs to wealth, are not safe from their mother: care must be taken to prevent her poisoning them. The crimes of Medea and Procne are no mere figments of ancient tragedy; they are repeated in Rome today. A cold-blooded, mercenary crime is far worse than one committed in passionate excitement. Alcestis will find no wife to imitate her, but modern Clytaemnestras will be found in every street. The means used may be different, but the husband dies all the same.*

628. paelice, 'a rival', seems to be an invidious name given by the stepmother to the first wife. **privignum,** the son of a husband by a former marriage. **iam iam** here = *nunc* of l. 659. **fas est,** i.e. the accepted morality of the day does not condemn it.

629. vos is a climax: her own children even are not safe; their father is dead and the mother eludes the care of their guardian, wishing to secure their property for herself.

For **ego** P reads *equo*, which perhaps points to *quoque* as the true reading; the Schol. explains: *hoc videris quid sit, quod et suis infestae sunt filiis.*

630. 'Guard your lives and distrust every dish.'

631. livida refers to the effects of the poison: see n. to *nigros...maritos*, I. 72.

632. illa quae peperit, ἡ τεκοῦσα, 'their own mother'. The emperors regularly took the precaution of having a *praegustator*; cf. Tac. *Ann.* XII. 66; Suetonius mentions a report that Claudius was poisoned by his own 'taster' at a public banquet (*Claud.* 44).

633. **papas** is found only in inscriptions elsewhere: it cannot mean 'fond father' (so Weidn.), as (1) the *pupilli* are orphans, (2) the Latin equivalent of our 'dada' is *tăta*. We should expect *paedagogus*, which the metre will not allow; probably this is a child's name for the same.

634. 'You say, I invent this, and satire puts on the high buskin; and going beyond the limits of satire and the rules of my predecessors, I cry out in frenzy' etc. See Introduction, p. xxxiv.

The position of **scilicet** would be more usual if it changed places with *fingimus* below: the meaning is like that of *at enim* in prose.

636. Sophocles is mentioned as the typical tragic poet: cf. Mart. v. 30. 1 *Varro, Sophocleo non infitiande cothurno*.

638. **nos utinam vani,** sc. *essemus*, 'would that mine were an idle tale!' **Pontia** is twice mentioned as a poisoner by Martial, IV. 43. 5, VI. 75; cf. also II. 34. 6 *o mater, qua nec Pontia deterior*. The Scholiast says she was a daughter of Publius Petronius, who was condemned for a conspiracy against Nero. She was convicted of poisoning her sons and committed suicide. A well-attested instance of this crime is mentioned by Statius (*Silv.* V. 2. 76–97): a modern bard, in complimenting a friend, would not recall the fact that the friend's mother had been executed for attempting to poison her own son. The Crispinus there addressed was a *pupillus* (l. 629).

For **feci,** 'I am guilty', see n. to *fecisset*, IV. 12.

639. **aconita:** for this poison, cf. I. 158.

640. **tamen** belongs to *peregi*: cf. VIII. 272. Written at length, the sentence would be: *quamquam ipsa peregi, tamen peregi*: cf. Ovid, *Fasti*, II. 311 *aurea pellebant nitidos umbracula soles, | quae tamen Herculeae sustinuere manus*: i.e. though the hands were those of Hercules, yet they held up the parasol.

641. For the omission of the verb, here *sustulisti* or of like meaning, see n. to I. 1: so *sustulissem* must be supplied in the next line.

643. 'Let us take the word of the tragic poets for all the tales of fierce Medea and Procne'; both are stock examples of mothers who killed their children.

644. **conor,** sc. *dicere*: cf. II. 2 *quotiens aliquid de moribus audent* (sc. *dicere*); X. 175. **et,** 'also'.

647. If a full stop is placed after **monstris** (so Büch.), *illarum* must be understood with *monstris*, as the modern crimes of which Juv. speaks are *summa monstra* just as much as the

crimes of Medea and Procne. The omission, however, of *illarum* is so harsh that it seems necessary to connect *minor...monstris* with what follows. The difficulty was early felt; for in MSS. of the 9th and 10th cent. an *et* is found between *sexum* and *rabie*, which certainly simplifies the construction. The meaning is that rage is a less horrible motive for crime than avarice.

648. **hunc sexum:** cf. Virg. *Aen.* v. 6 *notumque furens quid femina possit*; Tac. *Ann.* I. 4 *muliebris impotentia* (want of control).

649. **ut saxa...recedit,** 'like boulders torn from the heights, where a mountain sinks and its side falls back and the slope hangs tottering': when masses of rock fall from the top, the mountain sinks lower and the side seems to recede.

651. **non tulerim,** 'I cannot endure'; cf. II. 24 *quis* (= *nemo*) *tulerit Gracchos de seditione querentes?* VII. 140, VIII. 30. **computat,** 'calculates' her gains by the deed.

652. **spectant,** 'they see on the stage'. **mariti,** i.e. *non sua.*

654. Pope had this line in mind when he wrote: 'Not louder shrieks to pitying heaven are cast, | When husbands or when lap-dogs breathe their last.'

655. The daughters of Danaus, Eriphyle and Clytemnestra all alike slew their husbands.

656. **mane,** 'of a morning', i.e. any day. **vicus,** 'street', is properly a division of the city, which was divided into *regiones,* and these into *vici.*

657. **refert** is used for *interest*; cf. v. 123, XI. 21; Mart. VIII. 38. 7 *refert sis bonus an velis videri.* The word is convenient metrically; but even prose-writers, e.g. Seneca and Quintilian, use the two apparently without distinction.

Clyt., the daughter of Tyndareus, killed Agamemnon with 'a stupid, senseless axe', which required both hands to hold it; more refined methods are in use now, says Juv.

659. **res agitur,** 'the business is done with the tiny lung of a toad'; for **rubeta,** see n. to I. 70.

660. **et ferro,** 'with the steel as well'. **Atrides,** 'her Agamemnon'.

661. Mithridates, king of Pontus, was conquered by Lucullus, Sulla, and Pompey successively. It was believed that this king had secured his life against poison by taking it constantly in small doses: cf. XIV. 252; Mart. v. 76 *profecit poto Mithridates saepe veneno, | toxica ne possent saeva nocere sibi*; Celsus, v. 23. 3 *nobilissimum (antidotum) est Mithridatis quod cotidie sumendo rex ille dicitur adversus venenorum pericula tutum corpus suum*

reddidisse: he then gives the prescription which contains an immense number of ingredients.

Pontica should be translated as an epithet of **regis,** not of **medicamina**: Juv. would have written *Pontici*, if metre had not prevented him. He elides *quantulum* once (VI. 151) and *plurimum* once (XIV. 73), but never elides a word metrically equivalent to *Pontici*.

SATIRE VII

THE TROUBLES OF LITERARY MEN

1–35. There is no prospect for men of letters except the emperor's patronage. Distinguished poets have been forced by imminent starvation to take up most prosaic occupations; even this is better than to live by perjury. But now there will be a different state of things, when the emperor is eager to reward merit. Otherwise, poets might as well burn their writings, for the rich will give them nothing but praise. Meanwhile, the poet grows old in poverty, and learns to curse his genius.

It is urged by Friedländer that this introduction has no real connection with what follows. First, whereas the introduction insists on an improvement in the position of literary men, owing to the emperor's interest in them, the body of the satire goes on to depict the present position of all this class as utterly hopeless. Secondly, whereas the satire deals with literary men of every kind, the introduction speaks only of poets and poetry. Friedl. concludes that the satire was written under Trajan, and the introduction added when Hadrian ascended the throne.

The argument is not convincing. Juv. is careful to point out that the emperor is the only patron literature can expect (*tantum* l. 1, *solus* l. 2); and one swallow does not make a summer. The attack on the selfish nobles, which pervades the whole satire, is not inconsistent with the preliminary compliment to the emperor, a compliment based mainly on a sense of favours to come.

1. 'All the prospects of literature and all the inducements to it depend on the Emperor.' In the absence of other means of dating the satire, there has been much discussion as to which emperor is meant, as one in particular seems plainly referred to. See Introduction, p. xvi.

2. **tempestate** = *tempore*: see n. to IV. 140.

3. **respexit,** 'has taken pity on'; cf. Cic. *in Verr.* II. 3. 26
*quid praetereo? an illud ubi caves tamen Siculis et miseros respicis
aratores?* It is difficult to say whether **iam** is to be taken with
cum or **celebres**; in the latter case, it means 'quite'; see n.
to III. 206.

4. **furnos,** 'bake-houses'; the poets turn bakers or bath-
keepers.

7. **atria,** 'auction-rooms': that the profession of auctioneer
was more lucrative than poetry, appears from Mart. VI. 8
praetores duo, quattuor tribuni, | *septem causidici, decem poetae,* |
cuiusdam modo nuptias petebant | *a quodam sene. non moratus
ille* | *praeconi dedit Eulogo puellam.* Martial abounds in similar
complaints of the poverty of poets.

8. **umbra:** see n. to l. 105.

9. **ames,** 'you must put up with'; perhaps an imitation of
στέργειν in this sense. The context shows that Machaera was
an auctioneer; that the profession was not held in high repute,
appears from III. 157.

10. **commissa auctio,** 'the strife of the auction'; cf. *pugna
commissa,* V. 29: the verb is used with any object that suggests
competition, e.g. *ludos, certamen, iudicium, bellum* etc.

12. In the course of his business as auctioneer, the poet has
to sell tragedies by his brother-poets: the authors here men-
tioned are not otherwise known. For Tereus as the subject of a
tragedy, cf. VI. 644.

Alcithoen, found in some MSS., must be read for the un-
intelligible *Alcitheon* of P: Alcithoe and her sisters were changed
into bats for slighting the worship of Bacchus: cf. Ovid, *Met.*
IV. 1 ff.

13. Cf. XVI. 30. **sub iudice,** 'before the court'; so *sub iudice
lis est.*

14. **faciant** = *hoc faciant,* 'do so'; a common use; see
Munro on Lucr. IV. 1112. The mood is to be noted: in Cicero
quamquam is always followed by a verb in the indicative, except
where the verb itself has a conditional or potential force; this
rule is relaxed in silver Latin; Juv. never has the indic. after
quamquam, sometimes the subj. (cf. X. 34, XI. 295), sometimes a
participial phrase: see n. to IV. 60.

equites Asiani is a sarcastic phrase, a knight being properly
styled *eques Romanus,* while in the case of a senator the adjective
is unnecessary. These men have become Roman knights by
sordid and disgraceful occupations, but they first came to
Rome as slaves from the East: cf. Mart. X. 76. 3 *de Cappadocis*

eques catastis. The *equites*, originally a division of the citizen
army, had long ago become a class of wealthy business men, and
were now being organised into an imperial civil service; see nn.
to I. 58, IV. 32. A law of Tiberius required that they should be
able to show free birth for three generations; but, in practice,
this was freely ignored, the emperors setting the example by
giving the *anulus* to their own freedmen. If a man possessed his
quadringenta, he had little difficulty in assuming the knight's
privileges; hence Martial's sneer *dominae munere factus eques*
(VII. 64. 2).

Asia, the Roman province, does not include Cappadocia or
Bithynia.

15.   **Bithyni:** the first syll. is long elsewhere: so x. 162, xv. 1;
hence some edd. transpose *B.* and *Asiani*; Büch. proposes
faciantque equites Bith. Also, as Bithynia is quite distinct from
Galatia, the insertion of *et* after *Bithyni* (so Weise) seems de-
sirable.

16.   'Whom New Gaul sends across the sea with bare feet',
lit. ankles. A horde of Gauls crossed over into Asia about
280 B.C. They terrified and plundered the unwarlike Asiatics
till they were defeated by Attalus, ruler of Pergamum, and at
last settled in a district which was called from them Galatia or
Gallograecia. Galatia became an imperial province under
Augustus, 25 B.C.

traducit is used metaphorically elsewhere in Juv.: see n. to
VIII. 17.

nudo talo refers to the custom explained on I. 111.

17.   **studiis,** 'literature', as l. 1: especially poetry.

18, 19.   The language is purposely rather high flown. The
laurel was supposed to inspire the Delphian priestess, and
therefore to be the source of the poet's inspiration too.

20.   **hoc agite,** 'be diligent'; cf. v. 157; the opposite is
aliud agere, 'to be careless, inattentive'.

21.   **ducis,** 'of the emperor': so IV. 145; common in Martial
and Statius.

indulgentia: the regular use of this word in Pliny's letters
to Trajan shows that it had become a technical term for 'im-
perial favour'. The Papal Chancery, inheriting the word from
the imperial court, made it famous in another sense.

22.   **rerum,** 'fortunes'.

speranda: so A. E. Housman (*Class. Rev.* III, 201) for
spectanda of P, and *expectanda* of other MSS.: the Scholiast
seems to have read this; and cf. Mart. III. 38. 1 *quae te causa*

trahit vel quae fiducia Romam, | *Sexte? quid aut speras aut petis inde? refer.*

sperare is 'to look for'; *spectare*, 'to look at'.

23. **croceae membrana tabellae,** 'the parchment of the yellow page', i.e. the yellow page of parchment. The form of book here mentioned should be noted. Until near the end of the 1st cent. the Roman book was regularly a roll of papyrus with a stick (*umbilicus*) in the middle. The writing was in columns (σελίδες, *paginae*), like our newspapers, but the length of the roll, from top to bottom, was only a fraction of its length from side to side. Such a roll had generally a parchment cover (*membrana*). But about this time, a different material and different shape came into use. The material was *membrana*, the skin of an animal, and the form resembled that of our books, which was always used by the Romans for writing letters. This parchment was much stronger and more durable than papyrus, and could be written upon on both sides. As *pagina* is used for the column in the papyrus roll, so *tabella* is used for the page of the parchment book. In Martial's *Apophoreta* (Bk. XIV) where presents are arranged in pairs, to be raffled for, one being valuable and the other not, books of parchment and papyrus rolls come alternately; there is no doubt that the former were more valuable. Cf. Mart. I. 2. 3 *hos eme quos artat brevibus membrana tabellis*; idem, XIV. 184–92.

(Friedl. supposes that Juv. refers here, not to a parchment book, but to tablets covered with parchment, which were certainly used for rough drafts: cf. Mart. XIV. 7 *pugillares membranei. esse puta ceras, licet haec membrana vocetur:* | *delebis, quotiens scripta novare voles*; and Quintilian (X. 3. 31) recommends persons with weak sight to write their oratorical exercises on *membrana* rather than wax tablets. And this is more probable if—what is not quite certain—parchment was still a rare and costly material.)

24. **impletur:** *implentur*, read by P and the Scholiast, can only be translated if *membrana* is the nom. plur. of a form *membranum*, which is not found elsewhere till much later.

25. **Veneris...marito** = *Vulcano*, i.e. 'to the flames': for this indirect kind of description, which Quintilian (VIII. 6. 29) calls *antonomasia*, cf. *generum Cereris* for *Plutonem*, X. 112, and see n. to III. 25. Of this Telesinus nothing certain is known.

26. **positos...libellos,** 'leave them lying for the worms to bore holes in': the books are shut up in *scrinia*; for the meaning of **pertunde,** see n. to V. 131.

27. Cf. Mart. IX. 73. 9 *frange leves calamos et scinde, Thalia,* *libellos,* | *si dare sutori calceus ista potest.*

vigilata proelia, 'battles you have sat up to write', an epic poem being in question; cf. Stat. *Theb.* XII. 811 (addressing his own poem) *o mihi bis senos multum vigilata per annos* | *Thebai;* Ovid, *Ars,* II. 285 *vigilatum carmen.*

28. **cella** is properly the room of a slave.

29. **venias,** 'may come forward' M. **hederis et imagine,** 'a bust wreathed with ivy': thus Ovid writes in banishment to his friends (*Trist.* I. 7. 1) *si quis habes nostris similes in imagine* *vultus* (i.e. a bust of me,) | *deme meis hederas, Bacchica serta,* *comis:* so it appears that the ivy was natural, not sculptured. *hederae* does not refer to any particular prize, but, as ivy was sacred to Bacchus, the ivy-wreath was associated especially with poets, as the laurel-wreath was with conquerors: cf. Virg. *Ecl.* 8. 12 *hanc sine tempora circum* | *inter victrices hederam tibi* *serpere laurus.*

imagine macra: the bust is lean, because the poet is so too from his privations. It was the custom to place the busts of famous poets, either living or dead, in public and private libraries: so Mart. IX. *praef. Stertinium, clarissimum virum, qui* *imaginem meam ponere in bibliotheca sua voluit.* Asinius Pollio set the example of this good custom, when he founded the first public library at Rome.

31. **laudare:** cf. I. 74 *probitas laudatur et alget.*

32. The poet, when too old to be a merchant, soldier, or husbandman, repents too late his choice of a vocation.

35. **Terpsichoren** = *Musam,* the special function of this Muse not being considered; thus Clio stands for poetry l. 7.

odit may be transl., 'curses'; cf. III. 214, VI. 272; Pliny, *Paneg.* 33 *nemini impietas...obiecta, quod odisset gladiatorem.*

36–97. *Your rich patron, himself a poet, will do nothing for you* *that costs money. If you wish to give recitations of your works,* *he will lend you a disused house in a distant part of the town and* *send his freedmen to applaud; but he will pay none of the neces-* *sary expenses. And yet we persist in writing poetry! Real genius* *can never be nurtured under such conditions; to be a great poet,* *a man must be in easy circumstances. Virgil's poem would not* *be what it is, had Virgil been a needy man. The rich man can* *be generous to his mistress; he can even afford such fancies as* *keeping a tame lion; but for a poet he has nothing. For a rich* *poet, like Lucan, praise is enough; but starving poets want pudding as well. Statius was popular: all Rome ran to hear him recite; but he made nothing by it. An actor is a better patron than the great nobles of our time. In the days of Maecenas it was a different thing.*

36. **accipe nunc artes,** 'now I will tell you the devices' of patrons. **conferat** implies assistance in money; cf. III. 51.

37. Augustus placed Greek and Latin libraries in the temple of Apollo on the Palatine. There was a second library in another temple, also on the Palatine, built by Livia to Augustus and known as the *novum templum*; cf. Mart. XII. 3. 7 (to his book) *iure tuo veneranda novi pete limina templi | reddita Pierio sunt ubi templa choro*; from this it appears that the library in the latter was dedicated to the Muses. Thus Juv. may refer here to both the great public libraries: the poet, in order to court his patron, turns his back on them and their presiding deities.

38. Antiquity is the only ground on which he concedes superiority to Homer: cf. Sedigitus, *ap.* Gell. XV. 24 *decimum addo causa antiquitatis Ennium.*

40. **recites:** see n. to III. 9; for the constr., cf. VI. 470. **Maculonis** is the reading of P which Büch. keeps: he supposes this to be some private house which could be used, and was used, for the purpose; this being the general custom, until Hadrian built an *Athenaeum* for recitations. That it was unusually generous in a rich man to lend *his own* house for the purpose, appears from Pliny, *Epp.* VIII. 12. 2 *domum suam recitantibus praebet.* Juv.'s readers may have been as familiar with Maculo's house as with Hispulla's bulk (XII. 11), though both allusions fall flat to us.

On the other hand, it is clear from the Scholiast's note, that he read *maculosas*; and some edd. prefer this (or *maculonsas* Bywater), though the epithet is unexampled of a house. But it should be observed that this involves a further departure from P; for **haec** (l. 41) requires that one definite house should be mentioned here. Consequently, if *maculosas* is read, *ac*, the reading of the inferior MSS., must be substituted for *haec.*

41. **longe** cannot stand for *diu*, though *longum* sometimes does: it means 'at a distance' and is to be taken with **domus:** cf. Mart. III. 58. 51 *rus hoc vocari debet an domus longe?* (a town-house out of town). This connection of adv. with noun is due to the absence, in Latin, of a pres. partic. of *esse.* The want of an article is a further complication, so that the adv. may have to represent a noun in any case, *nunc* standing for οἱ νῦν ὄντες,

τοὺς νῦν ὄντας etc.: cf. Tac. *Hist.* I. 10 *palam* (his public life,
τὰ δημόσια) *laudares, secreta male audiebant*; Ovid *Met.* I. 19 (of
the primeval chaos) *frigida pugnabant calidis,...sine pondere*
(= τοῖς ἄνευ βάρους οὖσιν) *habentia pondus*.

servire, 'to be at your disposal': so, of a country-seat, Mart.
x. 30. 28 *o ianitores vilicique felices!* | *dominis parantur* (are
bought by their owners) *ista, serviunt vobis*.

42. **sollicitas,** i.e. of a besieged city: **portas** must of course
refer to a city. The house is securely barricaded, as if to prevent
all ingress, because, if it were in a remote and lonely part of the
town, it would be necessary to take precautions against burglars;
cf. III. 303 and 304.

imitatur, 'is like': the word is often used of inanimate things,
and then cannot imply conscious imitation: cf. Mart. I. 43. 6
imitata breves Punica grana rosas.

43. **scit dare,** 'he knows how to give', i.e. he is willing to
give: cf. Persius, I. 53 *calidum scis ponere sumen,* | *scis comitem
horridulum trita donare lacerna*. The freedmen sit at the end of
the benches, in order to distribute the sound.

44. **comitum:** see n. to I. 119. For this function of freed-
men and clients, cf. Mart. III. 46. 7 *quidlibet in causa narraveris,
ipse tacebo:* | *at tibi tergeminum mugiet ille* (i.e. *libertus meus*)
σοφῶς. People who attended courts and recitations, and ap-
plauded in hope of a dinner, were known as Σοφοκλεῖς (i.e.
'σοφῶς' καλοῦντες), or *laudiceni*, lit. Laodiceans (so Livy,
XXXIII. 18. 3 and Inscriptions), but really meaning *qui laudando
cenam captant*: 'good-cheerers' will render the latter play on
words.

45. **regum:** see n. to I. 136. *tantum* is understood as the
antecedent of **quanti.** For this and other expenses of a recita-
tion, cf. Tac. *Dial.* 9 (of Saleius Bassus) *cum toto anno, per
omnes dies, magna noctium parte unum librum...elucubravit,
rogare ultro et ambire cogitur ut sint qui dignentur audire, et ne id
quidem gratis: nam et domum mutuatur et auditorium exstruit
et subsellia conducit et libellos dispergit*.

46, 47. The room is a large one: immediately in front of the
reciter is a place corresponding to the *orchestra* in a theatre,
and here are placed cushioned chairs (*cathedrae*); further back
come benches (*subsellia*), and then, rising towards the roof,
tiers of seats supported on scaffolding (*tigilla*); the scaffolding is
hired for the occasion, and the chairs also have to be returned.

47. **posita** would be *disposita* in prose.

48. **hoc agimus,** 'give ourselves up to this', i.e. poetry; see

n. to l. 20. The business of ploughing the sand is proverbially unprofitable: see n. to l. 157.

50. **si discedas,** 'if you try to give it up'; *recedere* is commoner in this sense. The subj. may be due, not to the conditional form of the sentence, but to the indefinite use of the 2nd person, 'you' being = 'anyone': cf. Munro on Lucr. I. 327.

ambitiosi...mali, lit. 'of a mischief which seeks for publicity', i.e. 'of morbid love of display': for *mali*, cf. *malum*, VI. 109.

52. **cacoethes** may be transl. 'itch': it is the medical term for a malignant growth of the nature of cancer but curable.

53. **publica vena,** 'hackneyed Muse': the metaphor is probably from water in the earth, not from metals in a mine: cf. Ovid, *ex Pont.* II. 5, 21 *ingenioque meo vena quod paupere manat, | plaudis; ibid.* IV. 2. 20 *et carmen vena pauperiore fluit.*

54. **expositum,** 'trite'; the word is used in the sense of *in medio positum,* δεδημευμένον. **deducere** is a metaphor from weaving: cf. Hor. *Epp.* II. I. 225 *tenui deducta poemata filo.* The word seems not to be used, like *producere* (cf. XV. 166) or *procudere,* of forging metal.

55. Another metaphor, from coining. For **feriat,** cf. the title of the Commissioners for the Coinage, III *viri A.A.A.F.F.* i.e. *auro, argento, aeri flando feriundo*: we say 'to strike', in the same sense.

communi moneta, 'at the mint of commonplace': 'mint' is derived from *moneta,* the temple of Juno Moneta having been the Mint at Rome. For **triviale,** cf. Quint. I. 4. 27 *litterarii paene ista sunt ludi* (an elementary school) *et trivialis scientiae.*

56. I can point to no such poet now, says Juv., as there is none; but I feel there might be in a happier age.

57. Lewis quotes Ovid's excuses for the falling off in his poetry after his banishment to Tomi, *Trist.* I. I. 39 *carmina proveniunt animo deducta sereno; ibid.* 41 *carmina secessum scribentis et otia quaerunt; ibid.* 43 *carminibus metus omnis obest.*

58. **impatiens** seems to be used with the meaning of *expers.* For this idea of a poet's life, cf. Tac. *Dial.* 9 *poetis... deserenda officia utque ipsi dicunt, in nemora et lucos, id est, in solitudinem secedendum est; ibid.* 13 *me vero dulces, ut Vergilius ait, Musae, remotum a sollicitudinibus et curis...in illa sacra illosque ad fontes ferant.*

60. **thyrsum...contingere:** the thyrsus is the ivy-bound staff of Dionysus, which by its touch produced the frenzy of his worshippers; hence it is used for inspiration generally; cf.

Lucr. I. 922 *acri | percussit thyrso laudis spes magna meum cor*;
Ovid, *Trist.* IV. I. 43 *sic ubi mota calent viridi mea pectora thyrso,
| altior humano spiritus ille malo est.*

62. Horace has had a good meal when he writes a fine lyric;
the allusion is to Hor. *Carm.* II. 19. 5 *euhoe! recenti mens trepidat
metu, | plenoque Bacchi pectore turbidum | laetatur.* εὐοῖ is the cry
of the Bacchanals; whence Dionysus is called *Euhius* and *Euhan.*

64. **dominis...feruntur**, 'are borne along by the lords of
Cirrha and Nysa': Apollo is the lord of Cirrha, the port of
Delphi; Nysa was a favoured haunt of Dionysus; cf. Mart. IV.
44. 3 *haec iuga quam Nysae colles plus Bacchus amavit*: its
position was disputed.

dominis is probably dat. of the agent after a pass. verb: a
common constr. in poetry: cf. H r. *Epp.* I. 19. 3 *carmina quae
scribuntur aquae potoribus.* Cicero could not say *scribuntur nobis
multa*, but could say *scripta sunt nobis*: see Madv. on Cic. *de
Fin.* I. II. For **feruntur**, cf. *Musae me ferant* quoted on l. 58.

Or *dominis* may be abl., like *Tisiphone*, VI. 29, Bacchus and
Apollo being regarded not as agents but as causes: see n. to I.
13, and add to references given there Ovid, *Fast.* VI. 99 *perierunt
iudice formae | Pergama* (Troy was overthrown by Paris).

65. **vestra** = *poetarum.* **duas...curas**, i.e. poetry and
poverty.

66. **paranda:** see n. to III. 224.

67. **attonitae:** there is a play on the double meaning of the
word 'inspired' and 'distressed'; so perhaps also in Mart.
VIII. 56. 17 (of Virgil when enriched by Maecenas) *excidit
attonito pinguis Galatea poetae*; for the second meaning of the
word, cf. IV. 77; Sen. *Epp.* 108. 37 *quid me potest adiuvare rector
navigii attonitus et vomitans?*

68. The allusion is to Virg. *Aen.* VII. 445–66 where the Fury
Allecto drives Turnus (*Rutulum* of I. 162) to madness.

69. For the idea that a poet's genius depends on freedom
from material cares, cf. Ovid, *Trist.* I. I. 47 *da mihi Maeoniden
et tot circumspice casus; | ingenium tantis excidet omne malis.*
Martial (VIII. 56) attributes the Aeneid to the wealth given to
Virgil by Maecenas. **puer**, 'a slave'.

desset: for the tense, see n. to *liceret*, IV. 85.

70. I.e. his descriptions would not be what they are.

71. **surda** is passive in sense, 'unheard'; so *caecus* often
means 'unseen', e.g. as an epithet of *primordia* in Lucretius.

poscimus: *postulare* is regularly used in this constr. in
Plautus and in prose.

72. Of Rubrenus Lappa nothing is known: Atreus sounds like the name of a tragedy, or else *cothurno* might apply as well to epic poetry: Martial twice (v. 5. 8, VII. 63. 5) speaks of *cothurnatus Maro*. The 'tragedians of old', if Latin writers, may be Ovid and Varius, or, of a more ancient time, Ennius and Pacuvius.

73. **alveolos:** cf. v. 88.

pignerat, 'pawns'; i.e. in order to live while writing his tragedy, he pawns his bits of furniture and clothes.

74. Numitor stands for the *dives avarus* of l. 30. **quod mittat,** 'to give away'; see n. to III. 45. Quintilla is the name of Numitor's mistress: cf. III. 133, IV. 20.

78. **nimirum,** 'of course'; sarcastic.

79. 'Lucan may repose in his park, adorned with statues, and find fame enough': the poet Lucan (A.D. 39–65) belonged to a wealthy family which took its rise from Corduba in Spain; his father was Annaeus Mela, and his two uncles were the philosopher Seneca, and Junius Gallio, so named from adoption, who refused, when pro-consul in Achaia, to hear the complaint of the Corinthian Jews against Paul (Acts xviii. 12). Lucan had been long dead when this was written; he, with all his relatives, lost his life in consequence of Piso's conspiracy in A.D. 65.

hortis: see n. to I. 75: they were commonly adorned with statues.

80. Serranus and Saleius Bassus were both epic poets of some fame, and both died young, before the publication of Quintilian's *Institutio* (ca. A.D. 94): they are mentioned together there X. 1. 89 and 90. Some light is thrown on the circumstances of Bassus by Tacitus, *Dial.* 9 *omnis illa laus...ad nullam certam et solidam pervenit frugem...; laudavimus nuper ut miram et eximiam Vespasiani liberalitatem, quod quingenta sestertia Basso donasset.* The date of the Dialogue is A.D. 75.

tenui: see n. to *tenues*, III. 163.

82. 'The town runs to hear his pleasant voice and that popular poem, the Thebais': this poem, in twelve books, is the chief work of P. Papinius Statius (A.D. 45–96), a Neapolitan: see n. to l. 27. During the twelve years he spent over his poem, he sometimes, as Virgil had done, gave recitations of parts he had finished; he speaks himself of giving recitations, but apparently not from the *Thebais*: see Stat. *Silv.* v. 3. 215 (to his father) *qualis eras, Latios quoties ego carmine patres | mulcerem, felixque tui spectator adesses | muneris; ibid.* 2. 160 he speaks of giving recitations from his *Achilleis*.

84. **diem,** sc. *recitando.* **captos afficit** = *capit et afficit.*

86. **fregit...versu,** 'his lines have brought down the house', L.: the damage to the furniture is due to the demonstrations of his hearers, not to the energy of his own declamation, in spite of 1. 13.

87. It is not known from any other source that Statius wrote for the stage: his *Agave* would deal with the same matter as the *Bacchae* of Euripides. We have here a clear statement that money was paid for plays: but that poets were paid by their publisher is by no means certain. Martial once or twice refers a friend, who wishes to get his poetry for nothing, to the publisher; but this is not conclusive, and it is remarkable that he never complains of the publisher's meanness to himself. It seems probable that poets depended entirely on the munificence of patrons.

intactam, 'virgin'; cf. Mart. 1. 66. 7 *virginis pater chartae*: a double meaning is intended, as if Paris were a *leno.*

Paridi: cf. VI. 87; this was a famous dancer (*pantomimus*) of Domitian's time, who was put to death in A.D. 83 for an intrigue with the empress, Domitia; Martial wrote an epitaph for him (XI. 13). It seems to have been customary for artists to take the name of some famous predecessor: the chief *pantomimus* of Nero's reign, executed in A.D. 67, was also called Paris; and there were others of the same name in the next century.

The art of the *pantomimus* consisted entirely in dramatic gesticulation. The subjects of the *fabulae salticae* were generally taken from mythology, e.g. the murder of Pentheus by his mother. A large chorus, accompanied by an orchestra, sang the *fabula*, which consisted of a string of lyric odes, dealing with the most important episodes of the story. Meanwhile, the *pantomimus*, alone upon the stage, represented the action by gesticulation, without uttering a word. This was by far the most popular spectacle in theatres of the Roman Empire. Even parts of the Aeneid were so represented: cf. Suet. *Nero*, 54 *voverat...proditurum se...histrionem* (as a dancer) *saltaturumque Vergili Turnum.* It is difficult for us to realise the power of expression shown by the best *pantomimi*: cf. Quint. XI. 3. 66 *saltatio frequenter sine voce intelligitur*: Tacitus (*Dial.* 26) mentions as a phrase of the time that *histriones* are said *diserte saltare.* It must be remembered that, in ancient dancing, the arms were more important than the legs.

88, 89. 'Paris bestows on many rank in the army as well, and

puts, for six months' service, the ring round poets' fingers.'
All the officers of the legion, called *tribuni*, received the privi-
leges of equestrian rank; and Claudius instituted a custom of
making sinecure appointments to the tribunate, in order that
the holder might become an *eques*, without any military service:
cf. Suet. *Claud.* 25 *instituit...imaginariae militiae genus, quod
vocatur supra numerum, quo absentes et titulo tenus fungerentur.*
A six months tenure of the rank was sufficient to ennoble the
holder: cf. Pliny, *Epp.* IV. 4 (to Sosius Senecio) *hunc rogo seme-
stri tribunatu splendidiorem...facias.* Martial became an *eques*
by this back-door, probably by gift of Titus: cf. Mart. III. 95.
9 *vidit me Roma tribunum, | et sedeo qua te suscitat Oceanus* (i.e.
I can sit undisturbed in the knights' seats in the theatre).
Hence **aurum semenstre** = *aurum semenstrium tribunorum*
and means 'knighthood gained by a command for six months'.

These appointments were nominally conferred by the em-
peror alone; but Paris has such influence at court, that he can
secure his friend's promotion.

90. **histrio** = *pantomimus*, which is inadmissible in dactylic
metre: the word properly means 'actor', but is restricted, under
the Empire, to the meaning 'dancer'—a proof of the exclusive
popularity of this kind of spectacle.

91. Barea (cf. III. 116) and Camerinus (cf. VIII. 38) are taken
as typical aristocrats.

magna atria curas: Martial too considers this an unsatis-
factory profession: e.g. III. 38. 11 *'atria magna colam.' vix tres
aut quattuor ista | res aluit; pallet cetera turba fame.*

92. Pelopea and Philomela are both names of *fabulae salticae*
which Paris may have represented. The meaning is that the
actor grants access to the equestrian *cursus honorum*: for
praefecti and *tribuni*, see n. to *curam cohortis*, I. 58.

The comment of the Scholiast on this l. is as follows:
propter hunc versum missus est in exilio (sic) *a Claudio Nerone.*
For Juvenal's exile and its cause, see Introduction, pp. xii,
xiii.

93. **haut...invideas,** 'one need not envy': i.e. he is not so
well off after all. For the constr., cf. Cic. *de Off.* III. 110 *haud
facile quis dixerit:* there is no difference in meaning, in this
constr., between the pres. and aor. subj.; cf. v. 139. The l. is
parenthetical.

94, 95. Cf. v. 108 ff. This complaint of the decay of patronage
might be illustrated at any length from Martial: cf. esp. VIII.
56. 5 *sunt Maecenates, non deerunt, Flacce, Nerones,* and Mart.

quoted on v. 109. C. Proculeius was a friend of Augustus, and is mentioned by Horace (*Carm.* II. 2. 5) as a generous man; Paulus Fabius Maximus was (like Cotta; see n. to v. 109) a patron of Ovid; Lentulus cannot with certainty be identified among the many Lentuli Cornelii.

97. **pallere:** this was the natural result of the sedentary and exhausting life of a poet: cf. Hor. *Epp.* I. 19. 17 (of his imitators) *quod si | pallerem casu, biberent exsangue cuminum*; hence Martial's jest on a man who confused cause and effect: VII. 4 *esset, Castrice, cum mali coloris, | versus scribere coepit Oppianus.*

vinum…Decembri: all classes allowed themselves some relaxation and enjoyment at the Saturnalia (17–19 Dec.); hence Mart. XIII. 1. 4 *ebria bruma.* The poet alone abstains, for the purpose of study; another reason, perhaps, was that the Saturnalia seems to have been the publishing season for light literature.

98–104. *Historians are in no better case than poets. Their labour and research is more exacting; the material of their many rolls is more expensive. But their remuneration is quite inadequate.*

98. **historiarum:** history was considered by the ancients to have more affinity to poetry than we should allow; here they together represent literature: cf. Quint. X. 1. 31 (of history) *est proxima poetis et quodammodo carmen solutum, et scribitur ad narrandum non ad probandum.*

99. **perit:** so P for *petit*, the reading of the other MSS.: though *petit* makes sense, *perit* is certainly right: it gives the passive of the common phrase *perdere oleum*: cf. Cic. *ad Att.* II. 17. 1 *haec non deflebimus, ne et opera et oleum philologiae nostrae perierit*; Mart. II. 1. 3 *at nunc succinti quae sint bona disce libelli; | hoc primum est brevior quod mihi charta perit.* **hic** = *in hoc.*

100. 'For there is no end to it, and every one of you goes on to his thousandth column, and, as the roll swells, the account runs up for the quantity of paper.' **millensima pagina:** rolls were not manufactured of more than a certain size, 200 *paginae* (σελίδες) being the extreme limit of a prose-book; in this case the 1000th column would be about the end of the fifth roll (*volumen* or *liber*). Livy's history (*Livius ingens*, Mart. XIV. 190) consisted of 142 rolls.

surgit, where we should say 'is turned over', is accounted for by the shape of the roll.

101. **crescit,** 'lengthens out': cf. Mart. v. 6. 15 *nigris pagina crevit umbilicis.* Papyrus was not costly, except in very large quantities: Mart. 1. 66. 3 speaks of *tomus vilis.*

102. **ingens rerum numerus...,** 'the huge extent of the subject and the rules of the composition'.

103. **inde** = *a labore.*

104. For the **acta,** see n. to VI. 483: a slave would be employed to read the official Gazette aloud; so at Trimalchio's table (Petron. 53) *actuarius...tanquam urbis acta recitavit*: this parody of the Gazette begins with the date and includes the births of slaves, accounts, the crucifixion of a slave, a fire, etc.

The slave would of course not be paid for reading this: but Juv. means that rich men have poor taste in literature and would, to use a modern comparison, prefer 'a copy of the Times to all the works of Thucydides'.

105–49. *You say that men of letters are a self-indulgent set and don't deserve anything better. Lawyers are busy enough; do they make large incomes? No; though, in the presence of a creditor or a possible client, they pretend they do, as a fact one jockey gets as much as a hundred barristers. The lawyer pleads, till he bursts his lungs, before a stupid jury, and gets, as the reward of success, some second-rate wine or food. If he gets a guinea for four appearances, the attorney claims his share. A rich lawyer can name his own fee. So, in order to seem rich, and to keep up appearances, many lawyers ruin themselves by extravagance. Eloquence alone will attract no clients, unless you wear a fine ring and are attended by a retinue of slaves and clients. Eloquence is resented in a poor man; he had better retire to the bar of Africa or Gaul.*

105. This applies to poets as well as historians: cf. Ovid, *Am.* I. 9. 41 *ipse ego segnis eram discinctaque in otia natus: | mollierant animos lectus et umbra meos.* The **lectus** (or *lectulus*) here mentioned was used by the student while reading or writing, and is different from the *l. tricliniaris* (v. 17) and the *l. cubicularis* (VI. 268); cf. Sen. *Epp.* 72. 2 *quaedam* (some kinds of writing) *lectum et otium et secretum desiderant.*

umbra, 'the cloister', suggests a life of retirement: cf. l. 173 and Hor. *Epp.* II. 2. 77 *scriptorum chorus omnis amat nemus et fugit urbes, | rite cliens Bacchi somno gaudentis et umbra.* So we speak of a man living 'in cotton-wool'.

106. Juv. here replies to the imaginary objector. **civilia officia,** 'practice at the bar': a common phrase at this period:

see Quint. x. 3. 11, *ibid.* 7. 1, XI. 3. 22, XII. 2. 6; Pliny, *Epp.* VI.
32, IV. 24. 3 *alius exercitum regit, illum civilibus officiis principis
amicitia exemit.* The last example shows how 'civil' became
opposed to military life. Here the epithet suggests that the
barrister is useful to society in a way the literary recluse is
not.

107. **libelli**, 'briefs': cf. Mart. v. 51 (of a lawyer) *hic qui
libellis praegravem gerit laevam*: the word may stand for many
kinds of documents, e.g. 'petitions' (XIV. 193). Quintilian (X. 7.
31) uses it for the notes of a speech, which a barrister holds in
his hand in court.

108. **ipsi...sonant**, 'they talk big themselves' of their
gains: the opposing clause begins at *veram*, l. 112.

sed, 'and', does not contradict, but emphasises, the previous
statement: see n. to IV. 27.

109, 110. The passage is explained thus by Madvig (*Opusc.*
II. 179): a lawyer brags most of his gains, either before a creditor
of his own, or before a litigant who comes, with a huge account-
book, to press for payment of a debt: the latter 'gives a fillip'
to the advocate's power of lying, as there is a chance of his
getting the brief. **acrior** is practically = *acrius* the adv.: he is
the cause of greater keenness in the barrister. Jahn brackets the
l. Friedl. (after the Schol.) supposes l. 110 to refer to the
barrister's banker who is doubtful of his client's credit.

110. **dubium** suggests a difficult case, **grandi** a rich client—
both of which circumstances promise a large fee.

codex is the book in which debtors' *names* and their debts are
entered; hence *nomen* is 'debt' and *nomina facere* 'to lend
money'; cf. Sen. *de Ben.* I. 1. 2 *nomina facturi diligenter
patrimonium et vitam debitoris inquirimus.* The book is generally
called *Kalendarium* in prose: cf. Sen. *Epp.* 87. 7 *magnus Kalen-
darii liber volvitur*; Mart. VIII. 44. 11 *centum explicentur paginae
Kalendarum.*

111. The man's lungs are compared to a blacksmith's
bellows, as by Hor. *Sat.* I. 4. 19 *at tu conclusas hircinis follibus
auras,* | *...ut mavis, imitare.*

112. **conspuiturque sinus**, 'and they spit upon their
breasts': this was a charm, to obviate the evil effects of boasting
and to propitiate Adrasteia: cf. Theocr. VI. 39 (Polyphemus,
after admiring his own reflection) ὡς μὴ βασκανθῶ δέ,, τρὶς εἰς
ἐμὸν ἔπτυσα κόλπον: Petron. 74 (Trimalchio, of the lady who has
been lucky enough to become his wife) *at inflat se tanquam rana,
et in sinum suum non spuit.*

This seems better than the other explanation, that they splutter in their eagerness; the Scholiast gives both.

messem continues the metaphor of l. 103.

113. **patrimonia**, 'fortunes'; see n. to XII. 50.

114. **parte alia** = *inde*, 'on the other side'; cf. VI. 437 where it is opposed to *inde*.

russati...Lacertae: for *russati*, see n. to XI. 198. Lacerta is an *auriga* or *agitator* in the circus: the name (which may have been borne by many successive *aurigae*: see n. to *Paridi*, l. 87) is found on a lamp with the figure of an *auriga*. Martial also complains (X. 74. 2) of the disproportion between his gains, as a *salutator*, and those of a circus-driver, *quamdiu salutator | ante-ambulones et togatulos inter | centum merebor plumbeos die toto, | cum Scorpus una quindecim graves hora | ferventis auri victor auferat saccos?*

115. The l. is a parody of the contest for the arms of Achilles, as described by Ovid, *Met.* XIII. 1 *consedere duces, et vulgi stante corona | surgit ad hos clipei dominus septemplicis Aiax.* This contest was a stock subject for declamation in the schools; so that Ajax stands for *declamator*; see n. to X. 83. For **consedere,** cf. Livy, IV. 15. 2 *se ad causam cognoscendam consedisse*; Pliny, *Epp.* VI. 31. 9 (of Trajan on the bench) *consederat auditurus.* **duces:** the Greek chieftains in the original *iudicium* are here the presiding magistrates of the court.

116. **pro libertate:** a common case in Roman courts, called *assertio in libertatem*; proof was given, that a man detained as a slave was really free; the procedure is well illustrated by Mart. I. 52. 4 (of his books, the authorship of which is being claimed by a plagiarist) *si de servitio gravi queruntur, | assertor venias satisque praestes, | et, cum se dominum vocabit ille, | dicas esse meos manuque missos.*

bubulco iudice: Quintilian often reminds the pleader that the jury he wishes to convince may consist of unlearned men, e.g. XII. 10. 53 *cum...laturi sententiam indocti saepius atque interim* (sometimes) *rustici* (*sint*); idem, IV. 2. 45. Of the method by which the *decuriae* of *iudices* were selected under the Empire, little is known.

117. **rumpe...tensum:** see n. to *captos afficit*, l. 84.

118. A success of a pleader in the courts was indicated by palm-branches stuck up at his house-door: cf. Mart. VII. 28. 5 (to a lawyer) *sic fora mirentur, sic te palatia laudent, | excolat et geminas plurima palma fores.* By **scalarum** it is suggested that the lawyer here spoken of is a poor man living in a garret.

119. Mart. (IV. 46) gives a similar account of the scraps of food and bits of furniture which Sabellus, a *causidicus*, thinks himself very lucky to get as a fee.

120. **bulbi**, 'roots', probably a kind of onion; *epimenia* is Greek for the Lat. *menstrua*.

121. **Tiberi devectum:** all the wines brought *down* the Tiber to Rome were poor, *Vaticanum* being a common instance of bad wine; the choice wines from Campania and elsewhere came *up* the river.

122. **egisti**, 'have spoken in court'; *actio* is often 'a speech in court', *actor*, 'a pleader'.

aureus unus, 'a single guinea': this was a gold coin, first struck by Augustus, worth 100 sesterces or 25 denarii.

123. **pragmaticorum:** 'the attorneys' claim their share for legal information supplied to the pleader. Quintilian (XII. 3. 4) says of them, *tela agentibus subministrant*. The two professions were not so sharply distinguished as in England: the *pragmatici* were often barristers who could not learn to speak and therefore made a special study of law; cf. Quint. *ibid.* 9 *plerique, desperata facultate agendi, ad discendum ius declinaverunt.*

124. Aemilius is a noble and therefore receives the full fee permitted by law (*quantum licet*). The most famous law concerning the fees of advocates was the *lex Cincia* of 204 B.C., making any fee illegal: this was repealed by Claudius who allowed a maximum fee of 10,000 sesterces = 100 *aurei* (Tac. *Ann.* XI. 5–7): there was further legislation under Nero to protect litigants (Suet. *Nero*, 17). Pliny (*Epp.* VI. 23. 1) lets us know that he spoke without remuneration. Quintilian discusses the question in a commonsense way (XII. 7. 8): he objects chiefly to a bargain made beforehand, but asserts the duty of the client to show his gratitude practically.

et, 'and yet': so I. 74, XIII. 91; see n. to I. 93.

125. **huius** goes with *vestibulis*, not with *currus*.

126. **quadriiuges**, i.e. a triumphal statue of some ancestor; under the Empire no one but the emperor could celebrate a triumph. Equestrian statues of great lawyers seem to have been common: cf. Mart. IX. 68. 6 *causidicum medio cum faber aptat equo*.

127, 128. There are serious difficulties in this description, obviously sarcastic, of the armed equestrian statue of a lawyer. A spear-shaft should be straight; why is it **curvatum** here? M. explains that the shaft bends when poised for the throw; but it would be strange to represent this in a statue. Friedl. suggests that *curv.* is used for *rotatum*, 'brandished', but gives no similar instance.

lusca offers a second difficulty: various explanations are given, e.g. that the statue has one eye closed, as taking aim; that the statue is seen in profile; that Aemilius may have been *luscus* himself. From the point of view of art, all these are unsatisfactory. Friedl. suggests that the eye-balls were, as often, represented by coloured stones, one of which has fallen out.

If we accept the last view, perhaps *curvatum* may be 'crooked' 'out of the perpendicular': Juv. then means that the statue is badly wrought and in bad repair.

129. By imitating the display of the rich and noble (**sic**), poor and plebeian lawyers are ruined; the earthen pitchers try to swim with the brass pots. That a barrister was expected to make some display of wealth, is shown by Pliny, *Epp.* VI. 32. 1 *Nonio Celeri, cui ratio civilium officiorum* (see n. to l. 106) *necessitatem quandam nitoris imponit.*

conturbat, sc. *rationes*, 'becomes bankrupt'; cf. Mart. VII. 27. 10 *conturbator aper*, 'boar that makes me bankrupt', i.e. white elephant.

130. **rhinocerote,** an oil-flask (*gutus*) made of rhinoceros horn.

131. T. mobs the bath with a crowd of clients who are muddy from attending him through the streets.

132. **iuvenes Maedos,** 'stout Maedian slaves': the Maedi were a tribe in the west of Thrace, who were used, like Syrians and Liburnians, as *lecticarii.* **longo** suggests that the litter is a *hexaphoron* or *octophoron.*

133. He buys all the most costly articles for sale in the Forum in the Via Sacra, the Bond Street of Rome: for *murrina*, see n. to VI. 156.

134. **spondet,** 'gets him credit'.

stlattaria: this very rare word is apparently derived from *stlatta* (Gell. X. 25. 5), a kind of ship used by pirates, and means either 'seaborne', i.e. imported in a ship of this kind, or 'deceptive' (so the Schol.), from the wiles adopted by such vessels.

135. **vendit,** 'makes him sell', i.e. gets him practice; cf. Hor. *Epp.* II. 1. 75 *totum ducit venditque poema* [*versus concinnior*].

136. **amethystina:** 'amethyst' was a particular shade of purple, especially valued; cf. Pliny, *Nat. Hist.* IX. 135 *amethysti colos eximius ille*; it was obtained by mixing the juice of the purple-fish with that of another shell-fish, the *bucinum.*

138. Display is necessary for success; but the extravagant way of life at Rome carries the lawyer too far and ruins him.

140. **dederit** is aorist, like δοίη ἄν: *det* (= διδοίη ἄν) would mean just the same. Martial also declares that eloquence alone will not bring success: III. 38. 3 *causas, inquis, agam Cicerone disertior ipso:* 5 *egit Atestinus causas et Civis; utrumque | noras; sed neutri pensio tota fuit.*

141. **an**, 'whether', where *num* or *ně* would be used in classical Latin, *an* meaning 'or'. But in silver-age Latin this meaning of *an* is common: cf. VI. 387, 567, 591; VII. 162; XIII. 203; XV. 89.

142. **comites** and **togati** both refer to free clients, wearing the *toga*; cf. I. 96.

143. **agebat**, 'spoke in court'; cf. l. 122. For **conducta**, see n. to VI. 352; for **sardonyche**, n. to VI. 381. In the modern comic opera, the impecunious barrister had 'a ring that looked like a ruby', as part of his stock-in-trade.

146. A poor ill-dressed barrister cannot be permitted to make any sensation in the court. It was customary for the barrister during the peroration (*epilogus*) of his speech to excite sympathy for his client by bringing forward weeping relatives. Quint. VI. 1. 41 shows that mishaps sometimes occurred to spoil the effect: *nec ignotum quid Glyconi...acciderit. huic puer, quem is productum* (cf. *producere* here) *quid fleret interrogabat, a paedagogo se vellicari respondit.* The custom was common in Athenian courts also; cf. Aristoph. *Wasps*, 568 τὰ παιδάρι᾽ εὐθὺς ἀνέλκει· | ...τὰ δὲ συγκύψανθ᾽ ἅμα βληχᾶται· κἄπειθ᾽ ὁ πατὴρ ὑπὲρ αὐτῶν | ὥσπερ θεὸν ἀντιβολεῖ με τρέμων τῆς εὐθύνης ἀπολῦσαι.

147. 'All would resent eloquence in a Basilus'; for the constr., cf. II. 24 *quis tulerit Gracchos de seditione querentes?* Aorist and present are identical in meaning; see n. to l. 140.

148. It appears that the study of Roman law and literature was already considerable in Africa; Fronto and Apuleius, both names of great importance in Roman literature, belong to a rather later epoch.

150–214. *Those engaged in teaching are no better off than lawyers: to teach oratory is duller and no more lucrative than speaking in court. Pupils refuse to pay their fees and lay the blame for their own stupidity upon their teachers. Many a professor throws aside the stock subjects of sham debate in the schools, and engages for once in a real contest in the law-courts for the pittance unfairly denied him. A teacher of music is better paid than he. A rich man must have his splendid house, with its bath and covered walk; he must have his expensive cook; the payment to his son's instructors is the most trifling item in his expenditure. It is true that some men, Quintilian for instance, have made a fortune as professors; but that is exceptional, a mere freak of omnipotent fortune. For most men this way of life is a bitter disappointment, as recent examples show. In former times a teacher was feared and respected by his pupils; in our days they take the rod to him.*

The rest of the satire deals with the position of teachers, *grammatici* and *rhetores*. Though the instruction of the *grammaticus* came first, the *rhetor* is first dealt with here, the transition from the law-court to the school of rhetoric being an easy one. For the respective duties of *grammaticus* and *rhetor*, see nn. to l. 230, I. 15 and 16, VI. 450.

150. **declamare**, 'to speak in the schools', whereas *causas agere* is 'to speak in court'; cf. Mart. II. 7 *declamas belle, causas agis, Attice, belle*. It was common for adult Romans also to practise *declamatio*: see Introduction, p. xii.

ferrea, 'much-enduring'; cf. I. 31. Vettius is a *rhetor*, otherwise unknown.

151. One of the stock-exercises in the schools was to deliver invectives against tyrants or panegyrics upon tyrannicides. It might be supposed that the imperial government would not approve of this practice as a regular part of education; but the tyrant of the schools was too fantastic and unreal a creation to be taken seriously.

numerosa, 'crowded'; cf. X. 105: in classical Latin 'melodious': cf. Ovid, *Trist.* IV. 10. 49 (where he is speaking of the great poets of his youth) *tenuit nostras numerosus Horatius aures*.

152. The meaning of the l. cannot be positively fixed, for want of exact knowledge of the routine followed in such a class. **sedens** is probably masculine, 'a pupil' being understood as subject. Some of the less advanced students read their speeches without standing; others, more experienced, stand up to declaim; but all repeat precisely the same arguments, while 'killing their tyrant'.

153. **perferet**, 'will rehearse'; cf. VI. 392. **cantabit:** this word, like *decantare*, suggests trite repetition.

versibus, 'lines', of prose as well as poetry; cf. Pliny, *Epp.* III. 5. 12 (of his uncle) *memini quendam ex amicis, cum lector quaedam perperam pronuntiasset, revocasse et repeti coegisse, huic*

avunculum meum dixisse 'intellexeras nempe?' cum ille adnuisset,
'cur ergo revocabas? decem amplius versus hac tua interpellatione
perdidimus.'

154. 'Till, like hash'd cabbage served with each repast,
     The repetition kills the wretch at last.'   (Gifford)
The stale repetition is more fatal to the teacher than to the
tyrants. **crambe repetita** = ἑωλοκρασία: the Scholiast quotes a
proverb, δὶς κράμβη θάνατος.

155. **color**, 'line of defence'; n. to VI. 280. **causae genus:**
cf. Quint. III. 9. 6 *sed ante omnia intueri oportet, quod sit genus
causae.* **summa quaestio**, 'the cardinal point of a cause',
called also by Quint. *causae cardo* (V. 12. 3, XII. 8. 2).

156. **diversae**, 'from the opposite side'; cf. XIII. 136. Quint.
uses *ex parte adversa* (II. 4. 28), and *ex diverso* (V. 13. 1) in this
sense. **forte** seems weak but has better authority than *a parte*.
A refutation of the arguments which the adversary is likely to
use (*occupatio*), occurs commonly in ancient speeches: cf.
Quint. V. 13. 44 *ridiculum est...prius cogitare quid responderi
quam quid ex diverso dici possit.* For the metaphor of **sagittae,**
cf. *tela* quoted on l. 123.

157. Ovid also speaks of the difficulty teachers have in getting
their fees paid, *Fast.* III. 829 *vos, turba fere censu fraudata, magistri.*

158. 'Do you dun me for your fee?': *appellare* is regularly
used in this sense, but the constr. is usually a personal accus.:
cf. Mart. VII. 92. 3 *appellat rigida tristis me voce Secundus*; Juv.
IX. 64 *appellat puer unicus* 'my one slave duns me' for his wages.
Seneca uses the same constr., *Dial.* IX. 11. 3 *appellaverit natura
quae* (neut.) *prior nobis credidit.*

159. **laevae...mamillae:** as generally in Latin, the heart
is the seat, not of feeling, but of intellect.

160. **salit:** the word is often used of the beating of a pulse;
cf. Ovid, *Met.* X. 289 (of Pygmalion's statue coming to life)
saliunt temptatae pollice venae. Juv. means that the youth is
duri ingenii, dull, stolid. **Arcadico** may be translated 'bucolic',
the Arcadians being proverbial in Greece for ἀγροικία. So
Quintilian (II. 8. 7) says dryly of pupils in the school, *nonnulli
rus fortasse mittendi.*

mihi: the *rhetor* himself is supposed to be the speaker.

Once a week there is a formal exhibition in the school,
attended by anxious parents, when each pupil delivers a
suasoria or speech purporting to be made by some historical
personage at some important crisis of his career: see n. to I. 16.
Hannibal was commonly used for this purpose; cf. X. 167.

161. **dirus** is the common epithet of Hannibal, e.g. Hor. *Carm.* II. 12. 2 with Page's note; here it has a secondary sense, i.e. cursed by the *rhetor*.

162. **deliberat** is the technical word used in the *lemmata* of these *suasoriae*; cf. Quint. III. 8. 19 *deliberat C. Caesar, an perseveret in Germaniam ire, cum milites passim testamenta facerent.*

It must be noted that *an petat urbem . . ., an . . . circumagat . . . cohortes* does not refer to an alternative before H. at one time, but to two quite different occasions, the latter being in 211 B.C., five years after Cannae, when he offered battle under the walls of Rome: see n. to VI. 291; cf. Livy, XXVI. 11 *instructis utrimque exercitibus in eius pugnae casum, in qua urbs Roma victori praemium esset, imber ingens grandine mixtus utramque aciem turbavit . . . et postero die . . . acies instructas eadem tempestas diremit.*

164. **a,** 'after' and 'in consequence of'; to be taken after *madidas*, not after *circumagat*.

165. The *rhetor* speaks, saying he would give anything to oblige the boy's father to listen as often to the wearisome speech. The final clause (*ut . . . audiat*) is to be taken closely both with the imperative clause (*quantum . . . accipe*), and with *quid do*, which is added asyndetically. Transl. 'bargain for any sum you please, and I'll pay it on the spot—I'd pay anything for his father to hear him as often' as I do myself. **quid do ut** . . . is apparently a colloquial idiom: cf. III. 184 *quid das* (i.e. you give much) *ut Cossum . . . salutes?*

Some edd. keep *quod* (relative) *do* of the inferior MSS.; this, like many of their variants, is simpler, but weak. Merry (*Class. Rev.* IX, 29) proposes *quiddam* (included as a MS. reading in Jahn's *apparatus*); but surely *partem* must have been used in this sense.

167. **sophistae:** this name for *rhetores* is uncommon in Latin. Quintilian (XI. 3. 127) uses *antisophista* for 'a rival declaimer'; and Dio Cassius uses σοφιστής as the Greek equivalent of *rhetor* (e.g. LXVII. 12); but Lucian (*Merc. Cond.* 25) has γραμματικὸν ἢ ῥήτορα.

168. The professors engage in an actual lawsuit to recover their fees, abandoning the imaginary crimes and criminals of the schools, among whom the *raptor* (abducer of a maiden) was a common figure.

veras lites are opposed to the themes of *controversiae* in the schools; see n. to I. 16: for the complaint of their unreality, cf. Tac. *Dial.* 31 *fictis nec ullo modo ad veritatem accedentibus controversiis*; Petron. 1 *hoc tantum proficiunt, ut cum in forum* *venerint, putent se in alium orbem terrarum delatos. et ideo ego adulescentulos existimo in scholis stultissimos fieri, quia nihil ex his quae in usu habemus aut audiunt aut vident.*

169, 170. The reference is to the strange world of the *controversiae*—in which poison played a large part, as well as unlikely cures, and in which ingratitude was 'actionable'.

171. **rudem:** see n. to VI. 113.

172. **diversum,** 'different': see n. to III. 268.

173. **rhetorica...umbra,** 'the retirement of a professor,' i.e. learned leisure. In Petron. 2 a *rhetor* is called *umbraticus doctor*: the *lux* of the senate or a law-court is often opposed to the *umbra* of the schools.

174. **tessera:** *t. frumentariae* or *nummariae* were tokens given to all citizens at Rome whose names were upon the roll (*incisi*): see n. to X. 81: with these they could get a fixed quantity of corn, by repairing at a stated time to the *porticus Minucia*, a great hall with many entrances, where the corn was distributed. These tickets could be bequeathed by will, transferred, or (as here) sold: a person buying one was said *emere tribum*: indeed *tribus*, having ceased to have any political significance, is now commonly used in connection with poor-law relief: cf. Stat. *Silv.* III. 3. 99 *evolvit quantum Romana sub omni | pila die, quantumque tribus...poscant* (i.e. he reckons the amount required to pay the armies in every clime and provide the Roman corn-supply).

175. **quippe,** 'for': see n. to XIII. 26. **merces** is commonly used for the δίδακτρον paid to a *rhetor*; so l. 157.

tempta, 'try by experiment' M.; a strange word here. The constr. is to be noted: the imperative, followed by a verb in the fut. (*scindes*), is equivalent to the protasis of a conditional sentence: 'if you find out...then you will tear up...': cf. I. 155. So, in English, 'ask, and ye shall receive'. Cicero often uses the constr., but never, as silver-age writers often do, inserts *et* between the two clauses: cf. XVI. 29 and 31.

176. Chrysogonus (VI. 74) is a singer, Polio (*ibid.* 387) a player on the *cithara*: Juv. means that music-teachers get much higher fees.

177. **artem,** 'hand-book' of rhetoric: see n. to VI. 452. Theodorus, whose name is used as we might use Blair's, was a native of Gadara, a famous professor of rhetoric at Rhodes, and instructor of the emperor Tiberius. He founded a school, known as *Theodorei*, and left writings behind him (Quint. III. 1. 17 and 18). **scindes** is Jahn's certain correction of *scindens*.

178. **sescentis,** sc. *parantur*: 600,000 sesterces = £6,000. For **porticus,** see n. to IV. 5.

181. **hic** = *in porticu*, 'under cover'. Some edd. bracket the line, which is certainly weak, but not more so than many in Juv.

182. **parte alia:** the whole refers to one house, but the dining-hall (*cenatio*) is at some distance from the *porticus* and so disposed as to catch all the winter sunlight (*algentem solem*). A house of Pliny's, which he calls a *villula* (*Epp.* II. 17), had no less than three *cenationes*, several *porticus*, and a bath with the usual accessories.

Numidarum: 'Numidian' marble is that now known as *giallo antico*, a red-yellow stone; the entrance-hall of the National Gallery is adorned with fine slabs of this marble, brought from Tunis.

184. **domus,** sc. *constat*. A *structor* to arrange the table (see n. to V. 120) and a cook are indispensable.

185. **condit:** *condiat* was proposed by Lachmann, but this, though more suitable to the mood of *componat*, is metrically improbable, as much as *ludium* VI. 82. For *qui condit*, cf. *qui dispensat* l. 219, and *qui pingit* IX. 146.

186. **Quintiliano:** M. Fabius Quintilianus, about A.D. 35–95, was brought to Rome by Galba in 68 and appointed by Vespasian professor of Latin Rhetoric with a salary of 100,000 sesterces (£1,000). He had also a large practice at the Bar. He wrote the *Institutio Oratoria* in his retirement, before the death of Domitian in 96. His name is used here as the type of a *rhetor*; though Juv. goes on to point out that his good fortune was an exception.

187. **ut multum** = and a great deal too: cf. Mart. X. 11. 5 *donavi tamen, inquis, amico milia quinque | et lotam ut multum terque quaterque togam.*

188. **filius,** i.e. the education of his son. Cf. Pliny, *Epp.* IV. 13. 5 (where he tells how he urged his fellow-citizens of Comum to spend more on teachers for their children, and less on building, travelling, and foreign luxuries).

189. **novorum,** 'strange', 'unusual'.

190. **transi,** 'pass over', 'ignore': see n. to III. 114.

felix (= ὁ εὐδαίμων) is the subject, the other adjectives being predicates. Juv. heaps upon the lucky man, such as Quintilian was, all the qualities attributed by the Stoics to their *sapiens*. Quintilian (VI. *praef.*) estimates his own fortune very differently: §5 *ego vel hoc uno malo* (the loss of his wife) *sic eram afflictus ut*

me iam nulla fortuna posset efficere felicem: he is lamenting the
death of both his sons in boyhood.

192. 'He wears on his foot the crescent fastened to the black
leather'; i.e. he wears the *mulleus* or patrician shoe, which had
an ivory crescent sewed upon the instep: this *lunula* was per-
haps, like the *bulla*, an amulet. Even the children of *patricii*
wore this shoe: cf. Stat. *Silv.* v. 2. 27 *sic te, clare puer, genitum
sibi curia sensit | primaque patricia clausit vestigia luna.* As the
mulleus was of red leather, **nigra aluta** refers probably to the
straps (*corrigiae*), which were black; so *nigris pellibus* Hor.
Sat. I. 6. 27. The prep. in *subtexit* may be compared with ὑπό
in ὑποδεῖσθαι, ὑπολύεσθαι etc.

The l. does not mean 'the lucky man is a senator'; for the
calceus senatorius had apparently no *lunula*.

193. **iaculator** refers to the sports of the Campus, to suc-
cess in which the Romans attached much importance; cf. Mart.
VII. 72. 1 (to a lawyer) *gratus sic tibi, Paule, sit December, |
...seu, quod te potius iuvat capitque, | ...sic palmam tibi de
trigone nudo | unctae det favor arbiter coronae.* This particular
exercise is seldom mentioned at this date, but was popular
earlier: cf. Hor. *Carm.* I. 8. 12 *trans finem iaculo nobilis expedito*;
Ovid, *Ars*, III. 383 *sunt illis celeresque pilae iaculumque trochique |
armaque.*

194. **si** = even if.

195. **sidera:** Juv. frequently expresses the Stoical belief
that the fortunes of men are influenced by the aspect of the
planets at their birth; cf. IX. 33 *nam, si tibi sidera cessant, |nil
facies*; XVI. 4 and 5.

196. **vagitus, rubentem:** cf. Tolstoi, *La Guerre et la Paix*,
I, 363 *quelque chose de petit et de rouge vagissait dans les bras de
la sage-femme.*

197. Juv. is referring to an actual case, that of Valerius
Licinianus, a senator and distinguished orator, who had been
praetor. He was banished for intrigue with a Vestal virgin, and
started as a professor of rhetoric in Sicily under Trajan, be-
ginning his first lecture with this sentence, *quos tibi, Fortuna,
ludos facis! facis enim ex senatoribus professores, ex professoribus
senatores* (Pliny, *Epp.* IV. 11. 1).

Quintilian was, in a sense, *de rhetore consul*, as he received the
ornamenta consularia by favour of T. Flavius Clemens, husband
of Domitilla, Domitian's sister.

199. P. Ventidius Bassus, often quoted as an example of
astonishing vicissitudes, was led as a captive in the triumph of

Cn. Pompeius Strabo in the Social War 89 B.C., was consul in 43 B.C., and triumphed himself over the Parthians in 38.

For Servius Tullius, cf. VIII. 259.

202. corvo...albo: cf. *nigro cycno*, VI. 165. **quoque**, 'even'.

203 is a δέ clause, opposed to the previous l.; this has been made clearer by lightening the stop after *albo*. **sterilis cathedrae:** Mart. I. 76. 14 has the same phrase.

204. Lysimachus, according to the Schol., was an Athenian rhetor, who hanged himself. Secundus Carrinas was banished from Rome by Caligula; the text shows that he poisoned himself in poverty at Athens.

205. et hunc, 'Secundus also', as well as Lysimachus.

206. gelidas...cicutas: cf. Ovid, *Am.* III. 7. 13 *gelida... cicuta*: the effect produced by hemlock is made a quality of the plant itself.

207. terram: there is an ellipse of *dent*, as often in prayers and imprecations. The l. is a poetic expansion of the inscription common on graves, S. T. T. L., i.e. *sit tibi terra levis*.

208. spirantis, 'fragrant': generally with an accusative.

210. Achilles was taught to play the lyre by the Centaur Cheiron. **grandis** is a regular epithet of Achilles in this connection: e.g. Hor. *Epod.* 13. 11 *nobilis ut grandi cecinit Centaurus alumno*.

211. cantabat, 'had music-lessons'. **et cui,** i.e. 'so respectful that he...'; *et* connects the following adjectival clause with *metuens virgae*; for a similar constr., cf. V. 54, XIV. 52. *cŭi* is a pyrrhic, as in III. 49. **tunc,** 'in those days', as opposed to ours.

213. caedit, 'flogs', 'thrashes': i.e. the roles are reversed nowadays.

214. This Rufus, a *rhetor* whom his class maltreated, was nevertheless admired by them for his eloquence, so that they often called him, being presumably a Gaul, the Allobrogian Cicero.

For **quem** the inferior MSS. have *qui*, of which two explanations are given: (i) that Rufus criticised Cicero as a barbarous orator, (ii) that he accused Cicero of dealings with the Allobrogian envoys at the time of Catiline's conspiracy.

215–43. *The schoolmaster is treated even worse than the professor: he is worse paid, and his fees are docked before they reach him. But he must submit to this, in order to get some return for his* *hard work and early hours. Parents are unwilling to pay, and yet they require a tremendously high standard of knowledge from the teacher: he must have all history, all literature at his fingers' ends. The morals also of the boys are under his charge; and for all this, he receives, at the end of the year, as much as a successful actor.*

215. gremio, 'to the pocket': *sinus* is the common word. Celadus is an unknown *grammaticus*. Q. Remmius Palaemon was the most distinguished of his profession; cf. Suet. *de Gramm.* 23 *docuit Romae ac principem locum inter grammaticos tenuit.* We are told (*ibid.*) that he had an income of 400,000 sesterces (£4,000) from his school; but this was no doubt exceptional.

216. grammaticus = in teaching literature: see n. to l. 230.

217. aera = δίδακτρα, 'fee'.

218. acoenonoetus: the *paedagogus*, who nibbles off something from the schoolmaster's fees, is said *communi sensu carere*, 'to have no sympathy': see n. to VIII. 73. There is no reason to alter the text to *acoenonetus* (ἀκοινώνητος), 'refusing to go shares'; for the latter, see Pliny, *Epp.* III. 9. 8 with Mayor's note.

Mr Lendrum would translate 'without an idea in his head'; and ἀκοινονόητοι is used much in this sense by Cicero in a jest preserved by Gellius (XII. 12. 4).

219. qui dispensat, the *dispensator* or 'cashier', who pays money out on his master's account.

220. The schoolmaster must not stand on his dignity, but must let himself be beaten down like a small tradesman. **inde** = *a mercede.*

222. School began very early in Rome; Mart. more than once complains that his sleep is broken by the noise from schools: e.g. IX. 68 *quid tibi nobiscum est, ludi scelerate magister,* | *invisum pueris virginibusque caput?* | *nondum cristati rupere silentia galli,* | *murmure iam saevo verberibusque tonas.*

223. sedisti: the master sat, the pupils stood (*stabant* l. 226) round him. **sederet:** the subj. is consecutive.

224. obliquo...ferro, 'to card wool with slanting comb of steel'; no mechanic begins his work so early.

225. dummodo non: *non* is rare for *ne* after *dummodo* or *dum* 'provided that': but cf. Ovid, *ex Pont.* I. 1. 14 *dummodo non sit amor*; Stat. *Theb.* XI. 751 (Creon to Oedipus) *occursu dum non pia templa domosque* | *commacules*; Pliny, *Paneg.* 27 *non alat dum non occidat*; Quint. IX. 4. 58, X. 3. 7.

lucernas: each boy brought a lamp with him—which shows that they began early. Cf. Pattison's *Casaubon*, p. 95, Henri de Mesmes (at Toulouse in 1545) describes himself as going to school at 5 a.m., 'with our big books under our arms, our portfolios and lanterns in our hands'.

227. **Flaccus...Maroni:** the fate which Horace deprecated for his poems (*Epp.* I. 20. 17) soon overtook them: both he and Virgil were school-books as early as the first century. The teacher read aloud (*praelegebat*) and commented on (*enarrabat*) the author in hand; that the boys themselves held copies of the book is proved, I think, by Quint. II. 5. 4 *praelectio...in hoc adhibetur ut facile atque distincte pueri scripta oculis sequantur.* The books would soon be blackened by smoke from the boys' lamps (*fuligo*).

Friedl. thinks that busts of the poets, hung up in the schoolroom, and not books are meant. The Schol. is against him.

228. **cognitione tribuni,** 'an investigation of the magistrate': we might say 'a county-court summons'. The jurisdiction of the *tribuni plebis* in cases of this kind is not mentioned elsewhere.

229. **inponite** and *exigite* l. 237 are ironical; the meaning is much the same as if Juv. had said *inponitis* and *exigitis*.

230–6. All the learning required of a *grammaticus* is here mentioned: there is a precisely similar list given by Quintilian, I. 2. 14 *grammaticus...de loquendi ratione disserat, quaestiones explicet* (ll. 232–6), *historias exponat, poemata enarret.*

230. **verborum regula constet,** 'be faultless in his grammar'; *v. r.* = *lex loquendi* of VI. 453 where see n.; he must avoid all *barbarismi* and *soloecismi*; see n. to VI. 456. For *constet*, see n. to VI. 166.

231. **historias,** an extension of the proper province of a *grammaticus*: cf. Quint. I. 4. 2, I. 8. 18.

233. He must be able to answer offhand the most puzzling questions (*quaestiones explicare*). There were three *thermae*, great public bathing-establishments, in Rome at this time, those of Agrippa, Nero, and Titus. The *balnea Phoebi* would be a smaller bath, kept by a man of that name, like the baths of Lupus and Gryllus mentioned by Martial.

234. The Romans had a surprising interest in minute and useless knowledge of this kind: so the emperor Tiberius (Suet. *Tib.* 70) *grammaticos...eiusmodi fere quaestionibus experiebatur, quae mater Hecubae, quod Achilli nomen inter virgines fuisset, quid Sirenes cantare sint solitae.* For Quintilian's opinion of such

learning, see n. to VI. 451. Anchemolus and his step-mother occur in the *Aen.* X. 389; Acestes entertained Aeneas in Sicily and supplied him with wine, *Aen.* V. 73 ff. The point of the conundrums seems to be that they have no answers.

235. **annis:** for the case, see n. to VI. 183.

237. **mores,** sc. *puerorum.* **ducat** is a metaphor from working in wax, clay, or other plastic material: cf. R. Ascham, *Schole-master*: 'for the pure cleane witte of a sweete young babe is like the newest wax, most hable to receive the best and fayrest printing'.

238. **cera:** portrait-busts were often made of wax; cf. Mart. VII. 44. 2 *cuius adhuc vultum vivida cera tenet.* They were the ancient substitute for photographs: so Ovid. *Remed.* 723 (advising a lover how to conquer his passion) *si potes, et ceras remove,* 'if you have the heart, hide away her portraits too'.

239. **ipsius** is apparently used here by hypallage for *ipse*: cf. VIII. 138 where *ipsorum* is used, though much less harshly, for *ipsa.* It cannot = *totius.* **ne turpia ludant,** 'to prevent indecent tricks'.

242. **inquit,** 'says he', i.e. any parent; cf. III. 153.

243. **victori:** it seems impossible that this can refer to a successful chariot-driver in the Circus, as l. 114 and Mart. there quoted show that he received much larger sums than the *grammaticus* could ever hope for. The Schol. explains it of actors in the theatre, who were not allowed to receive a reward of more than five *aurei.* But it still remains uncertain in what sense an actor could be *victor* in a Roman theatre, where there were not competitions between plays.

SATIRE VIII

ON NOBLE BIRTH

1–38. *There is no advantage in a long line of distinguished an-cestors, if their descendant and representative is vicious and effeminate, and a disgrace to his forefathers. Virtue is the only true nobility, and blue blood must come second to honour and goodness. Weakness and wickedness are only more conspicuous when attached to a great and famous name.*

1. **stemmata,** 'family-trees'. In a Roman noble's house the *atrium* was adorned by *imagines,* wax portrait-masks of deceased ancestors, each preserved in a separate cabinet (*armarium*) with

an inscription (*titulus*) below. The *ius imaginum* was restricted
to those who had held curule office. Whether these masks were
placed upon busts, is uncertain; if so, they could be detached,
as they were worn by actors in the funeral procession of any
member of the family. Cf. Pliny, *Nat. Hist.* XXXV. 6 *apud
maiores in atriis…expressi cera voltus singulis disponebantur
armariis, ut essent imagines quae comitarentur gentilicia funera,
semperque defuncto aliquo totus aderat familiae eius qui unquam
fuerat populus. stemmata vero lineis discurrebant ad* (to cor-
respond with) *imagines pictas.* The last sentence indicates that
the *imagines* were distinct from the *stemma*, the latter being a
genealogical chart (cf. *generis tabula*, l. 6), which was so called
because the names on it were illuminated with painted garlands.
The same distinction between *imagines* and *stemma* appears in
Sen. *de Ben.* III. 28. 2 *qui imagines in atrio exponunt et nomina
familiae suae…multis stemmatum inligata flexuris in parte prima
aedium collocant.* (In distinguishing *imagines* from *stemma*, and
in the quotations to prove the distinction, I follow Mr Lendrum
in *Hermathena*, VI. 360.)

quid faciunt? 'are useless', 'absurd': the use of *facere*
is idiomatic: cf. Mart. VII. 64. 6 *quid facit infelix et fugitiva
quies?*

Ponticus is again addressed ll. 75 and 179; the name may be
invented that the satire may be addressed to some one in
particular: see n. to Postumus, VI. 28.

2. censeri, 'to be appraised by', i.e. to be valued for; again
l. 74; a common silver-age idiom; cf. Tac. *Dial.* 39 *itaque eius
modi libri extant, ut ipsi quoque qui egerunt non aliis magis ora-
tionibus censeantur*; Pliny, *Paneg.* 75 *quisquis paulo vetustior
miles, hic te commilitone censetur* (i.e. the most remarkable thing
about him is, that he has served with Trajan).

picti vultus m. are the *imagines* above spoken of.

3. in curribus, i.e. a triumphal statue; cf. VII. 125.

Aemilianos: the termination shows that a member of the
gens Aemilia had been adopted into another *gens*, as Octavius
became Octavianus, when adopted into the *gens Iulia*. A son of
Aemilius Paulus, when adopted by P. Cornelius Scipio, added
Aemilianus to the name of his adopting father; in after years he
was the conqueror of Carthage, the second of the *duo fulmina
belli*.

4. iam dimidios, 'mutilated by time': cf. XV. 5. **humeros,**
lit. 'as to the shoulders'; 'minus the shoulders', we might say.

5. The emperor Galba was of a very ancient family, whose antiquity he exaggerated so far, *ut imperator stemma in atrio proposuerit, quo paternam originem ad Iovem, maternam ad Pasiphaam Minois uxorem referret* (Suet. *Galb.* 2). If, as Marquardt supposes, the *stemma* was identical with the *imagines*, are we to believe that he had a wax mask of Jupiter on his wall? The name of the god on a family-tree would be much more permissible.

6. **generis tabula** = *stemmate*, a sort of key to the *imagines* to which Juv. is returning after the mention of statues.

7. The repetition of **Corvinum** and the weakness of **posthac** have led some critics to suspect interpolation; for *posthac*, Withof ingeniously suggested *posse ac*. In some MSS. the l. is omitted, in others *Fabricium* is found for *Corvinum*. But Juv. may well have written the l. as P gives it, with all its diffuseness.

multa...virga, 'to claim kinship through many a branch with...': *virgae* are the *lineae* which connect one part of the *stemma* with another: so *ramus* Pers. III. 28. For this sense of **contingere,** cf. XI. 62 and Suet. *Galb.* 2 *Neroni Galba successit nullo gradu contingens Caesarum domum.*

8. **fumosos:** cf. Mart. VIII. 6. 3 *argenti fumosa...stemmata;* the *imagines*, being kept in the *atrium*, 'the blackened room', naturally got blackened with smoke in course of time.

9. **quo,** 'what is the use of...?' **effigies** is probably accus., governed by some verb understood: cf. l. 142, XIV. 135 and Hor. *Epp.* I. 5. 12 *quo mihi fortunam* with Wilkins' note.

11. **Numantinos:** this cognomen was conferred on Scipio (see n. to l. 3) after his conquest of Numantia 134 B.C.: the Numantini are present in the form of masks or statues.

12. **duces,** 'they as generals', i.e. Scipio and others.

13. The *gens Fabia* was traditionally descended from Hercules and Vinduna, daughter of Evander; one of the chief glories of the family was the conquest of the Allobroges by Q. Fabius Maximus 121 B.C., who afterwards assumed the *cognomen* of Allobrogicus. His son, who was conspicuous only for his vices, is probably alluded to in what follows.

magna...ara: the *ara maxima Herculis*, a religious monument of great antiquity situated near the Circus and the Tiber, was naturally of special interest to his reputed descendants.

15. **Euganea** is merely an ornamental epithet: the flocks of Altinum in that district were celebrated, but no softer than flocks elsewhere.

16. **Catinensi:** Catina or Catana, near Etna in Sicily, supplied the pumice-stone which was used by the effeminate to

remove all hair from the body and limbs: within limits this practice was permissible, and even expected: cf. Sen. *Epp.* 114. 14 *alter se plus iusto colit, alter plus iusto neglegit; ille et crura, hic ne alas quidem vellit.*

17. **squalentis** = *hirsutos.*

traducit, 'caricatures': this common silver-age use of *traducere*, 'to make an exhibition of', 'to parody', is perhaps derived from the custom of marching prisoners in mockery through the streets of Rome at a triumph: comp. Livy, XXXVI. 40. 11 *cum captivis nobilibus equorum quoque captorum gregem traduxit*, with idem, II. 38. 3 *vestras coniuges, vestros liberos traductos per ora hominum?* The latter, metaphorical, sense occurs first in Livy and becomes the commoner in silver-age Latin; cf. II. 159 *heu! miseri traducimur*; XI. 31; Mart. VI. 77. 5 *rideris, multoque magis traduceris, Afer, | quam nudus medio si spatiere foro.* The noun *traductio* is used in the same sense by Seneca, *Dial.* III. 6. 4 *damnatos cum dedecore et traductione vita exigit.*

18. The **imago** (see n. to l. 1) of a criminal would not be suffered to stand with those of his ancestors, but be destroyed: cf. Tac. *Ann.* II. 32. 2 *tunc Cotta, ne imago Libonis exsequias posterorum comitaretur, censuit.* By a somewhat similar custom, *praenomina* were abandoned in certain *gentes* because of the crimes of those who had borne them: cf. Suet. *Tib.* 1 *gens Claudia...Luci praenomen consensu repudiavit, postquam e duobus gentilibus praeditis eo alter latrocinii, caedis alter convictus est.*

19. **cerae:** see n. to l. 1.

20. **virtus** is subject, **nobilitas** is predicate: so Tennyson ''tis only noble to be good'.

21. L. Aemilius Paulus, the conqueror of Perseus, and Drusus, brother of Tiberius, were both famous for high character as well as military successes. Which of the Cornelii Cossi is here meant, is uncertain.

22. **hos,** i.e. *mores*: as of more importance, they must have precedence even of the *imagines.*

23. **illi** is not contrasted with *hos* above, but refers to the same noun *mores*. **virgas:** see n. to l. 136: the lictors went in front of the magistrate: cf. Cic. *in Verr.* II. 5. 22 *quaeret quamobrem fasces praetoribus praeferantur.*

24. **prima,** 'in the first place'.

animi bona, 'virtues', is a phrase derived from philosophical terminology, where 'goods' are divided into different classes:

cf. Sen. *de Ben.* v. 13. 1 *sunt animi bona, sunt corporis, sunt for-tunae.*

sanctus: '*sanctus* is stainless in all relations of life, impervious to any degrading influence whatever' (Nettleship).

26. **agnosco procerem,** 'then I recognise you as a noble'; this noun is rare in the singular. The Iunii Silani and Cornelii Gaetulici, both noble families, were connected by adoption. Juv. means, 'if you are virtuous, then your noble birth is allowed full value'.

27. **Silanus,** sc. *es.*

29. **Osiri invento:** cf. the Schol. *populus Aegypti invento Osiri dicit,* εὑρήκαμεν, συνχαίρομεν: the phrase is quoted by Seneca, *Apocol.* 13. 4 (of Claudius' victims when they meet him in Tartarus) *cum plausu procedunt cantantes,* εὑρήκαμεν, συγχαίρ-ωμεν. For the worship of Egyptian deities at Rome, see nn. to VI. 532-4.

30. **dixerit** = ἂν εἴποι: it is aor. optative in a conditional clause. The omission of the verb (*est*) in the relative clause is unusual.

32. A new point: a dwarf is sometimes called 'Goliath' in mockery, or a negro 'Snowball'; it would not be pleasant to bear a great name, implying great qualities, on similar terms.

nanum: the Romans had a perverted fancy for pets of this kind, dwarfs (*nani, pumili*) and idiots (*moriones, fatui*); that Augustus did not share this taste, was thought remarkable; cf. Suet. *Aug.* 83 *pumilos atque distortos et omnis generis eiusdem ut ludibria naturae malique ominis abhorrebat.* Atlas is the Titan who supports the sky on his shoulders.

33. **pravam,** 'crooked', often used of physical defect; cf. Hor. *Sat.* I. 3. 48 *pravis fultum male talis.* **extortam** is rare in the sense of *distortam,* as in Suet. quoted above.

34. **Europa** is typical of beauty and stature.

35. **levibus:** they have lost their hair from mange, and, unable to crawl about, try to extract some oil from a dry lamp; yet we continue to call them, as if in mockery, by the names of the fiercest wild beasts, which have ceased to be appropriate.

36. **adhuc** goes with *quid* and = *aliud*: in silver-age Latin *adhuc* means 'in addition': for Cicero's use, see n. to VI. 502.

37. **fremat,** 'roars': used esp. of lions: see n. to XIV. 247.

38. **sic,** 'in the same way', i.e. in an ironical sense: a necessary correction for the *sis* (or *si*) of MSS., which need not be expressed.

Creticus, a cognomen in the family of the Caecilii Metelli, gained after the conquest of Crete 62 B.C. The Camerini were a distinguished family of the *gens Sulpicia*. They stand here for typical aristocrats.

39–70. *This advice is addressed to Rubellius Blandus, who prides himself upon his lineage and despises the humbly born. And yet we see that the world's work, in peace and war, is generally done by men of no ancestry, while the aristocrat is of no practical use at all. In the case of animals blood alone goes for little: the stock of the most illustrious sires, if they lose their races, come to ignoble uses. So the aristocrat, if we are to admire him, must not depend entirely on his ancestors.*

39. Rubellius Blandus is taken as a type of noble birth and nothing more. A man of this name was married in A.D. 33 to Julia, grand-daughter of Tiberius; they had a son, Rubellius Plautus, who was murdered by Nero's orders in A.D. 62: cf. Tac. *Ann.* XIII. 19, XIV. 22, 57 ff. Juv. must mean one of this family, possibly a son or brother of Rubellius Plautus.

40. **tumes** cet.: cf. Tac. *Hist.* I. 16 (*Neronem*) *longa Caesarum serie tumentem*: Nero and R. Plautus were related to Augustus in exactly the same degree, Drusus, son of Tiberius, being grandfather of Plautus.

41. **feceris ipse:** cf. Ovid, *Met.* XIII. 140 *genus et proavos et quae non fecimus ipsi,* | *vix ea nostra voco.*

42. If **conciperet** is used strictly, the person addressed must be a brother of Plautus.

fulget: cf. Sen. *Medea,* 209 *quondam nobili fulsi patre* | *avoque clarum sole deduxi genus.*

Iuli: the *gens Iulia* claimed descent from Iulus, son of Aeneas.

43. 'And not a woman who weaves for hire beside the wind-swept embankment.' The **agger** was a defensive earthwork built by Servius Tullius to protect Rome on the east; it extended from the Esquiline to the Colline Gate, flanking the Quirinal and Viminal Hills. The citizens found it an agreeable and breezy (*ventoso*) place for walks: cf. Hor. *Sat.* I. 8. 15 *licet... aggere in aprico spatiari*; fortune-tellers frequently established themselves there (cf. VI. 588); and performing animals might be seen (cf. V. 153). In this passage, the probable meaning is not that women wove in the open air, but that they lived in houses abutting on the embankment, and worked for their living, either in their own houses, or as 'hands' in a *textrinum*.

46. **Cecropides:** Cecrops was the mythical founder of

monarchy at Athens, and there is also an allusion to the boast of the Athenians that they were αὐτόχθονες, indigenous to the soil of Attica.

vivas...feras, 'long life to you! may you long enjoy the happiness of such a descent!' For this use of *longa*, see n. to *longo* VI. 561.

47. **Quiritem:** the sing. of this noun is uncommon: but cf. Ovid, *Am.* I. 7. 29; *ibid*, III. 14. 9; *Trist.* II. 569.

The three professions which follow are often coupled as paths to advancement: cf. XIV. 191 ff.; Livy, XXXIX. 40. 5 *ad summos honores alios scientia iuris, alios eloquentia, alios gloria militaris provexit.*

49. **nobilis** is used as a noun; cf. *dives avarus*, VII. 30. **togata,** 'civilian': in contrast with the soldier below.

50. The noble requires not only a barrister to speak for him but a *iuris consultus* to explain the knotty points (*nodos*) of law which arise in his case; the common people will supply him with both.

51. **hic,** 'another' plebeian: Weidn.'s suggestion, *hinc* (= *a plebe*), is ingenious. Plebeians serve their country as soldiers, on the east and west frontiers of the Empire, the Euphrates and the Rhine. **iuvenis,** 'as a soldier'; cf. the common use of *iuventus = exercitus.*

52. **aquilas:** the eagles stand for the legions themselves.

53. **trunco,** 'limbless', not 'mutilated'.

Hermae: cf. Grote's *History of Greece*, chap. 58: 'the Hermae, or half-statues of the god Hermes, were blocks of marble about the height of the human figure. The upper part was cut into a head, face, neck, and bust; the lower part was left as a quadrangular pillar, broad at the base, without arms, body, or legs. They were distributed in great numbers throughout Athens.'

54. **quippe,** 'for': see n. to XIII. 26. **vincis,** 'you are superior'.

55. **tua...imago,** 'while you are a living statue'; though alive, which the marble is not, you are just as useless.

56. **Teucrorum proles** = *Troiugena* of I. 100 where see n. **muta,** i.e. without articulate speech: sometimes a noun, e.g. XV. 143.

58. **sic** = *quod fortis est.*

facili...circo, 'to whom, an easy winner, falls many a hotly-contested palm, and for whom the shout of victory swells in the roaring Circus'. **facili** is dat. agreeing with *cui*; *plurima* may be taken with *victoria* as well as with *palma*. For the noise in the Circus, cf. XI. 197.

61. **fuga**, 'speed'; cf. Virg. *Aen.* I. 317 *volucremque fuga praevertitur Hebrum*.

pulvis is the cloud of dust raised on the course (*aequor*) by the horse as he runs: cf. Mart. VI. 38. 7 *acris equi suboles magno sic pulvere gaudet* (where perhaps *magno* is a gloss, the right reading being *circi* which had *magni* written over it; then *magni* ousted *circi* from the text and was made to agree with *pulvere*).

62. **venale**, sc. *est*, 'come to the hammer'. **pecus** is contemptuous. Coryphaeus and Hirpinus are the names of famous race-horses; cf. Mart. III. 63. 12 *Hirpini veteres qui bene novit avos*.

63. **rara**, 'seldom': the adj. is much oftener used in Latin than the adv.: so also *frequens*. **iugo:** the horses in the Circus-races were always driven, not ridden.

64. **ibi**, 'in their case', although a worthless noble may be respected. M. explains, 'in the Circus': but cf. XI. 176.

66. **epiraedia:** the word is mentioned by Quintilian (I. 5. 68) as a hybrid, *epi* being Greek and *raeda* Celtic, though the whole compound is used as a Latin word: the meaning is usually given as 'traces'; but *ducunt* seems to suggest some kind of carriage or cart.

67. **nepotes**, = *posteritas*, is subject to *iubentur* and *ducunt*: old edd. generally read *Nepotis* which most MSS. give, but the reading of P is certainly right: M. quotes an inscription in which a horse (Hirpinus: cf. Mart. quoted above) is called *nepos Aquilonis*. The *pistrinum* was often turned by asses, sometimes, as a punishment, by slaves.

68. **privum aliquid da**, 'instance something of your own'; cf. Ovid quoted on l. 41. *privum* is Salmasius' certain correction of *primum* which all MSS. give. The same corruption is found in the MSS. of Lucr. III. 372, where Bentley restored *privis* for *primis*.

69. **titulis**, 'the inscription' on your statue or bust; see n. to I. 130.

70. **illis**, i.e. *maioribus iuis*.

71–145. *Enough said to a selfish aristocrat like Rubellius Blandus: you, Ponticus, will not, I hope, be satisfied to depend on the distinctions of your ancestors. Fulfil all the duties of a good citizen, and tell the truth at any cost to yourself. Some day you will become governor of a province; there is temptation there, but*

*I hope you will pity the poor provincials and restrain your
avarice. It is true that earlier depredations have stripped the
provinces almost bare: the great harvests of a Verres have left
only slender gleanings for later rulers. And remember not to
provoke the wretched too far; what may be done safely in the
East may provoke armed resistance in the West. What I say is
said in earnest: if you are a good governor and insist on good
conduct in your subordinates, then you may make unchallenged
any pretensions you please to noble birth; but if you are cruel
and rapacious, nobility only makes your wickedness more con-
spicuous. If you are a forger and a profligate, I think nothing of
the distinctions of your ancestors.*

It must be admitted that this long digression about provin-
cial misgovernment has little connection with the main subject
of the satire. There is good evidence that the provinces were
better treated under the Empire than under the Republic; see
Furneaux's Tacitus *Ann.* Introduction, p. 101; Hardy's
Pliny's Correspondence with Trajan, pp. 29–49. In spite of this,
scandals occurred; and Juv., as a satirist, makes the most of
them.

71. The language (**fama tradit**) seems to show, what we
should expect, that the youth addressed is no longer living. The
description does not agree with what Tacitus says of Rubellius
Plautus: cf. esp. *Ann.* XIV. 22. 3 and 57. 5.

72. **Nerone propinquo**, 'his kinship to Nero': see n. to l. 40
and Tac. *Ann.* XIII. 19. 3 *Rubellium Plautum, per maternam
originem pari ac Nero gradu a divo Augusto.*

73. **ferme**, 'as a rule': cf. XIII. 236.

sensus communis, 'feeling for others', the recognition of
the bond between all men, which is expressed in Terence's line,
homo sum; humani nihil a me alienum puto. Not unlike is the
sense of *civilitas*: see n. to v. 112. The meaning is slightly
different in Hor. *Sat.* I. 3. 66 *communi sensu plane caret*, where
'ordinary tact' (Palmer) gives the sense. A third meaning is
found in Lucr. I. 422 *corpus enim per se communis dedicat esse |
sensus*, where it is a transl. of αἴσθησις πᾶσι κοινή,, 'perception
which all men share'. Lastly there are passages where *communis
sensus* represents our 'commonsense', something opposed to a
loftier but less practical wisdom, e.g. Sen. *de Ben.* I. 12. 3 *nemo
tam stultus est ut monendus sit ne cui...mittat vestimenta
aestivā brumā, hiberna solstitio: sit in beneficio sensus communis:
tempus, locum observet.* See n. to VII. 218.

74. **fortuna,** 'rank'; cf. XI. 176; Sen. *Dial.* IV. 21. 7 *nonne vides ut maiorem quamque fortunam maior ira comitetur?*

76. **laudis** is governed by *nihil*, and denotes the good action itself as well as the praise won by it.

77. **ne...ruant...tecta** may be transl. 'or the roof may fall in....'.

78 is very abrupt as a separate statement. I therefore follow Beer (*Spicilegium*, p. 73) in changing the stop after *columnis* to a comma, and reading *desideret* for *desiderat*: a variant, *discinderet*, in the margin of P, gives some confirmation to this reading; and the passage gains decidedly by the co-ordination of the two similes.

(Mr Lendrum would keep the text, explaining the abruptness by the fact that l. 78 is proverbial; he compares Aristoph. *Wasps*, 1291 εἶτα νῦν ἐξηπάτησεν ἡ χάραξ τὴν ἄμπελον.)

viduas...ulmos: see n. to VI. 150: this metaphor is constant in Latin: the elm is here called *vidua* because the vine has fallen from it, just as the plane is called *vidua* by Martial (III. 58. 3) and *caelebs* by Horace (*Carm.* II. 15. 4), because it was never used to train the vine upon: for *vidua* means both 'maid' and 'widow'; cf. IV. 4.

79. **estŏ:** but *estō* l. 164 where the *o* is in arsis.

81. **rei,** here, and often in Plautus, an iambus, is in Lucretius always a spondee or one long syllable: see Munro on Lucr. I. 688: the form occurs only here in Juv. and is very rare in verse after Lucr.: *fidĕi* occurs XIII. 6.

Phalaris is the tyrant of Agrigentum, famous for the brazen bull in which he roasted his victims alive; cf. VI. 486.

82. **admoto:** this verb is regularly used of applying means of compulsion: cf. Hor. *Carm.* III. 21. 13 *tormentum ingenio admoves*; Livy, XXVII. 43. 3 *metus tormentorum admotus.*

83. **pudori,** 'honour'; again XVI. 34, but a rare sense of the word: M. quotes Pliny, *Epp.* II. 4. 2 *debes famam defuncti pudoremque suscipere*, where the two words seem almost synonymous.

84. **vivendi causas,** 'what makes life worth living'; cf. XI. 11 and Pliny, *Epp.* I. 12. 3 *plurimas vivendi causas habentem, optimam conscientiam, optimam famam, maximam auctoritatem.*

85. **perit,** 'is dead already'; perfect, not present.

86. **Gaurana** = *Lucrina*: the *mons Gaurus* overhung the Lucrine lake where the best oysters were bred: see n. to IV. 140.

Cosmus was a famous perfumer, often mentioned by Martial.

toto mergatur, 'immerse his whole body': *toto*, by a

common idiom, goes rather with the subject of the sentence than with the word with which it is grammatically connected; cf. Mart. IV. 22. 4 (of a woman in a bath) *lucebat, totis cum tegeretur aquis*, where *totis* should be translated as *tota*. There is no adverb formed from *totus*, which partly accounts for the idiom.

88. **accipiet:** see n. to x. 26: P has *accipiat*, but the subj. seems impossible here, and the reading of the worse MSS. is to be preferred: in l. 91 P has *mandat* where *mandet* must be right.

89. **sociorum:** again ll. 99 and 136: this name is often somewhat loosely applied to the inhabitants of subject provinces, who are properly *stipendiarii*.

90. 'You see mere empty bones, with the marrow sucked out': the robbers of the provinces, not content with stripping all the flesh off the bones, have sucked the marrow out as well.

ossa rerum, lit. 'bones of power': *rerum* is to be compared with the common *res Romana* 'the Roman state', *rerum potiri* etc. *regum*, 'of the subject kings', is an easier reading but has poor authority. **vacuis exucta medullis:** the meaning would be more accurately expressed by *vacua exuctis medullis*: cf. [Sen.] *Herc. Oet.* 1230 *malum | hausit medullas; ossibus vacuis sedet.*

92. **fulmine,** 'thunderstroke': the word is used by Ovid of his banishment by Augustus, *Trist.* I. 1. 72 *venit in hoc illa fulmen ab arce caput*; by Mart. of a condemnation by Domitian, VI. 83. 3 *nam tu missa tua revocasti fulmina dextra*; and by Statius (of the same case) *Silv.* III. 3. 158 *venturi fulminis ictus*: it is used of bereavement in Livy, XLV. 41. 1 *quae duo fulmina domum meam...perculerint, non ignorare vos arbitror*, and by Stat. *Silv.* II. 1. 30.

93. **Capito:** cf. Tac. *Ann.* XIII. 33: Cossutianus Capito, legate of Cilicia, was accused by the province before the senate A.D. 57, and condemned for extortion; he was however restored to his place in the senate, four or five years afterwards, by the influence of his father-in-law Tigellinus (*Ann.* XIV. 48. 2).

Of **Numitor**, perhaps not a governor but a *procurator*, nothing else is known.

94. **piratae Cilicum** = *p. piratarum*, Cilicia being the head-quarters of the nest of pirates whom Pompey rooted out in 67 B.C.: cf. Lucan, III. 228 (of the Cilicians in Pompey's host) *itque Cilix iusta, non iam pirata, carina.*

The robber is condemned, but the impoverished provinces are no richer for their success; cf. I. 50. The next 30 lines are decidedly irrelevant.

95. Chaerippus must be a Cilician who took part in accusing the extortionate governor; he has nothing but rags left, and must get an auctioneer (*praeco*) to sell these, that he may get bread. **pannis** is dat. after *praeconem*.

96. The names Pansa and Natta seem not to refer to actual occurrences but to stand for 'Governor A' and 'Governor B'. Condemnation has not even a deterrent influence.

97. **naulum** is either 'your passage-money' to Rome, where the provincials must go, to plead before the senate, or 'Charon's penny', the coin placed in the mouth of the dead: see n. to III. 267. The latter meaning has more point; but no instance is given of ναῦλος alone with this meaning.

100. Cf. Cic. *in Verr.* II. 4. 46 (of Sicily before Verres' government) *domus erat ante istum praetorem nulla paullo locupletior, qua in domo haec non essent...patella grandis cum sigillis et simulacris deorum, patera...turibulum.*

101. **Spartana...Coa:** the purples dyed at Sparta and at Cos were famous; cf. Hor. *Carm.* II. 18. 7 *Laconicas...purpuras; ibid.* IV. 13. 13 *Coae...purpurae.* For *Coae vestes*, see n. to VI. 260.

102. Parrhasius was one of the most famous of Greek painters, and flourished about 400 B.C.; the other artists here mentioned were the most famous Greek workers in marble and metal during the 3rd and 4th cent. B.C. Their names are often mentioned together: cf. Mart. VIII. 51. 1 *quis labor in phiala, docti Myos anne Myronos?* | *Mentoris haec manus est an, Polyclite, tua?*

103. **vivebat:** the statues are so wrought that they seem alive. **ebur:** the chief works of Pheidias, his statues of Zeus at Olympia and of Athena at Athens, were of ivory and gold.

104. **labor,** 'workmanship', as in Mart. quoted above. Mentor was especially famous as an engraver of cups: Juv. says that few side-boards (*mensae*) in the provinces lacked a specimen of his art.

105. The l. as it stands in the MSS. is unmetrical, wanting a syllable; hence *istinc* is read for *hinc* by some edd. The adverbs are identical in meaning, each being = *a divitiis provinciarum*: so *illi* and *hos* l. 23.

Cn. Cornelius Dolabella, governor of Cilicia, was accused in 78 B.C. by M. Scaurus of extortionate conduct, and condemned. Another Dolabella was impeached for similar conduct in Macedonia by the young Julius Caesar in 77 B.C. Hence some edd. read *Dolabellae*, which the l. might scan by hiatus: see n. to X. 281.

Antonius: this is the uncle of Cicero's enemy, a monster of rapacity: he was prosecuted, also by Caesar, in 76 B.C. for plundering Greece and exiled in 59 B.C. for plundering Macedonia, where he was governor 62 B.C.

106. **Verres:** see n. to III. 53: as propraetor of Sicily, 73–70 B.C., he robbed the inhabitants of their wealth and the temples of their statues (hence *sacrilegus*), on such an immense scale, that his name became typical of such crimes; cf. II. 25 *quis caelum terris non misceat et mare caelo,* | *si fur displiceat Verri?*

navibus altis | **occulta,** 'smuggled in the hold of their ships'; *altis* here suggests depth, not height. The last syllable of *occulta* is lengthened before *sp* following, a licence almost unknown to the Augustan poets: Martial, however, lengthens a short vowel before *sp* (II. 66. 8), before *st* (V. 69. 3), and before *pr* (*Epig. Lib.* 28. 10). Horace sometimes has a short vowel before *st* and *sc*, but only in the Satires: cf. *Sat.* I. 10. 72 with Palmer's note.

107. **de pace:** *pace = pacatis gentibus, triumphare de aliquo* being the usual constr.; in the triumph over a conquered people, the spoils were displayed in the procession; these robbers in time of peace celebrate more triumphs than successful generals, in the sense that they bring back more booty; cf. Cic. *in Verr.* II. 4. 124 *honestius est reipublicae nostrae ea quae illis* (i.e. *Graecis*) *pulchra esse videantur, imperatorem nostrum* (Marcellus) *in bello reliquisse quam praetorem* (Verres) *in pace abstulisse.* Cf. Pelham's *History of Rome*, p. 169: 'the spoils of peace were richer than those of war, and were more easily won'.

108. **nunc** is opposed to **tunc** l. 100, '*but* now'.

111. **aedicula,** 'a little shrine' in a private house, not 'a temple': M. quotes Petron. 29 *grande armarium in angulo vidi, in cuius aedicula erant Lares argentei positi Venerisque signum marmoreum.* In many Pompeian houses the *Lararium* is found in the form of a little shrine in the *atrium.*

111, 112. The awkwardness of expression is remarkable. It has been proposed to omit both lines: Büch.'s *iam* for *nam* is attractive, getting rid of the repeated 'for'.

The meaning is: these poor pickings are the most you can get now, and represent all the wealth there once was.

113. **unctam,** 'scented'; the epithet suggests luxury and softness; cf. *madidum Tarentum*, VI. 297.

114. **despicias** is conditional, 'you would be right in despising them'.

resinata: *resina* is the gum of fir-trees, much used in antiquity for removing hair from the body and limbs; see n. to l. 16.

116. **horrida...Hispania:** cf. Mart. x. 65. 6 (where he contrasts himself, a rugged Spaniard, with an effeminate Corinthian) *tu flexa nitidus coma vagaris, | Hispanis ego contumax capillis; | levis dropace tu cotidiano, | hirsutis ego cruribus genisque.*

117. **latus,** 'shore', 'coast': Illyricum is the north-eastern coast of the Adriatic, one of the chief recruiting-grounds of the imperial armies.

messoribus = *Afris:* cf. v. 118: of the regular Roman corn-supply, Africa sent two-thirds, Egypt the rest.

118. 'By whom the citizens are fed, and need think of nothing but the races and the theatre': Juv. elsewhere (x. 81) says that the populace of Rome has only two objects of desire, bread and the races; here he says that they need not trouble themselves about the first of these.

saturant = *saturam faciunt:* this verb has generally an abl. of the thing as well as an accus. of the person: but cf. XIV. 166; Lucan, VIII. 506 *senatus, | cuius Thessalicas saturat pars magna volucres.* For **circo,** see n. to III. 223.

vacantem = *ut vacare possit. vacare alicui rei* is 'to have leisure for something', and then, by a slight extension of meaning, 'to give leisure to a thing, to devote yourself exclusively to it'; hence the French *vaquer à;* the latter is the meaning here: both meanings are seen in Mart. XI. 1. 6 *nec Musis vacat, aut suis vacaret,* 'he has no time for poetry: if he had, he would *spend it* in writing poetry himself'.

119. **dirae,** 'monstrous', because it causes starvation at Rome.

120. For the depredations of Marius in Africa, see n. to I. 49. **tenuis,** 'needy' (so III. 163), may be *proleptic,* expressing the result of his treatment. **discinxerit** = *spoliaverit,* as it was customary to carry money in the belt.

122–4. The pleonasm is remarkable; hence Lachmann was for striking out l. 124; this would be an improvement, but we have seen how frequent repetitions are in Juv.; and it is quite in his manner to repeat in an epigrammatic form exactly what he has just said: cf. XV. 110–12.

125. **quod modo proposui** refers back to ll. 87–94. **sententia,** 'a thing said for effect': the word originally represents γνώμη, a maxim or general statement dealing with human life or

action; but in the rhetorical schools the meaning was considerably modified, and Quintilian defines *sententiae* as *lumina...
praecipueque in clausulis posita*, 'striking phrases, especially at the end of a period'. Cf. also Tac. *Dial.* 22 (of Cicero's style) *pauci sensus apte et cum quodam lumine* (i.e. *sententia*) *terminantur*. Thus the meaning is akin to our 'epigram': cf. Sen. *Epp.* 100. 8 *subiti ictus sententiarum*. Silver-age literature suffers from perpetual straining for *sententiae*. Juv. has many such; *spoliatis arma supersunt* in l. 124 is a *sententia*.

Quintilian deals at length with this important item of rhetorical equipment (VIII. 5); he defends them in moderation, but considers they were over-done in his time: §34 *ego vero haec lumina orationis velut oculos quosdam eloquentiae credo: sed non oculos esse toto corpore velim:* §2 (*sententiae*) *minus celebratae apud antiquos nostris temporibus modo carent.*

verum est: truth is opposed to a *sententia*, which aims only at rhetorical effect: cf. Petron. 1 *sententiarum vanissimo strepitu.*

126. **vobis** is addressed to all who, like Ponticus, hope some day to rule a province.

folium...Sibyllae, i.e. a divine prophecy and therefore a certain truth: for the Sibyl, see n. to III. 3: her oracles, said to be written on palm-leaves, were preserved in the Capitol, and to consult them was the special duty of the college of *quindecemviri.*

127. **cohors** is the technical word for a provincial governor's staff: cf. Catull. x. 10 *respondi id quod erat, nihil neque ipsis | nec praetoribus esse nec cohorti.*

tribunal = legal decisions.

128. **acersecomes,** *intonsus,* properly an epithet of Apollo, is here = *puer capillatus,* a favourite slave, who is allowed to interfere with the course of justice. **coniuge:** in the trials for *repetundae* mentioned by Tacitus and Pliny, the governor's wife is in several cases put on trial; cf. Tac. *Ann.* III. 33 *cogitarent ipsi, quoties repetundarum aliqui arguerentur, plura uxoribus obiectari,* and Furneaux's note.

129. **conventus,** 'districts': to facilitate the administration of justice, a province was divided into a number of districts or circuits, called *conventus:* the word is also applied to the meetings held at these centres; the governor was said *conventum agere:* cf. Suet. *Galb.* 9, *Iul.* 30.

curvis unguibus: again XIII. 169.

130. **Celaeno,** i.e. like a Harpy: cf. Virg. *Aen.* III. 211 *quas dira Celaeno | Harpyiaeque colunt aliae.*

131. The apodosis begins at **tu licet.** Picus was the son of Saturn and father of Faunus, the first in the line of Laurentine kings. **licet** = you have my leave to...; *per me* is often added with this meaning. For such fictitious pedigrees, see n. to l. 5.

132. **Titanida pugnam,** 'the warrior Titans', who rebelled against Zeus in the beginning of the world's history.

133. **Promethĕā:** for the quantity, see n. to III. 266. Prometheus made the first men out of clay, so that no human ancestor could go so far back; see n. to IV. 133.

134. **libro:** such a book as Hesiod's *Theogonia* seems to be indicated.

135. **rapit,** sc. *te.*

136. **virgas** = *fasces:* these were the bundles of rods, with an axe (*securis*) inserted, which were carried by lictors before the higher Roman magistrates, as a sign of their executive power. The repetition of *si te delectant* from l. 131 is careless writing.

137. **lasso lictore** is an abl. absol., but may be translated as a nom. coordinate with *h. secures:* their bluntness is due to constant use.

139. **facem praeferre:** cf. Sallust, *Iug.* 85. 23 *maiorum gloria posteris quasi lumen est, neque bona neque mala eorum in occulto patitur.* **pudendis** is neut., 'your shameful actions'.

142. **quo mihi,** 'how does it profit me...'; cf. Hor. *Sat.* I. 6. 24 *quo tibi, Tilli, | sumere depositum clavum?* There is an ellipse of some such verb as *prodest.* The logic is better brought out by reversing the clauses in English: 'how does it profit me that your ancestor built the temples, in which', etc.

falsas...tabellas: for signatures to wills, see n. to III. 82; wills were often witnessed in a temple (cf. Mart. x. 70. 7 *nunc ad luciferam signat mea gemma Dianam*) and then deposited there for safety. For *falsas,* 'forged', see n. to I. 67: in this case the forger visits the temple, where the will is kept, and substitutes a fraudulent document with forged seals. His father's statue stands, apparently, in the temple.

144. **quo, si...:** the expression is awkward: Juv. means 'what is your nobility to me, if etc.'

145. Cf. Hor. *Sat.* II. 7. 55 *caput obscurante lacerna.*

Santonico, i.e. made by the Santones in Gaul: Mart. (XIV. 128) applies the epithet to *bardocucullus.*

146-82. *Here is another instance of misconduct in a man of noble birth. Lateranus, the consul, drives himself along the public*

*roads. He does this by night; when he ceases to hold office, he
will act coachman in the light of day. He visits the night-houses,
an honoured guest in that low company. Nor has he the excuse of
youth: he is a man of full age who might be serviceable to his
country. Even a slave, who had such a passion for bad company,
would be punished; but aristocrats pardon such conduct in them-
selves.*

L.'s crime consists in driving himself instead of being driven:
cf. 1. 61 *nam lora tenebat | ipse*: the charge seems to us trifling,
but Juv. often attacks undignified conduct as fiercely as the
worst vices: cf. l. 220, and see Introduction, p. xxxviii.

146. He drives along the public roads, e.g. Appia and
Flaminia, and there passes the tombs of his ancestors: see n.
to 1. 171.

147. **carpento:** this was a two-wheeled, comfortable car-
riage, drawn by two horses.

Lateranus: cf. x. 17: Plautius Lateranus, when consul desig-
nate in A.D. 65, was implicated in Piso's conspiracy against Nero
and put to death. No Lateranus occurs in the Fasti of Nero's
reign; T. Sextius Lateranus was consul A.D. 94; but the re-
ference to Nero l. 170 shows that he cannot be meant here; so
Juv. probably refers to the consul designate, of whom Tacitus
speaks favourably (*Ann.* xv. 60).

148. **mulio consul:** all edd. before Büch. (1886), read
multo sufflamine c., no other reading being known; but the
Florilegium of St Gall (see Introduction, p. xlv), which in-
cludes this line, reads as in the text; and this is confirmed (1)
by a Schol. on l. 157, *quia mulio est qui consul fertur*, (2) by a
late grammarian, who, in order to show that *mulio* can be a
dactyl, quotes from Juv. *mulio consul*. The reading is certainly
right: *multo sufflamine* is meaningless. In P the true reading has
been erased. For the cause of corruption, see Introduction,
p. xlvi.

(A 9th-cent. MS. of Juv. in the British Museum (no. 15600)
reads *mulio sufflamine consul*: *mulio* being a correction, ap-
parently of *multo*.)

149. **quidem,** 'it is true'. **testes** is taken by M. to be nom.;
it may, however, be accus. agreeing with *oculos*. For **oculos,** cf.
M. Arnold's *Obermann* 'and on his grave with shining eyes | the
Syrian stars look down'.

150. **honoris,** 'office'.

153. **iam senis:** his age makes the friend more likely to be a

stickler for respectability, so that it would be only decent in L. to pass without recognising him.

virga...annuet, 'salute him with his whip': the phrase occurs in a different sense III. 317.

154. **hordea:** this plural, used by Virgil (*Georg.* I. 210), was regarded as doubtful Latin, and satirised by Bavius and Maevius, whom Virgil had ridiculed, in the following verse preserved by Servius, *hordea qui dixit superest ut tritica dicat.* Yet the poets continued to speak of 'barleys', undeterred by Bavius and Maevius.

155. **intrea,** 'till then', when all restrictions will be removed by his ceasing to hold office.

lanatas robumque iuvencum: the epithets are obviously archaic and sacrificial, like καρταίπους, the 'Delphic' word for 'ox' (Pind. *Ol.* 13. 81) and φερέοικος 'a snail', and ἀνόστεος 'a cuttle-fish' in Hesiod. All MSS. read *torvum* for the unfamiliar *robum*, except P which has the word erased; but the Scholiast preserves *robum* and explains it as *robustum* or *rufum*; the latter seems the true meaning. A similar epithet, *robius*, applied to victims, is quoted without explanation by Gellius (IV. 6) from a *senatus-consultum* of 99 B.C.

156. **more Numae:** all ancient religious ceremonies were attributed to Numa. The ceremony here meant is the *feriae Latinae*, at which the consuls celebrated a solemn sacrifice to Jupiter Latiaris on the Alban mount.

altaria is not used in the sing. by good writers; hence it may mean 'the altar'.

157. **Eponam:** *E. dea mulionum est* Schol. He insults Jupiter by swearing before his altar by such an obscure divinity. **facies** may refer to representations of Epona, daubed on the stable-walls.

158. It is difficult to see the force of **sed:** perhaps it opposes his private amusements to his solemn public duties; or it may be intensive, 'still more': see n. to IV. 27.

instaurare: the word means 'to renew' (e.g. *pugnam,* XV. 74), and hence (because the least informality made a repetition necessary) 'to celebrate solemnly', esp of the. *feriae Latinae*: e.g. Cic. *ad Q. fr.* II. 4. 4 *Latinae instaurantur.* Here it is generally transl. 'to visit'; but surely Juv. is using it in exactly the same sense as Cicero above; and *popinas* is added παρὰ προσδοκίαν (as in l. 172) with good ironical effect, where we should expect *Latinas.* It may be noted that **pervigiles** might be applied to the festival; cf. Lucan, V. 402 *vidit flammifera confectas nocte Latinas.*

popinas are low taverns, where food and drink could be got; a haunter of such places was called *popino*. Nero himself was fond of such haunts and might have met his consul-designate there: cf. Suet. *Nero*, 26 *post crepusculum statim adrepto pilleo vel galero popinas inibat*; Dio Cass. LXI. 8 ἔς τε καπηλεῖα ἐσῄει καὶ πανταχόσε ὡς καὶ ἰδιώτης ἐπλανᾶτο.

159. **adsiduo,** 'invariable'. *Syrophoenicia* is the name by which the Romans called Phoenicia, a part of the province of Syria.

160. The meaning of **Idumaea porta** is uncertain: some take it to be a mountain-pass in Palestine, *porta* being used like Πύλαι: but *incola* seems against this. Others suppose that one of the gates of Rome, or the arch of Titus, or some quarter in the city, was called by this name; the last hypothesis it the most probable.

161. **hospitis adfectu,** 'with hospitable air' he welcomes his distinguished guest. For the titles of honour, cf. Mart. IV. 83. 5 *sollicitus* (when you're in trouble) *donas, dominum regemque salutas.*

162. **Cyane** is the hostess, a foreigner like Virgil's *copa Syrisca*: her dress is tucked up (*succincta*) for the better performance of her duties: cf. Hor. *Sat.* II. 6. 107 *veluti succinctus cursitat hospes.*

164. **nempe,** 'but of course': again l. 180.

165. **turpiter audes:** cf. VI. 97. The number of times the same thing is here repeated, should be observed.

166. **cum prima...barba:** the *barbae depositio* (see n. to III. 186), when the beard was first cut, marked entrance upon manhood and was the occasion of a holiday: Trimalchio (Petron. 29) kept the first clippings of his beard in a gold box in the shrine of the Lares. As to the age, at which the ceremony was observed, the emperor Caligula removed the beard on the day when he assumed the *toga virilis*; and this was probably a common custom, the usual age being about 15 or 16.

168. **thermarum calices:** at all public baths, especially at the thermae, food and drink could be got: cf. Mart. XII. 70. 8 *sobrius a thermis nescit abire domum.* **inscripta lintea** are linen curtains or awnings, in front of the *popinae*, covered with advertisements of the entertainment within. The wording of these may be inferred from Sen. *Epp.* 21. 10 (of Epicurus' school) *cum adieris hortulos et inscriptum hortulis,* 'hospes, hic bene manebis, hic summum bonum voluptas est' cet. Cf. Mart. I. 117. 11 (of a bookseller's shop) *scriptis postibus.*

169. The 'rivers of Armenia and Syria' are the Euphrates and the Tigris. The Hister (Danube) separated Dacia from the Roman territory and, with the Rhine and Euphrates, formed the natural boundary of the Empire until the conquests of Trajan.

170. **Neronem** without an epithet (see n. to IV. 38) cannot mean Domitian; therefore T. Sextius Lateranus (see n. to l. 147) cannot be the consul here spoken of.

171. **aetas** = ἡλικία, 'time of life'. **mitte**, sc. *eum*, i.e. *Lateranum*. **Ostia**, 'to Ostia', acc. plur.; a form *Ostia, Ostiae*, is also found; Ostia was the port where a general would go on board in order to sail to his command abroad.

172. **legatum:** the general style of the governor of an imperial province was *legatus Augusti pro praetore*.

popina, παρὰ προσδοκίαν *pro castris* Schol.: see n. to l. 158: *magna* shows that his explanation is right.

173. **iacentem** etc., 'cheek by jowl with some assassin, hand and glove with sailors, thieves, and tramps': *iacentem* suggests, not helpless intoxication, but the position the Romans assumed at meals: cf. *iacetis*, v. 169; Sen. *Dial.* II. 33. 4 *iacebat conviva centesimus*.

174. **nautis:** the word seems to show that the scene passes in a *popina* at Ostia, not at Rome.

175. **sandapila** is a wooden bier, not a coffin, on which the bodies of the poor were carried out by *vispillones* for burial: Mart. couples *carnifices* and *vispillones* as the dregs of the populace, II. 61. 3 *triste caput fastidia vispillonum | et miseri meruit taedia carnificis*.

176. **tympana galli:** for the priests of Cybele and their music, see n. to VI. 515. This priest has been upset in the orgy and his tambourine is silent.

177. **aequa libertas,** 'liberty and equality'. For the different kinds of *lecti*, see n. to VII. 105.

178. **alius,** 'different': so *aliam aquam*, v. 52.

180. **Lucanos,** 'your estate in Lucania': the people of Lucania come to be used for the place: cf. Hor. *Carm.* II. 18. 14 *satis beatus unicis Sabinis* (nom. *Sabini*); Mart. VII. 31. 9 *quidquid vilicus Umber aut colonus | aut Tusci tibi Tusculive mittunt*; Pliny, *Epp.* III. 4. 2 *cum...in Tuscos excucurrissem* with Mayor's note.

Tusca ergastula, 'your chain-gang in Etruria'; *ergastulum* was a prison in which slaves were kept and set to work in chains on the land; cf. VI. 151. The hybrid word shows this institution was not Roman originally: it was brought to Italy from Sicily where the Carthaginians had introduced it. To be transferred at all from town to country service was a punishment for slaves: cf. Plaut. *Most.* 18 *cis hercle paucas tempestates, Tranio,* | *augebis ruri numerum, genus ferratile*; Hor. *Sat.* II. 7. 117 *ocius hinc te* | *ni rapis, accedes opera agro nona Sabino.*

181. **Troiugenae:** see n. to I. 100.

182. Cf. IV. 13. **cerdoni** = *plebeio*. The word is a Greek proper name (Κέρδων) usually borne by slaves or workmen: it occurs in Herondas *Mim.* 6 and 7 as the name, apparently a common one, of a shoe-maker, and is also given by Martial to one of the same trade (III. 16. 59 and 99) apparently as a proper name; it is found in Petron. 60 (where of two slaves one is called *Cerdo*, another *Lucrio*, a Latin equivalent), in Persius 4. 51, Juv. IV. 153 and here. In Latin it is clearly used as a contemptuous soubriquet for the class engaged in small trade and handicrafts, those whom Cicero (*pro Flacc.* 17) calls *sutores et zonarii*: cf. the French *épicier.*

Volesos: one of the great names of early Rome: the father of P. Valerius Publicola was so called; Publicola was present with L. Junius Brutus, when Lucretia killed herself (Livy, I. 58).

183–210. *Conduct like this in our nobles is bad enough, but worse remains behind: they have actually stooped to appear on the stage in low farces. The people too are to blame for their willingness to look on at such a disgraceful sight. Under Nero, when refusal might mean death, there might have been some excuse for the nobles; but even in our own time, when there is no compulsion, they still appear on the stage. A true man would prefer death to such a disgrace as this. There is only one downward step possible after this; and even that we see, when a Gracchus appears as a gladiator in the arena. And further, he chooses to appear as a retiarius, on purpose that his face may be uncovered and that he may be recognised by all the spectators. The antagonist, who fights with him, suffers disgrace worse than any wound.*

185. **Damasippe:** Horace (*Sat.* II. 3) has a character of this name, who has ruined himself by speculation in works of art; but Juv. is probably speaking of some otherwise unknown figure of Nero's time.

vocem…locasti: anyone, appearing as an actor or singer at public shows, was said *vocem locare praetori*; see n. to l. 194.

186. **sipario,** 'to the curtain': farces (*mimi*) were played on the front part of the stage, which was divided from the back by the *siparium*, which is therefore used as = *mimus*: cf. Sen. *Dial.*

IX. II. 8 *Publius, tragicis comiscisque vehementior ingeniis, quotiens mimicas ineptias et verba ad summam caveam spectantia reliquit, inter multa alia cothurno, non tantum sipario, fortiora, et hoc ait,* etc. Publius was a writer of mimes.

The 'Ghost' of Catullus was a 'noisy' *mimus*, perhaps owing to the screams of the characters to whom the ghost appeared. For *Phasma*, cf. Plautus' *Mostellaria* (the story of the *mostellum* or ghost), which was founded on a Greek original called Φάσμα.

Catullus was a well-known farce-writer of the Neronian age; cf. XIII. 111; Mart. v. 30. 3 *facundi scena Catulli*.

187. **Laureolus** was a famous highwayman, whose crucifixion was represented in Caligula's reign in a *mimus* which seems to have become a stock-piece: cf. Suet. *Cal.* 57 and Mart. *Epig. Lib.* 7. 4 *non falsa pendens in cruce Laureolus.*

velox, 'nimble': either (1) because activity was required to preserve the illusion, or (2) because Laureolus was a *fugitivus*: for the latter, cf. XIII. 111.

The Lentuli Cornelii were a high patrician family.

189. **ignoscas**, 'ought you to pardon'; see n. to *expectes*, 1.

14. **frons durior:** the forehead is regarded as indicating modesty or the want of it; cf. the common phrase *perfricui frontem*, 'I lay modesty aside': see n. to XIII. 242.

190. **triscurria:** the prefix intensifies the meaning, as in τρίδουλος, *trifurcifer* etc.

191. **planipedes**, 'barefooted': tragic actors wore a high boot (*cothurnus*), comic actors a low shoe (*soccus*); the actors in *mimi* wore no covering on the foot. The Fabii and Mamerci are given as types of noble birth; a Mamercus was the original ancestor of the *gens Aemilia*.

192. **alapas:** an invariable feature in the *mimus* was the blows bestowed upon the *stupidus*, a stock character in these pieces, who represented such parts as a deceived husband (see n. to VI. 276), and, like our Pantaloon, came in for all the kicks and blows; cf. Mart. v. 61. 11 *o quam dignus eras alapis, Mariane, Latini*. (The quotation shows that *Mamercorum* may mean 'dealt *by* the M'.)

quanti sua funera vendant etc. Much has been written on this difficult passage. The primary fact to realise is, that Juv. is not here speaking of nobles fighting as gladiators; that topic begins at l. 199; here he is dealing only with appearances on the stage. Two serious difficulties remain, in *quanti*, and *sua funera vendant*. With regard to the latter, such examples as *letum* (or *vitam*) *pacisci* seem quite unlike: *s. f. vendant* is not = *vitam*

suam redimant, for *vendant* must be used in the same sense as
in l. 193 (cf. l. 185 and VI. 380), where the meaning clearly
cannot be that life is gained as the price of degradation, as Juv.
expressly says there is now no compulsion, no fear of death.
Madvig (*Opusc.* p. 545) explains *sua funera* as *relliquias mortuas
tanti generis*: perhaps rather 'their moral suicide', an oxymoron
like *dignus morte perit* l. 85; after such a disgraceful compliance,
they cannot be said to be living. Accepting this, what does
quanti mean? what is the price spoken of? The logic of the
passage shows that *quanti* has the same syntactical relation to
vendant as *cogente Nerone* to *vendunt*. Therefore *quanti* refers
not to a sum of money paid to nobles for compliance, but to
the motive which compelled them, i.e. fear of death. The whole
passage may then be paraphrased: 'you say that the motive, for
which Nero's nobles sold their dead selves, was a powerful one:
but that is unimportant; for they do the very same under Trajan
and Hadrian, with no fear of suffering death from a tyrant; they
do it at the shows of a mere praetor, not an emperor'. (This
explanation of *funera* and *quanti* was given by Mr Lendrum,
Class. Rev. IV, 229.)

193. **cogente Nerone**: cf. Tac. *Ann.* XIV. 14. 5 [*Nero*] *ratus
dedecus molliri si plures foedasset, nobilium familiarum posteros
egestate venales in scaenam deduxit; quos fato perfunctos ne
nominatim tradam, maioribus eorum tribuendum puto*; the last
sentence shows how entirely alike is the view of Tacitus and
Juv. on this point. For similar appearances in public of nobles
before Nero's time, see Furneaux's note to Tac. *Ann. l.l.*

194. **celsi praetoris**: the management of all the public
shows in circus and theatre was transferred by Augustus from
the aediles to the praetors 22 B.C.; cf. VI. 380 *vocem vendentis
praetoribus*; X. 36; XIV. 257. In the theatre the praetor, as
president, sat 'throned on high' (*celsus*) on his *tribunal*, im-
mediately over the stage on the spectator's left, like the 'royal
box' in a modern theatre.

ludis, 'at the games', abl. not dat.: so *comitiis, sollemnibus* etc.

195. **tamen** is resumptive, from *vendunt* to *ludis* being a
parenthetical reference to Juv.'s own times: 'but suppose they
have to choose between execution and the stage'. The choice is
not between acting and fighting as a gladiator, but between
acting and being put to death. **inde**, 'on one side', **hinc**, 'on the
other'; so l. 65; the use of the two words in l. 105 is quite
different. For **gladios**, cf. IV. 96, X. 123.

196. **quid** is used in the sense of *utrum*, πότερον: the use of

quid X. 338 and *quisque* for *uterque* I. 41, is similar, though not such a marked deviation from ordinary usage. So in English, 'which' has ousted 'whether'.

sit = *velit esse*, as in phrases with *tanti est*: see n. to III. 54; the sense might be expressed by *nemini vita tanti est ut sit*, etc.

197. The 'jealous husband', and the *stupidus* (see n. to l. 192) were stock characters in the *mimus*; Thymele, mentioned l. 36, and Corinthus acted in such pieces. **collega** is sarcastic; for the serious use, cf. l. 253, XI. 92.

198. **citharoedo principe:** cf. *mulio consul*, l. 148. Nero made his *début* as a harper at the *ludi Iuvenales* of his own instituting A.D. 59: cf. Tac. *Ann.* XIV. 15 *postremum ipse scaenam incedit multa cura temptans citharam*; but this was a semi-private performance; in A.D. 65 he appeared publicly in Pompey's theatre as a *citharoedus*; cf. *ibid.* XVI. 4 *ingreditur theatrum, cunctis citharae legibus* (etiquette) *obtemperans*.

199. **ludus**, 'the gladiators' school', in which they were taught to fight by a *lanista*; also VI. 82, XI. 20. *ludus* is to be distinguished from *ludi*, 'a show'; a show of gladiators is often called *munus*, never *ludus*: see n. to III. 36. Here, as in II. 143 ff., the public appearance of men of rank in the arena is regarded as the worst feature of the age.

illic = *in illa re*, 'in that respect also'.

200. Gracchus, in the arena, refuses to wear any equipment which includes a vizor to cover the face, such as was worn by the *murmillones* and other gladiators who fought in heavy armour.

201. **clipeo,** 'a round shield', borne probably by the *secutor*; see n. to l. 210. The small round shield (*parma*) was peculiar to the *Threces*, the large square shield (*scutum*) to the *Samnites*.

falce supina, 'back-bent sickle': this is the *sica*, the short curved sword, which was the national weapon of the Thracians and was always borne by the gladiators called *Threces*.

202. **damnat,** 'he decides against', is not so strong as *odit*; cf. Sen. *Dial.* III. 16. 7 *bonus iudex damnat improbanda, non odit*; for *damnare*, cf. Mart. II. 64. 7 *si schola damnatur, fora litibus omnia fervent*; idem, VI. 32. 3 (of Otho's suicide) *damnavit multo staturum sanguine Martem*.

sed: see n. to IV. 27: it is used before the repeated verb in the sense of 'yes, indeed'; M. quotes Diderot *c'est un auteur de beaucoup, mais de beaucoup d'esprit*.

203. Gracchus appears as a *retiarius*, in a tunic, with head bare, with a net, trident and dagger as weapons; that the

retiarius, alone among gladiators, had no head-covering, appears from Suet. *Claud.* 34 *forte prolapsos iugulari iubebat [Claudius], maxime retiarios, ut expirantium facies videret.* There is a similar description of this appearance of a Gracchus II. 143 ff. *vicit et hoc monstrum tunicati fuscina Gracchi*, 'even this horror was surpassed by the trident and tunic of a Gracchus'.

204. The *retiarius* had for antagonist a *secutor*, whom he endeavoured first to entangle in his net and then to despatch with his trident and dagger.

205. **spectacula,** i.e. the spectators' seats: cf. VI. 61; Livy, XLV. I. 2 *cum in circo ludi fierent, murmur...tota spectacula pervasit.*

206. **tota...harena** to be taken with *fugit*; cf. II. 144 *lustravitque fuga mediam gladiator harenam.* **agnoscendus,** 'recognisable': in Cicero *agnoscendus est* always means *debet agnosci*, never *potest agnosci*: this use of the gerundive (e.g. *videndus*, 'visible') begins with Ovid, e.g. *Trist.* III. 4. 56: see Madv. on Cic. *de Fin.* I. 6.

207, 208. 'Let us be convinced, since the golden cord stretches from the throat of the tunic, and shoots out from the long shoulder-guard.' Some edd. put the comma after **tunicae;** but it is the golden *spira*, not the tunic, which forces the spectator to admit the painful fact. For the *tunica* of the *retiarius*, cf. II. 143 quoted above, and Suet. *Cal.* 30 *retiarii tunicati quinque numero...totidem secutoribus succubuerant; cum occidi iuberentur, unus resumpta fuscina omnes victores interemit.*

aurea: a proof that here is no common gladiator.

208. **porrigat:** some understand *tunica* as subject, but the verb required in that case would surely be *demittat*; the subject is **spira,** the coiled rope or cord by means of which the net was thrown and recovered: for the reservation of the subject till the end of the clause, cf. IV. 71, VI. 177. **galerus** is usually a head-covering worn as disguise: cf. VI. 120 *crinem abscondente galero*, and Suet. quoted on l. 158. The context shows this meaning to be impossible here, as Gracchus is determined to be recognised. Hence an explanation given by the Schol. is generally accepted: *galerus est umero impositus gladiatoris.* For it is known that a high guard of leather or metal was worn on the left shoulder by the *retiarius* to serve as a shield; in some ancient works of art this pad sticks up and out like a wing; hence *longo*. A good figure will be found in Baumeister's *Denkmäler*, p. 2097. This passage seems to show that the *spira* was attached to the *galerus*: neither net nor *spira* appears in the figure referred to: they are seldom represented in any work of art.

209. **omni,** 'than any': cf. III. 38.

210. The antagonist of the *retiarius*, called **secutor,** was armed with sword and shield, greaves, and a helmet with vizor. We need not believe that the *secutor* felt the indignity so acutely as Juv. would have us think: *sententia est; non verum est.*

211–30. *Nero, the last of the Caesars, was a monster of crime, not fit to live, far less to rule. He murdered his own mother; so did Orestes, but his motives were noble, and he did not stain himself with other crimes and degradations. Nero's vice and cruelty did not deserve punishment so richly as his public appearances as a singer. What would his ancestors have thought of distinctions so gained?*

212. Juv. may allude to a report which said that in the Pisonian conspiracy against Nero A.D. 65 it was intended to pass Piso over and place Seneca on the throne: cf. Tac. *Ann.* xv. 65 *fama fuit Subrium Flavum…destinavisse ut post occisum opera Pisonis Neronem Piso quoque interficeretur tradereturque imperium Senecae, quasi..claritudine virtutum ad summum fastigium delecto.* As it was, Seneca, who had tried in vain to guide and restrain his dangerous pupil, was forced to commit suicide.

213. **non una,** 'more than one': a common Latin idiom.

214. Cf. XIII. 155 and n. there: the traditional punishment for *parricidium* (the murder of any near relation) was that the criminal should be sewn up in a sack, with a dog, a cock, a snake, and a monkey; the sack was then thrown into the sea. Nero murdered his mother Agrippina, his wife Octavia, and others closely related to him.

215. **Agamemnonidae** = *Orestae*; he also was μητρο-κτόνος. **causa,** 'motive': cf. Quint. VII. 4. 8 *fortissimum est si crimen causa facti tuemur, qualis est defensio Orestis.* The case of Orestes was a favourite subject in the rhetorical schools.

216. **quippe,** 'for': see n. to XIII. 26. **deis auctoribus:** the express command of the Delphic oracle is repeatedly urged by Orestes: cf. Aesch. *Choeph.* 269 Λοξίου μεγασθενὴς | χρησμὸς κελεύων τόνδε κίνδυνον περᾶν.

217. **media inter pocula:** Juv. follows the Homeric account, *Od.* XI. 409 ἀλλά μοι Αἴγισθος τεύξας θάνατόν τε μόρον τε | ἔκτα σὺν οὐλομένῃ ἀλόχῳ, οἰκόνδε καλέσσας, | δειπνίσσας, ὡς τίς τε κατέκτανε βοῦν ἐπὶ φάτνῃ.

218. Orestes did not kill Electra or his wife, Hermione, whereas Nero killed Octavia and Antonia, his sister by adoption.

219. **coniugii** = *coniugis*; cf. the use of *mancipium*, 'a slave'; *custodiae*, 'prisoners'. **aconita:** cf. I. 158.

220. It is a mistake to suppose that any humour is intended in this climax of all Nero's crimes; to Juv., as to Tacitus, these public appearances as an artist were worse than any vice or cruelty. **cantavit** refers to his appearances as a tragic actor, not as a *citharoedus*: he chose at least one appropriate part, cf. Suet. *Nero*, 21 *inter cetera cantavit...Oresten matricidam.*

221. **Troica,** an epic poem by Nero describing the fall of Troy, recited by him in the poetical competition founded by himself A.D. 65. M. suggests that the ἅλωσις Ἰλίου, which Nero declaimed, while Rome was burning (Suet. *Nero*, 38), may have been an extract from this poem; but the ἅλωσις Ἰλίου seems to have been a tragedy.

Verginius: in the year 68 Julius Vindex, legate of Gallia Lugdunensis, revolted against Nero and offered the throne to Galba the future emperor, then governor of Hispania Tarraconensis; Verginius Rufus, legate of Upper Germany, marched against Vindex and utterly defeated him; though he fought for Nero then, he afterwards acquiesced in Nero's deposition, and hence is coupled here with Vindex and Galba. He refused the throne for himself on two different occasions, and died as a private citizen of the highest distinction A.D. 97. He was honoured with a public funeral, and a *laudatio* was pronounced over his remains by Tacitus, consul in that year.

223. **quod** for *quid* of MSS. is due to Madvig, who also removed the mark of interrogation after *Galba*, l. 222.

The antecedent of *quod* is *quid* in l. 221: for the form of sentence, cf. Cic. *Phil.* XII. 5 *quid autem non integrum est sapienti, quod restitui potest?*

225. **peregrina ad pulpita:** Nero made a progress in Greece which lasted a whole year A.D. 67–68; the calendar was altered so that all the great Greek games might fall within the same year; at all of them he entered as a competitor in various contests, and as a matter of course, gained the prize; he brought 1,808 crowns back to Rome.

226. **prostitui,** 'to prostitute himself'. **apium:** the prize at Nemea was a wreath of parsley (σέλινον).

228. **Domiti:** Nero belonged to the *gens Domitia*; until his adoption by Claudius A.D. 50, his name was L. Domitius Ahenobarbus, and after that Ti. Claudius Nero Caesar. The founder of the family is here meant; cf. Suet. *Nero*, 1 *Aenobarbi auctorem originis itemque cognominis habent L. Domitium.*

229. **syrma** (σύρμα from σύρειν) is the long trailing robe worn by tragic actors; cf. xv. 30; Thyestes, Antigone, and Melanippe (the subject of two lost plays of Euripides) were tragic parts acted by Nero.

230. **marmoreo...colosso:** as Nero's own colossal statue, 120 feet high, was of brass, a statue to one of his family may be meant. Cf. Suet. *Nero,* 12 *coronam...citharae a iudicibus ad se delatam adoravit ferrique ad Augusti statuam iussit.*

231–68. *Which was more truly noble, Catiline and his high-born confederates who plotted to burn Rome, or Cicero, the man of obscure birth, who thwarted their designs and so gained as much glory as Augustus at Actium? Was it to the plebeian Marius or to his noble colleague Catulus that the great victory over the Teutons was due? Were the Decii, who gave their lives for their country, men of noble birth? Servius Tullius was the son of a slave; but those who tried to betray the infant republic were the sons of Brutus, the first of the consuls.*

231. Cf. x. 287. Catiline belonged to the patrician *gens Sergia,* Cethegus to a patrician family of the *gens Cornelia;* with Lentulus they were the ringleaders in the conspiracy of 63 B.C.

natales, 'birth', 'ancestry', a silver-age use; *natalis,* 'birth-day'. For the apostrophe (*Catilina*), see n. to I. 50.

233. **nocturna:** adj. where English has adverb; cf. *alea pernox* l. 10. **paratis,** the reading of P (other MSS. have *parastis*), is clearly right: the pres. is dramatic and is continued in the account of Cicero's activity.

234. **bracatorum:** Gallia Narbonensis was earlier called *Bracata,* because of the dress of the inhabitants. **Senonum:** the Gauls who defeated the Romans on the Allia 390 B.C. and then took Rome and burnt it, were supposed to be descended from this tribe: see n. to XI. 113. The Schol. seems to have read *Allobrogum.*

235. Criminals guilty of arson were burnt alive in what was called the **tunica molesta:** the Christians, whom Nero accused of setting fire to Rome A.D. 64, were probably the first to undergo this punishment (Tac. *Ann.* xv. 44). The 'shirt of pain' was made of pitch and other inflammable materials, by which the criminal was consumed; cf. I. 155. Martial applies the phrase to the paper wrapper round frying fish, IV. 86. 8 (to his book) *nec scombris tunicas dabis molestas.*

236. Cicero was consul, with C. Antonius, in 63 B.C.

vexilla, 'bands', often used of a body of troops serving under a *vexilllum*, e.g. Tac. *Hist.* I. 70 *Britannorum cohortibus et Germanorum vexillis.*

237. **novus Arpinas:** the term *novus homo* was applied to any Roman who, like Cicero, was the first of his family to attain curule office. **ignobilis** is not a synonym, but refers more generally to the obscurity of his family: Cicero's early speeches frequently refer to the inferiority which the *nobiles* made him feel. Cicero and Marius were both natives of Arpinum in the Volscian country.

238. **municipalis eques,** 'a provincial knight': Cicero belonged by birth to the second order of the Roman aristocracy, who were small men at Rome though great men in the provincial towns; his own talents enabled him to bridge over the gulf fixed between an *eques Romanus* and a *senator*: he calls himself *equitis Romani filius* (*pro Muren.* 17).

galeatum: see n. to I. 169. **ponit ubique** = *disponit* of prose.

239. **attonitis,** 'the terrified citizens': see n. to IV. 77.

montes, read by the Schol., has been changed in P to *gente*, which the worse MSS. have: the seven hills of Rome are generally called *colles*, e.g. III. 71, IX. 131 (*stantibus et salvis his collibus*); but cf. Mart. IV. 64. 11 *hinc septem dominos videre montes* | *et totam licet aestimare Romam*: so *in omni monte* = *tota Roma*: cf. Stat. *Silv.* IV. 5. 33 *in omni vertice Romuli.*

240. **toga** is the dress worn in time of peace and hence is used for 'peace' itself: Cicero wrote of his own exploit, *cedant arma togae, concedat laurea laudi*, and defends the verse *in Pis.* 73. M. quotes Pliny's fine address to Cicero, *Nat. Hist.* VII. 117 *salve primus omnium parens patriae appellate, primus in toga triumphum linquaeque lauream merite, et facundiae Latiarumque litterarum parens.* That the Augustan poets pass over Cicero in silence, is a fact easily explained by their relation to Augustus: cf. Plut. *Cicero*, 49 πυνθάνομαι δὲ Καίσαρα (Augustus) χρόνοις πολλοῖς ὕστερον εἰσελθεῖν πρὸς ἕνα τῶν θυγατριδῶν (grandsons) · τὸν δὲ βιβλίον [ἔχοντα Κικέρωνος ἐν ταῖς χερσὶν ἐκπλαγέντα τῷ ἱματίῳ περικαλύπτειν· ἰδόντα δὲ Καίσαρα λαβεῖν καὶ διελθεῖν μέρος πολὺ τοῦ βιβλίου, πάλιν δ' ἀποδιδόντα τῷ μειρακίῳ φάναι, 'λόγιος ἀνήρ, ὦ παῖ, λόγιος καὶ φιλόπατρις': yet Livy did not fear to praise him (M. Senec. *Suasor.* 6. 22), and even Asinius Pollio allowed his greatness (*ibid.* 24). The writers of a later age, the younger like the elder Pliny, Quintilian, Martial, and Juvenal, are unanimous and enthusiastic in his praise. His oratorical style is criticised by a speaker in Tacitus' *Dialogue*, cc. 22 and 23.

241. **quantum in Leucade,** the reading of P, is unmetrical,

such a hiatus being impossible: the worse MSS. read *non*, which is not satisfactory in sense; as (1) *non* cannot be left unexpressed with the second *quantum*; (2) if it could, it would not convey Juv.'s meaning, which is that Cicero gained the *same* title as Augustus but more gloriously; *sed* alone shows this. Many corrections of *in* have been suggested, *vix, tum in, sub, unda*: the neatest is Mr S. G. Owen's *vi*: yet the sense required seems to be that of *armis* (which differs somewhat from *vi*, though often coupled with it) or *mari*, or *navibus*.

Leucas, an island on the coast of Acarnania, is 30 miles south of Actium, where Augustus established his power by defeating Antony 2 Sept. 31 B.C.; but the Roman poets are not very scrupulous about geographic accuracy and had metrical difficulties about Actium. So Philippi, the scene of the victory against Brutus 42 B.C., is not, as here stated, in Thessaly but in Macedonia; Pharsalia is in Thessaly, and the poets all followed Virgil (*Georg.* I. 489), who appears to identify the two battlefields.

242. Octavian first received the name *Augustus* from the senate and people 16 Jan. 27 B.C.

243. **gladio,** instrumental.

sed: both gained the same title, *but* the circumstances were different.

244. **libera,** 'while yet free', is strongly emphatic; the state was no longer free when Augustus received this title, 2 B.C.: cf. *Mon. Ancyr.* cap. 35 *tertium decimum consulatum cum gerebam, senatus et equester ordo populusque Romanus universus appellavit me patrem patriae.* Cicero, after the exploits of his consulship, was hailed by this name in the senate by Q. Catulus. Others had been so saluted before, in spite of Pliny quoted on l. 240.

245. Marius, as well as Cicero, was a native of Arpinum in the Volscian hills. For **alius** = *alter*, cf. IV. 138, X. 257.

246. **poscere mercedes,** 'to earn wages'; he was a *mercennarius*, not even owner of the land he tilled; cf. Mart. I. 55. 3 *hoc petit: esse sui, nec magni, ruris arator.*

247. **frangebat…vitem,** 'he had the rattan broken on his head', i.e. he served as a private soldier: for the phrase, cf. VI. 479 *hic frangit ferulas. vitis* is the switch of vine-wood which each centurion carried: cf. Mart. X. 26. 1 *Vare, Paraetonias Latia modo vite per urbes | nobilis et centum dux memorande viris.* Hence it is used XIV. 193 as = *centurionatus.*

248. **muniret,** by throwing up earth-works round it. **dolabra** is not so common as its diminutive, Dolabella: comp. Caligula from *caliga*: both are pet-names of the camp.

249. **tamen,** in spite of these obscure beginnings.

Cimbros: a horde of German invaders, Cimbri and Teu-
tones, swarmed over the Alps into Italy 102 B.C. and were
annihilated in the following year by Marius and Q. Lutatius
Catulus in the Raudine plain near Vercellae. **rerum,** 'to the
state': the use in l. 90 is not unlike.

250. **excipit,** 'faced', lit. 'is ready to receive'; a word taken
from field-sports: cf. Sen. *Dial.* III. 11. 2 *an tu putas venatorem
irasci feris? atqui et venientes excipit et fugientes persequitur, et
omnia illa sine ira facit ratio. quid Cimbrorum Teutonorumque tot
milia superfusa Alpibus sustulit...nisi quod erat illis pro virtute
ira?*

251. **Cimbros stragemque,** 'the heaps of slain Cimbri', a
form of hendiadys common in Latin poetry.

252. **maiora cadavera:** the great size of the Germans is
often remarked on; cf. Tac. *Germ.* 20 *in omni domo nudi ac
sordidi in hos artus, in haec corpora, quae miramur, excrescunt.*

253. Catulus triumphed together with Marius, but was
altogether thrown into the shade by his colleague, partly be-
cause M.'s victory at Aquae Sextiae the year before had made the
final triumph possible, but also, according to Mommsen, be-
cause of the desire for a democratic revolution in which it was
hoped that Marius might lead the way.

254. P. Decius Mus gave up his life to save the Roman army
in battle against the Latins 340 B.C.; his son, of the same name,
repeated the act of heroism in the battle of Sentinum against the
Samnites 295 B.C.: cf. XIV. 239. Their family was plebeian.

255. **totis** is used for *omnibus*, a licence which first appears in
silver-age poets: cf. VI. 61; Martial seems certainly to use *totis
diebus* in the sense of *tous les jours*: cf. II. 5. 1, IV. 37. 6, IV. 54. 3.

257. The Romans were warned by a dream of both consuls,
that the gods of the world below required the sacrifice of the
general commanding one of the two armies; whichever army lost
its general, would be victorious, but the other army was
doomed. Decius accordingly sacrificed his life, devoting him-
self in the following form: *pro re publica populi Romani Quiri-
tium...legiones auxiliaque hostium mecum deis Manibus Tellurique
devoveo* (Livy, VIII. 9. 8). The son repeated his father's words
and action, 45 years later (Livy, X. 28. 15).

258 was condemned by Dobree, acutely: the l. is weak in
itself, and its form (*enim*) is like an inserted explanation of what
precedes; but that Juv. had better not have written it, is no
proof that he did not.

259. **ancilla natus,** Servius Tullius; cf. VII. 199. **trabeam:** see n. to X. 35.

260. **fasces,** called by Livy (III. 36. 3) *insigne regium*, were not first instituted under the republic.

261. In transl., prefix 'whereas' to this sentence, which contrasts the merits of Servius Tullius with the treason of the nobly born: Latin prefers asyndeton, where Greek and English insert particles.

laxabant, 'intended to loosen'.

262. **iuvenes** is used for *filii*, for metrical reasons; see n. to III. 158. The sons of Brutus formed a plot to admit Tarquinius Superbus within the walls; the plot was discovered, and Brutus had his sons executed.

et quos: the English idiom is to omit 'and' here: cf. v. 54, VII. 211.

263. **dubia,** i.e. not yet firmly established.

deceret is used for *decuisset*; see n. to *liceret*, IV. 85.

264. These are the heroic names of the early republic—Horatius Cocles who 'kept the bridge', Mucius Scaevola who burned his hand in Porsena's camp to show his courage, and Cloelia, a hostage who left the Etruscan camp and got home by swimming the Tiber.

265. **imperii fines,** in apposition with *Tiberinum*; for the boundaries of the empire in later times, see n. to l. 169. **Tiberinus,** though often used as a noun, is properly the adj. of *Tiberis*, with which *amnis* is understood.

266. **crimina,** 'guilt', not 'accusation', as the epithet shows. The name of the slave was Vindicius.

267. **matronis lugendus:** the Roman matrons mourned Brutus and Valerius, as the fathers of the state, for a year; Juv. says they ought to have mourned Vindicius also, as he too saved his country.

268. **legum,** i.e. constitutional, as opposed to the arbitrary punishments of the kings. The expression is commoner in Greek than in Latin; thus the Thebans plead excuse for the action of their country while governed by an oligarchy, Thuc. III. 62. 4 οὐδ' ἄξιον αὐτῇ ὀνειδίσαι ὧν μὴ μετὰ νόμων ἥμαρτεν.

269–75. *Better be the good son of a bad father than the bad son of a great hero. Besides, the longest Roman pedigrees go back to a somewhat troubled fountain, the Sanctuary of Romulus.*

269. Among the Homeric chiefs, Thersites is the mean and

ignoble figure, Achilles (grandson of Aeacus) is the hero, who
fights in armour made for him by Hephaestus (Vulcan).

malo...sit: *volo, nolo,* and *malo* are followed by the subj.
without *ut,* when the subject of the dependent clause is different
from that of the main verb.

271. **producat,** 'begat': cf. VI. 241.

272. **tamen** belongs to the verb *deducis:* see Housman on
Juv. VI. 640. **ut,** 'although'. For **repetas,** cf. Mart. V. 35. 4
longumque pulchra stemma repetit a Leda ('traces back to L.' is
our idiom).

273. **asylo:** to people his newly founded city, Romulus
established a Sanctuary, where slaves, debtors, and criminals
could take refuge. The more respectable inhabitants were
simple shepherds; the others are better left undescribed, says
Juv. Cf. M. Seneca *Controv.* 1. 6. 4 *quemcunque voluerimus re-
volve nobilem: ad humilitatem pervenies.*

SATIRE X

THE VANITY OF HUMAN WISHES, AND THE FOLLY
OF HUMAN PRAYERS

'The common-place about the vanity of human wishes...has
found a place, in various shapes, in the works of many of the
greatest of all writers from the days of Solomon to our own...
Juvenal, perhaps, has preached this doctrine as vigorously as
anyone else—the more so, no doubt, because the whole con-
stitution of his mind exemplified that one-sided vehemence and
absence of subtlety which belonged to so many Latin writers...
When the illustrations appended to such a text are well chosen
and dramatically worked out, as in Juvenal's poem, the general
drift of the sermon derives a degree of weight from its separate
parts to which it is not fairly entitled. The true answer to such a
lesson is, that people are in reality less ambitious and more
successful than it assumes them to be' (Stephen's *Essays,* 1862,
p. 184).

This is the best known of Juvenal's satires, partly from John-
son's famous imitation of it. After long passing for a master-
piece, it has been roughly handled by some critics: see Friedl.'s
introduction. It is true that it is easy to pick holes in the argu-
ment: for instance, it was not the object of Hannibal's life to
avoid being defeated and die in his bed; but the power and
beauty of much of the declamation are suprior to all criticism.

1–53. *Of all mankind, how few can see clearly what is good for them and what bad. Men pray for eloquence, for strength, above all for wealth; but how many have lost their lives by the granting of their prayers. The rich man is a mark for the tyrant's envy; the poor man passes unmolested. Considering the irrationality of men, Democritus could not help laughing and Heracleitus weeping, whenever they mingled with their kind. Yet how little there was to laugh at in ancient Greece, compared with the absurd sights which the Roman streets present. Democritus, though born in a country of dullards, was a wise man: he laughed at his fellow-men, and, for his own part, defied Fortune.*

1. **Gadibus:** Gades (Cadiz) is considered as the world's western boundary: cf. Sen. *Nat. Quaest.* 1. prol. 13 *quantum est, quod ab ultimis litoribus Hispaniae usque ad Indos iacet* (i.e. how small the world is). **usque** is seldom used thus without *ad* except before the names of towns.

2. **pauci,** 'only a few': cf. ll. 19 and 111.

3. **illis,** dat.: cf. Hor. *Epp.* 1. 18. 5 *est huic diversum vitio vitium prope maius.* **multa diversa,** i.e. *mala*: the adv. shows that *diversa* is not used in its classical sense of 'opposite': see n. to III. 268.

4. **quid—cupimus,** 'when are our fears or desires based upon reason?' **ratione:** the abl. without *cum* or an epithet, is used like an adverb; so *iure, iniuria, ordine, silentio, vitio (creatus),* and a few other words.

5. **tam dextro pede,** 'so auspiciously': the origin of this metaphor, common in Latin, is shown by Petron. 30 *cum conaremur in triclinium intrare, exclamavit unus e pueris... 'dextro pede'. sine dubio paulisper trepidavimus ne contra praeceptum aliquis nostrum* (gen.) *limen transiret* (i.e. the guests were told to cross the threshold with the right foot first, for good luck). Hence, by a metaphor, the phrase is used of other actions than walking.

7. **optantibus:** *optare* is constantly used throughout the satire, as elsewhere, in the sense of 'to pray' or 'pray for': so ll. 115, 189, 289, 293, 346: the sense of 'to wish' is less common; cf. VI. 487.

ipsis, i.e. *dominis:* *ipse* was used colloquially in this sense by the slaves of a household, but there need be no allusion to that here.

8. **faciles,** 'compliant': cf. Mart. 1. 103. 4 *riserunt faciles et tribuere dei.* **toga,** 'in time of peace', is to be taken with

petuntur: for the contrasted terms, cf. Mart. I. 55. 2 *clarum militiae, Fronto, togaeque decus*.

9. **torrens**: cf. l. 128 and *sermo Isaeo torrentior*, III. 74.

10. **viribus** is to be taken twice, with *confisus* and with *periit*. **ille** is Milo, the athlete of Crotona, who met his death, 'wedged in the timber which he strove to rend'.

11. **periit**: for the quantity, see n. to *redit*, III. 174. That **lacerti** are 'muscles' rather than 'arms', is shown by Martial's phrase *omnibus lacertis* (V. 12. 3).

13. **cuncta** = all other.

patrimonia is not different from *census*: see n. to XII. 50.

14. **ballaena Britannica**: whales are much rarer in the Mediterranean than in the German Ocean, with which the Romans were now becoming familiar, after the victories of Paulinus and Agricola in Britain.

15. **temporibus diris**, 'in the reign of terror': used of Domitian's reign, IV. 80; cf. Mart. XII. 6. 11 (of the same reign) *sub principe duro | temporibusque malis*: Juv. refers to the suppression of the conspiracy of A.D. 65.

16. **C. Cassius Longinus**, a famous jurist, was banished by Nero to Sardinia. For Seneca's fate, see n. to VIII. 212; Tacitus also calls him *praedives et praepotens* (*Ann.* XV. 64. 6). Some at least of his immense wealth he had made over to Nero in his life-time, in the vain hope of escaping his fate thus. **hortos**, 'park': see n. to I. 75.

17. **Lateranorum**: see n. to VIII. 147. This house on the Caelian hill was confiscated, and is often mentioned later as an imperial residence. The site is now occupied by the church of St John Lateran.

18. **rarus**, 'seldom': see n. to VIII. 63; cf. Mart. V. 10. 9 *rara coronato plausere theatra Menandro* (where *rara* 'seldom' applies to *coronato* as well as to *plausere*).

cenacula, 'a garret': cf. Mart. I. 108. 3 *at mea Vipsanas spectant cenacula laurus*: all the rooms on the upper story of a Roman house were called *cenacula*: the plur. is used only for metrical convenience, the sing. being used in prose: cf. Livy, XXXIX. 14. 2 *cenaculum super aedes datum est*.

miles, i.e. in the character of executioner: soldiers, esp. of the praetorian guard, were used as a kind of executive police at Rome: cf. Tac. *Ann.* XI. 32 *ceteris passim dilabentibus adfuere centuriones, inditaque sunt vincla*; *ibid.* 37 (of Messalina's death) *adstitit tribunus*; *Ann.* XII. 22 *in Lolliam mittitur tribunus a quo ad mortem adigeretur*.

19. **pauca** and **puri** are both emphatic: your plate may be small in amount, and the workmanship may be plain, but still, etc. *purum argentum* is opposed to *a. caelatum*, 'embossed silver': cf. XIV. 62.

20. **nocte iter ingressus:** see n. to V. 55. The Roman traveller carried his plate with him: cf. Sen. *Epp.* 123. 7 *omnes iam sic peregrinantur, ut illos Numidarum praecurrat equitatus... omnes iam mulos habent qui crystallina et murrina* (see n. to VI. 155) *et caelata magnorum artificum manu portent: turpe est videri eas te habere sarcinas, quae tuto concuti possint.*

21. **ad lunam,** 'in the moon-light'; so *ad lumina*, 'by lamp-light', *ad lucem*, 'by day-light'. Cf. Lucan, VIII. 5 (of Pompey, flying from Pharsalia) *pavet ille fragorem | motorum ventis nemorum.*

22. In English this sentence should begin with 'but': it is a contrast. Cf. Sen. *Epp.* 14. 9 *nudum latro transmittit.* **cantabit,** 'will sing', i.e. will be light-hearted. **vacuus,** 'empty-handed'; *inanis* is commoner in this sense; cf. Mart. III. 58. 33 *nec venit inanis rusticus salutator.* After the civil war, the public roads, even in Italy, were beset by armed highwaymen; both Augustus and Tiberius found it necessary to plant military posts throughout the country, to keep down *grassatura* and *latrocinium* (Suet. *Aug.* 32; *Tib.* 37). See n. to III. 307.

23. **fere,** 'in most cases'. **vota,** 'object of prayer'.

25. **foro:** the bankers (*argentarii*) carried on their business in the forum and kept their clients' *arcae* there; hence *cedere foro*, 'to become bankrupt', XI. 50. **nulla,** where we naturally say 'never': so VIII. 219.

26. **illa:** for this use of *ille*, where there is no special emphasis, cf. V. 139, VI. 274. The emphatic word here is **tunc.**

sumes: for the idiomatic fut., cf. V. 60 *respice, cum sities*; VIII. 87 *provincia cum te | rectorem accipiet, pone irae frena.*

27. Cf. Sen. *Thyest.* 453 *venenum in auro bibitur.* For gold and jewelled drinking-cups, cf. V. 37–45; for Setine wine, see n. to 5. 33. **ardebit,** 'shall sparkle' M.; but the word also suggests *red* wine: cf. XI. 155.

28. **iam,** i.e. after what I have said: cf. Lucr. I. 907 *iamne vides igitur, paulo quod diximus ante cet.* A tradition, constantly recurring in Latin authors, represents Democritus as unable to restrain his laughter, and Heracleitus his tears, at the spectacle of human life; cf. Sen. *Dial.* IX. 15. 2 *Democritum potius imitemur quam Heraclitum. hic enim, quotiens in publicum processerat, flebat, ille ridebat.* The tradition is well illustrated by Horace

Walpole's favourite saying, 'Life is a comedy to those who think, a tragedy to those who feel.'

29. **de limine**, 'over his threshold'.

30. **auctor**, 'teacher': cf. Stat. *Silv.* II. 2. 113 *Gargettius auctor*, i.e. Epicurus. This is Heracleitus, not directly named after Juv.'s manner, the 'dark' philosopher of Ephesus (fl. 520 B.C.).

32. 'The real wonder is, what source supplied the other's eyes with sufficient moisture': *illi*, i.e. *Heraclito*, would give the required sense more simply.

33. **pulmonem**, 'his sides', we say; yet cf. Shak. *A. Y. L.* II. 7: 'My lungs began to crow like chanticleer.'

34. Cf. Hor. *Epp.* II. 1. 194 *si foret in terris, rideret Democritus* (if he saw the public shows). For the subj. after *quamquam*, see n. to VII. 14. **illis**, 'of the past': cf. *hos*, 'of the present', I. 89.

35. Cf. Sen. *Dial.* II. 12. 2 (*pueri*) *inter ipsos, magistratus gerunt et praetextam fascesque ac tribunal imitantur.* The objects mentioned are various insignia of high rank: the **toga praetexta,** bordered with purple, was worn by the curule magistrates; before them the **fasces** (see n. to VIII. 136) were borne, and the **tribunal,** a kind of scaffold, was set up to support their curule chairs. The **trabea,** a toga striped with purple, had been worn by the kings and was now the full-dress uniform of the knights, who wore it when reviewed by the censor. To be carried about the streets of Rome in a **lectica,** was a privilege originally confined to the wives of senators (cf. VI. 351), but now commonly usurped by men also (cf. I. 32).

36. Democr. would have had good reason to laugh, had he seen the *pompa Circensis*, with praetor or consul presiding in state. This procession, which preceded the races in the Circus, marched from the Capitol through the Forum and the Velabrum and round the Circus itself. At the head, surrounded by musicians and clients, came the president of the games, whether consul or praetor, driving in a chariot and wearing the dress of a triumphing general. Images of the gods and statues of the imperial family were borne along in the procession.

praetorem: the praetor becomes a consul in l. 41: either magistrate might be entrusted with *cura ludorum*; see n. to VIII. 194.

37. **sublimem:** cf. Livy, XXVIII. 9. 15 *iret alter consul sublimis curru* (i.e. let him triumph).

38. A magistrate presiding at the races, or a general cele-

brating a triumph, wore the *tunica palmata* embroidered with palms, and the *toga picta* of purple with gold embroidery: both parts of this *augustissima vestis* (Livy, v. 41. 2) belonged to Jupiter Capitolinus and were borrowed on each occasion from his treasury.

Sarrana = Tyrian, Sarra being one of the names of Tyre.

39. **aulaea:** the *toga picta* is so called because of its folds and great size.

coronae: a great wreath, of gold oak-leaves set with jewels, was held over the magistrate's head by a public slave, being too heavy to wear: cf. Mart. VIII. 33. 1 *de praetoricia folium mihi, Paule, corona | mittis.*

41. **quippe,** 'for': see n. to XIII. 26. **sibi...placeat,** 'to prevent the consul being conceited': cf. VI. 276. The same custom, in the case of a triumph, is mentioned by Pliny, *Nat. Hist.* XXXIII. 11 (gold rings used to be rare in Italy) *volgoque sic triumphabant, et cum corona ex auro sustineretur tergo, anulus tamen in digito ferreus erat aeque triumphantis et servi fortasse coronam sustinentis.* The origin of the custom, as well as of the lampoons sung by the soldiers round their general's chariot, may have been the wish to appease Nemesis, always most formidable to a man at the supreme moment of his fortune; cf. Munro's *Criticisms of Catullus*, pp. 75 ff.

43. **da nunc,** 'next imagine'. The bird is an eagle, which topped the ivory staff held by a triumphing general: when the Gauls took Rome 390 B.C., the senators awaited them in the Forum in triumphal dress, arrayed for death (see n. to III. 171) like the Spartans at Thermopylae, and M. Papirius *dicitur Gallo, barbam suam permulcenti, scipione eburneo in caput incusso iram movisse* (Livy, v. 41. 9).

44. **praecedentia...officia:** often called *anteambulones* in the iambic verse of Martial. *officia,* 'escort', denotes the attentive clients rather than the attention; cf. Pliny, *Paneg.* 76 (of Trajan when consul) *ipsius quidem officium tam modicum,...ut antiquus aliquis consul sub bono principe incedere videretur:* see nn. to VI. 203, III. 239.

45. **niveos,** i.e. *togatos:* see n. to I. 96. Cf. Mart. IX. 49. 8 (of his *toga*) *possis niveam dicere iure tuo* (i.e. it is as cold, if not as white, as snow).

46. The clients have all got their *sportula* already, having been paid in the morning, and each has it buried away in his purse; so in I. 95 (where see n.) the dole is distributed in the morning. **defossa** suggests buried treasure. **loculi** is a portable

purse as opposed to *arca*, 'a strong box'; cf. I. 89. P reads *loculos*, the other MSS. *loculis*: cf. Livy, VIII. 10. 12 *in terram defodi*.

47. tunc quoque, even at a time when there was no gorgeous praetor to be seen.

48. prudentia, 'wisdom'. Democritus, a native of Abdera and a contemporary of Socrates, was a traveller and a man of science, whose name is chiefly famous in connection with the atomic theory.

50. vervecum, 'block-heads': Abdera, a town on the south coast of Thrace, although the birth-place of Protagoras and other famous men as well as Democr., became a proverb for the stupidity of its inhabitants; cf. Mart. X. 25. 4 (if you think so and so) *Abderitanae pectora plebis habes*, i.e. you are very stupid. For the view that climate affects intellect, cf. Hor. *Epp.* II. I. 244 *Boeotum in crasso iurares aere natum* and Wilkins' n. there.

51. nec non et: the same pleonasm was used, III. 204.

53. 'Bade her go hang, and shook his fist at her': the 'fist' is an equivalent, not a translation, the ancients using the fingers, especially the middle finger, to convey opprobrious insults to an adversary: cf. Sen. *Dial.* II. 5. 2 *lacrimas evocant nomina parum grata auribus et digitorum motus*. Hence it was known as *infamis, impudicus*: cf. Mart. VI. 70. 5 *ostendit digitum, sed impudicum,* | *Alconti Dasioque Symmachoque*, 'he shows, not his wrist but his fist, to all the doctors'; so in French *il montra le poing au ciel*.

54–113. Is it true that even innocent objects of prayer are useless or even harmful? Great place is exposed to great envy and often brings ruin on its possessors. Consider the instance of Sejanus, how his fall from the second place in the world was hailed with joy by the very populace, which used to worship him. A little change of fortune, and he might have crushed Tiberius and been emperor himself; but now all are eager to trample on the fallen. Would it be better to be Sejanus, the possessor of boundless power and wealth, and then to fall from that height, or to pass safely through life as a poor little provincial magistrate? Sejanus, at least, was wrong when he prayed for wealth and power; and so was Crassus, and Pompey, and Julius Caesar. For tyrants seldom die in their beds.

54, 55. All MSS. read *aut perniciosa petuntur*, where half a foot is wanting. Some word must be inserted; and, as *vel* is often expressed in MSS. by a symbol like *t*, it may have fallen out after *aut*. Many further corrections have been suggested;

that of Mr H. Richards (*Class. Rev.* II, 326) is here adopted. Juv. asks 'are we then to deem those things superfluous or even (*vel*) baneful, for which men offer innocent prayer to heaven?' He answers by showing from examples that power (ll. 56–113), eloquence (114–32), success in war (133–87), long life (188–288), and beauty (289–345), all legitimate objects of prayer, have all brought ruin on their possessors. He then asks, *nil ergo optabunt homines?* For the form of the sentence, Mr Richards compares 1. 158 *qui dedit ergo tribus patruis aconita, vehatur | pensilibus plumis atque illinc despiciat nos?*

Büch. inserts *quae* before *perniciosa*, and puts a question after both *petuntur* and *deorum*; he also attaches the couplet to the preceding paragraph, with which it seems to have no connection.

55. fas est, 'it is allowable, permissible'; cf. 1. 58. 131; VI. 628; X. 257: such things as power and eloquence are recognised as proper objects of prayer, which can be asked for *aperto voto* (Pers. 2. 7): there are objects of prayer which cannot be avowed; cf. Sen. *Epp.* 10. 5 *quanta dementia est hominum! turpissima vota dis insusurrant: si quis admoverit aurem, conticescent.*

genua incerare deorum: this refers to the custom, common in antiquity, of writing a prayer or vow on a wax tablet and placing it on a temple wall or the knees of a divine image; such a *votiva tabella* is generally mentioned as a sign of gratitude for prayer answered, but was also used in the way here described; cf. XII. 100. The knees were constantly grasped, or appealed to, by suppliants. Hence the Homeric ταῦτα θεῶν ἐν γούνασι κεῖται (await the decision of heaven).

57. mergit, 'they are wrecked by...'; the verb is active, as XIII. 8.

58. pagina honorum = *tituli*; see n. to I. 130. *pagina*, lit. 'column', not 'page': see n. to VII. 23.

descendunt statuae, 'down come the statues': the order of words should be kept. See n. to VIII. 18. It was a regular thing for the people to pull down and destroy the statues of a fallen favourite or dethroned emperor; cf. Tac. *Hist.* III. 85 *Vitellium ...coactum...cadentes statuas suas contueri*; Suet. *Dom.* 23 (after his death) *senatus adeo laetatus est ut...non temperaret, quin...scalas inferri clipeosque et imagines eius coram detrahi et ibidem solo affligi iuberet.*

sequuntur: cf. I. 164.

59, 60. Even the chariot and horses of a triumphal statue are not spared. **caballis:** cf. III. 118; but the word seems to have no satirical purpose here; it may already have been the colloquial equivalent for *equus*, which it finally displaced in the Romance languages, as the French *cheval* shows. The number of statues erected to Sejanus was enormous: the senate voted the erection of one in the theatre of Pompey (Tac. *Ann.* III. 72. 5); in the camp of every army, except in Syria, his statue was found near that of the emperor (Suet. *Tib.* 48).

62. **adoratum,** 'worshipped', in the literal sense; cf. Dio Cass. LVIII. 4. 3 καὶ τέλος καὶ ταῖς εἰκόσιν αὐτοῦ ὥσπερ καὶ ταῖς τοῦ Τιβερίου ἔθυον.

63. **Seianus:** L. Aelius Sejanus, the son of a Roman knight, born at Vulsinii in Etruria, became prefect of the praetorian guards, and finally the *alter ego* of Tiberius; at last the emperor discovered his treachery, and his utter ruin followed; he was executed 18 Oct. A.D. 31. Seneca also quotes him as the chief example of the instability of great power, *Dial.* IX. 11. 11 *honoribus summis functus es: numquid aut tam magnis aut tam insperatis aut tam universis quam Seianus? quo die illum senatus deduxerat, populus in frusta divisit.* Johnson, in his imitation of this satire, takes Cardinal Wolsey as the modern Sejanus. **secunda:** second only to the emperor himself.

64. The statue mentioned here is obviously of metal, not marble; that in Pompey's theatre was of bronze.

65. **laurus:** the outer door of a house was decorated with laurel on occasions of public or private rejoicing: cf. VI. 79 (for a marriage) *ornentur postes et grandi ianua lauro*; IX. 85 (for a birth) *foribus suspende coronas*; XII. 91.

66. Victims must be sacrificed to Jupiter in gratitude for the emperor's safety. If the victim, which should be white, has any dark spots, they must be chalked over (**cretatum**): the phrase, which, according to the Schol., comes from Lucilius, is surely sarcastic here; perhaps παρὰ προσδοκίαν for *auratum* (see n. to VI. 28).

ducitor unco: cf. XIII. 245; the body of an executed criminal was dragged from prison to the Gemonian steps by the *uncus* of the executioner, and there exposed for three days to public insult.

67. A conversation between two citizens begins here. **quae labra** refers to S.'s haughty expression: cf. *clauso labello*, III. 185; Quint. XI. 3. 80 *naribus labrisque non fere quidquam decenter ostendimus, tametsi derisus iis, contemptus, fastidium significari solet.*

70. **delator:** see n. to III. 116. The practice was at its worst in the reign of Tiberius, and no one had encouraged *delatores* more than Sejanus himself.

71. Dio Cassius (LVIII. 9) describes the scene in the senate when this letter was read and gives a summary of it. It was long,

and ambiguous after the usual manner of Tiberius even when he wished to make his meaning clear; but when the reading was over, the senators understood what was required of them.

72. **Capreis:** see n. to l. 93. **bene habet** = καλῶς ἔχει, 'all right': cf. Prop. V. 11. 97; Livy, VI. 35. 8, VIII. 6. 4, 9. 1 and 35. 3, XXXIX. 50. 8; Sen. *Herc. Fur.* 1040; idem, *Oed.* 1020; Stat. *Theb.* XI. 557, XII. 338. There are several instances in the elder Seneca. Terence (*Phorm.* 429) has *bene habent tibi principia.* In all the instances in Livy, it is, as here, colloquial.

Part of what follows is plainly a reflection of Juv.'s own, interrupting the conversation: it seems to begin at *sed quid* etc. The *turba Remi* are to be distinguished from the persons of the dialogue, who, as is shown by their anxiety, are men of some position.

73. **Remi:** *Romuli* would no doubt be preferred, if it were equally convenient metrically: in hendecasyllabic verse Catullus (49. 1) calls Cicero *disertissime Romuli nepotum. Quirini* is often used for *Romuli*, e.g. VIII. 259.

74. **Nortia,** the Etruscan goddess of Fortune, worshipped especially at Volsinii, the birth-place of Sejanus: hence **Tusco,** 'her Tuscan'.

75. **secura,** 'off his guard', is part of the supposition; Tiberius was not in any sense *securus*: cf. Suet. *Tib.* 63–6.

76. **diceret** = ἐκάλει ἄν, 'would be calling'.

77. **suffragia nulli vendimus:** the great constitutional change, by which the people ceased to be an element of the constitution, and the election of magistrates was transferred from the popular assemblies to the senate, was effected by Tiberius A.D. 14: cf. Tac. *Ann.* I. 15 *tum primum e campo comitia ad patres translata sunt: nam ad eam diem, etsi potissima* (the consular elections?) *arbitrio principis, quaedam tamen studiis tribuum fiebant. neque populus ademptum ius questus est nisi inani rumore*—a sufficient proof of their unworthiness. Caligula professed to make some attempt to restore their electoral rights (Suet. *Cal.* 16).

78. **vendimus,** substituted παρὰ προσδοκίαν for *damus* with an ironical reference to the bribery which was almost universal at elections in the latter days of the republic: indeed the same word (*ambitus*) meant 'canvassing' and 'bribery'.

effudit curas, 'has cast off its burdens', i.e. of public affairs: cf. Tac. *Hist.* I. 1 *inscitia reipublicae ut alienae*: the point is not that the people are anxious to have Sejanus rule over them, but they are indifferent and will acquiesce in any accomplished fact.

79. **imperium,** 'power', the administrative authority wielded by all the higher magistrates, at Rome and, with less restriction, beyond the city. **se continet,** 'narrows its field'.

80. **duas...res:** see n. to VIII. 118.

81. **panem:** the system of supplying corn below the market price to the populace at Rome, was introduced by Gaius Gracchus 123 B.C. Later demagogues, especially Clodius, went further and supplied it for nothing (cf. Cic. *pro Sest.* 55). The consequence was that poor citizens flocked from all parts of Italy to the capital, where they lived as idle and dangerous paupers. Julius Caesar reduced the number of corn-receivers from 320,000 to 150,000 (Suet. *Iul.* 41); but even he could not abolish the system, which demoralised the recipients and burdened the public revenue. Under the Empire the corn was brought chiefly from Egypt and Africa (see n. to VIII. 117), under the superintendence of the *praefectus annonae* (see n. to IV. 32): for the system of distribution, see n. to VII. 174.

circenses, sc. *ludos,* 'races', rather than 'games': see nn. to III. 223, X. 36 (*pompa Circensis*), XI. 198. The conversation, interrupted l. 72, is here resumed.

perituros...multos: cf. Tac. *Ann.* VI. 19 (*Tiberius*) *cunctos, qui carcere attinebantur accusati societatis cum Seiano, necari iubet. iacuit inmensa strages, omnis sexus, omnis aetas, inlustres, ignobiles.* Suet. (*Tib.* 61) speaks of twenty executions in one day as something exceptional; so the language of Tac. seems exaggerated.

82. **magna est fornacula,** 'the furnace is a big one', a colloquial way of expressing that 'many will find it hot'. The diminutive has lost all diminutive sense, when such an epithet as *magna* can be applied to it; cf. μέγα θηρίον Hom. *Od.* X. 171.

pallidulus: see n. to *lividulus,* XI. 110.

83–5. Madvig (*Opusc.* I. 44) explains this as follows: Bruttidius was in the habit of declaiming in the schools, where one of the stock subjects of debate (*controversiae,* see n. to I. 16) was the *armorum iudicium* between Ajax and Ulysses: cf. VII. 115: B. had made a poor speech on the side of Ajax, and the speaker here, who is not really a friend, says ironically, 'how I fear that Ajax, beaten in his case, is punishing B. for his bad speech'; the real cause of B.'s pallor is of course his fear of being involved in the downfall of Sejanus.

But there is no proof that Bruttidius spoke in the schools: he is mentioned as a historian by Seneca (*Suasor.* 6. 20) not as a declaimer. And it is simpler to understand Tiberius by Ajax, such a use of Homeric names being common (see n. to *Automedon*, I. 61), and to suppose that the speaker is expressing his fear that the emperor may punish freely in resentment for the danger he has run. Just as Ajax killed the cattle, so Tiberius, like Ajax when he lost the prize (*victus Aiax*), may slaughter the citizens, under the impression that (*ut*) he has been ill-defended. Consequently the two speakers, in order to prove their zeal, make haste to kick the prostrate body of the emperor's foe: this is the form their *defensio* takes.

This Bruttidius is probably Bruttidius Niger, who, as aedile A.D. 22, prosecuted Silanus for *maiestas* (Tac. *Ann.* III. 66. 2). The altar of Mars was in the Campus Martius.

86. **in ripa,** sc. *Tiberis*: the body of Sejanus, as those of many other criminals, was thrown into the river after being exposed.

87. **ne quis,** sc. *servorum*. **in ius,** 'into court': cf. Mart. I. 103. 11 *in ius, o fallax atque infitiator, eamus.*

90. **salutari:** this does not refer to greetings in the streets but to the swarm of citizens, high and low, who thronged S.'s reception-rooms in the morning as *salutatores*. **habere** = κεκτῆσθαι, 'to possess': cf. XIV. 207.

91, 92. **illi...illum,** 'one', 'another': again l. 196: so *hic... hic* I. 46 and 47.

91. **curules,** sc. *sellas*, which the worse MSS. give for *summas*; but the word can be understood: M. quotes Stat. *Silv.* III. 3. 115 *fasces summamque curulem | frater...tulit.* The consulship is meant, for which the favour of Sejanus was a necessary passport (Tac. *Ann.* IV. 68. 2).

93. **angusta:** *augusta* of P seems a mere slip; 'imperial' is here a feeble epithet, while there is point in the ruler of the world choosing a 'narrow' peak to perch on. In A.D. 27, under the influence of Sejanus, Tiberius withdrew permanently from Rome and never again entered it alive: he settled in the island fortress of Capreae (Capri) off the coast of Campania, where he is said to have abandoned himself to shameless profligacy. Cf. Suet. *Tib.* 40 *Capreas se contulit, praecipue delectatus insula septa undique praeruptis immensae altitudinis rupibus et profundo mari.*

94. **grege Chaldaeo:** see n. to VI. 553. Tiberius was *addictus mathematicae* (Suet. *Tib.* 69), having learned the science at Rhodes from Thrasyllus (see n. to VI. 576), whom he brought back to Rome and kept with him until his death A.D. 36.

certe, 'at any rate' you want executive power, if not outward show. **pila, cohortes,** 'the cohorts and javelins': this refers to the praetorian guard, consisting of ten cohorts, each 1,000 strong; they were quartered in barracks built for them by Tiberius in front of the *porta Viminalis*. Sejanus was their commander (*praefectus praetorio*). For *pila* as their weapon, cf. Mart. x. 48. 2 *pilata...cohors*.

95. **egregios equites,** 'distinguished knights': *vir egregius* is the common style of an equestrian magnate after the 1st cent., just as a senator is called *vir clarissimus*. Juv. here means that class among the *equites* who were selected by the emperor to enter on a career as *procuratores Augusti*: see n. to I. 58: they are constantly mentioned by imperial writers as *equites inlustres, insignes, splendidi*. Sejanus himself had been one of them. They are naturally mentioned here together with the praetorian guard; for the *eques*, who was to become a *procurator*, might perform his preliminary military service, either with an army abroad, or in the capital as *tribunus cohortis vigilum*, then *tribunus cohortis urbanae*, and lastly *tribunus cohortis praetoriae* : these also were *equestres militiae*.

Lewis and Friedl. are of opinion that the mention of these knights is irrelevant here, where Juv. is speaking only of military power, and that the cavalry attached to the praetorian cohorts, or else the *speculatores*, must be meant; but what of the epithet *egregios*? Their difficulty vanishes, when it is remembered that these civil functionaries had to begin their career by military service.

castra domestica, 'a barrack at your own disposal': the meaning is that S., by virtue of his position, could treat the praetorian guard as a part of his own household.

97. **posse,** 'the power to do so': cf. Ovid, *Heroid.* XII. 75 *perdere posse sat est, si quem iuvet ipsa potestas*. The clause is a good instance of a *sententia* (see n. to VIII. 125).

sed quae...malorum, 'but no glory and success is worth having, if prosperity must be followed by an equal measure of calamity': a more literal translation would be cumbrous: for this constr. with *tanti est ut...*, cf. VI. 178 and see n. to III. 54.

98. Cf. XIV. 314. The repetition of *mensura* l. 101 is somewhat slipshod.

99. **trahitur,** sc. *unco*; cf. l. 66. **praetextam:** see n. to l. 35. As *praefectus praetorio*, and therefore a knight, Sejanus had no right to this dress; but, by an extraordinary mark of favour, he received the *ornamenta praetoria* A.D. 20 which conferred the right; and in the year of his death A.D. 31 he was actually consul

together with Tiberius (Suet. *Tib.* 65). **sumere** is regularly
used of putting on clothes; cf. III. 172 *nemo togam sumit.*

100. Fidenae, Gabii and Ulubrae were little towns in Latium
which even in Horace's time had become proverbial for loss of
population and for grass-grown streets: cf. Hor. *Epp.* I. II. 7
Gabiis desertior atque | Fidenis vicus; ibid. 29 *quod petit* (i.e.
happiness) *hic est, | est Ulubris, animus si te non deficit aequus.*

potestas, 'magistrate', used in a concrete sense like ἀρχή: the
Italians still call a magistrate *podestà.*

101. Magistrates of all ranks are said *ius dicere:* cf. Cic. *ad
Att.* v. 15. 1 *quippe, ius Laodiceae me dicere, cum Romae A.
Plotius dicat,* Plotius being *praetor urbanus* when Cicero was
governor of Cilicia. The provincial magistrate has a humble
sphere of jurisdiction and is chiefly employed in seeing that the
shopkeepers don't cheat their customers by short measures and
false weights: cf. Pers. I. 129 *sese aliquem credens, Italo quod
honore supinus | fregerit heminas Arreti aedilis iniquas.*

102. For **aedilis** as the title of a provincial magistrate, see
n. to III. 179.

105. **numerosa:** see n. to VII. 151.

106. **tabulata:** cf. III. 199. **altior,** 'from a greater height'.
The sentiment is a commonplace of all ages: cf. Hor. *Carm.* II.
10. 10 *celsae graviore casu | decidunt turres.*

107. **et...ruinae,** 'and that the sheer descent of the totter-
ing building, once o'erthrown, might be measureless'. **inpulsae**
lit. 'pushed', i.e. sent over the edge: cf. Cic. *pro Cluent.* 70
praecipitantem impellamus. The word often means 'to over-
throw', 'to strike down' in silver-age writers, especially Lucan,
e.g. I. 149 (of Caesar) *impellens quidquid sibi summa petenti |
obstaret, gaudensque viam fecisse ruina;* idem, III. 440 *procumbunt
orni, nodosa impellitur ilex.* **praeceps** is here a neuter noun:
cf. Sen. *Epp.* 94. 73 *exanimantur et trepidant, quotiens despexerunt
in illud magnitudinis suae praeceps.* For **ruinae,** see n. to XI. 13.

108. **Crassos...Pompeios,** i.e. men like Crassus and
Pompey: cf. *Maecenatibus,* XII. 39.

109. Julius Caesar is certainly meant, who with Crassus and
Pompey formed the first triumvirate; Crassus was killed in
battle against the Parthians at Carrhae 53 B.C., Pompey was
murdered off the Egyptian coast 48 B.C., and Caesar in Pom-
pey's theatre 15 March 44 B.C.

flagra, 'his lash', as if they were slaves: cf. v. 173.

Friedl. quotes Plut. *Caes.* 57 δεδεγμένοι τὸν χαλινὸν...
δικτάτορα...αὐτὸν ἀπέδειξαν διὰ βίου.

110. Cf. Sen. *Epp.* 95. 3 *illos, quos honores nulla non arte atque opera petiti discruciant*, perhaps remembered by Juv.

111. **magna...vota,** 'prayers for greatness'.

112. **generum Cereris,** i.e. *Plutonem*; see n. to VII. 25.

113. **sicca,** 'bloodless'.

114–32. *Eloquence too, which every schoolboy longs and prays for, is often a snare to its possessors; witness the greatest orator of Greece and of Rome. If Cicero had confined himself to poetry, he would have been safe; it was his splendid eloquence that cost him his head. Demosthenes, also, came to a violent end, and would have been wiser to stick to his father's trade of blacksmith.*

Cf. Seneca (Haase, III. 454) *si muti fuissent Cicero et Demosthenes, et diutius vixissent et lenius obiisent.*

114. **ac famam:** so the inferior MSS.: P, followed by Büch., reads *aut famam.* Cf. Tac. *Ann.* I. 8. 3, where Nipperdey reads *ac* for *aut* of MSS. A distributive sense of *aut* is found in silverage Latin: e.g. Tac. *Ann.* XI. 38 *iugulo aut pectori* (i.e. now to her throat, now to her breast): but such a sense is impossible here, because the orator's fame is inseparable from his eloquence.

It has already been pointed out that eloquence was the object towards which the whole of Roman education was directed, the instruction of the *grammaticus* being considered merely preparatory to the formal rhetorical training which followed, and completed the curriculum: see nn. to I. 15 and 16.

115. The Quinquatrus was an annual feast of Minerva, 19–23 March, especially observed by teachers and scholars, she being the goddess of wisdom: edd. quote Ovid, *Fasti*, III. 809 *fiunt sacra Minervae, | nomina quae iunctis quinque diebus habent; ibid.* 815 *Pallada nunc pueri teneraeque orate puellae. | qui bene placarit Pallada, doctus erit.*

totis, i.e. during all the five days of the feast.

116. **Minervam** is generally explained as 'learning', and the *as* as the boy's fee to his teacher. This explanation seems to be due to the epithet *parcam.* But the evidence is against such a low fee as one *as:* Horace mentions 8 *asses* a month as the fee in an elementary country school (*Sat.* I. 6. 75). On the other hand, an *as* is often mentioned as the amount of a *stips* or contribution to a god's treasury: *as* and *stips* are even used as synonyms: thus the saying of Augustus, reported by Quint. VI. 3. 59 *noli timere, quasi assem elephanto des*, reappears in Suetonius (*Aug.* 53) with *assem* changed to *stipem.* Therefore it seems likely that the *as* is neither a fee nor a present to the master, but a *stips*

offered by the little boy to the goddess of wisdom, who can make him wise. Cf. Petron. 88 *quis unquam venit in templum et votum fecit, si* (on condition of) *ad eloquentiam pervenisset?* (where the context shows that this was a right thing to do). A difficulty remains in the epithet **parcam.** The Schol. explains *vilioris pretii fictile Minervae signum*: I would rather transl. 'economical', the quality, which properly belongs to *asse*, being transferred to the goddess: see n. to XII. 82; cf. Mart. VIII. 33. 11 (of a *strena* or new-year's gift, in the form of a gilded date and a *stips*) *hoc linitur sputo Iani caryota Kalendis, | quam fert cum parco sordidus asse cliens.*

adhuc, 'as yet'; when he is older, he will give more.

117. The boy is followed to school by a little slave who carries his little box of books. **capsa** is a circular box of beechwood used for the transport of books: cf. Catull. 68. 36 *capsula me sequitur*: how it differed from a *scrinium*, is uncertain: see Rich's *Companion*, p. 588. In the country boys carried their own slates and satchels; not in Rome (Hor. *Sat.* I. 6. 74).

118. **perit:** for the long final syll., see n. to III. 174.

119. **leto dedit,** 'laid to rest': an old Roman formula here used ironically: see Mommsen's *History*, I. 78 n.: 'The solemn announcement of the funeral of a citizen ran in the words *"ollus Quiris leto datus".*'

120. **ingenio** is dat.; 'genius had its hand and head cut off'. When Cicero was murdered by Antony's order 43 B.C., his head and right hand were cut off and fastened upon the Rostra in the Forum, where the heads of political victims were often exposed, as on Temple Bar in London as late as 1773: cf. Boswell's *Johnson* (1874), I, 470.

122. 'O happy Rome, born in my consulship'; i.e. Rome would have died but for Cicero's suppression of Catiline. The quotation is a notorious verse of Cicero's, probably from his epic poem in three books *de suo consulatu*: it was a famous mark for the banter of ancient critics, and is certainly a very ugly line. Cicero must have intended the assonance; but the effect is far from pleasing. His poetry is generally spoken of with contempt, yet a verse translation of Aratus, written by him in boyhood, found a great poet to admire it; see Munro on Lucr. V. 619.

123. Juv. refers to Cic. *Phil.* II. 118 *contempsi Catilinae gladios, non pertimescam tuos* (i.e. *Antoni*).

124. **ridenda poemata malo,** i.e. they are better, as being safer. The argument will not bear much scrutiny, and might be refuted from Juv. himself VIII. 83 and 84, or by Scott's stirring

verses: 'Sound, sound the clarion, fill the fife! | To all the
sensual world proclaim: | one crowded hour of glorious life | is
worth an age without a name.'

125. Cicero's Philippics are the fourteen speeches against
Antony, so named after Demosthenes' invectives against Philip.
Of these the most famous, the second, was never spoken: it was
written in answer to an attack by Antony in the senate 19 Sept.
44 B.C. For the fame of these speeches, cf. Tac. *Dial.* 37 *nec
Ciceronem magnum oratorem P. Quinctius defensus aut Licinius
Archias faciunt; Catilina et Milo et Verres et Antonius hanc illi
famam circumdederunt.*

126. **volveris** = art read: cf. VI. 452; XV. 30: *evolvere* is
commoner in prose: cf. Sen. *Dial.* XII. 1. 2 *cum omnia claris-
simorum ingeniorum monumenta...evolverem.* For the form of the
ancient book, see n. to VII. 23. **a prima proxima,** 'next to the
first', i.e. second: cf. l. 247; Ovid, *Trist.* V. 8. 38 *haec sunt a
primis proxima vota meis.*

illum: Demosthenes, when the struggle against Macedonia
became hopeless, took poison in the temple of Poseidon at
Calauria 322 B.C.

128. **torrentem:** cf. l. 9. **moderantem frena** seems to be
imitated by Milton, *P.R.* IV: 'Thence to the famous orators
repair, | those ancient, whose resistless eloquence | wielded at
will that fierce democratie.' **theatri:** in the 5th cent. B.C. the
theatre of Dionysus was used on special occasions for meetings
of the *ecclesia* instead of the Pnyx; in later times it was regularly
used.

130. Demosthenes' father was not a blacksmith but a wealthy
manufacturer of swords: he died when D. was seven years old.

133–87. *Men thirst after military glory, as the greatest prize life
can offer, and will undergo any toil and any danger in pursuit of
it. Yet ere now men have ruined their country from this ambition.
And how short-lived is their fame! how often the end is defeat
and humiliation! Hannibal, after raging resistless all over the
world, came to be slighted by a petty king, and poisoned himself
at last. Alexander sighed for more worlds to conquer, but soon
found he must be content with six feet of earth. Xerxes set nature
at defiance, when he invaded Greece, and punished the very
winds and seas for thwarting his will; but in what a plight did
he return!*

133. **exuviae,** 'spoils', lit. 'strippings': cf. Stat. *Theb.* II.
725 *nunc tibi fracta virum spolia informesque dicamus | exuvias.*

truncis...tropaeis: the epithet refers to the simplest form of a trophy, erected on the field after a victory: this was the stump of a tree, stripped of the leaves and most of the branches, and then covered with captured weapons and pieces of armour: cf. Suet. *Cal.* 45 *truncatis arboribus et in modum tropaeorum adornatis*; Sen. *Thyest.* 659 *affixa* (sc. *quercui*) *inhaerent dona, vocales tubae,* | *fractique currus...victaeque falsis axibus pendent rotae.*

135. **curtum temone,** 'shorn of its pole'; for the constr., cf. *puppe minor* (Lucan, II. 717), *capite truncum* (Mart. VII. 20. 15).

136. **arcu,** a triumphal arch, such as those still remaining in the Roman Forum, with 'downcast captives' sculptured upon it: cf. Suet. *Dom.* 13 *arcus...tantos ac tot extruxit, ut cuidam Graece inscriptum sit,* 'arci' (i.e. ἀρκεῖ, 'enough arches').

137. **humanis maiora,** 'too great for man', i.e. divine: in a different sense XIII. 221. **ad hoc,** 'to this end', i.e. military glory.

138. **barbarus,** i.e. neither Greek nor Roman. Plautus, translating from Greek originals, often uses the word as = *Romanus* for comic effect; so *Mostell.* 828 *pultifagus opifex... barbarus,* where see Lorenz. But later writers use it in the same sense as here, e.g. Quint. V. 10. 24 *nec idem in barbaro, Romano, Graeco, probabile est.*

induperator: cf. IV. 29: the form is not ironical here but required by the metre; it is common in Lucr. but never found in the Augustan poets, nor in Martial's dactylic verse; Mart. has *imperator* twice, and Catullus often, in iambic verse.

141. **ipsam,** 'for herself'.

142. **olim,** 'again and again'. Seneca (*de Ben.* v. 16) gives a list of Romans who turned their arms against their country— Coriolanus, Catiline, Marius, Sulla, Pompey, Caesar, and Antony.

143. **gloria,** 'ambition': cf. Sen. *Epp.* 94. 65 *quid Caesarem in sua fata pariter ac publica inmisit? gloria et ambitio.* For **tituli,** see n. to I. 130; such an inscription may be on the base of a statue, or beneath an *imago* (VIII. 69), or, as here, on a tomb; cf. VI. 230; Sen. *Dial.* x. 20. 1 *misera subit cogitatio laborasse ipsos in titulum sepulcri.*

145. Cf. Mart. x. 2. 9 *marmora Messalae findit caprificus.*

147. 'If you lay Hannibal in the scale, how many pounds will you find in the greatest of commanders?' His ashes (cf. l. 172) are meant: cf. Ovid, *Met.* XII. 615 *iam cinis est; et de tam*

magno restat Achille | nescio quid, parvam quod non bene compleat urnam; [Seneca] *Herc. Oet.* 1767 *ecce vis totam Hercules | complevit urnam. quam leve est pondus mihi, | cui totus aether pondus incubuit leve.*

148. **non capit** = οὐ χωρεῖ,, 'cannot contain'. The boundaries here assigned to Africa are naturally not those of modern geography: Juv.'s Africa is bounded on the west by the Moorish Ocean, i.e. the Atlantic off the coast of Mauretania; it stretches eastwards as far as the Nile, and southwards as far as Aethiopia.

150. **rursus**, lit. 'backwards', i.e. southwards: cf. XII. 76: *admota* must be supplied again here.

aliosque elephantos: this is the reading of the inferior MSS.; P has *altos*, but with a gloss *praeter Indicos* which shows *alios* to be its original reading: for the confusion of *t* and *i*, see Introduction, p. xlvi. It means, 'a second race of elephants', *alius* being used for *alter* as VIII. 245. There were elephants in the west in Mauretania and Gaetulia (l. 158), and also in the far south beyond Syene: cf. XI. 124.

151. **Hispania:** Hannibal was commander of the Carthaginian forces in Spain 221 B.C.; he crossed the Pyrenees 218 B.C. and the Alps in October of the same year.

153. **aceto:** Livy (XXI. 37. 2) relates that H. blasted the rocks by pouring vinegar on them when heated by fire; and Pliny mentions it as a common process in the Spanish mines, where H. and many of his soldiers must have seen it. Calcareous rocks would be dissolved by vinegar; it seems doubtful whether heat would add to the effect. H. would have plenty of vinegar, this mixed with water being the regular drink of soldiers, as of the common people (III. 292).

155. Perhaps a reminiscence of Lucan, II. 657 (of Caesar) *nil actum credens cum quid superesset agendum.*

156. **Subura:** see n. to V. 106.

157. **facies**, 'sight', not 'face', a common silver-age use of the word: cf. VII. 137; Pliny, *Paneg.* 35 *memoranda facies*; 56 *decora facies*; 71 *insolita facies*; 82 *foeda facies, cum populi Romani imperator alienum cursum...sequeretur.* Pliny in each case omits the verb *est*: so here. Virgil often has the word in this sense with a gen.

quali digna tabella, 'how fit for caricature!'

158. Hannibal crossed the Apennines on his sole surviving elephant in the spring of 217 B.C., and lost an eye, from disease, in marching through the country flooded by the Arno (Livy, XXII. 2. 10).

159. **vincitur:** his first and last defeat was at Zama 202 B.C. by the elder Africanus.

161. **sedet** suggests the attitude of a suppliant: see Liddell and Scott under καθέζεσθαι.

praetoria regis: *praetorium is any large house or palace*: cf. I. 75. H. fled from Carthage 193 B.C., fearing to be given up to Rome; he went to the East and was living at the court of Prusias I, king of Bithynia, when Flamininus, after the defeat of Antio-chus, required the surrender of Rome's inveterate enemy. Escape being impossible, he took poison, which he carried about in a ring (*anulus*, l. 166), and died at the age of 76, 183 B.C.

162. **vigilare:** Juv. is transferring to Bithynia the customs of Rome, where a *salutator* had often to wait until the great man chose to get up: cf. v. 19; Sen. *Dial*. x. 14. 4 *suum somnum rumpunt ut alienum expectent* (to wait for the end of the great man's sleep).

163. **miscuit,** 'turned upside down': see n. to VI. 283.

165. **Cannae,** in Apulia, was the site of the greatest victory of H. over the Romans 216 B.C. The mention of Cannae and the emphatic position of *anulus* make it probable that Juv. refers to the story that the messenger who bore the news of Cannae to Carthage, confirmed his tidings by pouring out in an immense heap before the senators the gold rings of the Romans killed in the battle (Livy, XXIII. 12. 1).

166. **i et...curre:** see n. to VI. 306: *nunc* is generally found between the two imperatives.

167. 'To suit the taste of schoolboys and become the subject of their speeches': it is really their own composition that *placet pueris*: cf. Petron. 4 (eloquence is taught ready-made to boys, with the worst results) *quodsi paterentur laborum gradus fieri... ut persuaderent sibi nihil esse magnificum quod pueris placeret, iam illa grandis oratio haberet maiestatis suae pondus*. For the exer-cises in the rhetorical school, see n. to I. 16; for Hannibal as the subject of a *suasoria*, see n. to VII. 161. The fate that has befallen H.'s name is satirically, or rhetorically, represented as the object of his exertions.

Johnson takes Charles XII of Sweden as the modern Hanni-bal, and ends his paragraph with the familiar couplet, 'He left a name, at which the world grew pale, | to point a moral and adorn a tale.'

168. Cf. XIV. 313; Lucan, x. 456 (of Caesar) *hic, cui Romani spatium non sufficit orbis*.

Alexander the Great was born at Pella, and died at Babylon in his 33rd year, 323 B.C. Juv. means that a native of the little town of Pella might have been content with possessing a single world.

169. **limite** = *propter limitem*: abl. of cause: cf. VI. 272.

170. For Gyarus (or Gyara) as a place of exile, see n. to I. 73; for Seriphus, n. to VI. 564. Such rocky islands are often called *scopuli*: e.g. Tac. *Hist.* I. 2 *plenum exiliis mare, infecti caedibus scopuli.*

171. Babylon, famous for the brick walls built round it by Semiramis, is described rather than directly named: cf. Ovid, *Met.* IV. 57 *dicitur altam | coctilibus muris cinxisse Semiramis urbem.*

172. '**sarcophagus**, "flesh-eating", is an epithet of *lapis*, *lapis Assius*, a stone chosen for coffins as hastening decay' (M.).

173. **creditur olim**, 'it has long been believed', = *iam dudum creditur*: see n. to III. 163. So M.; but it is also possible to take *olim* with *velificatus* in the sense of 'long ago'. Juv. evidently disbelieved that such a canal had been made through the peninsula of Mount Athos, but modern travellers have found indubitable remains of it. Modern investigation has repeatedly justified Herodotus against his ancient critics: his account of the canal will be found VII. 22, of the bridge *ibid.* 33 ff.

175. **audet**, sc. *dicere*: see n. to VI. 644.

constratum classibus does not refer to the bridge of boats (see l. 176) but only to the size of the fleet: cf. Livy, XXXV. 49. 5 (of Antiochus, who made no bridge) *rex...consternit maria classibus suis.*

176. This refers to the bridge over the Hellespont made by Xerxes, preparatory to his invasion of Greece. The same meaning would be given more simply by *impositas mari rotas*, *rotae* being the chariots which passed over.

177. For rivers drunk dry by the invading host, cf. Herod. VII. 21 κοῖον δὲ πινόμενόν μιν ὕδωρ οὐκ ἐπέλιπε, πλὴν τῶν μεγάλων ποταμῶν; *ibid.* 108. The saving-clause shows that the historian, though unwilling to omit this picturesque detail, felt some qualms about it.

178. **prandente**, 'breakfasting', is sarcastic: cf. Mart. IV. 49. 3 *qui scribit prandia saevi | Tereos.* Xerxes did not require an ἄριστον but only a δεῖπνον from the cities he passed through, as is shown by the witty remark of the man of Abdera (Herod. VII. 120).

madidis...alis: Sostratus, an unknown poet, who ap-

parently wrote on the invasion of Xerxes. Friedl. suggests that
he may have been a candidate for the prize for Greek poetry at
the last *agon Capitolinus* (see n. to VI. 387) in A.D. 126. The
meaning of *madidis alis* has been much debated: the most pro-
bable interpretation is that of the Schol., *quia omnes qui cum
sollicitudine recitant, necesse est ut alae eis sudent.*

179. **ille,** i.e. Xerxes.

180. **solitus,** 'he who had been wont'. When the bridge over
the Hellespont was made, it was broken by a storm; Xerxes
ordered 300 lashes to be inflicted on the sea and fetters to be
sunk into it, with an insulting address (Herod. VII. 35). The
historian adds that he 'had heard' that Xerxes sent branders also
to brand the sea.

181. **barbarus** may imply either cruelty or folly: as Xerxes
may be charged with both, it is difficult to say which Juv. means
to convey here. In the narrative of Herodotus (VII. 35 ἐνετέλλετο
ῥαπίζοντας λέγειν βάρβαρά τε καὶ ἀτάσθαλα), βάρβαρος means 'non-
Hellenic', i.e. irrational.

Aeolio: cf. V. 101.

183. 'It showed leniency, it is true, that he did not think the
god deserved branding as well.' Weber altered the text (to
sane; quid? non...credidit?) to avoid a contradiction of Hero-
dotus: perhaps rightly: but H. tells this detail with some hesita-
tion, and Juv. need not have been careful to follow that narrative.

184. **servire:** as Apollo served Admetus. All the punish-
ments mentioned above are those of slaves; cf. XIV. 19 *flagellis*,
23 *catenae*, 24 *inscripti*.

185. **sed,** returning after a digression. **nempe,** 'surely', in
the answer to a question; so l. 110, VIII. 180.

187. **totiens** may be taken both with *optata* and *exigit*: 'such
the penalty which glory so often prayed for has so often
exacted'.

188–288. *Long life is another object for which all men pray. They
forget the countless miseries of old age, the physical deformity
which makes the old repulsive to their nearest and dearest, the
lessened power of enjoyment, the decay of all the faculties, the
constant aches and ailments, and, worse than all, the loss of sense
and reason. If the mind is still unimpaired, the old man only
suffers worse from successive bereavements which carry off the
younger generation before him. Think of Nestor, mourning over
the length of life which made him survive his own son. Think of
Priam: how happy for him, if he had died earlier while Troy*

still stood firm. He survived to see the general overturn and died
a violent death; Hecuba lived longer, and her fate was even worse.
In Roman history, Marius and Pompey survived their own
prosperity; for them length of life was sheer misfortune.

It has been suggested that Shak. refers to this passage in
Hamlet's reply to Polonius (Act II, Scene 2): 'the satirical rogue
says here that old men have grey beards, that their faces are
wrinkled, their eyes purging thick amber and plum-tree gum,
and that they have a plentiful lack of wit, together with most
weak hams: all which, sir, though I most powerfully and
potently believe, yet I hold it not honesty to have it thus set
down.' Swift takes some traits from Juv. for his own much more
repulsive picture of the Struldbrugs (*Gulliver's Travels*, part II,
chap. X): 'at ninety they lose their teeth and hair; they have at
that age no distinction of taste, but eat and drink whatever they
can get, without relish or appetite. The diseases they were
subject to still continue, without increasing or diminishing. In
talking they forget the common appellations of things and the
names of persons, even of those who are their nearest friends
and relations.'

189. **recto vultu:** cf. *recta facie*, VI. 401; but here 'the un-
distorted face' surely denotes youth, as opposed to the *malae
labantes* (Suet. *Aug.* 99) or *pendentes genae* (l. 193) and *pallor* (cf.
l. 229 and Sen. *de Ben.* VII. 27. 3) of old age.

solum hoc...optas, 'you pray for this and for nothing else';
solum goes only with the second clause.

191. **ante omnia,** i.e. physical deformity is only the first of
many miseries; for a different use, cf. XI. 192.

192. The repetition of **deformem** is remarkable: it does not
seem effective and is perhaps accidental.

194. **Thabraca** (now Tabarca), a town on the coast of
Numidia.

198. **una...facies,** 'but old men all look alike'.

cum voce...membra = *et vox et membra*: these and the
following nominatives are subjects to a phrase understood, such
as *mala sunt senectutis*.

199. **madidi...nasi:** cf. VI. 148.

202. **captatori...Cosso:** see nn. to III. 129, IV. 19. The
will-hunter is not easily disgusted, but even he must draw the
line here.

209. **partis,** i.e. faculty, that of hearing.

210, 211. **quae cantante** cet., 'what pleasure has he in

singing, though the singer be a rare performer, or (if he is present) when Seleucus is the harpist, and those etc.': there seems to be an ellipse such as is supplied in the transl. Juv. is enumerating the different kinds of music, so that l. 212 may allude to the gorgeous dresses worn by *tibicines* in the theatre: cf. Hor. *Ars*, 214 *luxuriem addidit arti | tibicen traxitque vagus per pulpita vestem.* Seleucus is not certainly mentioned elsewhere.

213. **magni...theatri:** there were at this time two large theatres in Rome, the *Marcellianum* and *Pompeianum* (Mart. II. 29. 5, X. 51. 11), as well as the smaller theatre of Balbus: the *Pompeianum* held over 17,000 spectators.

214. In all ancient theatrical entertainments, music played a part; in the theatre of Juv.'s age, a very large orchestra, chiefly of wind-instruments, was essential: see n. to *Paridi*, VII. 87. Seneca (*Epp.* 84. 10) declares that there are more musicians in the modern orchestra than there used to be spectators in the theatre.

216. **quot nuntiet horas:** the Romans reckoned time either by sun-dials (*solaria*), or by water-clocks, which, however, were different from the *clepsydrae* used in the law-courts. Their reckoning of time was essentially different from ours in this respect, that their hours, being each $\frac{1}{12}$th of the solar day, were of unequal lengths at different times of the year, being longer in summer and shorter in winter. This placed peculiar difficulty in the way of constructing clocks. Not carrying a watch, the Roman set a slave to watch a sun-dial or water-clock and report the hour: cf. Pliny, *Epp.* III. 1. 8 *ubi hora balinei nuntiata est*; Mart. VIII. 67. 1 *horas quinque puer nondum tibi nuntiat.* Hence 'to ask the time' is *mittere ad horas* (Cic. *Brut.* 200).

217. **gelido iam in corpore:** cf. Sen. *Epp.* 67. 1 *iam aetas mea contenta est suo frigore; vix media regelatur aestate.*

218. **agmine facto:** cf. III. 162.

221. **Themison:** Seneca (*Epp.* 95. 9) and Celsus (III. 4) mention a famous physician of this name: Juv. is probably speaking of a contemporary of the same name. For Greek physicians, see n. to III. 58. **autumno,** the season of most sickness at Rome; see n. to IV. 56.

222. **socios, 'partners'** in business: this Basilus seems to be a different person from the *causidicus* of VII. 145.

226 is repeated from I. 25: see n. to X. 365. Such repetitions occur also in Horace.

gravis...sonabat, 'rasped'; for the age up to which the beard was cropped, see n. to VI. 105.

227. **debilis,** 'ailing': it often means 'maimed', and is then opposed to *integer*.

232. **ieiuna:** this detail is not necessary to the simile but is kept from Hom. *Il.* IX. 323 ὡς δ' ὄρνις ἀπτῆσι νεοσσοῖσι προφέρῃσιν | μάστακ' ἐπεί κε λάβῃσι, κακῶς δ' ἄρα οἱ πέλει αὐτῇ (where Achilles is speaking of his toils on behalf of the Achaeans).

omni, 'any': a commonsense after the comparative: cf. VIII. 209, X. 303.

233. **maior:** *maius* (sc. *damnum*) would be more accurate; and such a corruption could be easily accounted for.

236. **saevo,** 'unnatural', is a metrical equivalent for *impio*.

237. Edd. generally take **suos** as an epithet of **heredes,** *sui heredes* being a term of Roman law, and including a wife who is *in manu*, sc. *mariti*, and children who are *in potestate*, sc. *patris*. But Juv. was not bound to bring in this legal term; and it is much more natural to take *suos* (his own flesh and blood) as object of *vetat*, and *heredes esse* as predicate. *heres esse* is common for 'to inherit'.

vetat: *vetare* and *iubere* are often used of decisions expressed in a will or other legal document: cf. Mart. IX. 87. 3 *affers nescio quas mihi tabellas, | et dicis 'modo liberum esse iussi | Nastam— servolus est mihi paternus— | signa'*; Pliny, *Epp.* IV. 10. 1 *Modesto, quem liberum esse iussi.*

238. Phiale is the old man's mistress.

tantum...oris: transl. 'so prevailing the breath of her artful mouth'.

240. **ut,** 'though'.

241. **funera natorum:** that the father should bury the children seemed to the ancients a tragic inversion of the order of nature: cf. Mart. I. 114. 3 *condidit hic natae cineres nomenque sacravit | quod legis Antullae, dignior ipse legi. | ad Stygias aequum fuerat pater isset ad umbras.*

aspiciendus: cf. the use of ἐπιδεῖν, for which see Thompson's n. on Plat. *Gorg.* 473 C.

242. **sororibus,** i.e. the ashes of his sisters.

245. Quintilian, when both his sons had died young, asks (VI. *praef.* 4) *quis mihi bonus parens ignoscat,...si quis in me alius usus vocis, quam ut incusem deos superstes omnium meorum?* Cf. Suet. *Tib.* 62 *felicem Priamum vocabat, quod superstes omnium suorum extitisset*; Trevelyan's *Macaulay*, II, 417: His imagination was deeply impressed by an old Roman imprecation, which he had noticed long ago in a Gallery of Inscriptions: '"ultimus suorum moriatur"; an awful curse!'

246. **rex Pylius,** Nestor: cf. Hom. *Il.* I. 250 τῷ δ' ἤδη δύο μὲν γενεαὶ μερόπων ἀνθρώπων | ἐφθίαθ'. . . μετὰ δὲ τριτάτοισιν ἄνασσεν.

247. **vitae,** 'of long life'. **a cornice secundae:** cf. Hor. *Sat.* II. 3. 193 *Aiax, heros ab Achille secundus; a* is lit. 'reckoning from'. The long life of the crow was proverbial, e.g. Hesiod *fragm.* ἐννέα τοι ζώει γενεὰς λακέρυζα κορώνη | ἀνδρῶν ἡβώντων.

249. **iam dextra conputat,** 'begins to reckon on his right hand': the ancients counted units and tens on the fingers of the left hand, and hundreds on those of the right: hence Juv. means that Nestor has reached 100 years, having lived through three *saecula,* or generations of men, each reckoned at 30 years.

251. No such scene is found in Homer; in the Odyssey (III. 111 ff.) Nestor alludes to the death of his son at Troy.

252. **stamine,** lit. 'thread', i.e. of life, spun for him by the Parcae; cf. III. 27, XIV. 249.

253. **Antilochi barbam ardentem,** 'the beard of A. on the pyre'; cf. Sen. *Epp.* 99. 27 *ne illo quidem tempore quo filius ardet...cessare pateris voluptatem.* The 'beard' shows that, according to Roman customs, he was still in the prime of life: see n. to VI. 105.

256. **Peleus,** sc. *dicit* or *queritur.*

257. **alius,** 'another', i.e. Laertes, the old father of the wandering Ulysses: cf. I. 10.

cui fas...lugere, 'whom nature bids lament' M.: cf. Stat. *Theb.* XII. 79 *fas sit lugere parenti.* Friedl. takes *fas* as = *fatum,* but the other instances in Juv. (quoted on l. 55) are not in favour of this. *lugere* is 'to mourn as dead': cf. III. 279; and the point seems to be, that though it is generally *nefas* to mourn for the living, it was not so in the case of Ulysses, who could not be supposed to be still alive.

natantem, 'on the sea', does not refer particularly to the famous swim to the Phaeacian coast: the word is often used of ships.

258. Priam is the next example of the misery of living too long; cf. Suet. quoted on l. 245. Edd. point out that the whole description is taken apparently from Cicero *Tusc. Disp.* I. 85.

259. **Assaracus,** son of Tros, was the brother of Priam's grandfather, Ilos.

magnis, 'splendid', *sollemnia* being used as a noun. **funus,** 'his dead body': see n. to IV. 109.

260. **reliquis:** this convenient word is never used by Virgil, Horace, or Ovid, who considered it a tetrasyllable beginning with 3 shorts; Lucretius uses it in the form *rellicuus,* treating it

as the Greeks treated ἀθάνατος: see Munro on Lucr. 1. 560. Juv. uses it v. 149, XIV. 36; Martial only once (1. 49. 41) in the form *relicum*.

reliquis f. cervicibus = the other brothers on their shoulders.

261. **ut...palla:** the clause defines and amplifies *Iliadum lacrimas*; for similar clauses, cf. v. 2, VIII. 42. Yet there is some awkwardness here, as this detail ought to be exactly parallel with *Hectore funus portante*: and it is possible that *ut* may be 'when', as the subj. is required by the conditional form of the whole sentence. Cassandra and Polyxena were two of Priam's daughters. For rending of clothes as a sign of mourning, cf. XIII. 132.

263. **diverso,** 'different'; see n. to III. 268.

264. **audaces carinas,** 'daring keels': for the same phrase used of ships in general, cf. Stat. *Silv.* III. 2. 1.

265. **contulit,** sc. *Priamo.* **vidit** = ἐπεῖδε: see n. to l. 241.

267. **tiara** here is perhaps not a crown, but the national head-dress of the Phrygians, a kind of fez; see n. to VI. 516.

268. **ruit,** 'was felled' by Pyrrhus; cf. Virg. *Aen.* II. 506-58.

270. **ingrato:** gratitude for the ox's labour is due from the plougher, not the plough; but such a transference of epithet is common in all poetry: so *audaces carinas* above. **fastiditus,** 'discarded'.

271. **utcumque hominis,** 'was human at any rate'; *utcumque,* like *quicumque,* is in Cicero a *relative* adverb and must be followed by a verb; this use, as an *indefinite* adverb, dates from Livy: see n. to III. 156.

torva is not an epithet of *uxor* but part of the predicate. Legend said that Hecuba was changed into a dog: cf. [Seneca] *Agam.* 745 *circa ruinas rabida latravit suas,* | *Troiae superstes, Hectori, Priamo, sibi.*

272. **vixerat:** the plpf. is apparently used with the meaning of an aorist; cf. VI. 281. Examples of the plpf. thus used occur in many Latin poets and abound in Martial: see n. to XV. 16.

273. **nostros,** i.e. examples from Roman history. For **transeo,** see n. to III. 114. Mithridates, king of Pontus, and Croesus of Lydia, were both monarchs who lost their thrones.

275. **spatia,** perhaps 'lap', a metaphor taken from the Circus; cf. Sen. *Dial.* IX. 9. 3 *non in cursu tantum circique certamine sed in his spatiis vitae interius flectendum est* (you must take the inside course): see n. to III. 223. Solon's interview with Croesus, which modern historians pronounce to be mythical, is related in Herod. I. 29 ff.

276. Marius appears elsewhere (VIII. 245 ff.) to point a different moral: he tried to hide from Sulla's pursuit 88 B.C. in a swamp at Minturnae, in the south of Latium on the river Liris, but was captured and imprisoned. When permitted to escape, he went to Africa and lived a beggar in a hut on the ruins of Carthage. It is irrelevant to Juv.'s purpose to mention that he lived to return to Rome a conqueror and to hold a seventh consulship.

278. **hinc** = *a longa vita*. For **quid**, see n. to VI. 284.

279. The climax formed by **Roma** is remarkable: it is significant of the estimate formed by the Romans of their imperial city.

280. **circumducto**: a triumphal procession made a long circuit through the city; the chariot of the general was preceded by the principal captives in chains; for the general's state, see nn. to ll. 38–43.

281. **bellorum pompa**, sc. *circumducta*: this refers to the procession generally, and perhaps especially to the spoils displayed, *pompa* meaning both 'show' and 'procession'. The last syll. of *pompa* is unelided; *hiatus*, not uncommon in Juv., is commonest at this place in the verse: so III. 70; VI. 274, 468; XII. 110; XIV. 49; XV. 126.

animam exhalasset opimam, 'he had breathed forth his life in glory': cf. Sen. *Epp.* 101. 14 *invenitur aliquis qui velit... per stilicidia emittere animam quam semel exhalare?* The epithet *opimus* is properly applied to *spolia*, hence, as by Horace (*Carm.* IV. 4. 51), to *triumphus*, *decus* etc., and here more boldly to the soul or life of the *triumphator*.

282. **Teutonico...curru**, i.e. the triumph over the Teutones and Cimbri; see nn. to VIII. 249–53.

vellet = was about to: the future tense in English (I will) is exactly parallel. Had Marius died as soon as his triumph was over, he would have been, in Tacitus' words, *felix opportunitate mortis*.

283. The reflections about Pompey also are drawn from Cicero *Tusc. Disp.* I. 86. He had a dangerous fever at Naples 50 B.C.: cf. Cic. *ad Att.* VIII. 16. 1 *municipia vero deum* (sc. *Caesarem putant*), *nec simulant ut cum de illo* (i.e. Pompey) *aegroto vota faciebant*.

284. **optandas**, 'for which he should have prayed', whereas in fact all the prayers sought to banish it.

285. **igitur**: see n. to VI. 210. This position of *igitur* is the rule in Sallust and Tacitus.

urbis is added to imply that the fortunes of Rome were bound up with those of Pompey, that her freedom died with him.

286. **caput abstulit:** his head was cut off by order of Ptolemy's advisers, on the Egyptian coast 29 Sept. 48 B.C.

287. The worst of traitors suffered less than he whom Cicero (*Phil.* II. 54) calls 'the light and glory of the Roman Empire', and Seneca (*Dial.* VI. 20. 4), *decus firmamentumque imperii.* Lentulus and Cethegus (see n. to VIII. 231) were strangled in the Tullianum, by order of the Senate; Catiline fell fighting bravely in battle at Pistoria, early in 62 B.C.

caruit: see n. to VI. 564.

288. **iacuit,** 'lay' on the battle-field.

Catiline forms the climax: to Juv., as to Virgil (*Aen.* VIII. 668), he is the chief of sinners: the comparative innocence of 'political' crimes is a modern discovery.

toto, 'unmutilated'.

289–345. *Every mother prays for beauty for her boys, and still more for her girls: but the fate of Lucretia shows that beauty is often a deadly danger. A handsome boy too has dangers in his path, which the ugly and deformed escape. If he leads a dissolute life, his punishment may be severe and his character will deteriorate; even if he is chaste, the examples of Hippolytus and Bellerophon show how dangerous beauty is. A modern instance is supplied by Gaius Silius, whose beauty Messalina coveted. If he refused to comply with her wishes, he had to die that day; if he complied, death could not be long delayed, when the outraged emperor heard the tale; in any case, he had to die.*

289. **optat:** see n. to l. 7; cf. Sen. *Epp.* 60. 1 *etiamnunc optas quod tibi optavit nutrix tua aut paedagogus aut mater? nondum intellegis, quantum mali optaverint?* The exquisite story in Herodotus (VI. 61) may be compared, of the nurse who carried an ugly child every day to the temple of Helen, ὅκως δὲ ἐνείκειε ἡ τροφός, πρός τε τὤγαλμα ἵστα καὶ ἐλίσσετο τὴν θεὸν ἀπαλλάξαι τῆς δυσμορφίης τὸ παιδίον.

290. **Veneris:** she is the giver of *venustas*, beauty and charm.

fanum videt: see n. to l. 116.

291. **usque ad delicias votorum,** 'even to fancifulness in her prayers', i.e. praying even for trifles: *deliciae* often has the meaning 'airs and graces' (see n. to VI. 47) and sometimes the sense of *delicatus homo* (see n. to IV. 4); the idea is the same here. The word might denote either a feeling in the mind of the person

praying, or a quality in the object prayed for; in the present instance, the latter seems more likely. The meaning is, that mothers are not content to ask simply for beauty; they want something more out of the common, specifying, for example, the colour of hair or eyes they wish their children to have: cf. Sen. *de Ben.* IV. 5. 1 *neque enim necessitatibus tantummodo nostris provisum est: usque in delicias amamur* (i.e. the gods not only provide us with necessaries but show their love by providing us with luxuries as well); idem, *Epp.* 90. 16 *simplici cura constant necessaria: in delicias laboratur* (where the two things are opposed in just the same way). In the present instance, *forma* may be called the *necessitas*, but mothers are not content with that, unless they secure the *deliciae* (superfluities) as well.

inquit: see n. to III. 153.

292. pulchra...Diana = in her daughter's beauty: taken from Hom. *Od.* VI. 102 οἵη δ'Ἄρτεμις εἶσι...γέγηθέ δέ τε φρένα Λητώ: or from Virg. *Aen.* I. 502 *Latonae tacitum pertemptant gaudia pectus.*

293. Lucretia is the wife of Collatinus, who killed herself when outraged by Sextus Tarquinius: cf. Livy, I. 57 ff.

294. Rutila is no doubt an actual person, whose deformity was familiar to Juv.'s contemporaries; Virginia is the beautiful girl, whom her father killed to save her from the lust of Appius Claudius: cf. Livy, III. 44 ff.

295. suam: P has *suum* which Büch. keeps, supplying *gibbum* in the sense of *papillas*, which seems impossible. The writing is somewhat careless, but *faciem* is the word which must be supplied from l. 293.

autem, 'again', 'on the other hand'; a good and classical use of the word.

296. miseros...parentes | semper habet, 'keeps his parents in constant misery'; cf. XIII. 194. *sollicitum habere aliquem* is a common idiom: cf. Terence, *Haut. Tim.* 461; Cic. *ad Fam.* VI. 13. 3; *Cato*, 66; Livy, VIII. 29. 1, X. 11. 9, XXVIII. 25. 8, XXXVIII. 30. 6. Yet in this passage it is possible that *habet* has its common meaning: cf. Mart. V. 22. 13 *semper inhumanos habet officiosus amicos.*

297. rara...pudicitiae: edd. quote Ovid, *Heroid.* 15 (16), 288 *lis est cum forma magna pudicitiae.* **adeo** applies to *rara* only and is here used like *tam*: cf. XIII. 59.

298. licet, 'although': the apodosis is in l. 304.

horrida, 'plain', 'simple': this and *sordidus* are often eulogistic epithets.

299. Sabinos: see n. to *Sabina*, VI. 164.

300. modesto sanguine: for blushing as a sign (sometimes deceptive) of modesty, cf. XI. 154, XIII. 242; Suet. *Dom.* 18 *fuit vultu modesto ruborisque pleno.*

304. viro: so Jahn for *viros* of P; the acc. might stand: cf. Quint. IV. 4. 6 *procuratorem tibi esse non licuit*: but the sing. seems necessary. Transl. 'still he may not keep his manhood'.

305. improbitas, 'impudence': cf. IV. 106.

310. i nunc et...: see n. to VI. 306. For **iuvenis,** supplying the oblique cases of *filius,* cf. III. 158, VIII. 262.

312. publicus, 'promiscuous'.
The reading in the text is that of P which Büch. keeps and explains thus: *debet poenas quascumque metuit ex ira mariti.* The Latin is doubtful (e.g. the sense of the gen. *mariti* and the order of words), and the phrase excessively awkward. Rigalt's *iratis maritis* is too bold; Madvig's emendation, *ira sibi,* is the simplest which gives a good sense: the injured husband's anger is not satisfied until it has exacted sufficient punishment.

313. nec...astro Martis = *nec astrum* ('his luck') *eius felicius erit astro Martis*: for the constr., cf. III. 74 *sermo Isaeo torrentior.* For *astrum,* cf. *sidus,* VII. 200. Ares, the lover of Aphrodite, was caught in a cunning snare, laid by the injured husband Hephaestus (Hom. *Od.* VIII. 266 ff.).

315. dolor is regularly used for the resentment of a deceived husband or wife; cf. Mart. X. 41. 1 *veterem, Proculeia, maritum | deseris atque iubes res sibi habere suas. | quid, rogo, quid factum est? subiti quae causa doloris?*

324. immo, 'nay', contradicting the statement implied in the previous question.

325. grave propositum, 'his temperate resolve'. M. Hippolytus and Bellerophon are the Josephs of pagan mythology; the former was tempted by Phaedra his stepmother, wife of Theseus; the latter by Sthenoboea, wife of his host Proetus. Both resisted; and in each case the wife accused the innocent man to her husband. Hippol. lost his life in consequence; Beller. was set to accomplish a number of dangerous tasks which he did successfully.

326. haec is shown by the next l. to refer to Phaedra, daughter of the 'Cretan' Minos, though, according to correct usage, it should refer to the less remote instance, i.e. the temptress of Bellerophon: hence Haupt proposed *hac* agreeing with *repulsa,* the abl. of cause: if *hac* be read, *Sthenob.* is subject to *erubuit.* But the correction is needless: some silver-age writers, especially Quintilian, use *hic*,...*ille* freely for 'the former', 'the latter'.

ceu fastidita, 'like a woman scorned': the real motive for rejection was not *fastidium* ('unwillingness to be pleased': cf. Mart. III. 76. 1 *fastidis, Basse, puellas, | nec formosa tibi sed moritura placet*), but the *grave propositum* of the young men.

repulsa, 'because of the rebuff', is abl. of a noun and is governed by *erubuit*, not by *fastidita*.

327. **Cressa**, i.e. Phaedra, daughter of Minos, king of Crete.

excanduit, 'flared up', i.e. 'burst into a rage': not used by Cicero with a personal subject, but very common in writers of the Neronian period, e.g. Petronius and Seneca.

328. **se concussere**, 'shook with passion': *concuti* is used of any violent agitation, due to anger, fear, laughter etc.: cf. Sen. *Dial.* IX. 2. 3 *magnum et summum est deoque vicinum, non concuti*.

330. **suadendum**: *ei* must be supplied, as the antecedent of *cui*.

The story is told at length by Tacitus *Ann.* XI. 12 and 26 ff. Messalina, wife of Claudius, conceived a guilty passion for C. Silius, the consul designate, himself a married man, and actually went through a formal ceremony of marriage with him: the death of both followed A.D. 47.

331. **optimus**: the only instance in Juv. of a dactylic word forming the 2nd foot of the verse; rare in Virgil, very common in Lucr.: see Munro's *Lucretius*, II, p. 13.

formosissimus: cf. Tac. *Ann.* XI. 12 *Silium iuventutis Romanae pulcherrimum*.

332. **patriciae** is used in the sense of *nobilis*: the *gens Silia* was plebeian: see n. to I. 24. **rapitur** is properly applied to the abduction of a woman; see n. to VII. 168.

334. **flammeolo**: the diminutive is used for metrical purposes; see n. to VI. 225.

genialis, sc. *lectus*: for an empress it is covered with purple (*Tyrius*). The *lectus genialis* (also called *adversus* because it stood opposite the door) was not used as a bed, as is shown from the fact that it stood in the *atrium*, but was placed there as a symbol of marriage: cf. Hor. *Epp.* I. 1. 87 *lectus genialis in aula est* (= if he is married), | *nil ait esse prius, melius nil caelibe vita*: but it might be used as a seat: cf. Prop. V. 11. 85 *seu tamen adversum mutarit ianua lectum, | sederit et nostro cauta noverca toro.* **hortis,** probably the gardens of Lucullus on the Pincian Hill: in order to get them Messalina had caused the death of their owner, Valerius Asiaticus (Tac. *Ann.* XI. 1). Tac. does not say that the marriage took place there, but that on the return of Claudius to Rome from Ostia, Messalina took refuge in these gardens and was killed there.

335. **ritu...antiquo:** the transference of the dowry to the husband was part of the ancient legal ceremony; the amount of the dowry (in this case and VI. 137 a million sesterces or £10,000; in II. 117 *quadringenta* or £4,000) is not material.

336. **signatores** are friends who attend the ceremony (*officium*) in order to sign the marriage-contract (*tabulae*); see nn. to III. 82, VI. 25. The **auspex** gave the sanction of heaven to the nuptials, taking the omens before the marriage and performing a sacrifice as part of the ceremony. On this occasion, as on others in private and public life under the Empire, divination was practised by the inspection of entrails, not by the flight of birds, though the diviner was still called *auspex*, not *haruspex*. Messalina was determined to have everything done in proper form (**legitime**); cf. Tac. *Ann.* XI. 27 *adhibitis qui obsignarent ...illam audisse auspicum verba*; Suet. *Claud.* 26 *quam, cum comperisset C. Silio etiam nupsisse, dote inter auspices consignata, supplicio affecit.*

(Mr Lendrum suggests that the business of the *auspex* was not to sacrifice but to speak words of good omen: he compares Tac. *l.l. auspicum verba* with Juv. II. 119 *signatae tabulae, dictum 'feliciter!'*)

337. **tu** must refer to Silius; the line is somewhat awkwardly inserted here.

338. **quid** is used for *utrum*: see n. to VIII. 196.

339. **ante lucernas,** 'before lamps are lit', i.e. before dark: similar expressions for denoting time (e.g. περὶ λύχνων ἁφάς, Herod. VII. 215; περὶ πρῶτον ὕπνον, Thuc. II. 2; περὶ πλήθουσαν ἀγοράν, Xen.; βουλυτόνδε, Hom.) were commonly used by the ancients in default of a more accurate method of reckoning hours. They may still be heard in country districts; e.g. in Scotland 'the milkin' o' the kye' is a well-understood hour of the day.

340. **admittas,** 'commit'; so generally; cf. l. 255 and n. to XIII. 1.

341. Cf. Tac. *Ann.* XI. 30 (Narcissus to Claudius) '*an discidium tuum nosti? nam matrimonium Silii vidit populus et senatus et miles*'. The μετεωρία and ἀβλεψία of Claudius (cf. Suet. *Claud.* 39) were conspicuous in this scene of his life.

343. **tanti,** sc. *est*, 'is worth it'; the phrase is used in its original form (see n. to III. 54) and might be completed by *ut obsequaris*.

344. **quidquid,** 'whichever course'.

345. **praebere** (or *porrigere*) **cervicem,** with or without *gladio*, is regularly used by imperial writers for 'to be executed': cf. Sen. *Epp.* 82. 9 *non dubitabo porrigere cervicem*; *ibid.* 12 *evocatus ad mortem iussusque praebere cervicem.*

346–66. *Is all prayer then a mistake? The wisest plan is to let the gods give us what they think good for us; they love us and know our wants better than we do ourselves. But if pray you must and offer sacrifice, pray for wisdom and health, for courage and endurance. These are things which you can give yourself. We fear Fortune and call her divine, but the wise man can dispense with her favours.*

346. Juv. now returns to the point from which he started in ll. 54 and 55.

347. **permittes,** 'you will leave it to...': cf. Hor. *Carm.* I. 9. 9 *permitte divis cetera.* **ipsis,** i.e. without prayer from us.

353. **pueri,** sc. *futuri sint.*

354. **et** 'both' may be followed by *que* 'and' in silver-age Latin. The accumulation of diminutives (*sacellis, candiduli, tomacula*) is intended to suggest disrespect for the customary method of prayer with offerings.

355. **divina,** either 'a feast for gods' M., or 'presaging', in allusion to divination of the future by the *viscera* which are sarcastically called 'sausages'; for the latter sense, cf. *divinat*, IV. 124, and Mart. III. 71. 2 *non sum divinus sed scio quid facias.*

356. **mens...sano:** a slight alteration of the common prayer for *bona mens* and *bona valetudo*; cf. Petron. 61 (when wine is brought in, the guests) *omnes bonam mentem bonamque valetudinem sibi optarunt.*

358. **spatium vitae,** 'long life', as l. 188.

extremum inter munera, 'the last (i.e. least) among nature's gifts'. Some take *extremum* with *spatium*, and render 'reckoning life's close one of kind Nature's boons'; but cf. Sen. *Epp.* 4. 4 *nulli potest secura vita contingere, qui inter magna bona multos consules* (i.e. *annos*) *numerat.*

359. **quoscumque:** for the indefinite use of this pronoun, see n. to III. 156.

360. We should desire to obtain the Stoic ἀπάθεια, i.e. freedom from the passions (πάθη, *perturbationes*): of these the Stoics reckoned four, sickness of mind, fear, lust, and pleasure: they are discussed at length by Cicero, *Tusc. Disp.* IV.

361. **Herculis aerumnas:** cf. Circ. *de Fin.* II. 118 ...*Her-*

culis perpeti aerumnas. sic enim maiores nostri labores non fugiendos tristissimo tamen verbo aerumnas etiam in deo nominaverunt. Hercules was much in the mouths of the Stoics, and was a kind of patron-saint of the Cynics, as an embodiment of their watch-words, αὐτάρκεια, ἐλευθερία, and παρρησία.

labores: the repetition of this word should be noticed, and also the rhyming endings of three consecutive lines.

362. **Sardanapalli:** Assur-bani-pal, a king of Assyria, was a proverb in antiquity for luxury and effeminacy: cf. Cic. *de Fin.* II. 106; *Tusc. Disp.* V. 101.

363. Cf. Sen. *Dial.* VII. 4. 2 *honesti cultor, virtute contentus... qui nullum maius bonum eo quod sibi ipse dare potest noverit.* M. points out that **monstrare** is technically used of physicians' prescriptions: it is also used of a teacher of any kind: cf. VI. 261, XIV. 208.

365, 366 are repeated with the omission of two words XIV. 315, 316, where see n.

365. **numen,** 'sacredness', 'divinity'; cf. Mart. *Epig. Lib.* 30. 7 *numen habet Caesar; sacra est haec, sacra potestas.* **si sit prudentia** = but for human folly. **nos,** 'it is we who...'. For the sense, M. quotes Sen. *Epp.* 85. 2 *prudens beatus est et prudentia ad beatam vitam satis est.*

366. **facimus...deam:** cf. Mart. VIII. 24. 5 *qui fingit sacros auro vel marmore vultus,* | *non facit ille deos; qui rogat, ille facit.* See n. to XIII. 86.

SATIRE XI

SOME REFLECTIONS ON EXTRAVAGANCE IN EATING, FOLLOWED BY AN INVITATION TO DINNER

1-55. *A poor man becomes a general laughing-stock when he orders a dinner fit for a millionaire. Such tastes united with poverty bring men to the gladiators' school. They pawn their remaining property all to pay for one dainty dish. In a rich man extravagance is permissible; but in all circumstances of life, one should remember to 'cut one's coat according to one's cloth'. Old age is more dreadful than death to the poor man of extravagant tastes. The rakes' progress is marked by regular stages: they cut a dash at Rome on borrowed money, then go bankrupt, and are off to enjoy the oysters at Baiae. Not a blush will you see on their faces; their one regret is to miss the races for the year.*

1. **Atticus,** a notoriously rich man, is probably Ti. Claudius Atticus, father of Herodes Atticus; he became immensely rich by the discovery of a treasure on his estate in Attica, and was twice consul. **lautus,** 'princely': the adj. is regularly applied to *cena, convivia, dapes,* and then to persons who eat good dinners and give them; and it is used, more rarely, of magnificence in other respects: cf. *merces lautissima,* VII. 175; *praetoris lauti,* XIV. 257.

2. **Rutilus,** unknown, but evidently a poor man.

maiore cachinno: cf. III. 100.

3. **Apicius:** for this famous gourmand, see n. to IV. 23.

omnis, nom. plur.

4. All places of public resort ring with the absurd doings of Rutilus. **convictus** seems used for *convivia,* 'society', with especial reference to society round a dinner-table; **thermae** were not merely baths but also convenient places for eating, drinking, exercise, and conversation; **stationes,** 'lounges', are any places in the open street or in public buildings, where people met in groups (*circuli*) for talk or business. For similar enumerations, cf. Mart. VII. 76. 2 *per convivia, porticus, theatra;* V. 20. 9 *campus, porticus, umbra, virgo* (the aqueduct), *thermae,* | *haec essent loca semper.*

5. **de Rutilo,** sc. *loquuntur,* 'are busy with R.'.

6. **galeae** = military service: cf. VII. 33.

ardet is the conj. of Guietus for *ardenti* of P and many other MSS. It seems possible that *ardenti* should be kept, and *dumque* struck out: for in other passages (e.g. XII. 110) hiatus has led to corruption of the text. For *ardenti sanguine fertur,* cf. Sen. *Dial.* X. 19. 2 *nunc, dum calet sanguis...ad meliora eundum est.*

fertur, 'he rushes on' in his course of extravagance: cf. VI. 648; Lucan, IV. 268 *miles...casurus in hostes* | *fertur;* Sen. *Herc.* 185 *at gens hominum fertur rapidis* | *obvia fatis.* The other rendering, 'he is said', is tame, and the ellipse of *esse,* with the *fut.* participle, is harsh.

7. 'Not indeed compelled, but also not prevented, by the tribune.' When a Roman citizen was about to hire himself as a gladiator, he was obliged to signify his intention to one of the *tribuni plebis,* who apparently had power to sanction or annul such a contract (*auctoramentum*). **cogente** need not imply that the tribune had power to compel a man to take this step; it may refer to the fact that many *had been* compelled by tyrannical emperors: cf. VIII. 193 *nullo cogente Nerone.*

nec should be noticed: in classical prose *ne...quidem* would be used; but in the silver age *nec* is used for both senses of *ne...quidem*, (1) 'not even', (2) 'also not', where there is no comparison between the things negatived. Compare Mart. II. 36. 1 *flectere te nolim sed nec* (also not) *turbare capillos*, with Sen. *Epp.* 5. 3 *non splendeat toga; ne sordeat quidem*. In either passage, 'not even' would imply that one habit was worse than the other; whereas the writers put them on the same level, as equally blameworthy. Cf. too Sen. *de Ben.* IV. 9. 2 (we ought to choose carefully the recipients of our benefits) *quia ne agricolae quidem semina harenis committunt*, 'farmers also do not plough the sand': 'not even' would imply that farmers know less of agriculture than other people. This *nec* may be transl. 'neither'; but must be distinguished from 'neither' preceding 'nor', which is entirely different, as different as οὔτε from οὐδέ.

8. 'Soon to sign the conditions and tyrannous terms of the trainer': the reference is to the oath taken by gladiators, given most fully in Petron. 117 *in verba Eumolpi sacramentum iuravimus: uri vinciri verberari ferroque necari, et quicquid aliud Eumolpus iussisset. tanquam legitimi gladiatores domino corpora animasque religiosissime addicimus*. The quotation shows that *verba* is used of the oath dictated by the *lanista*, not of the words of command used in training.

regia implies that the gladiator contracted himself out of his legal rights, so that the *lanista* became *rex* over him, i.e. a person whose actions were not limited by any law.

For nobles fighting as gladiators, cf. VIII. 199; and, for this end to a career of extravagance, cf. Sen. *Epp.* 99. 13 *aspice illos iuvenes quos ex nobilissimis domibus in harenam luxuria proiecit*.

9. **elusus**, 'bilked'.

10. **creditor**, 'money-lender'; *creditum* is the regular word for 'money lent', 'a loan', *credere*, 'to lend money'; cf. Mart. III. 15 *plus credit nemo tota quam Cordus in urbe. | cum sit tam pauper, quomodo? caecus amat*. **macelli:** see n. to VI. 40.

11. **vivendi causa:** cf. VIII. 84.

12. **egregius** is adv. of the comparative; a form not elsewhere found: there is no possible ambiguity here, or it would not be used.

The expression is somewhat irregular for *eo egregius cenat quo miserior est*.

13. **casurus** is syntactically an adj. joined by *et* to the other adj. *miserrimus*. **iam...ruina**, 'the falling building already letting in the light': the man's fortunes are compared to a house

about to fall and so dilapidated that light enters through the
cracks: so Waller 'The soul's dark cottage, battered and de-
cayed, | lets in new light thro' chinks that time has made', i.e.
the old are wiser. Cf. II. 78 (to a noble, wearing a transparent
garment) *Cretice, perluces*; Plaut. *Rudens*, 101 *villam integundam
intellego totam mihi; | nam nunc perlucet ea quam cribrum crebrius.*
ruina is not 'the fall' but 'the falling house': cf. X. 107.

14. **interea,** i.e. until the crash comes.

gustus, 'flavours', i.e. dainty dishes.

omnia: three at least of the elements are meant: earth, air,
and water are ransacked for beasts, birds, and fishes; cf. Seneca
quoted on v. 95.

15. **interius si adtendas,** 'a careful scrutiny will show
that...'; for the idiomatic subj. of the 2nd pers., see n. to III.
102. So in Horace (*Sat.* II. 2. 23) the gourmand prefers a pea-
cock to a fowl, because it costs more. Cf. Sen. *Nat. Quaest.* IV.
13. 4 *luxuriae nihil placere potest nisi carum.*

16. **ementur,** perh. 'will have to be bought': but the fut.,
read by P alone, is peculiar after *iuvant*: *emetur* (l. 36) after the
imperative is normal; and it seems possible that the fut. here is a
copyist's error due to the fut. below.

18. 'By pawning silver dishes or breaking up the bust of a
mother.' **oppositis,** sc. *pignori*: cf. *vasa novissima,* VI. 356.
imagines were made of metal as well as of marble; in this case
the son breaks up an *imago* of his mother, presumably of silver.
For **fracta** used of metal, cf. *frangebat pocula,* l. 102. Some take
fracta as an epithet, 'cracked', and supply *opposita* again with
imagine: but in that case a thing of value is coupled with
something worthless.

19. **quadring. nummis,** 400 sesterces (= £4). Seneca
speaks of a single meal costing one thousand times as much,
Epp. 95. 41 *quid est cena sumptuosa flagitiosius et equestrem censum
consumente?* For a mullet costing £50, see n. to IV. 15.

20. **fictile** is emphatic, their silver dishes having gone to
pay for the dainty. **miscillanea ludi,** 'the mess of the gla-
diator's school'; the phrase, only found here, suggests some-
thing like our 'resurrection-pie'. The ruined spendthrift is sure
of a sufficient quantity of food while training in the *ludus*,
though the quality leaves much to be desired.

21. **refert,** 'it makes a difference': see n. to VI. 657.

paret, 'buys': see n. to III. 224.

22. **est:** *haec eadem parare* must be supplied as subject.

Ventidio, a rich man unknown.

24. The meaning is, that knowledge of geography is not so important as common sense; to know the comparative height of hills, is less useful than to realise the comparative depth of purses.

25. **omnibus** is used where *ceteris* would be more exact.

hic: it is common in Latin (and the regular rule in Greek) to substitute a demonstrative pron. for a repeated relative; cf. XV. 169 and 170; Cic. *de Fin.* I. 42 *extremum bonorum..., quod ipsum nullam aliam ad rem, ad id autem res referuntur omnes* with Madv.'s note.

27. **sacculus,** like *loculi*, is a small portable purse, very different from a large iron-bound strong-box (*arca*). γνῶθι σεαυτόν: this famous motto, engraved upon the temple at Delphi, was ascribed to one or another of the Seven Sages, and also, from internal evidence, to Apollo himself: cf. Cic. *de Fin.* V. 44 *aliter nosmet ipsos nosse non possumus. quod praeceptum quia maius erat quam ut ab homine videretur, idcirco assignatum est deo.* This is the meaning of *e caelo descendit*: cf. II. 40 *tertius e caelo cecidit Cato.*

28. **figendum** need not be taken with *pectore*: cf. V. 12.

29. **sacri:** the epithet seems rare, though ἡ ἱερὰ σύγκλητος is common in Greek: Ovid (*ex Pont.* IV. 9. 17) applies *sanctus* to the senate: Martial (VIII. 66. 10) speaks of the consulship as *sacros honores*. A common phrase for the senate in prose is *amplissimus ordo*.

in parte senatus esse, 'to take a share in the senate', i.e. to be a member of it: cf. l. 101; Livy, VI. 37. 4 *nunquam plebem in parte pari rei publicae fore*; XXXI. 1. 1 *me quoque iuvat, velut ipse in parte laboris ac periculi fuerim, ad finem belli Punici pervenisse.* There are well-known examples of persons who, though eligible as senators, were content to remain knights, e.g. Maecenas, Vespasian in youth (Suet. *Vesp.* 2 *latum clavum...diu aversatus est*), Ovid (*Trist.* IV. 10. 35 *curia restabat. clavi mensura coacta est:* | *maius erat nostris viribus illud onus*), and Lucan's father, Annaeus Mela.

30. **neque enim:** so I. 89 and often, *neque* having in this phrase the sense of *non*. The reference is to the *armorum iudicium*: see n. to VII. 115.

31. **in qua,** 'wearing which', cf. *in tunica*, X. 38: the arms were adjudged to Ulysses: cf. Ovid, *Met.* XIII. 383 *fortis...viri tulit arma disertus.*

se traducebat, 'made an exhibition of himself'; see n. to VIII. 17.

32. **magno discrimine,** 'of important issues', abl. of quality.

33. **protegere,** i.e. in the character of *patronus*.

te consule, 'examine yourself': not to be confused with the same words in VIII. 23. **qui,** instead of *quis* in the indirect question, is very exceptional and is probably due here to a wish for euphony.

Cf. Quint. VI. 1. 45 *quare metiatur ac diligenter aestimet vires suas actor* (the barrister) *et quantum onus subiturus sit, intellegat*; Sen. *Dial.* v. 7. 2 *quotiens aliquid conaberis, te simul et ea quae paras quibusque pararis ipse metire.*

34. **C. et M. buccae,** 'or a mere mouther like Curtius and Matho': Curtius is unknown; a *causidicus* Matho occurs I. 32, VII. 129.

37. For the price of mullets, see n. to IV. 15: **gobio** = the price of a gudgeon.

40. **rebus,** sc. *paternis*; the *res* are enumerated below.

41. **argenti gravis,** 'solid silver plate': cf. Seneca quoted on XII. 44.

42. **exit,** 'leaves the family': a technical legal term for the alienation of property.

43. The spendthrift is an *eques*; as a bankrupt, he is obliged to lay aside the *anulus*, the sign of his rank: cf. Mart. VIII. 5 *dum donas, Macer, anulos puellis, | desisti, Macer, anulos habere.* A Crepereius Pollio is mentioned as deeply in debt IX. 6–8.

44. **funus acerbum:** we use the same metaphor, when we speak of a 'ripe' old age. Juv. is no doubt thinking of the famous instance of Apicius: see n. to IV. 23.

45. **luxuriae** is dat.: cf. the use of *ingenio*, X. 120.

46. **gradus,** 'the stages', in such a man's progress.

conducta, 'borrowed': cf. Hor. *Sat.* I. 2. 9 *omnia conductis coemens obsonia nummis*; but *conducere* is seldom used of money, though *locare* often is.

47. **dominis,** sc. *pecuniae*, 'the owners of the money', i.e. the lenders. **inde,** 'next'.

48. **faenoris auctor,** 'the money-lender': see n. to XIII. 2.

49. **vertere solum:** the phrase is used of insolvent debtors who make, as we say, 'a moonlight flitting': cf. Petron. 81 *conturbavit et...solum vertit. vertere* is an aor. of frequency. For the Lucrine oysters, see n. to IV. 140; and for Baiae, n. to XV. 46.

50. **cedere...foro,** 'to be bankrupt', lit. to retire from the forum, where all money business was carried on; see n. to X. 25.

Cf. Sen. *de Ben.* IV. 39. 2 *appellare debitorem ad diem possum,
et, si foro cesserit, portionem* (a dividend) *feram.* **iam,** 'now-a-days'.

51. **Esquilias:** cf. III. 71 and see n. to v. 77. **ferventi,**
'bustling'.

53. **anno:** for the case, cf. l. 72 and see n. to VI. 183.

circensibus: this form of *ludi*, the races, could be witnessed
only in Rome itself; see n. to III. 223.

54. **sanguinis gutta** = an apology for a blush; cf. X. 300.

55. **ridiculum et fugientem:** cf. Thuc. III. 83 καὶ τὸ
εὔηθες, οὗ τὸ γενναῖον πλεῖστον μετέχει, καταγελασθὲν ἠφανίσθη.

56–182. *You, Persicus, who are to dine with me today, will see
whether my practice squares with my theory. The dinner will be a
plain one, supplied entirely by my farm and garden. In the good
old times, even such a meal would have been a feast; and some of
the guests, once consuls or dictators, would put down their spade
an hour earlier than usual, to share it. In the days of Cato the
censor, food and furniture alike were simple; our soldiers, when
they took Greek cities, broke up master-pieces of immortal artists
to adorn their shields and helmets. Heaven was nearer us in those
days. A modern Roman cannot eat his dinner, unless the table
rests on ivory; silver even is not good enough. But in my house
you will find no scrap of ivory, no elaborately trained carver; I
have only country boys, in plain dress, to wait at my table. For
entertainment, you will have verses of Homer and Virgil recited,
but no Spanish dancing-girls; that kind of amusement may do for
the rich and splendid, but does not suit humble folks like me.*

56. **numquid,** 'whether'; rare in indirect questions, very
common in direct; cf. II. 51 *numquid nos* (we women) *agimus
causas?* Martial has at least twenty-five instances of the latter,
none of the former. For an instance, cf. Sen. *Epp.* 20. 3 *observa
itaque, numquid vestis tua domusque dissentiant.*

57. **Persicus,** an unknown friend of Juv., must be a different
man from the Persicus mentioned III. 221.

non praestem vitae tibi moribus et re, 'I do not carry
out for you in my habits of life and in deed': *tibi* is added by
Büch., the word before *moribus* being erased in P: most edd.
read *vita vel* with the inferior MSS. *vitae moribus* seems unusual
for *vita et moribus*.

58. **si laudem,** 'by praising': the sentence is conditional,
depending on *praestem.* **pultes:** this porridge, made of spelt
(*far*), was for long the national food of the Romans, whom

Plautus, accordingly, calls in jest *pultifagi barbari*: see n. to x. 138.

59. dictem: the frequentative is used with the meaning of the simple verb. The l. describes the master ordering his dinner, not eating it: cf. Cic. *pro Rosc. Am.* 59 (of an accuser in court) *ita neglegens esse coepit, ut...etiam puerum vocaret, credo, cui cenam imperaret.*

60. promissus, 'engaged': cf. Petron. 10 *ad cenam promisimus* (we are engaged to dine out); Seneca quoted on XIII. 213. **habebis,** 'you will find in me'.

61. Euandrum, i.e. a hospitable but not a wealthy entertainer: the legend represented him as leaving Arcadia to settle on the Palatine Hill, where he successively entertained Hercules and Aeneas: cf. Virg. *Aen.* VIII. 359–65: Juv. has in mind esp. l. 364 *aude, hospes, contemnere opes et te quoque dignum | finge deo, rebusque veni non asper egenis.*

Tirynthius: according to one account, Hercules was born at Tiryns, and he is constantly called *Tirynthius* by the Latin poets.

illo: for the meaning and position of the pron., cf. v. 139, XIII. 73.

62. contingens, 'related to'; cf. VIII. 7: Aeneas was the son of Venus.

63. Aeneas was drowned in the river Numicius, Hercules was burnt on Mount Oeta: both were deified.

64. fercula, 'the dinner', as distinguished from the *mensae secundae* (dessert), which begin l. 72; there is no mention of a *gustus* or *promulsis*: see n. to v. 147. **nullis ornata macellis,** 'that no market supplied': *ornata* suggests more elaborate display than *instructa* would.

65. That Juv. had a farm of his own at Tibur (Tivoli) is shown by the mention of the *vilica* below.

66. inscius herbae, i.e. unweaned.

69. asparagi: cf. v. 82. The *vilica*, wife of the *vilicus* or bailiff, puts her spinning aside to gather asparagus: Mart. III. 58. 20 *avidi secuntur vilicae sinum porci* shows her in another occupation, but elsewhere (IX. 60. 3, X. 48. 7) as picking vegetables and flowers.

70. faeno: cf. Mart. III. 47. 14 *tuta faeno cursor ova portabat*: they were wrapped in hay for safe transport.

71. matribus, sc. *ovorum*, i.e. *gallinis*: cf. Mart. VII. 31. 1 *raucae chortis aves et ova matrum.*

72. parte anni, 'during a good part of the year': perhaps *pars* is used for *dimidia pars*, as in Mart. II. 24. 6, III. 86. 1: see Munro on Lucr. II. 200. From the vintage until the date of this dinner (10 April) would be just about six months. The ancients had various ways of preserving grapes in a fresh state.

74. Picenum was famous both for apples and for olives.

76. **autumnum,** somewhat boldly used for 'their autumn taste'. Many fruits were considered less wholesome off the tree than when they had been stored for a time. This dinner took place in April; see n. to l. 193. **posuere:** cf. *posito,* l. 69; *ponenda,* III. 56.

77. **iam,** 'quite', modifies the adj. only; see n. to III. 206.

78. The ordinary dinner of the great men of old was much inferior to what I offer you. Manius Curius Dentatus, conqueror of Pyrrhus 274 B.C., was one of the most famous examples of the old Roman simplicity and frugality, one of Milton's 'men so poor, | who could do mighty things'. **horto,** a garden for vegetables: see n. to l. 75.

79. **ipse:** in this case personal service is as creditable as it was infamous in the case of Lateranus VIII. 147.

80. **in magna...compede:** the reference is to the chain-gang labour of the *ergastulum*; see nn. to VI. 151, VIII. 180.

81. The slave, sent to work in the country, as a punishment, hankers after the flesh-pots of the town: cf. Hor. *Epp.* I. 14. 21 *fornix tibi et uncta popina | incitunt urbis desiderium.*

82. **rara...crate,** 'from the open wicker-work' attached to the ceiling which served as a *carnarium*: for a figure, see Rich's *Companion s.v.* Juv. seems to have in mind the reception of the gods by Baucis and Philemon in Ovid, *Met.* VIII. 646 *quodque suus coniunx riguo collegerat horto | truncat holus foliis. furca levat ille bicorni | sordida terga suis nigro pendentia tigno.*

83. **moris erat:** this gen., found also in *consuetudinis*, is generally explained as similar to that in *stulti est nolle sumere quae di porrigant.* The nominative (*mos est*) is commoner before the silver age.

diebus: dative.

84. **natalicium,** 'only on birthdays': for the adverbial use of the adj., see n. to *aestivum,* I. 28.

85. **nova...carne,** 'fresh meat', in addition to the salt pork (*lardum*); in ancient times a sacrifice was practically synonymous with a feast, as Homer shows, e.g. *Od.* III. 1 ff.

87. **castrorum imperiis:** i.e. he had commanded an army in the field.

89. **erectum,** 'on his shoulder'. **domito:** a favourite meta-

phor, e.g. Virg. *Aen.* IX. 605 *aut rastris terram domat*: especially appropriate here where *monte* and *ligonem* are παρὰ προσδοκίαν. Cf. Sen. *Epp.* 86. 5 (describing the elder Scipio's simple bathroom) *in hoc angulo ille Carthaginis horror...abluebat corpus laboribus rusticis fessum. exercebat enim opere se, terramque, ut mos fuit priscis, ipse subigebat.*

90. **Fabios:** cf. VIII. 14; Cato is the famous censor, the 'sour John Knox' of antiquity; for the Scauri, cf. VI. 604; Fabricius, who fought against Pyrrhus, was censor 275 B.C. and was severe against extravagance; cf. IX. 141 *argenti vascula puri, | sed quae Fabricius censor notet.*

92. This refers to the quarrels between the censors of 204 B.C., M. Livius Salinator and C. Claudius Nero, related by Livy, XXIX. 37.

93. *habendam* of P is kept by Büch.: but it gives so harsh a construction that it is probably a mere slip of the scribe's pen.

94. *Oceani* is read by the worse MSS., but the reading of P should be kept, *Oceanus* being often used as an adj., as in *mare Oceanum.*

testudo: the wooden *lecti* were overlaid with tortoise-shell: cf. VI. 80 *testudineo...conopeo* (in a curtained cradle adorned with tortoise-shell); Mart. XII. 66. 5 *gemmantes prima* (the best) *fulgent testudine lecti.* For a parrot's cage thus adorned, cf. Stat. *Silv.* II. 4. 11.

95. **Troiugenis:** see n. to I. 100. **fulcrum,** 'head-rest', a part of the *lectus*; see n. to VI. 21 and below.

96. **nudo latere** is syntactically an adj., parallel with *parvis*, and qualifying *lectis*; cf. III. 48 *mancus et extinctae...dextrae.* The couches (*lecti tricliniares*) in old days were small; their sides were unadorned, i.e. were not inlaid with silver or tortoise-shell; the only adornment was the *frons aerea.* This latter is identical with the *fulcrum*, and formed the end of the frame-work on which the pillows were placed; even in a rude age the *fulcrum*, though not inlaid with tortoise-shell, was covered with bronze and had its characteristic ornament.

97. There are many extant specimens of this ornament in bronze, the head and shoulders of a mule or ass, decorated with a garland of vine-leaves, legend connecting this animal with the discovery of the vine and the worship of Bacchus. Transl. 'but the sides of their small couches had no inlay, and the head-rest of bronze displayed but the common head of an ass with a garland'.

98. **ad quod,** 'beside which'.

ruris alumni: *rustici infantes* Schol. M. understands the guests to be meant, and there is good authority for the heartiness with which the old Roman heroes could unbend at such times; cf. Hor. *Sat.* II. 1. 73 (of Scipio and Laelius) *nugari cum illo* (Lucilius) *et discincti ludere, donec | decoqueretur holus, soliti.* Yet *lascivi* (playful) seems to suit children better: it implies very active motion (e.g. *lascivos leporum cursus*), which would hardly be a treat for a man who had been digging all day (l. 89). And Juv. takes an interest in children's amusements: cf. v. 138–45; IX. 60 *rusticus infans | cum matre et casulis et conlusore catello*; XIV. 167 *casae, qua...infantes ludebant quattuor.*

99. 'Well, their meals were as plain as their house and furniture.'

100. Livy dates the appreciation of Greek art among his countrymen from the taking of Syracuse 212 B.C.: cf. XXV. 40. 2 *inde primum initium mirandi Graecarum artium opera.* He attributes the introduction of foreign luxuries at Rome to the soldiers of Manlius who triumphed over the Asiatic Gauls (see n. to VII. 16) 187 B.C.: cf. XXXIX. 6. 7 *luxuriae peregrinae origo ab exercitu Asiatico invecta in urbem est. ii primum lectos aeratos* (cf. l. 96)...*et monopodia* (cf. *pes* l. 128) *et abacos* (cf. III. 204) *Romam advexerunt.* Sallust agrees that Asia demoralised the Roman soldier but dates the decline from Sulla's command there 87–83 B.C.: cf. Sall. *Cat.* 11. 6 *ibi primum insuevit exercitus populi Romani amare, potare, signa, tabulas pictas, vasa caelata mirari.*

mirari, 'to appreciate': cf. Livy and Sallust above.

101. Polybius was present at the sack of Corinth 146 B.C., and describes how he saw Roman soldiers using priceless pictures, thrown from the walls to the ground, as tables to gamble on. **praedarum in parte reperta,** 'which they found in their share of the booty'.

102. **artificum:** the most famous *caelatores* are enumerated, VIII. 102–4. The soldier broke up a cup chased by Mentor, to adorn the trappings of his horse or his helmet with a detached piece of ornament.

103. **phaleris:** see nn. to l. 109 and XVI. 60.

104. The helmet represents the wolf that suckled Romulus and Remus, her two nurselings, and their father the god Mars. Virgil represents the wolf and the twins as sculptured on the shield of Aeneas (*Aen.* VIII. 630); and both from literature and the remains of ancient art, we see that it was a common practice to engrave these figures, especially upon weapons and armour.

simulacra: the plur, is used for metrical reasons.

105. **imperii fato:** the destiny of Rome required, for its fulfilment, that the wolf should lose her fierceness.

geminos...Quirinos, i.e. Romulus and Remus; so *Castores* is sometimes used for Castor and Pollux.

106. The third figure is the god Mars, who in the works of art depicting this myth is always represented as naked (except for a *chlamys* floating behind him), armed with shield and spear, and generally as hanging in the air, on his way to visit Rhea Silvia: for a figure, see Baumeister's *Denkmäler*, p. 886.

There is a difficulty in the ablatives *clipeo* and *hasta*. Some make *nudam* govern them; but the existing representations are against this, and *venientis* is very feeble by itself. Nor can any real parallel be given for *clipeo* = *cum clipeo*. Hence some emend: e.g. Merry reads *clipeoque nitentis* (*fulgentis* is read by the inferior MSS.); Müller, *in clipeo.*

I believe the text to be sound, and that Juv. is using **venientis** in its well-established sense of 'attacking'; cf. l. 113 and Munro on Lucr. III. 833 '*venio* is continually used by Livy for the hostile advance of soldiers'. If *venientis* has this sense, the ablatives are common instrumentals like *Gorgone* (= *clipeo*), XII. 4. No doubt, the artist represented Mars as borne on the wings of love to Rhea Silvia; but Juv. wishes to emphasise the other aspect of the god, swooping down as a destroyer on the doomed enemies of his Roman posterity.

108. The pottery of Etruria is often mentioned, as cheap and common. For **farrata,** which = *pultes,* see n. to l. 58.

109. Perhaps a reminiscence of Livy, XXII. 52. 4 (of the spoil taken by Hannibal at Cannae 216 B.C.) *praeter equos virosque et si quid argenti—quod plurimum in phaleris equorum* (cf. l. 103) *erat; nam ad vescendum facto perexiguo, utique militantes, utebantur— omnis cetera praeda diripienda data est.*

110. I.e. a man must be covetous indeed, to covet the simple contrivances of those days. **lividulus:** cf. *sordidula,* III. 149; *inprobulum,* V. 73; *pallidulus,* X. 82; *rancidula,* XI. 135; here and in some of the other instances, the diminutive sense seems latent; but such forms are appropriate to colloquial language and hence to *satura.*

111. **praesentior,** 'more near to help us'; cf. III. 18. The story of this divine φημή is told by Livy, V. 32. 6: it was heard at night near the temple of Vesta at the foot of the Palatine, and told the hearer to warn the magistrates that the Gauls were coming; but no attention was paid to it until it came true.

112. **audita,** sc. *est.*

113. **litore ab Oceani:** this is a rhetorical ornament taken apparently from Livy, V. 37. 2 *invisitato atque inaudito hoste ab Oceano terrarumque ultimis oris bellum ciente.* Part of Gaul was bounded by the Ocean; but, by Livy's own account, these Gauls had been settled in Etruria for 200 years. For **venientibus,** 'attacking' (cf. *bellum ciente* in Livy), see n. to l. 106. This invasion took place 391 B.C.

114. The stop at *peragentibus* is due to Madvig, who explains **his** as = *hac voce et huiusmodi signis.* The stop was formerly placed after *nos, vox* being taken as subject to *monuit*; which left *his* unintelligible.

116. **fictilis:** the oldest statues at Rome were exclusively of baked clay; Etruria was the home of the art, and Tarquinius Priscus had the first image of Jupiter made at Veii for the Capitol. For the sentiment, cf. Sen. *Epp.* 31. 11 *cogita illos* (i.e. *deos*) *cum propitii essent, fictiles fuisse.*

violatus: cf. III. 20.

117. **domi natas,** 'home-grown': the phrase is often applied to things one need not buy: cf. Petron. 38 *nec est quod putes illum quicquam emere; omnia domi nascuntur.* The tables were of common walnut or beech, not of the foreign *citrus.*

118. **hos:** P has *hoc* which Büch. keeps; and this may be right, if **ad usus** can mean, like *in usum,* 'for use' as opposed to 'ornament': cf. Sen. *Dial.* IX. 1. 7 *mensa non varietate macularum conspicua...sed in usum posita.* Yet the plur. is generally used with an adj., such as *multos, varios, hos* etc.; and *hoc* seems superfluous. The proximity of *lignum* would account for the corruption of *hos.*

122. **unguenta,** 'scents': these and the roses (cf. XV. 50) belong especially to the *comissatio* which followed the *cena*: see n. to IV. 108. For **orbes,** see n. to *mensas* I. 75.

123. **ebur et...pardus** = *eburneus pardus*: a hendiadys.

sublimis, 'rampant'.

124. **dentibus,** 'tusks': for this kind of table-rest, cf. Stat. *Silv.* IV. 2. 38 (of the tables in Domitian's palace) *Indis innixa columnis | robora Maurorum* (i.e. *orbes citrei* upon ivory supports).

porta Syenes: Syene (Assouan) on the Nile is so called, as being an outpost on the Roman frontier: cf. Stat. *Silv.* IV. 4. 63 *portae | limina Caspiacae*; Tac. *Ann.* II. 61 *Elephantinen ac Syenen, claustra olim Romani imperii.* Until Trajan's time, Syene was the most distant part of the empire: cf. Mart. I. 86. 6

tam longe est mihi quam Terentianus, | *qui nunc Niliacam regit Syenen.*

125. In addition to Mauretania and Aethiopia, the two breeding-grounds of elephants in Africa already mentioned, x. 150, India is here spoken of as a third source of the supply.

obscurior: yet Lucan, IV. 678 says *concolor Indo* | *Maurus.* In point of fact the Moors must have been darker: cf. v. 53.

126. **Nabataeo...saltu:** the Nabataei lived in part of Arabia Petraea, and there are no elephants in Arabia: the geography is thus a little inaccurate.

belua: cf. x. 158. The elephant does not really shed its tusks either by sticking them in the ground (so the Schol.) or in any other way.

127. **hinc,** i.e. *ab ebore.* **orexis:** see n. to VI. 427.

128. **pes,** sc. *mensae*: cf. Livy quoted on l. 100: mere silver is not fine enough.

129. **anulus...ferreus:** iron finger-rings, once universal (see Pliny quoted on x. 41), were now a sign that the wearer was a man of no rank, belonging to neither of the *ordines*: cf. Stat. *Silv.* III. 3. 143 (of Vespasian conferring knighthood on the father of Claudius Etruscus) *in cuneos populo seduxit equestres* | *mutavitque genus laevaeque* (sc. *manus*) *ignobile ferrum* | *exuit, et celse natorum aequavit honori.*

130. **qui me sibi comparat,** i.e. who draws comparisons between my possessions and his own. **res** are not, I think, 'dishes' (so Friedl.) but 'circumstances' in the sense of 'wealth'; cf. Mart. I. 55. 4 *sordidaque in parvis otia rebus amat.*

131. **adeo,** 'for indeed': for this absolute use of *adeo*, see n. to III. 274. It modifies the whole clause, not merely *nulla*: and the clause explains *exiguas*: i.e. my circumstances are narrow, *for* I have no scrap of ivory.

132. **tessellae** (usually *tesserae*) are κύβοι, 'dice'. **calculi** are 'counters', with which the Romans played games resembling our draughts and backgammon.

133. **quin,** 'nay'. The verb (*sunt*) is understood.

cultellorum: knives have been used for eating since immemorial times; forks, which one might expect to be mentioned here, are a comparatively modern invention: Coryate in his *Crudities* (published 1611) speaks of them as a novelty which he had observed in Italy, and says that, when he tried to introduce the custom in England, his friends called him *furcifer*. For forks, the ancients used fingers; cf. Mart. v. 78. 6 *ponetur digitis tenendus ustis* | ...*coliculus.*

136. **nec:** as well as no ivory knife-handles, you will *also* have no carver; see n. to l. 7.

structor (cf. v. 120) is used for *scissor*, his business being to carve, as well as to arrange the table.

137. **pergula,** 'booth', is an out-building in front of a house, with a roof but no wall facing the street; the narrow Roman streets were made still narrower by erections of this kind, which were constantly used as shops, or workshops, or schools. In such a *pergula*, Trypherus (τρυφερός = *delicatus*) taught the art of carving by means of wooden models. **omnis pergula,** 'the whole booth', i.e. all the scholars.

139. **Scythicae volucres,** 'pheasants', usually called *Phasianae aves*; the river Phasis, on the east of the Euxine, just south of the Caucasus, was supposed to be the original home of this bird.

140. **lautissima cena** is in apposition with all the preceding nominatives. For the separation of the adj. and noun, cf. *aurea*, VIII. 207.

142. **Afra avis** is the 'guinea-fowl', constantly mentioned with the pheasant as a costly delicacy. Juv. means that these dainties are not to be found in his house, for the slaves to steal.

143. **noster,** i.e. the lad that waits on me. **omni tempore,** 'all his days': or can this be a colloquial expression for 'entirely'?

144. **ofella,** a dimin. of *offa*, is a simple dish that needs no carving: cf. Mart. x. 48. 15 *quae non egeant ferro structoris ofellae.*

145. **plebeios calices,** not *crystallina* or *myrrhina*, for which see nn. to VI. 155 and 156.

146. **incultus:** cf. Sen. *Dial.* IX. 1. 7 *placet minister incultus et rudis vernula.*

a frigore tutus, i.e. warmly dressed, not half-naked as the *flos Asiae* (v. 56) might be, to display his beauty. Seneca (*Dial.* X. 12. 5) speaks of the *tunicae diligenter succinctae* of rich men's pages.

148. **et magno,** sc. *pretio*: cf. III. 166: *petitus* is understood again with *magno*. Büch. keeps the reading of P, *in magno*, and connects it with *cum posces, poculo* being understood; but the sense so obtained is unsatisfactory: for why should the guest speak Latin only when he wished a *large* glass? The Schol. understood it of the price paid for the slave. For expensive slaves, see n. to v. 60. **latine,** i.e. not in Greek; cf. VI. 185 ff.

149. **tonsi rectique capilli**: whereas the pages of the rich are regularly called *capillati* or *comati*, long hair being considered essential, and it was often artificially curled: cf. *acersecomes*, VIII. 128.

For plain cups and cup-bearers with plainly dressed hair, cf. Sen. *Epp.* 119. 14 *nam si pertinere ad te iudicas quam crinitus puer et quam perlucidum tibi poculum porrigat, non sitis* (you are not thirsty).

154. **ingenuus** is 'free-born', 'of gentle birth': so III. 131: but it means further 'frank', 'open': cf. II. 16 *verius ergo | et magis ingenue Peribomius*. The second is the meaning here, but the next line is suggested by the first meaning.

155. **ardens** refers to the bright colour: cf. X. 27; Stat. *Theb.* V. 438 *chlamys ardet*.

purpura is used for *praetexta*, the *toga* with purple border, worn by all free-born boys, *ingenui*, (and girls) until they put on the *toga virilis*: see n. to I. 78.

159. **diffusa**, 'bottled': see n. to V. 30.

161. The wine therefore comes from the country near Tibur; it is no foreign or Campanian vintage.

162. **Gaditana**, 'the songs of Gades': cf. Mart. III. 63. 5 *cantica qui Nili, qui Gaditana susurrat*. Spanish dancing-girls, especially from Cadiz, afforded an entertainment, which, though not considered respectable, was popular at some dinner-tables: so Pliny writes to a friend who had declined an invitation *Epp.* I. 15. 2 *audisses comoedos vel lectorem vel lyristen vel, quae mea liberalitas, omnes: at tu apud nescio quem ostrea, vulvas, echinos, Gaditanas maluisti*. Their dances were in this case accompanied by a chorus of singers.

165, 166 are found in different places in different MSS., and are omitted in some. Lewis suggests that the right place for them is immediately before l. 171, and that *spectent*, not *spectant*, should be read; cf. *fruatur* l. 173.

166. 'A sight which anyone would be ashamed to describe in their presence.'

171. **non capit**, 'is too small for': cf. X. 148 and Mart. VII. 27. 9 (to a large boar sent him as a present) *ad dominum redeas; noster te non capit ignis*.

174. **vocibus** refers to the Spanish songs.

libidinis arte, 'refinements of sensuality'.

175. 'Who makes the circles of Laconian marble slippery with wine he spits out', i.e. who has a dining-room floored with marble: cf. Sen. *Dial.* IX. 1. 8 *domus etiam qua calcatur pretiosa*.

A favourite kind was the green marble of the Eurotas-valley; this floor is inlaid with round pieces of it.

A man is said *pytissare* (πυτίζειν), when he takes a little wine into his mouth and then spits it out, by way of testing it. For this unpleasant practice, cf. Ter. *Haut. Tim.* 457 *nam ut alia omittam, pytissando modo* (merely) *mihi | quid vini absumpsit 'sic hoc,' dicens 'asperum, | pater, hoc est: aliud lenius sodes vide.'*

176. **ibi**, 'in his case'. **fortunae**, 'rank'; cf. VIII. 74.

177. **mediocribus**, 'men of moderate rank': cf. VI. 582 and *modicis*, v. 108.

178. **faciunt:** Büch. keeps *faciant* of P; but I feel that *cum* here is 'when', not 'though', and doubt whether Juv. would use the *pres.* subj. with this meaning. Jahn read *faciunt* with some MS. authority.

179. **ludos**, 'entertainment'; see n. to v. 157.

180. **cantabitur**, 'will be declaimed'. Recitations of all kinds were a common entertainment at a *cena*; but such an *acroama* might cease to be entertaining, if the host produced a thick roll of his own composing; Martial says, as a final inducement to accept a dinner-invitation, *nil recitabo tibi* (XI. 52. 16). Seneca tells of Calvisius Sabinus, a rich freedman of no education, that he gave a literary flavour to his dinners by means of his slaves; of these one knew Homer by heart, another Hesiod, while nine others could each repeat one of the nine Greek lyric poets; each slave had cost his master 100,000 sesterces (£1,000) (Sen. *Epp.* 27. 5).

For comparisons between Homer and Virgil, cf. VI. 436; though unfruitful, they were inevitable and were made even before the Aeneid was published: so Prop. III. 26. 66 *nescio quid maius nascitur Iliade*. Quintilian adopts the view of Domitius Afer who, in answer to his question, 'what poet is next to Homer?' replied, *secundus est Vergilius, propior tamen primo quam tertio*. But Quintilian's own remarks that follow, show how clearly he felt the differences which make comparison hopeless (Quint. X. 1. 86). Some acute appreciation of both poets will be found in Myers' *Classical Essays*, pp. 132–7.

182. The poetry is so good, that it will not matter if the slave who recites it is not a master of elocution: cf. Pliny quoted in VII. 153.

183–208. *But now throw aside all your cares and worries, and bring a light heart to your dinner. The races are going on in the Circus; and the shouts of triumph I hear show that the popular colour has won. It is quite appropriate for a smart young man to bet and shout in the Circus with a smart young woman at his side: I prefer to sit in the sun, in an easier dress. Today is a holiday, and you can go earlier to the bath than usual. Yet a life, all holidays, would soon pall.*

183. So Horace, inviting Torquatus to dinner, *Epp.* I. 5. 8 *mitte levis spes et certamina divitiarum | et Moschi causam; cras nato Caesare festus | dat veniam somnumque dies.*

184. **quando licebat...?**: so Büch. after P; other edd. read *licebit* with the other MSS. and explain *quando* as = *quandoquidem* as III. 21. The difficulty in P's reading is, that we should expect *licuit*.

185. **non faenoris ulla mentio,** sc. *sit*: therefore the negative should, according to rule, be *ne*, not *non*, but there are many exceptions to the rule. It may fairly be said that here and in some other instances (e.g. Hor. *Epp.* I. 18. 72 *non ancilla tuum iecur ulceret ulla*), *non* does not negative the whole sentence but a part of it, here the word *ulla*; see Nixon, *J. Philol.* VII, 54; but this explanation will not serve in Cic. *pro Cluent.* 155 *a legibus non recedamus.* For exceptions, though far more frequent in poetry, are found also in the best prose. Quintilian (I. 5. 50) brands *non feceris* for *ne feceris* as a solecism; but uses *non* himself, with a hortative pres. subj. in the 1st and 3rd persons e.g. I. 1. 19 *non ergo perdamus primum statim tempus.* See too Palmer's n. on Hor. *Sat.* II. 5. 91.

186. I.e. dismiss from your mind any suspicions of your wife.

191. **frangitur illis:** *illis* is dative of the agent. A slave belonging to Vedius Pollio broke a *crystallinum*, and was ordered by his master to be thrown to feed his lampreys: but Augustus, who was dining with Pollio, had all the *crystallina* broken in his presence and the fish-pond filled up (Seneca, *Dial.* v. 40. 2 and 3).

192. **perit,** 'is lost': passive in meaning, being the classical equivalent of *perditur*: see n. to I. 18.

ingratos...sodales, 'the ingratitude of friends', the worst heart-ache of all: cf. Mart. XII. 34. 8 *si vitare voles acerba quaedam | et tristes animi cavere morsus, | nulli te facias nimis sodalem.*

193. **Megalesiacae...mappae:** when the image of Cybele, the μεγάλη μήτηρ, was brought from Pessinus 204 B.C., games were instituted in her honour, lasting from 4 to 10 April. On the last day there were races (*circenses*): in these the signal for the

chariots to start was given by the presiding magistrate dropping a *mappa*; hence *mappae* is here used as = *ludorum*.

spectacula is used in the sense of *spectatores* (cf. VIII. 205) and is subject to *colunt*.

194. **Idaeum:** see n. to III. 137. **sollemne** is a noun and governs *mappae*.

triumpho is boldly used for *triumphanti*: for the praetor's state when presiding at games, see nn. to X. 36–45.

195. **praeda caballorum,** 'eaten up by the horses', i.e. he spends all his fortune in paying the *factiones* and rewarding the successful *aurigae*: cf. Mart. X. 41. 4 (Why does Proculeia divorce her husband?) *dicam ego: praetor erat. | constatura fuit Megalensis purpura centum | milibus, ut nimium munera parca dares*; v. 25. 7 *hoc, rogo, non melius...quam non sensuro dare quadringenta caballo, | aureus ut Scorpi nasus ubique micet?* Is there a reference here to the story of Actaeon, who was *praeda canum suorum?*

pace...plebis, 'without offending the people': so *pace tua, vestra,* etc. Romans were proud of the immense population of their city and might be offended at the statement that the Circus could hold it all. Under Vespasian the Circus held 250,000 spectators; it was enlarged by Trajan, and in the 4th cent. 385,000 persons found accommodation. The population of Rome cannot be certainly estimated; Friedländer is of opinion, judging partly by the number of those who received distributions of corn, that it was between one million and a half and two millions, at the end of the 1st century.

197. **capit,** 'holds': *non capere* is commoner: cf. l. 171.

fragor aurem percutit: cf. Sen. *Epp.* 83. 7 *ecce circensium obstrepit clamor.* We do not know where Juv.'s house was; but it need not have been very near the Circus, in order to hear the noise.

198. **quo...panni,** 'from which I infer the success of the green jacket'. The horses and drivers in the Circus were provided by wealthy companies (*factiones*), with whom the *munerarius*, whether praetor or consul, had to make a contract. There were four *factiones*, each possessing its own stud and training-stables, and also a colour, which was displayed on the chariot and person of the driver. They competed against one another, one chariot of each colour taking part in each race. The colours were red (*russata*, sc. *factio*), white (*alba*), green (*prasina*), and blue (*veneta*). Domitian's attempt to add a gold and a purple colour was abortive. By degrees the red and white became less

important; and the real struggle lay between the blue and the green. The partisanship of the populace was incredibly enthusiastic: see n. to v. 143: at Constantinople in A.D. 532 a riot arose in this way which cost 30,000 lives. The Christian converts, when rebuked by their pastors for their excessive interest in the competition, pleaded in excuse that Elijah had gone up to heaven in a chariot! For details of the races, see n. to III. 223.

eventum: cf. Pliny, *Epp.* V. 20. 2 *egi pro Vareno, non sine eventu.* The word generally means 'result', whether favourable or not.

viridis...panni: the *pannus* (the word is constantly used in this connection) is the tunic of the *auriga*, as Pliny shows, *Epp.* IX. 6 *favent panno, pannum amant...tanta gratia, tanta auctoritas in una vilissima tunica.* The colour of this *factio* is elsewhere usually called *prasinus*, lit. leek-coloured.

199. **attonitam,** 'stricken': cf. IV. 77: a stronger word than *maestam*: cf. Sen. *Dial.* XI. 15. 5 *non solum maestum sed etiam attonitum.*

200. **pulvere:** a high wind blew great dust-clouds in the faces of the Roman soldiers at Cannae 216 B.C.: cf. Livy, XXII. 46. 9 *ventus...adversus Romanis coortus multo pulvere in ipsa ora volvendo prospectum ademit.*

201. **consulibus:** L. Aemilius Paullus fell on the field; C. Terentius Varro survived and was thanked by the senate, *quod de republica non desperasset.*

spectent iuvenes: cf. l. 165.

202. **sponsio,** 'betting': cf. Mart. XI. 1. 15 *sed cum sponsio fabulaeque lassae | de Scorpo fuerint et Incitato.*

cultae, 'smart', 'well-dressed': cf. VI. 352.

adsedisse puellae: in the theatre and amphitheatre, but not in the Circus, the sexes were separated: cf. Ovid, *Trist.* II. 283 *tollatur circus: non tuta licentia circi est; | hic sedet ignoto iuncta puella viro.*

203. 'But let my wrinkled skin drink in the spring sunshine': the day of this dinner is 10 April: see n. to l. 193. Sitting in the sun (*apricatio*) was thought good for the health, especially of old men, such as Juv. now was; cf. Mart. X. 12. 7 *totos avida cute combibe soles.*

204. **effugiatque togam:** cf. I. 96, III. 171; the *toga* was the compulsory dress of all citizens in the Circus and amphitheatre: cf. Suet. *Aug.* 40 *negotium aedilibus dedit ne quem posthac paterentur in foro circove nisi positis lacernis togatum consistere.* But it was unpopular, being hot and heavy and also expensive

to keep clean; so we find Martial contrasting *tunicata quies* with *opera togata*, and rejoicing in his Spanish retirement that the *toga* is there unknown (XII. 18. 17).

balnea: dinner was preceded by the bath, as a matter of course. **salva fronte,** 'without a blush', *frons* being used as = *pudor*, because modesty is shown by blushing: cf. VIII. 189; and see n. to XIII. 242.

205. **solida hora,** 'a full hour': so with *dies*, and with adverbs of amount, e.g. *solidum centiens*, 'full ten million sesterces'. **supersit:** for the mood, see n. to VII. 14.

206. **ad sextam,** 'till noon': *sexta hora* may be either a period of time, lasting an hour, or a point of time: here it is the latter. The date being soon after the equinox, the hour is about 60 minutes; in mid-winter it would, according to the Roman reckoning, be much shorter, in mid-summer; much longer: see n. to X. 216. The *cena* usually began at the ninth hour, and the public baths (*thermae*) were not opened until an hour before; but on a holiday they may have opened earlier, or Persicus may have resorted to one of the many private baths in Rome.

quinque diebus, 'for a whole week', is not to be understood to refer to the festival of the Megalesia which lasted, not five, but seven days; it is used colloquially for 'some considerable time': so Hor. *Epp.* I. 7. 1 *quinque dies*; *Sat.* I. 3. 16 *quinque diebus*.

208. **voluptates commendat,** 'makes pleasure more pleasant': cf. Mart. IV. 64. 25 *rus...commendat dominus* (the villa is made pleasanter by the character of its owner).

SATIRE XII

WELCOME TO A FRIEND ON HIS ESCAPE FROM SHIP-WRECK: AND SOME REFLECTIONS ON LEGACY-HUNTERS

1–82. *I keep today as a holiday and offer to the gods such victims as my means will allow, in gratitude for the safe return of my friend, Catullus. He has encountered many dangers at sea, first fire, when his ship was struck by lightning, and then shipwreck. When the hold was full of water, and things looked desperate, he had resolution to throw all his most precious possessions overboard, to lighten the ship; though many a man nowadays would rather lose his life than his property. Lastly, he cut his mast loose, a desperate remedy. But at length the wind fell and the sun shone out, and he ran safely into the harbour of Ostia.*

1. **natali...die:** cf. Hor. *Carm.* IV. 11. 17 (*dies*) *iure sollennis mihi sanctiorque | paene natali proprio. natalis* without *dies* often has this sense; for *natales*, see n. to VIII. 231. The Romans, even after childhood, kept their birthdays (cf. XI. 84) as sacred especially to the *genius*.

Neither Corvinus nor Catullus (l. 29) is otherwise known.

2. **caespes,** an altar of turf, erected outside a temple: burnt-offerings were not placed on altars inside Roman temples. Private persons could make offerings at the Capitol, under fixed conditions, as at other temples: cf. X. 65.

3. Each of the three Capitoline deities, Jupiter, Juno Regina, and Minerva, is to have a victim.

niveam: white victims were offered to the gods of heaven (X. 66), black to the infernal gods.

4. **vellus** = *ovis*: so Browning speaks of 'our many-tinkling fleece'. She who 'fights with the Moorish Gorgon' is Pallas or Minerva, *regina bellorum virago* as Statius calls her (*Silv.* IV. 5. 23). She bears the *aegis*, her shield, in the centre of which is set the head of the Gorgon, Medusa; hence the *aegis* itself is called *Gorgon*, e.g. Mart. VI. 10. 11 *sic breviter posita mihi Gorgone Pallas.*

Maura: legend placed Medusa in Libya: the epithet is sarcastic here, 'the negress'.

5. The calf, reserved for Jupiter, is dragged to the altar unwillingly by a rope. **petulans** is perhaps used in the sense of *petulcus*.

7. **quippe,** 'for: see n. to XIII. 26.

8. **mero:** the victim was dedicated, before being slain, by the pouring over its horns, of wine, incense, and *mola salsa* (hence *immolare*).

10. **res ampla domi:** see n. to *res angusta domi*, VI. 357.

similis affectibus, 'to correspond with what I feel'; see nn. to VI. 214, XV. 150.

11. **Hispulla** must have been a familiar figure in the streets of Rome; see n. to *Procula*, III. 203.

13. **laeta,** 'rich', 'luxuriant', a constant epithet of such words as *fruges, pabula, segetes.* Clitumnus is a river in Umbria, the meadows on whose banks supplied a breed of white oxen which are repeatedly mentioned as the choicest victims.

The animal is first called *sanguis*, and then *cervix*, much as an elephant is called *ebur*, l. 112: this is hardly tolerable in English, so transl., 'whose blood (i.e. race) proclaims the rich pastures of Clitumnus, and whose neck needs a tall sacrificer to strike the blow'.

For victims so described, cf. Ovid, *Fasti*, IV. 415 *apta iugo cervix non est ferienda securi*; Stat. *Theb*. IV. 446 *quaecunque gregum pulcherrima cervix,* | *ducitur*; ibid. XI. 284 *hostia, nate, iaces ceu mutus et e grege sanguis*. (If seems as if *sanguis* was in the sense of *victima*.)

14. The process of slaughtering is described by Seneca, *Dial*. I. 6. 8 *tenui ferro commissura cervicis abrumpitur, et cum articulus ille qui caput collumque committit incisus est, tanta illa moles corruit*. Hence **grandi** does not mean that the man must be strong, but tall enough to reach up to the animal's neck.

ministro: the priest's attendants are properly called *popae*.

16. **incolumem,** 'alive'.

19. **nube una:** cf. C. Bronte, *Shirley*, p. 336, 'there is only one cloud in the sky; but it curtains it from pole to pole'.

21. **attonitus,** 'thunderstruck' only in the metaphorical sense: cf. Seneca quoted on VII. 67.

nullum conferri etc.: cf. XIV. 19 which shows that *conferri* has here the sense of *comparari*; so XIII. 144.

22. **velis ardentibus,** 'a fire at sea'.

23. **poetica...tempestas,** 'a storm in a poem': so Seneca speaks of a *mimicum naufragium*, 'a shipwreck on the stage' (*Dial*. IV. 2. 5). A *poetica tempestas* with all possible details will be found in [Seneca] *Agamem*. 483–599. Juv. seems a little sceptical of his friend's account of his disasters.

25. **quamquam sint:** for the mood, see n. to VII. 14.

27. **votiva...tabella,** a picture, representing the danger escaped, which is hung on the temple wall in gratitude for deliverance; many such may be seen in Italian churches. Especially those saved from shipwreck showed their gratitude in this way: edd. quote Cic. *de Nat. Deor*. III. 89 *nonne animadvertis ex tot tabulis pictis, quam multi votis vim tempestatis effugerint in portumque salvi pervenerint?* A different use of such a picture is mentioned XIV. 302.

28. Painters 'are supported by Isis', because they get their bread by painting *votivae tabellae* for her temple: she was supposed to have great power to help, when a ship was sinking.

31. **alternum puppis latus,** 'first one side of the ship and then the other': cf. Petron. 113 *ecce iam ratem fluctus evertet.*

32. **arboris incertae:** *arbor* is constantly used for 'mast' in Latin poetry, and even in prose: cf. Petron. 114 *non arbor erat relicta, non gubernacula, non funis aut remus*; Pliny, *Epp*. IX. 26. 4 *cum stridunt funes, curvatur arbor, gubernacula gemunt*. It probably means 'mast' here, and, if so, is a gen. of quality to be

taken with *alveus*, 'with tottering mast', an abl. absol. being inserted between the two epithets, to show why the mast was tottering: so Madvig: cf. XI. 96. Others suppose, on slight authority, that *arboris* is 'the ship', and so take *puppis* above as 'stern', governing *arboris*. Lachmann read *arbori incertae*, the dat. being governed by *conferret opem*; but such a hiatus though found elsewhere (e.g. Virg. *Ecl.* 3. 79 *valĕ inquit Iolla*) is not found in Juv. or his contemporaries.

cani, 'white-haired', and therefore experienced.

33. **rectoris,** often used for *gubernatoris*; but in prose *navigii* is usually added.

decidere…cum ventis, 'to compound with the winds', a commercial phrase used of the debtor, who, unable to pay in full, bargains for a quittance by paying a part of his debt (*portio*): cf. Mart. IX. 3. 5 *conturbabit Atlas, et non erit uncia tota | decidat tecum qua pater ipse deum* (i.e. Jupiter will not be able to pay 1s. 6d. in the pound): it is often, as here, metaphorically used: so Sen. *Dial.* XI. 12. 1 (to Polybius who had lost a brother but had a wife and child and other brothers living) *pro omnium horum salute hac tecum portione fortuna decīdit*. The composition offered is here and elsewhere expressed in the ablative.

iactu is strangely used for *iacturā*: if the winds grant him life, he will surrender to them his goods.

34. **castora:** fable represented the beaver as biting off that part of his body for the sake of which he was hunted.

36. **testiculi:** the only hiatus in this part of the verse in Juv.

adeo…inguen, 'for well aware is he of the drug he carries in his groin': see n. to III. 274.

37. From the description it appears probable that Catullus was a merchant, what Seneca calls *institor delicatarum mercium* (*de Ben.* VI. 38. 3): it seems unlikely that he was carrying all the luxuries described (e.g. *mille escaria*) for his own use.

39. **Maecenatibus,** i.e. men like Maecenas, fops and voluptuaries; see n. to I. 66.

40. **alias…pecus,** 'other garments, of which the wool was dyed on the sheep's back (*ipsum*) by the quality of the splendid pasture': *pecus* is lit., the flock that bore the wool. The wool of the sheep in Baetica had a natural tinge of a golden colour, as Martial repeatedly states, e.g. XIV. 133 *lacernae Baeticae. non est lana mihi mendax, nec mutor aheno* (the vat): *| sic Tyriae placeant, me mea tinxit ovis*.

41. **egregius fons,** i.e. the river Baetis (Guadalquiver): cf. Mart. XII. 98. 1 *Baetis,…aurea qui nitidis vellera tinguis aquis*.

Seneca neatly describes the Peneus, which had similar properties, as *gratuitus infector* (i.e. it saves the dyer).

43. argentum, 'silver-plate': *lances* and *cratera* are in
apposition. **mittere**, 'to throw into the sea'.

44. Parthenio is said by the Schol. to be the name of the
engraver (*caelator*): plate was often named after the engraver
and his name stamped upon it (cf. Sen. *Dial*. IX. 1. 7 *argentum
grave rustici patris sine ullo nomine artificis*); so we read in Martial of *Gratiana* (sc. *vasa*), and *Septiciana*, in Pliny of *Clodiana*
and *Furniana*. No engraver called Parthenius is known.

Friedl. translates, 'for Parthenius', and supposes P. to be the
well-known *cubicularius* of Domitian, whose name, as that of a
famous connoisseur, adds value to the plate: cf. l. 47. But it is
strange that Martial should never mention P.'s plate, if so
famous.

The Romans set an especially high value on plate or jewellery
that had belonged to illustrious possessors (*per multas dominorum
elegantium successiones civitati nota* Sen. *Dial*. IX. 1. 7). Some of
the pieces of connoisseurs at this time had a *stemma* which taxes
our powers of belief: so Horace (*Sat*. I. 3. 91) speaks of plate
that had belonged to Evander and Sisyphus: and a statuette of
Hercules by Lysippus, belonging to Nonius Vindex, is said by
Martial (IX. 43) and Statius (*Silv*. IV. 6) to have belonged successively to Alexander, Hannibal, and Sulla. Berenice's diamond
(VI. 156) is a case in point: its value was increased, because so
famous a personage had worn it.

urnae is a measure of capacity, holding 24 *sextarii* or about
3 gallons: cf. VI. 426.

45. Pholus was a Centaur, who used an immense *crater* as a
weapon in the famous fight at a feast between the Centaurs and
Lapithae. Martial substitutes the name of Rhoecus, VIII. 6. 7
hoc cratere ferox commisit praelia Rhoecus | cum Lapithis.

Fuscus' wife must have been notorious at Rome for her love
of wine.

46. bascaudas: *vasa ubi calices lavabantur* Schol. A gloss explains the word by *conchas aereas*: M. says 'rather our *basket*'.
For the thing was British: cf. Mart. XIV. 99 *barbara de pictis veni
bascauda Britannis*, where it comes in a long list of *calices* of
different kinds. Rich *s.v.* supposes it to be a basket and compares the Welsh *basgawd*. Perhaps it was a cup with wicker
cover: for cf. Mart. II. 85. 1 *vimine clusa levi niveae custodia
coctae*.

47. caelati, sc. *argenti*: cf. *multum flammarum*, III. 285

quo: for the constr., cf. Lucan, IV. 380 *non auro murrhaque bibunt.*

'The cunning buyer of Olynthus' is Philip of Macedonia, who took Olynthus 347 B.C. by bribing the chief citizens: he said that 'no fort was impregnable into which an ass, laden with gold, could make its way'.

49. **argento** is not 'money', but 'plate': see n. to XIV. 291.

50. **patrimonia,** 'wealth', not necessarily inherited, as this passage shows: cf. VII. 113, X. 13.

51. **vitio,** i.e. *avaritia.*

52. **rerum utilium:** some see in this a sarcastic reference to the luxurious objects named above; but the natural meaning is that, as well as the costly cargo, most of the food and stores (*utensilium* in prose) were thrown out.

nec, 'not even': see n. to v. 3. **damna** is nom.

54. **reccidit:** so *reccidere* Lucr. I. 857 and often in the poets. **summitteret,** the regular word for lowering the mast, is here used ironically with *ferro* (= *securi*).

55. **se explicat angustum,** 'he finds a way out of his straits': the partic. of *esse* is understood with *angustum:* in fact *angustum = ex angustiis:* cf. Cic. *pro Cael.* 67 *qui se nunquam profecto, si in istum locum processerint, explicabunt.*

ultima: cf. XV. 95: the plur. is unusual when a gen. in the sing. follows: but cf. Lucan, VIII. 665 *ultima mortis.*

56. **factura,** 'that must make': for another silver-age use of the fut. partic., see n. to VI. 277.

57. **i nunc et...committe:** see n. to VI. 306.

58. **digitis,** 'finger-breadths': to both Greeks and Romans this was the smallest unit of linear measure, of which there were sixteen in a foot. The acc. is commoner to express dimensions, e.g. Cic. *Acad.* II. 58 *ab hac mihi non licet transversum, ut aiunt, digitum discedere.*

59. **taedae,** 'of pine-wood', is governed by *digitis:* this is the reading of P, restored by Büch. who quotes from Dio Chrys. τριδάκτυλον αὐτοὺς σῴζει ξύλον πεύκινον: cf. XIV. 289. *taeda* (the old reading) must be supplied as subject to *sit.*

60. **mox,** 'in future'. M. says 'when on board'; but surely Juv.'s point is that axes, to cut loose your rigging, ought to be provided with other indispensable stores before your next voyage.

reticulis et pane: the *reticula* are to hold the bread, as appears from Hor. *Sat.* I. 1. 47 *ut si | reticulum panis venalis inter onusto | forte vehas humero.*

ventre lagonae, 'big-bellied flasks'; so *Montani venter,* IV. 107.

61. It is doubtful whether **aspice** can mean 'provide', though *circumspice* (cf. VIII. 95) is so used, and *prospicere,* e.g. Sen. *Dial.* IV. 35. 1 *arma nobis expedita prospicimus, gladium commodum et habilem.* Hence Jahn proposed *respice.* Yet Livy often uses *aspicere,* 'to examine', 'to review', e.g. XLII. 5. 8 *Ap. Claudium legatum ad eas res aspiciendas componendasque senatus miserat*: and such a sense would be appropriate here.

sumendas, 'for future use': see n. to XIV. 268.

63. **vectoris** is perhaps governed by *fatum,* not by *tempora*; if so, *que* is attached to the second word in the sentence, by a licence not uncommon in the poets. The predicate of *fatum,* i.e. *prosperum erat,* is understood.

euro is used generally for 'wind': the winds are called *austri* l. 69.

65. **staminis** is governed by *lanificae* which is coupled by *et* with the other epithet *hilares.* For *staminis,* cf. X. 252: those for whom the Parcae spin white wool, are fortunate, while black wool means the opposite fortune: cf. Mart. VI. 58. 7 *si mihi lanificae ducunt non pulla sorores | stamina.*

68. **vestibus,** in default of sails which had disappeared with the mast.

superaverat: this verb is used in the best Latin with the sense of *superesse.*

69. **velo prora suo:** some ancient ships had a small fore-sail, called *dolon,* which could be set, not on the mast, but on the prow. *prora,* which as subject is used for *navis,* has its restricted sense in this part of the sentence.

70. **gratus Iulo,** etc.: Ascanius (or Iulus) left Lavinium, which Aeneas had founded and named after Lavinia, in posses-sion of Lavinia, his stepmother, and founded Alba Longa, beneath the Alban Mount.

71. **Lavino:** the town is generally called *Lavinium,* but an adj. *Lavinus* is found (e.g. Virg. *Aen.* I. 2), derived from a name *Lavinum* which is used here. Virgil has both *Lāvina and Lăvini.*

72. **sublimis apex,** i.e. *Mons Albanus.* This portent of a white sow with a litter of thirty (cf. VI. 177) was twice foretold to Aeneas, by Helenus (*Aen.* III. 389) and by Tiberinus, *ibid.* VIII. 43 *littoreis ingens inventa sub ilicibus sus, | triginta capitum fetus enixa, iacebit, | alba, solo recubans, albi circum ubera nati. | hic locus urbis erit, requies ea certa laborum, | ex quo ter denis urbem redeuntibus annis | Ascanius clari condet cognominis Albam.*

The number of pigs presaged the number of years before the founding of Alba.

73. mirabile sumen: cf. Virg. *Aen.* VIII. 81 (of this sow) *mirabile monstrum*. Büch. keeps *miserabile* of P, which the Schol., who recognises both readings, explains *quod miseratum sit Phrygibus*: for adjectives in *-bilis* with active meaning, see Munro on Lucr. I. 11: but there are no other certain examples in Juv., and it seems more probable that the copyist's eye wandered to l. 67.

In Ovid, *Heroid.* XII. 99 all MSS. read *facinus miserabile* except the best (Puteanus) which has *f. mirabile*: so it seems there was some tendency to confuse the two words.

sumen is a satirical equivalent of *feta sus*.

74. numquam visis, 'never seen', sc. elsewhere. It is not meant that the Trojans never really saw such a thing but that such sights are rare. The phrase is peculiar but seems to be an equivalent for *invisitatis*, a word which Livy often uses of extraordinary sights, e.g. XLV. 42, 12 *naves invisitatae ante magnitudinis*; and (without *ante*) V. 35. 4 *cum formas hominum invisitatas cernerent*.

75. The port of Rome was Ostia, at the mouth of the Tiber. The river-deposit, which has left the ancient harbour over a mile inland, made frequent operations and improvements necessary. The *portus Augusti* was constructed by Claudius, two miles to the north of Ostia, and connected with the Tiber by a canal. Cf. Suet. *Claud.* 20 *portum Ostiae extruxit, circumducto dextra sinistraque bracchio et ad introitum profundo iam salo mole obiecta...congestisque pilis superposuit altissimam turrem in exemplum Alexandrini Phari, ut ad nocturnos ignes cursum navigia dirigerent.*

moles: in Suet.'s description, *moles* denotes the breakwater or artificial island on which the lighthouse stood, outside the two great piers: but here *moles* is used for all the artificial structures which formed the harbour: the verb (**intrat**) shows this.

76. Tyrrhenamque pharon, 'and the Pharos of the Tyrrhene sea': the light-house island of Alexandria (see n. to VI. 83) gave its name to similar structures elsewhere; cf. Stat. *Silv.* III. 5. 99 (of a light-house off Campania) *Bacchei vineta madentia Gauri | Teleboumque domos* (Capri), *trepidis ubi dulcia nautis | lumina noctivagae tollit pharus aemula lunae.* A light-house is *phare* in French, and Browning uses the same word in English.

bracchia are the two great piers which run out to sea and

finally bend inwards again: for *rursum*, cf. X. 150. They were
built with *returns*, as we say.

78. **non sic,** i.e. far less, the *portus Augusti* being entirely
artificial (*manu factus*).

79. **sed,** 'to return', the digression about harbours being at
an end. **trunca,** i.e. dismasted. **magister,** 'the skipper': cf.
IV. 45.

80. **interiora:** *Traianus portum Augusti restauravit in melius,
et interius tutiorem, nominis sui, fecit* Schol. on l. 76. Trajan
added an inner basin to Claudius' harbour, which, being
smooth as a pond, is here called *stagna*: a plan of the whole is
given by Rich *s.v. portus*.

Baianae...cumbae: even the light pleasure-boats of Baiae
might sail about there; they generally plied in the land-locked
waters of the Lucrine; cf. Mart. III. 20. 19 *an aestuantes iam
profectus ad Baias | piger Lucrino nauculatur in stagno?*

81. **vertice raso,** in performance of a vow which was often
made by those in peril at sea: cf. Petron. 103 (where some
passengers are shaving their heads as a disguise) *unus forte ex
vectoribus, qui acclinatus lateri navis exonerabat stomachum
nausea gravem, notavit sibi ad lunam tonsorem intempestivo
inhaerentem ministerio, execratusque omen, quod imitaretur nau-
fragorum ultimum votum, in cubile reiectus est.*

82. **garrula,** as an epithet of *pericula* is very bold: in sense
it is either an epithet of *nautae* (cf. *parcam*, X. 116) or an adverb
with *narrare* (cf. *montana*, VI. 5), 'spin long yarns of dangers
escaped'. Juv. would not use *garruli*: see n. to *Pontica*, VI. 661.

83–130. *I bid my slaves make all preparations for the sacrifice;
this completed, I will make offerings to my household gods as
well, and decorate my house. Nor have these rejoicings any
interested motive; Catullus has children, so that I cannot expect
to inherit from him. As a rule it is only the childless who have
such attentions paid them; no sacrifice is too costly for the greedy
will-hunters to offer, if a childless rich man is even slightly ill.
If elephants were to be bought, they would not hesitate to sacrifice
one, or a slave, or even, like Agamemnon, a daughter. A legacy
is certainly worth more anxiety than a fleet of a thousand ships.
For the rich man may recover, and, in gratitude for this wonder-
ful devotion, may make the sacrificer of a daughter his sole heir.
Long life and great wealth to the will-hunter! and may he love
none, and none love him!*

83. **pueri,** 'slaves': so l. 117.

linguis...faventes, 'with reverent lips and hearts': a sacrifice was commonly preceded by the injunction *favete linguis* (= εὐφημεῖτε): Sen. *Dial.* VII. 26. 7 says that this was a command, not for words of good omen, but for silence, *ut rite peragi possit sacrum nulla voce mala obstrepente.*

84. **delubris** are the shrines of the three deities on the Capitol, in front of which the three altars of turf (l. 85) are erected: see n. to l. 2. **farra** is the same as *mola salsa*: it was sprinkled over the victim's head, the altar, and the sacrificial knives.

86. **sequar,** i.e. I will follow you to the Capitol where the altars are erected and duly perform the more important (*quod praestat*) sacrifice, then returning home for the lesser rites.

praestat: this absolute, personal use of *praestare* is rare: it occurs in a letter of Asinius Pollio to Cicero: Cic. *ad Fam.* X. 32. 4 *sed de illo plura coram: nunc, quod praestat, quid me velitis facere, constituite.*

87. **coronas,** 'wreaths'; **serta** above are 'festoons': cf. IX. 137 *o parvi nostrique Lares, quos ture minuto* | *aut farre aut tenui soleo exorare corona.*

88. **cera:** images and busts of wax, especially of the dead, are constantly mentioned; but it is pointed out that such a material is unsuitable for the Lares which stood before a fire. M. shows that wax was used as a varnish, and the Schol.'s explanation (*incerata signa deorum*) seems to point to this. The wax is called *fragilis* (crumbling), because it must be melted for this purpose.

89. **hic,** i.e. *domi*. The worship of other gods was often united with that of the *Lares*; Jupiter, Juno, Vesta and perhaps other deities were, as *Penates*, the object of domestic worship.

90. **violae:** cf. Pliny, *Nat. Hist.* XXI. 27 *violis honos proximus* (after roses and lilies), *earumque plura genera, purpureae, luteae, albae*: so Virgil calls them both *pallentes* 'yellow' (*Ecl.* II. 47) and *nigrae* (*ibid.* X. 39). It seems doubtful whether their modern representative is the pansy or the stock.

91. **erexit,** 'has put forth', as if they grew there. For the custom of adorning house-doors, see n. to X. 65.

92. **operatur,** 'worships', i.e. takes part in the rites. The pres. of this verb is extremely rare, and apparently unknown in classical Latin: the particip. *operatus* is constantly used with a present sense. The lighting of lamps by daylight is the sign of a religious ceremony: cf. Sen. *Epp.* 95. 47 *accendere aliquem lucernas sabbatis prohibeamus, quoniam nec lumine di egent et ne homines quidem delectantur fuligine.*

93. You must not suspect me of *captatio*: see nn. to III. 129, IV. 19.

94. **altaria** is here a true plural: see n. to VIII. 156.

95. **heredes** are his children: Catullus has the *ius trium liberorum*.

expectare, 'to wait and see': cf. VI. 274; Mart. IV. 40. 8 *expecto, Postume, quid facias.*

97. **sterili:** a friend is *sterilis* when his wife is *fecunda*: cf. V. 140 *carum sterilis facit uxor amicum.* **verum** is used where *immo* would be commoner. **inpensa** is probably a noun, but may also be the participle: cf. *inpendat* above.

cŏturnix: for the quantity, see Munro on Lucr. I. 360.

98. **patre,** 'one who has children': the opposite of *orbus.* **cadet,** 'will be sacrificed'; cf. l. 113.

sentire calorem, 'to be affected by the heat': cf. Sen. *Dial.* V. 21. 1 (of a river) *cum sensit aestatem et ad minimum deductus est.*

99. **coepit:** the sing. verb is followed by two subjects with a plur. adj.: Büch., to avoid this, puts a comma after *Pacius*, and takes *orbi* as gen. governed by *porticus*: but M. quotes exact parallels, e.g. Cic. *in Verr.* II. 4. 92 *dixit hoc apud vos Zosippus et Ismenias, homines nobilissimi.*

Galitta is originally a pet-name for *Galla*; this name occurs Pliny *Epp.* VI. 31. 4. Juv. has *Pollitta* (dimin. of *Polla*), II. 68.

100. The whole length of a *porticus*, either in the rich man's house or (more probably) in a temple, is covered with placards, on which are written the vows which mercenary friends undertake to perform, in case the invalid gets well; cf. Mart. XII. 90. 1 *pro sene, sed clare, votum Maro fecit amico,* | *cui gravis et fervens hemitritaeos* (a kind of fever) *erat,* | *si Stygias aeger non esset missus ad umbras,* | *ut caderet magno victima grata Iovi.*

legitime, 'in due form': cf. X. 338. **libelli** are *tabellae votivae*; see n. to X. 55.

But such a sense of *libellus* is rare: and *tota tabellis* might easily become *tota bellis* and be 'corrected' to *tota libellis.*

102. **quatenus** = *quandoquidem*, 'since'; only here in Juv.; cf. Mart. V. 19. 15 *quatenus hi non sunt, esto tu, Caesar, amicus.* They offer a hecatomb of oxen and not elephants, *because* these are not to be had. The **non** is superfluous when *nec* (= οὔτε) follows immediately.

103. **sub nostro sidere:** cf. Sen. *Dial.* X. 13. 7 *sub alio caelo natis beluis* (elephants).

104. **furva,** 'dark-skinned': the Moors, Ethiopians, and Indians: see n. to XI. 125.

105. **Rutulae arbores** may mean merely 'Roman trees': cf. VI. 637: but, as inscriptions show that the imperial herd of elephants was kept at Laurentum, Juv. may refer strictly to Ardea, the town of the Rutuli, which was not far from Laurentum.

107. **privato**, 'subject': cf. I. 16, VI. 114, XIII. 41. The possession of elephants was a privilege exclusively reserved for the emperors.

siquidem, 'for', as VI. 621; yet, as Hannibal and the Roman generals were *privati*, *sicut* or *quamquam* would be more logical: but the idea is that beasts, whose progenitors decided battles, have a right to be proud.

Tyrio = *Poeno*, Carthage being a colony of Tyre.

108. **Hannibali**: cf. X. 158. The *rex Molossus* is Pyrrhus who was the first to bring *boves Lucae* into Italy 280 B.C.

110. **partem aliquam belli**, 'a considerable part of an army', is in apposition with *cohortis*. **turrem** is a second object of *ferre* and co-ordinate with *cohortis*: cf. Lucr. V. 1302 *boves Lucas turrito* (= *turrigero*) *corpore...belli docuerunt vulnera Poeni* | *sufferre*; Livy, XXXVII. 40. 4 *addebant speciem* (sc. *elephantis*) *frontalia et cristae et tergo impositae turres turribusque superstantes praeter rectorem quaterni armati*.

belli: for the hiatus, see n. to X. 281. The reading of the worse MSS. (*bellique et*) is an attempt to get rid of the hiatus: see n. to XI. 6.

111. **nulla...mora per N.**: cf. VI. 333 *mora nulla per ipsam* | *quominus...summittat*. Novius is not to blame, but only circumstances, if the thing is not done. In Greek ἕνεκα γε is used just like *per* here: cf. Aristoph. *Ach*. 389 λαβὲ δ' ἐμοῦ γ' ἕνεκα = *nulla per me mora quin sumas*. Novius and Pacuvius Hister are *captatores*.

112. **ebur**, 'elephant': cf. *vellus*, l. 4, *sanguis*, l. 13: conversely *elephas* sometimes (as in Greek) = 'ivory'.

114. **horum**, i.e. *Larium*, the *Lares* being identified with the person in whose house they are worshipped.

115. **alter** is Pacuvius: cf. l. 125. **si concedas**, 'if permitted': see n. to III. 102.

116. The superl. (*pulcherrima*) is unusual after the positive: for the reverse order, cf. XIII. 13; Tac. *Ann*. I. 48 *foedissimum quemque et seditioni promptum*.

118. **vittae** and **infulae** (see Rich *s.vv*.) were placed on the head and neck of victims, whether animals or human beings: cf. Lucr. I. 87 (of Iphigenia) *cui simul infula virgineos circumdata*

comptus | ex utraque pari malarum parte profusast. Sacrificing
priests and Vestals (cf. IV. 9) also wore *vittae*.

nubilis: the age of Iphig. seemed to make her fate more
tragical: cf. Lucr. I. 98 *nubendi tempore in ipso | hostia concideret
mactatu maesta parentis.*

119. **altaribus:** see n. to VIII. 156.

120. **tragicae,** 'in the tragedy': so *poetica* l. 23.

furtiva piacula, 'the secret substitution': cf. Eur. *Iph. Taur.*
28 ἀλλ' ἐξέκλεψεν ἔλαφον ἀντιδοῦσά μου | Ἄρτεμις Ἀχαιοῖς. This
tradition was followed by Euripides, not by Aeschylus or
Sophocles.

121. **laudo meum civem,** 'bravo! my fellow-countryman',
cf. IV. 18. *meum civem*: the Greek idiom is the same (τὸν ἐμὸν
πολίτην), the word itself implying fellowship; *concivis* and
συμπολίτης are barbarisms.

nec comparo, etc.: I don't put the two things on a par, i.e.
I consider the inheritance more important: cf. l. 21, XIV. 19.

122. **mille rates,** the traditional number of Agamemnon's
fleet, though the Homeric catalogue reckons 1,186: cf. [Seneca]
Agam. 172 (the nurse to Clytaemnestra enraged at the loss of
her daughter) *sed vela pariter mille fecerunt rates.*

Libitinam evaserit: cf. Hor. *Carm.* III. 30. 6 *non omnis
moriar multaque pars mei | vitabit Libitinam: Libitina = mors,*
because her temple was the headquarters of the undertakers
(*libitinarii*); *ratio Libitinae* (Suet. *Nero,* 39) is 'the bills of
mortality'.

123. **tabulas,** 'his will', by which he had left his money to
others than Pacuvius.

nassae: Horace also compares the methods of will-hunters
with those of fishermen: *Sat.* II. 5. 44 *plures adnabunt tunni et
cetaria crescent*; cf. Mart. VI. 63. 1 *scis te captari, scis hunc qui
captat, avarum, | et scis, qui captat, quid, Mariane, velit...
'munera magna tamen misit'. sed misit in hamo; | et piscatorem
piscis amare potest?*

125. **breviter,** 'summarily': cf. *exiguis tabulis,* I. 68.

127. **iugulata Mycenis,** 'the butchery of the maid of
Mycenae': Agamemnon was king of Mycenae; and the daughter
of Pacuvius is called *an* Iphigenia.

128. **vivat...Nestora totum,** 'may his life be long as all
Nestor's was': cf. X. 246. For the expression, cf. XIV. 326 *sume
duos equites,* i.e. *censum duorum equitum.* Edd. quote Mart. X. 24.
11 *post hunc Nestora nec diem rogabo*; but the context there
seems to require Friedl.'s conjecture *post hoc.*

129. **rapuit Nero:** cf. Tac. *Ann.* xv. 45 (A.D. 64) *conferendis pecuniis pervastata Italia, provinciae eversae...inque eam praedam etiam di cessere, spoliatis in urbe templis:* idem, *Hist.* I. 20 *bis et viciens miliens sestertium* (£22,000,000) *donationibus Nero effuderat.*

For the form of this imprecation, cf. Mart. vi. 86. 5 *possideat Libycas messes Hermumque Tagumque | et potet caldam, qui mihi livet, aquam.* In both cases, the last petition makes the other advantages worse than useless.

130. M. traces this l. to its source in Cic. *Laelius,* 52 *nam quis est...qui velit, ut neque diligat quemquam nec ipse ab ullo diligatur, circumfluere omnibus copiis atque in omnium rerum abundantia vivere?*

SATIRE XIII

CONSOLATION TO A FRIEND, WHO HAS BEEN CHEATED OF A SUM OF MONEY

1–70. *A guilty man, though he may be acquitted in the law-court, always passes sentence against himself. You, Calvinus, have lost some money by a friend's dishonesty; but there are various reasons why you should not take it so to heart. In the first place, the sum is small; in the second, you are only suffering what many others suffer. Yet, in spite of your sixty years of life, you are as angry as if your misfortune was something serious, and something new. But every day produces far worse crimes than this. Good men are few; the times are bad. People laugh at you for your simplicity. This is no Golden Age when crime is an exception; on the contrary, in our days, it is the honest man who is a* lusus naturae.

1. **exemplo...malo,** 'setting a bad example'; abl. of accompanying circumstances: cf. Mart. I. 27. 5 *non sobria verba subnotasti | exemplo nimium periculoso:* so *fato* l. 104.

committitur: so *committunt,* l. 104. Elsewhere Juv. uses *admittere* in this sense (e.g. l. 237), *committere* meaning 'to entrust' (e.g. l. 125) or 'to join'. Cicero apparently uses *admittere* of the moral guilt of an action, *committere* of the injury to others: cf. *ad Fam.* III. 10. 2 *si quid a me praetermissum erit, commissum facinus et admissum dedecus confitebor;* and so Terence, *Phorm.* 415 *ut nequid turpe civis in se admitteret;* idem, *Adelph.* 682. But later writers seem to use the words indifferently: cf. Sen. *de Clem.* I. 23 *videbis ea saepe committi,* *quae saepe vindicantur...multo minus audebant liberi nefas ultimum admittere, quamdiu sine lege crimen fuit.* Martial always has *admittere scelus*, never *committere*.

2. **auctori**, 'the doer': cf. Mart. II. 72. 5 *auctorem criminis huius*: the word means 'originator', and is applied indifferently to the writer of a book, the giver of a present, the planter of a tree, the maker of a cup, the source of a report, the founder of a philosophic school, etc. It is said that *auctor* is never used absolutely for *scriptor*; but cf. Sen. *Dial.* XI. 11. 5.

prima est haec ultio: cf. Sen. *Epp.* 97. 14 *prima illa et maxima peccantium est poena, peccasse.*

quod, 'that', not 'because'.

3. **se iudice**, 'at the bar of his own conscience': his conscience is regarded as outside himself, which accounts for the abl. absol. referring to the subject of the sentence.

4. **praetoris...urna**: this probably refers either to the balloting for the names of the *iudices* before trying a criminal case, or to the urn in which the *iudices* in the praetor's court deposited their votes.

Servius gives a different explanation in a comment on Virg. *Aen.* VI. 431 *nec vero sine sorte datae, sine iudice sedes: | quaesitor Minos urnam movet*: he quotes this l. from Juv., and says it refers to a ballot before the praetor, to determine the order in which cases shall come on. Friedl. thinks Servius must be followed. But the context makes it clear that what influence secures is an acquittal, not merely an alteration of the time of the trial.

vicerit, 'has gained a verdict': see n. to *victrix*, I. 50; so *victoria*, 'an acquittal'.

urna is abl. of means.

5. This satire, like the last, is addressed to a person otherwise unknown.

6. **crimine**, 'crime': cf. I. 75. **sed**: the connection seems to be, 'it is a wrong, *but* there are alleviating considerations'.

7. **tenuis census**: hence *tenuis* is often used of men who are poor: cf. III. 163. The sum lost is stated l. 71.

8. **mergat onus** is a mixed metaphor, unless the man is compared to a ship, which would be very awkward after *iacturae*: cf. X. 57.

rara, 'seldom': see n. to VIII. 63.

9. **multis cognitus**: cf. XII. 26. **iam**, 'quite': see n. to III. 206.

10. **e medio...acervo**, 'drawn from the middle of For-

tune's store', i.e. a piece of average fortune, neither very good nor very bad: cf. *Ham.* II. 2: '*Guild.* On fortune's cap we are not the very button. *Haml.* Nor the soles of her shoe? *Rosenc.* Neither, my lord.'

12. **dolor,** 'resentment'. For the abl. **vulnere,** see n. to IV. 66.

13. **quamvis,** 'however', qualifies *levium* only: so III. 282 and Mart. V. 52. 7 *quamvis ingentia dona | auctoris pereunt garrulitate sui.*

15. **visceribus:** the ancients generally considered the *iecur* to be the seat of anger: cf. I. 45, VI. 648, *spumare*, which suggests the foaming mouth of a wild boar, is a bold word to use of the *viscera.*

reddat is subj., because it is the reason in C.'s mind for his wrath.

16. **depositum:** refusing to refund a sum of money entrusted for safe keeping was a common form of crime, as is shown by Pliny's famous letter about the Christians in Bithynia (A.D. 112): they said that they took an oath, *ne furta, ne latrocinia, ne adulteria committerent, ne fidem fallerent, ne depositum appellati abnegarent* (Pliny, *ad Traj.* 96. 7).

17. This l. affords an exact date for this satire, and an approximate date for the book of which it is the first. When a year is denoted by the name of one consul only, the name is that which has precedence in the *Fasti*. This condition is satisfied only by Fonteius Capito, consul in the year A.D. 67. Consequently, Calvinus was 60 years old in 127; and this satire was written in, or immediately after, that year. The Schol.'s notion that Juv. is speaking of himself in this l., is a mere blunder.

18. **rerum...usu,** 'experience'.

proficit is the reading of P, which Büch. keeps: yet the third person, quite natural in the l. above, is somewhat strained here; and P elsewhere has *t* where *s* is required: see n. to III. 201, Introduction, p. xlvii. The Schol. apparently read either *proficis usu*, or *proficit usus*; the inferior MSS. have *proficis.*

19. It is difficult to say whether **magna** belongs to *praecepta* or to *sapientia*: cf. Virg. *Georg.* II. 490 *felix qui potuit rerum cognoscere causas* (i.e. the philosopher)...*fortunatus et ille* etc. Perhaps the quotation is in favour of *magna* being fem. sing.

20. **sapientia** is 'philosophy', the *ars vivendi*, by which Democritus defied Fortune (X. 52). Nor must we despise, says Juv., the man, who, though no philosopher of the schools, has been taught the lesson of Stoicism by life, who is an *abnormis sapiens* like Horace's Ofellus (*Sat.* II. 2. 3).

22. **iactare iugum**, like πρὸς κέντρα λακτίζειν, is properly said of oxen ploughing: cf. VI. 208.

23. **cesset prodere** = *non prodat*, 'that it fails to bring to light': for the infin., which is common in Plautus, cf. Propert. I. 12. 1 *quid mihi desidiae non cessas fingere crimen?* Pliny, *Epp.* VI. 20. 10 *quid cessatis evadere?*

25. **gladio vel pyxide**, 'by the dagger or the bowl': *pyxis*, properly a bowl of box-wood, was a name given to bowls of a certain shape, whatever the material, in which drugs (II. 141) or cosmetics (Mart. IX. 37. 4) were kept.

26. **quippe**, 'for', as always in Juv.; in XII. 7 it may be used, as sometimes by Livy, like ἅτε to introduce a causal participle; but after Cicero the word is generally used as a synonym of *nam* or *enim*. Juv. has it generally as the first, sometimes as the second, word of the sentence.

numera = *si numeres*: see n. to VII. 175. This is Büch.'s reading indicated by P; other MSS. have *numero vix sunt*.

27. So Abraham thought there might be ten righteous men in Sodom (Gen. xviii), but there were not. Thebes in Boeotia had seven gates, each of which in Aeschylus' play is assailed by a separate champion: the Egyptian city Thebes was ἑκατόμπυλος: see n. to xv. 6. The Nile had seven mouths. Thus Statius, enumerating the gates of Thebes (*Theb.* VIII. 353), illustrates their number by the mouths of the Nile.

divitis, not because it carries gold like Tagus or Pactolus, but because the wealth of Egypt depends on its waters.

28. **nunc**: so P: *peior* is to be supplied with *aetas*. **saecula** is really sing. in meaning: see n. to VI. 24: and this makes its connection with *aetas* much less harsh. *nona* of the other MSS. is explained as referring to the 9th cent. of the city; but *aetas* naturally refers in this context to the ages known as golden, silver, etc.; cf. VI. 23 and 24; so *nona* would require eight ages, each named after a separate metal. But the longest of such lists, that of Hesiod, recognises only (1) a golden age, (2) a silver age, (3) a brazen age, (4) a heroic age, (5) an iron age. So that *sexta* would be needed here.

30. **nomen**, i.e. a name of the same kind as *aureum*, *argenteum*, *ferreum*: it would be flattering the age to call it even *ferrea*.

a, 'after', like ἐπί with gen.

31. **hominum divumque fidem**, 'the protection of gods and men', so often in Plautus *di, vostram fidem!* The same use of *fides* is seen in the phrase *in fidem populi Romani venire*. The

phrase was used in an excited protest, as appears from Sen. *Epp.* 15. 7 *usque eo naturale est paulatim incitari, ut litigantes quoque* (even people in a quarrel) *a sermone incipiant, ad vociferationem transeant: nemo statim Quiritium fidem implorat.*

32. **quanto** = *tanto quanto.*

agentem, 'when speaking in court': cf. VII. 122.

33. **sportula** stands for the clients, who receive the dole (see n. to I. 95): cf. Mart. VI. 48 *quod tam grande sophos clamat tibi turba togata, | non tu, Pomponi, cena diserta tua est*; Pliny, *Epp.* II. 14. 4 *sequuntur auditores, actoribus similes, conducti et redempti: manceps* (a contractor) *convenitur: in media basilica tam palam sportulae quam in triclinio dantur.* See also n. to VII. 44.

bulla dignissime = μέγα νήπιε: see n. to V. 164.

34. **veneres,** 'charms': cf. Sen. *de Ben.* II. 28. 2 *ille non est mihi par virtutibus nec officiis sed habuit suam venerem.*

35. **simplicitas** = εὐήθεια, 'innocence'.

37. **numen:** cf. X. 365. **rubenti,** with the blood of victims, over which the oath was taken; cf. l. 89.

38. **indigenae,** 'primitive man', the αὐτόχθονες: for the innocence of the golden age, before Saturn was dethroned by Jupiter, see VI. 1–24.

39. **sumeret:** the subj. would not be used here after *priusquam* by Cicero or his contemporaries: they would say *sumpsit.*

41. **Idaeis...antris:** Jupiter was born at Ida in Crete and 'was still a subject' there, so long as Saturn reigned. The language is satirical: hence *virguncula, privatus, taberna, prandebat.* All the nominatives from *virguncula* to *Vulcanus,* l. 45, have a verb understood, generally *erat* or *erant.*

44. **ad cyathos,** sc. *stabat:* Hebe, Hercules' wife, was cupbearer at the celestial banquets, until Ganymede was carried off by the eagle from Troy, to fill that office. Hephaestus did, according to Homer (*Il.* I. 597), act as cup-bearer on one occasion to the great amusement of the gods; but Juv. seems not to refer to this, but to the fact that he comes, smutty from the forge, to the feast and cleans himself up during it.

et is used for *nec*: cf. XV. 125.

iam siccato nectare, 'even after a draught of nectar', whereas the other gods washed before their meal. *siccato* means no more than *hausto,* though *siccare calicem* is more correct than *siccare vinum*: cf. V. 47; Sen. *Dial.* IV. 33. 5 *pocula...siccaret.* The Schol. seems to have read *saccato,* which would give a good meaning: 'even after the nectar was strained (i.e. ready for the table), Hephaestus was still engaged upon his toilet'.

45. **Liparaea...taberna:** see n. to I. 7. In Homer (*Il.*
XVIII. 414) Hephaestus duly cleanses himself before leaving the
forge to receive his visitor: σπόγγῳ δ' ἀμφὶ πρόσωπα καὶ ἄμφω
χεῖρ' ἀπομόργνυ | αὐχένα τε στιβαρὸν καὶ στήθεα λαχνήεντα.

46. **prandebat,** 'took his breakfast' by himself, has the same
satirical force as in X. 178.

47. The increase of gods may refer partly to the deified
emperors, but much more to the foreign deities of Egypt and
the East, whose worship had gradually become established
together with that of the Olympian gods. Before heaven became
so over-populated, Atlas had not such a heavy weight to support.

49. **aliquis,** i.e. Neptune or Poseidon.

sortitus: the three brothers, Zeus, Poseidon, and Pluto, cast
lots for sovereignty over the upper world, the sea, and the
lower world; hence the sea is called *sors secunda* as Neptune's
realm, and Hades *sors tertia*.

profundi, 'the deep sea'.

50. **Sicula...coniuge,** Proserpina, who was carried off from
Henna in Sicily by Pluto.

51. There were no punishments in Hades then, no Ixion
bound on a wheel, no Sisyphus rolling a stone, no Tityus torn
for ever by a vulture.

52. **regibus,** i.e. king and queen.

53. **admirabilis,** 'astonishing'; cf. *admiratio,* VI. 646.

55. **vetulo non adsurrexerat:** for the constr., cf. Sen. *Dial.*
IV. 21. 8 *longe...ab adsentatione pueritia removenda est. ...maio-
ribus adsurgat.* It is a complaint commonly heard in all ages,
that the young are not so respectful to their elders as they used
to be.

56. **cuicumque** is used as an indefinite pronoun: see n. to
III. 156.

licet, etc., although his parents were richer in the simple
things in which wealth then consisted.

57. **fraga...glandis:** cf. Ovid, *Met.* I. 104 (of the golden
age) *arbuteos fetus montanaque fraga legebant, | et quae deciderant
patula Iovis arbore glandes.*

59. **adeo** modifies *par,* just as *tam* does *venerabile* in the l. above.

61. **veterem...follem,** 'the old leather bag', in which the
money was originally handed over.

cum tota aerugine, 'rust and all': cf. VI. 171 *cum tota
Carthagine;* XIV. 61 *cum tota...tela:* the *aerugo* gathers of course,
not on the bag, but on the money.

62. **prodigiosa fides,** 'such honesty is a portent', and there-

fore needs that the lore of the *haruspices* be consulted, to tell us
what god needs to be appeased by sacrifice: cf. II. 121 (of the
portentous vices of the nobles) *o proceres, censore opus est an
haruspice nobis?* The **Tusci libelli** are probably the books in
which this learning was contained, called by Cicero *Etruscorum
libri haruspicini et fulgurales*; some explain as 'the calendar of
portents'. *extispicium*, with many other Roman customs and rites,
was adopted from Etruria: a certain Tages had the credit of
being *conditor artis*.

63. **coronata** = *immolanda*, as victims for sacrifice were
crowned with garlands.

64–70. All the portents here mentioned occur in Livy and
other ancient authorities. Juv. gives this long list in mockery;
and, a century earlier, Livy explains to his own sceptical age
why he records the portents of each year: XLIII. 13. 1 *et mihi
vetustas res scribenti nescio quo pacto antiquus fit animus, et quae-
dam religio tenet, quae illi prudentissimi viri* (the wise men of old)
*publice suscipienda censuerint, ea pro indignis habere quae in meos
annales referam.*

bimembris means either 'with superfluous limbs' (Livy
often records the portent of a *biceps puer*), or 'half-human', the
word being applied in this sense to the Centaurs: cf. Livy,
XXVII. 11. 5 *cum elephanti capite puerum natum.*

65. **miranti**: so all MSS. except P which has *mirandis*: see
Introduction, p. xlix. For *miranti*, cf. a modern couplet, 'The
finny brood grow salt by slow degrees, | and pickled salmon
swim *th'* astounded seas.'

68. **uva,** 'cluster', used by Virgil also of the shape assumed
by a swarm of bees when they settle on the branch of a tree,
Georg. IV. 558 *uvam demittere ramis.* It appears from Livy,
XXXV. 9. 4 that wasps were equally ominous.

70. Henry (*Aeneidea*, I, 368), in his argument, or proof, that
gurges is 'body of water', not 'whirlpool', compares this l.
with Silius XIII. 566 *vorticibus furit et spumanti gurgite fertur.*
miris, because consisting of milk: Porson read *miniis* to explain
the Schol. *aut lacteis aut sanguineis.* 'Pliny speaks of the atmos-
phere raining blood and milk, as not unfrequent (*Nat. Hist.* II.
147), where he also mentions showers of flesh, iron, wool, and
baked tiles' (Lewis). In the case of the iron shower, he adds,
haruspices praemonuerunt superna volnera (cf. l. 62).

71–85. *Others have been cheated of much larger sums than you
have, and in the same way. Our swindlers do not hesitate to*

*perjure themselves by all the gods of heaven, and by the lives of
their own children.*

71. **decem...sestertia** = £100.

72. **sacrilega:** the deposit, being protected by an oath,
becomes sacred (l. 15); and hence the stealing of it is sacrilege.

73. **hoc...modo,** i.e. by perjury of the trustee.
arcana, 'deposited without witnesses'.

74. **quam...arcae,** 'which the corner of a capacious strong-
box had hardly contained', i.e. almost too large for a capacious
strong-box, crammed full, corners and all: *arca* usually suggests
wealth: cf. XI. 26. For this sense of *capere* (= χωρεῖν), see n. to
X. 148.

75. **testes** is in apposition with *superos*: cf. VIII. 149.

76. **quanta voce,** 'how loudly': cf. Mart. VII. 72. 15 *et
quantum poteris, sed usque, clames.*

77. **neget:** the subject is ὁ ἀποστερῶν.

78. **Tarpeia** = *Iovis Capitolini*: cf. VI. 47. His oath is either
that he has not received the money, or that he has repaid it.

79. **framea** is the German name for a spear: cf. Tac. *Germ.*
6 *hastas, vel ipsorum vocabulo frameas, gerunt*: the spear of Mars
was called *cuspis,* II. 130 and *hasta,* XI. 106.

The *Cirrhaeus vates* is Apollo (see n. to VII. 64), the *venatrix
puella* Diana.

81. **pater Aegaei:** the attributes of Poseidon are transferred
to the Roman god.

83. 'And all the weapons contained in the armouries of
heaven': i.e. he swears by any other gods there are and by the
weapons they carry. **telorum** is governed by *quidquid*.

Juv. is mocking descriptions of the gods' weapons such as
occur in Epic poetry: a specimen will be found in Lucan, VII.
144–50.

84. Cf. Pliny, *Epp.* II. 20. 6 *facit hoc Regulus non minus
scelerate quam frequenter, quod iram deorum, quos ipse cotidie
fallit, in caput infelicis pueri detestatur* (i.e. he perjures himself
after swearing by his son's life). This boy died young.

et, 'also', modifies *pater est.* **comedam** is pres. subj., the
condition, *si mentior,* being understood: the swindler imprecates
on himself the doom of Thyestes, as a climax of horror: cf.
VI. 16.

85. **Phario...aceto:** Egyptian vinegar seems to have been
good: cf. Mart. XIII. 122 *acetum. amphora Niliaci non sit tibi
vilis aceti;* | *esset cum vinum, vilior illa fuit.* Juv.'s language is

purposely ludicrous: the perjurer would not have added these picturesque details himself.

86–119. Some think that human destiny is all a matter of chance, and that there is no ruler of the world; these have no scruple in perjuring themselves. Others believe in Providence, and yet brave its anger, preferring to undergo any punishment the gods can inflict rather than surrender their unlawful gains. These reflect also that 'the mills of God grind slowly', that their punishment may be long delayed, that it may never come at all. Hence they find courage to swear the false oath. They play a part like skilful actors, while the friend they have robbed reproaches the gods for their indifference.

86. 'Some refer all events to the accidents of chance', i.e. are atheists, *fortuna* being here opposed to divine agency. Juv. seems to take this view generally, though, as he indicates elsewhere (XIV. 315), Fortune was often personified as a deity, so that her worshippers found themselves in harmony, to some extent, with religion, thus 'reconciling opposites by the apotheosis of a negation' (Furneaux, Tac. *Ann.* I, 21). The Epicureans are especially meant: cf. Tac. *Ann.* VI. 22 *mihi...in incerto iudicium est, fatone res mortalium et necessitate immutabili an forte volvantur*, where the *fatum* and *necessitas* of Stoicism are opposed to the *Fortuna* of Epicureanism. As a matter of fact the Epicureans, though holding that the atoms which form a world coalesced by chance, were the strongest upholders of the reign of law in the universe: this was the great discovery by which Epicurus freed mankind from superstitious terrors (Lucr. I. 75–8).

These three theories of the world's government are often distinguished: cf. Sen. *Epp.* 16. 4 *quid mihi prodest philosophia, si fatum est? quid prodest, si deus rector est? quid prodest, si casus imperat?* The first is the Stoic view, the last the Epicurean, the second being that of orthodox laymen.

87. **mundum** is the sky, with the heavenly bodies. **nullo rectore** is abl. absol., *rectore* being = *regente*. **moveri** is middle.

88. **natura** is regarded from the Epicurean stand-point, as 'at one and the same time blind chance and inexorable necessity' (Munro): cf. Lucr. V. 76 *praeterea solis cursus lunaeque meatus | expediam qua vi flectat natura gubernans*; *ibid.* 107 *fortuna gubernans*.

vices lucis are night and day; **vices anni,** the seasons.

89. **tangunt:** see n. to XIV. 219.

90. **alius** represents a second class who believe and yet sin.

91. **hic,** 'the latter'.

et, 'and yet': see n. to VII. 124. **secum,** sc. *loquitur*, 'speaks thus in his heart'.

93. **feriat,** 'blast': so Teiresias lost his sight as a punishment for seeing Athene in the bath; an epigram of Martial (IV. 30) describes how a man, who had angled for some pet fish of Domitian at Baiae, lost his sight for his impiety. Blindness, being common in Egypt, was frequently ascribed to the wrath of the Egyptian goddess, Isis. The **sistrum** (σεῖστρον from σείω) is a metal rattle, the noise being made by rings running on rods: Isis is often represented as carrying it in her right hand: for a figure, see Baumeister's *Denkmäler*, p. 761: her votaries carried it in their processions. The Latin word is *crepitacillum*.

94. **abnego** = ἀποστερῶ, 'I refuse to repay'.

96. **sunt tanti,** 'are a price worth paying': for this constr., and for the other in which *est tanti* means 'is a prize worth gaining', see n. to III. 54.

locupletem...podagram, 'the rich man's gout', i.e. wealth with the gout, it being believed that this disease was unknown to the poor; cf. *locupletem aquilam*, XIV. 197. For *optare*, see n. to X. 7.

97. **nec dubitet,** 'would not even hesitate'. Ladas was the name of two famous Greek runners, who won prizes at Olympia; the name is chosen here, because such an athlete would feel more than others the lameness of gout.

si non eget Anticyra = *nisi insanus est*, the name of the city being used as an equivalent for hellebore, which grew abundantly there and, as a strong purgative and emetic, was believed to cure madness: cf. Hor. *Sat.* II. 3. 82 *danda est ellebori multo pars maxima avaris;* | *nescio an Anticyram ratio illis destinet omnem.* Of two cities, which bore this name, the more famous was in Phocis.

98. **Archigene** = 'a doctor': see n. to VI. 236. The abl. of Greek nouns, that have dat. in -ει, is often lengthened in Latin poets, e.g. Ovid, *Met.* X. 608 *Hippomenē victo*; but is sometimes short, e.g. Hor. *Sat.* II. 3. 193 *heros ab Achillē secundus* (Lachmann on Lucr. I. 739).

99. **esuriens:** the epithet which properly belongs to the winner, is transferred to the prize, which at Olympia was a wreath of olive-leaves. **Pisaeae:** Pisa, an ancient city of Elis, was

near the precinct of Olympian Zeus and is often identified with it by the poets.

100. A common thought in antiquity, as in the proverb, ὀψὲ θεῶν ἀλέουσι μύλοι, ἀλέουσι δὲ λεπτά, 'though the mills of God grind slowly, yet they grind exceeding small'; cf. M. Seneca, *Controv.* x. 6 *sunt di immortales lenti quidem sed certi vindices generis humani*: there the guilty man, for his own comfort, gives a different turn to the common saying.

102. **ad me venient,** 'will they get as far as me'.

sed et, 'and besides', leaving the slowness of punishment out of account.

103. **his,** 'such acts as mine'.

104. **diverso** has here its classical sense of 'opposite': see n. to III. 268.

105. **tulit:** an aorist of repeated action, of a thing that has often happened. In Greek φέρεσθαι would be used where we say 'carry off': in the absence of a middle voice, *ferre* has to do double duty.

107. **ad delubra:** in order to take an oath before the image of the god. For this custom, cf. Suet. *Iul.* 85 *apud eam* (the altar in the shape of a pillar erected where Caesar's body had been burnt) *longo tempore sacrificare...controversias quasdam interposito per Caesarem iureiurando distrahere perseveravit [plebs].*

108. **ultro,** 'actually': this word is often used, especially by Virgil and Tacitus, to express that some limit has been overstepped, contrary to what might have been expected: here the idea is, 'he might have consented to go, when pressed; but he does even more than that, and insists on going'. The word does not occur elsewhere in Juv.; but cf. Suet. *Iul.* 63 *ultro ad deditionem hortatus* (i.e. Caesar might naturally have made off before a superior force; but, far from that, he actually claimed their surrender to himself); Tac. *Hist.* I. 71 *Celsus constanter servatae erga Galbam fidei crimen confessus, exemplum ultro imputavit* (i.e. he actually claimed credit for the conduct which was charged against him). The conduct of Socrates on his trial is a typical case of such action; and a Latin writer, in describing it, would be likely to use *ultro*.

vexare, 'to give you no rest': this frequentative of *veho* seems never to mean 'to drag', though it sometimes means 'to squeeze', e.g. Suet. *Aug.* 53 *in turba...vexatus.*

109. **superest,** 'there is abundance of': cf. l. 237; Ovid *Trist.* III. 9. 17 *superest ingens audacia menti.* Lewis explains it

as = *adest*, 'supports', 'backs up', quoting Suet. *Aug.* 56,
where *superesset* is apparently used in this sense but without the
dat.; but it seems safer to take the meaning here which is
required in l. 237.

110. And yet Lamb attacks as a 'popular fallacy' the belief
that 'of two disputants the warmest is generally in the wrong'.
fiducia is predicate.

111. **Catulli:** see n. to VIII. 186.

112. **tu,** 'but you': the clause, opposed to *mimum agit ille*,
begins here: you are not disposed to applaud his dramatic
powers. **Stentora:** the reference is to Homer, *Il.* v. 785
Στέντορι εἰσαμένη μεγαλήτορι χαλκεοφώνῳ, | ὃς τόσον αὐδήσασχ' ὅσον
ἄλλοι πεντήκοντα.

113. **Gradivus Homericus,** 'Ares in Homer', Gradivus
being a name of Mars: cf. Hom. *Il.* v. 859 ὃ δ' ἔβραχε χάλκεος
Ἄρης, | ὅσσον τ' ἐννεάχιλοι ἐπίαχον ἢ δεκάχιλοι | ἀνέρες ἐν πολέμῳ. For
Homericus, cf. Sen. *Dial.* IX. 2. 12 *qualis ille Homericus Achilles*;
Cic. *pro Rosc. Amer.* 46 *senex ille Caecilianus* (in Caecilius'
play); *ibid.* 47 *adulescentem comicum* (in the comedy).

114. **nec** has the sense of *nec tamen*: cf. III. 102.

115. **aut,** 'or else'.

116. **carbone,** 'censer' M. **charta...soluta,** 'do we un-
fasten the paper and...', *tus* being carried to the shrine in a
screw of paper: cf. Mart. III. 2. 5 (to his book) *ne...turis
piperisve sis cucullus.*

117. **albaque porci omenta:** cf. X. 355.

118. **ut video,** 'as far as I see': so *quod video*, VI. 395, where
the context resembles this passage.

119. **Vagelli:** the name recurs XVI. 23, whence it seems he
was some dull speaker, who had been rewarded by a statue for
his exertions; cf. VII. 126, and n. there. That a client's gratitude
was often expressed in this way, is shown by Sen. *de Ben.* v. 8. 2
*nemo, quamvis pro se dixerit, adfuisse sibi dicitur nec statuam sibi
tamquam patrono suo ponit.*

120–73. *Well, will you accept such consolation as I can offer,
though I am no philosopher of the schools? Consider that yours
is no isolated case, that impudent fraud is rife on all sides: why
should you be more fortunate than your neighbours? Again, there
are worse crimes than that from which you suffer: some men are
killed by violence; others have their houses burnt; temples are
robbed, and images of the gods melted down; remember all the
cases of poisoning, and of parricide. A single day, spent in the*

police-court, will prove to you that you have no business to complain. A thing that happens constantly, ought not to excite surprise and resentment.

120. **accipe,** 'hear', i.e. I will tell you: cf. VII. 36. It was a main function of philosophy to console people under misfortune: thus we have *consolationes* (παραμυθητικοί) addressed by Seneca to his mother, Helvia, on his exile; to Polybius, Claudius' freedman, on the death of a brother; and to Marcia, daughter of Cremutius Cordus, on the death of her son. **contra,** 'in mitigation'.

121. **et qui,** 'even one who...', i.e. Juv. himself.

122. **a cynicis...distantia,** 'which differ from those of the Cynics to the extent of a shirt': i.e. the Cynic view of life is identical with Stoicism, but the Cynics are dissenters so far that they refuse to wear the χιτών (*tunica* or 'shirt'), which the Stoics did not discard. The αὐτάρκεια of Cynicism discarded every kind of luxury. Cynicism was never an organised school of philosophy; it was rather 'a practical protest of individuals against the passions, follies, and sins of a civilisation petrified in lifeless forms and doomed to destruction' (Bernays, *Lukian und die Kyniker*, p. 25).

tunica is abl. of amount of difference.

123. **suspicit,** 'looks up to', 'reverences': cf. Mart. XI. 56. 2 *vis animum mirer suspiciamque tuum?* The particip. *suspectus,* which is constantly used (cf. III. 222), never has this meaning.

horti: the Epicurean school was often called the Garden, as that of their rivals the Porch; the garden at Athens, bought by Epicurus for 80 minae (£320), was the hearth and home of his school; its exact site is unknown. The epithet (**exigui**) is meant to suggest the simplicity of the life led there: cf. XIV. 319; also Mart. VII. 69. 3 *magni senis Atticus hortus* (where the context shows that Plato is not meant). Seneca often speaks of the Garden as *hortuli.*

laetum plantaribus does not imply that Epicurus was interested in horticulture but that he was content with a vegetable diet.

It should be observed that philosophy is represented by Stoicism, Epicureanism, and Cynicism, Plato and Aristotle being now almost unknown, when only a few students professed to carry on the Academic and Peripatetic doctrines. The rise and vogue of the two later schools may be placed between 250 B.C. and A.D. 150. 'In the Roman world, the Stoic and

Epicurean systems divided between themselves the suffrages of almost all who cared to think at all' (Wallace, *Epicureanism*, p. 2).

124. **dubii...aegri**, 'critical cases', in which recovery is a doubtful matter. **medicis:** dat.: see n. to *lectore*, I. 13.

125. **venam committe:** *vena* is either the pulse, the importance of which in diagnosis was well understood by the ancients (cf. Celsus, III. 6; Quint. VII. 10. 10); or a vein in the arm from which blood is to be taken: cf. VI. 46. Philippus is unknown, but no imputation against his skill is conveyed: the best doctors may have unskilful apprentices.

127. **ostendis,** 'you can point to'. Juv. repeats what he has already said at great length (ll. 23–70).

129. **claudenda est ianua:** in ancient times the house-door was generally left open during the day, with a slave to watch it: cf. Plaut. *Most.* 435 *sed quid hoc? occlusa ianua est interdius*: it was shut as a sign of mourning: cf. Livy, XXXV. 15. 7 *per luctum regia clausa*; Tac. *Ann.* II. 82 *hos vulgi sermones audita mors* (of Germanicus A.D. 19) *adeo incendit ut...desererentur fora, clauderentur domus*.

132. **fingit** is emphatic: 'grief is never a sham when this happens'.

vestem diducere summam: for tearing the clothes as a sign of mourning, cf. X. 262 *scissa Polyxena palla*. When not much grief is felt, a man is content, says Juv., with tearing the upper border of his dress.

133. **contentus** is qualified by the negative contained in *nemo*: 'no one finds it enough to...'.

umore coacto: cf. Mart. I. 33 *amissum non flet, cum sola est, Gellia patrem;* | *si quis adest, iussae prosiliunt lacrimae*.

135. **cuncta...fora:** there were at this time five forums in Rome, all used for legal business: (1) *forum Romanum*; (2) *forum Caesaris*; (3) *forum Augusti*; (4) *forum Nervae* or *transitorium*; this was completed A.D. 98; hence Martial, who in 87 writes *in triplici foro* (III. 38. 4), in 98 speaks of *fora iuncta quater* 'four adjacent forums' (X. 51. 12); (5) *forum Traiani*, the most splendid of all, constructed A.D. 113. The precinct of the temple of Peace built by Vespasian was also sometimes called *forum Pacis*.

136–9. The fraudulent trustee originally gave a written acknowledgement: this is recited again and again by the claimants, but the trustee declares it is a forgery, though confronted by his own handwriting and the seal of his own ring on the document.

136. **diversa parte,** 'on the other side', i.e. by the claimant and his lawyers; cf. VII. 156. Madvig explains thus, and quotes Suet. *Iul.* 29 *cum videret...consules e parte diversa.*

137 recurs, with a slight change, XVI. 41.

supervacui is part of the predicate, and the meaning is the same as if the text ran, *vana dicunt esse chirographa et super-vacuum lignum*: this is suggested by the proximity of the two epithets: see n. to VI. 5.

ligni: the document is written on wooden tablets, covered with wax.

138. **littera,** 'hand-writing': the regular word in Cicero is *manus.*

gemma: the seal of a man's signet-ring was as good evidence as his hand-writing, and indeed less likely to be forged. Thus Suetonius tells us that Augustus used three signet-rings successively, for public and private documents, of which the first represented a Sphinx, the second, Alexander the Great, and the third, himself (*Aug.* 50).

139. **sardonychum:** for this gem in rings, see n. to VI. 381. The ring is kept in a case, being too precious to be worn on the finger.

140. **o delicias,** 'what fastidiousness!' i.e. how hard you are to please! See nn. to VI. 47, X. 291; here also the meaning conveyed is that something beyond the common lot is demanded.

141. **gallinae filius albae:** *proverbium vulgare* Schol. But it is not found elsewhere, and the expression is strange, considering how common white hens are. The explanations offered by edd. seem insufficient; but the context shows that in this phrase a 'white hen' means a very special sort of hen.

142. **nos,** 'while we are...', this clause also being dependent on *quia* and contrasted with *tu...albae.*

144. **si flectas** goes closely with *ferendam*: 'if you turn your attention'; for the mood, see n. to *poscas*, III. 102.

145. **latronem** implies robbery with violence.

sulpure...atque dolo, 'by a treacherous match'; for the hendiadys, see n. to VIII. 251, and, for *sulpure*, n. to V. 48.

146. The street-door is fired first, to prevent escape: cf. IX. 98 *candelam apponere valvis.* **colligit,** 'is nursing': cf. Lucr. I. 722 *hic Aetnaea minantur | murmura flammarum rursus se colligere iras*; Virg. *Georg.* I. 427 *luna revertentes cum primum colligit ignes.*

148. **adorandae robiginis** is an epithet of *pocula* repeated: see n. to III. 5.

149. **dona** is regularly used in the sense of ἀναθήματα: cf. Suet. *Iul.* 54 *templa deum donis referta*; idem, *Nero* quoted below.

151. Metal or wood was gilded by laying on thin plates of gold (*bratteae* or *laminae*), which could be prised off by the finger of a thief: cf. Martial quoted on v. 41.

153. 'Is he likely to stick at this, when he often melts down a whole Jupiter?': cf. Suet. *Nero*, 32 *templis compluribus dona detraxit simulacraque ex auro vel argento fabricata conflavit*. The l. is unsatisfactory, because, apart from the exaggeration of speaking of such a crime as common, this is clearly not the business of a *minor sacrilegus* but a greater crime than even the robberies in ll. 147–9. Hence Munro proposed *an dubitet? solitumst cet.*

155, 156. For this punishment, see n. to VIII. 214; the *locus classicus* for parricide and its *singulare supplicium* is Cic. *pro Rosc. Am.* 62–73.

deducendum, 'him who deserves to be launched', i.e. *parricidam*: the word is properly applied to a ship.

157. **haec** is probably fem. sing., in agreement with its predicate *pars*: cf. *hic est cibus* (these things are the food), XIV. 79: it refers to all the crimes mentioned above, which are said to be only a fraction of those actually brought to justice: for the constr., cf. III. 61; Sen. *Dial.* IV. 9. 3 *et quota ista pars scelerum est?* Mart. v. 65. 7 *ista tuae, Caesar, quota pars spectatur harenae?* (where *ista*, though fem. sing. by attraction, refers to the labours of Hercules).

custos Gallicus urbis: Rutilius Gallicus was *praefectus urbi* under Domitian, and Statius wrote a poem (*Silv.* I. 4) on his recovery from an illness A.D. 88: Gallicus died in 91 or 92 (Vollmer's *Silvae*, p. 10, n. 2). It should be noticed how Juv. here speaks of a personage of nearly 40 years back, as if he were a contemporary. See Introduction, p. xxxv.

The *praefectus urbi*, unlike most of the other *praefecti*, was a senator: see n. to IV. 77: all the city police were under his orders, and he had a large criminal jurisdiction in Rome and Italy. Seneca (*Epp.* 83. 14) also calls this official *custos urbis*.

158. There are so many cases that the magistrate sits from dawn to dusk, though it was usual to stop earlier: cf. Suet. *Aug.* 33 *ipse ius dixit assidue et in noctem nonnunquam*. As a rule, the Romans began work early (see n. to XIV. 190) and did no business in the afternoon.

159. For the view of human nature to be obtained in a

police-court, cf. Boswell's *Johnson* (1874), II, 342: 'Johnson, who had an eager and unceasing curiosity to know human life in all its variety, told me, that he attended Mr Welch' (who succeeded Fielding as Justice of the Peace for Westminster) 'in his office for a whole winter, to hear the examinations of the culprits; but that he found an almost uniform tenor of misfortune, wretchedness, and profligacy.'

160. **una domus:** either the office or private house, used as an office, of Gallicus: not (as Friedl.) any private house taken at random.

161. **dicere te...aude,** 'call yourself wretched, if you dare'.

162. M. finds the source of this passage in Sen. *Dial.* v. 26. 3 *ad condicionem rerum humanarum respiciendum est, ut omnium accidentium aequi iudices simus. iniquus autem est, qui commune vitium singulis obiecit: non est Aethiopis inter suos insignitus color, nec rufus crinis et coactus in nodum apud Germanos virum dedecet.*

tumidum guttur, 'the goître': cf. Shak. *Temp.* III. 3. 43: 'When we were boys, | who would believe that there were mountaineers | dew-lapped like bulls, whose throats had hanging at them | wallets of flesh?'

163. This national peculiarity is not noticed elsewhere: for Meroe, which still retains its name, see n. to VI. 528.

164. Cf. Tac. *Germ.* 4 [*Germanis*] *truces et caerulei oculi, rutilae comae.*

165. **madido...cirro,** 'with greasy locks twisted into horns': the phrase is a second epithet of *caesariem, flavam* being the first. These *cornua* are elsewhere often called *nodi*; the former name makes it probable that each German wore more than one, and that they were not at the back of the head.

166. **nempe quod...,** 'of course not, because...': the interrogative sentences which precede are virtually negative. **natura** = φύσις, 'outward form'.

167. 'To meet the sudden onset and clanging cloud of birds from Thrace': for the hendiadys, see n. to VIII. 251. The legend of pitched battles between cranes and the Pygmies is found in Homer: cf. *Il.* III. 7 (the cranes fly southward in autumn) ἀνδράσι Πυγμαίοισι φόνον καὶ κῆρα φέρουσαι, where the noise they make in the sky (**nubem sonoram**) is also mentioned: the Pygmies lived in Ethiopia.

169. **curvis | unguibus:** cf. VIII. 129.

172. **spectentur,** 'are witnessed', not 'may be witnessed': see n. to VII. 14.

173. **cohors,** 'army'. *Pygmaeus* is derived from πυγμή, 'a cubit'.

174-249. *'Is the wrong, then, to go altogether unpunished?' you cry. Suppose him put to death, your loss remains the same. And, if you say that still revenge is sweet, the great teachers of philosophy say otherwise. Revenge is the pleasure of a small mind. Leave your enemy to the worst of penalties, an evil conscience. Think how Glaucus was punished for a mere purpose to defraud. The life of a guilty man is one of fear and terror: he cannot, like other men, ask help of heaven in sickness. Further, his crime leads on to other crimes; at last he will be caught and suffer dire penalties here on earth, while you will rejoice, and admit that the gods are neither deaf nor blind.*

174. **peiuri capitis:** the spelling *peiurus*, found here in P, appears often in the Milan palimpsest of Plautus, which, dating from the 5th cent., carries great weight. *capitis = hominis*, as often, esp. in Plautus: see Lorenz on *Mostell.* 244. It need not mean that the swindler has sworn falsely by his head.

176. **nostro...arbitrio,** 'in the manner we wish'.

177. **illa...iactura,** 'the old loss', i.e. of the money.

178. **sed...sanguis:** there is no need to change the text or read *si* for *sed* with Weidn.; the emphasis falls entirely on *invidiosa*, and the sense is: suppose him put to death, still you will not get back your money, and you will earn hatred by taking such a revenge. **minimus** is strange: apparently it = *vel minimus*, 'even a few drops of blood'.

180. **at,** 'you say, that...': *at enim* is the common prose form, used exactly as ἀλλὰ νὴ Δία by the Greek orators.

181. **nempe,** 'I grant, that...'. **indocti,** sc. *dicunt* or *putant*: cf. XI. 5. The 'ignorant' are those who have not been trained in philosophy: *imperitus* is general in prose, in this sense: cf. Cic. *pro Mur.* 61 *apud imperitam multitudinem*: but the poet has to use a word that will scan.

182. **quantulacumque** is here a relative.

adeo: several uses of this word are to be distinguished in Juv. (1) It is like *tam* in meaning and usage, and qualifies a single word, esp. an adj.: *usque* often precedes it, in this case: cf. III. 84; V. 129; VI. 59 *adeo senuerunt* (= *senes facti sunt*) *Iuppiter et Mars?*; VI. 182; VIII. 183; X. 201, 297; XIII. 59; XV. 82. (2) It begins a clause which contains a statement by way of explanation: Juv. has this *adeo* thrice, III. 274 (where see n.); XI. 131; XII. 36. (3) It is used in a corrective sense, 'nay rather', like *immo*: so XIV. 234, where see n. This use is only a modification of the preceding.

The present passage is an instance of (3): **adeo** = *immo* and corrects the previous statement: for its position after a relative, cf. Sen. *de Ben.* IV. 17. 2 *nemo in amorem sui cohortandus est, quem adeo dum nascitur trahit*, which would become an instance of (2) by writing, *cohortandus est: adeo eum* cet.

184. Chrysippus, the second founder of Stoicism, and Thales, the first of the physical inquirers of Ionia, are strangely coupled with Socrates as teachers of the duty of forgiveness. Perhaps Thales is mentioned to show that philosophy from the earliest times inculcated this duty; more probably, Juv., who cared, as he says, little for philosophy, takes the first names that come into his head.

mite...ingenium: cf. IV. 82.

185. **dulci** refers to the honey made on Hymettus. The **senex** is Socrates, who was 70 years old when accused 399 B.C. by Meletus, Anytus, and Lycon.

186. **cicutae:** cf. VII. 206.

187. **nollet** has the sense of *noluisset*: see n. to IV. 85.

188. Büch. puts the comma after *exuit*, and makes *omnes* the object of *docet*: the improvement seems doubtful.

189. **sapientia**, 'philosophy': cf. l. 20, and Hor. *Epp.* I. 1. 41 *virtus est vitium fugere et sapientia prima | stultitia caruisse*.

quippe, 'for': see n. to l. 26. The logic would be clearer, if *plurima...sapientia* were marked as a parenthesis.

191. **collige**, 'infer': cf. XI. 198. In classical prose *argumentis*, or a similar word, is often joined with *colligere* in this sense. For spondaic verses in Juv., see n. to IV. 87.

193–5. Cf. Lucr. III. 1018 *at mens sibi conscia factis | praemetuens adhibet stimulos terretque flagellis*; Sen. *Epp.* 97. 15 *consentiamus mala facinora conscientia flagellari, et plurimum illi tormentorum esse, eo quod perpetuo illam sollicitudo urget ac verberat*.

194. **habet attonitos**, 'keeps in terror': cf. Sen. *Epp.* 110. 5 *vana sunt ista, quae nos movent, quae attonitos habent*; and see n. to X. 296. **surdo**, 'unheard': see n. to VII. 71.

197. Caedicius is unknown: a pleader of the same name is mentioned XVI. 46. For Rhadamanthus, one of the judges in Hades, cf. Virg. *Aen.* VI. 566 *Gnosius haec Rhadamanthus habet durissima regna, | castigatque auditque dolos subigitque fateri*.

198. **nocte dieque:** cf. III. 105: the general prose form is *dies noctesque*. **suum**, 'against oneself'.

199. Herodotus (VI. 86) puts in the mouth of Leotychides the story of Glaucus, the son of Epicydes—how he consulted the oracle at Delphi whether he should fraudulently retain a sum of money entrusted to him, and was punished for this, though he did restore the money. The story seems irrelevant here, as it has nothing to do with the penalties of conscience.

200. **quondam,** 'in the future'; the priestess advised him to take the false oath, since τὸ μὲν αὐτίκα κέρδιον; but added some pregnant remarks about the 'child of the Oath' which frightened Glaucus into honesty.

dubitaret, 'he was inclined to': the verb, in this sense, is generally followed by *an* and subj.; but cf. Cic. *ad Att.* XII. 49. 1 *cum dubitet Curtius consulatum petere*; Tac. *Ann.* IV. 57. 5 *dubitaverat Augustus Germanicum... rei Romanae imponere.* The phrase seems always to be used of an intention which was not carried out.

201. **iure tueri | iurando:** for the same separation, cf. Hor. *Sat.* II. 3. 179 *iure | iurando obstringam ambo.*

203. **illi:** the use of this pronoun, instead of *sibi*, is an irregularity, of which I know no other instance: ἐκείνῳ might be so used in Greek.

204. **tamen,** although he restored it.

205. **probavit,** 'his fate proved...'.

206. Herodotus' words are (*l.l.*) ἐκτέτριπται πρόρριζος ἐκ Σπάρτης.

207. **quamvis...gente,** 'who traced descent from his stock, however far back'; i.e. even those of remote collateral descent were blotted out.

208. **sola voluntas:** so some MSS., while others have *saeva* or *scaeva*: *sola* seems necessary. P has *saeva voluptas*; the latter word is found in many MSS., but is a mere blunder which may have led to the further corruption.

210. **crimen,** 'the guilt'.

cedo...peregit, 'how then, if he has carried out his purpose?' The same constr., a rare one, was used VI. 504, where see n. *cedo si* is literally, 'tell me your view, in the case that...'.

211. *est* is understood with **perpetua.**

212. **ut morbo,** 'as if in sickness'.

213. **crescente cibo:** a symptom of want of appetite often mentioned in the ancients, that food seems to grow in the mouth: cf. Sen. *Epp.* 82. 21 (of the Spartans breakfasting at Thermopylae, when Leonidas told them they would sup in Hades) *non in ore crevit cibus, non haesit in faucibus...: alacres et ad prandium illi promiserunt* (they accepted his invitation) *et ad cenam.*

sed vina: *sed* may mean 'yes, and...': see n. to v. 147. But the conjecture *Setina* is extremely attractive, Setine and Alban being two of the most famous wines: see n. to v. 33.

214. **senectus:** used of wine, v. 34.

215. **ostendas** = 'if you produce'.

216. **acri...Falerno,** 'by *the* bitter Falernian': the article is important; for the better wine is Falernian, the only vintage which could rival Alban and Setine; but the very sight of it makes him screw up his face as if he were drinking it. That the very choicest and oldest wine should be bitter, is unlike our notions of wine; but cf. Sen. *Epp.* 63. 5 *in vino nimis veteri ipsa nos amaritudo delectat.*

218. **iam quiescunt,** 'begin to rest'.

220. **sudoribus** properly refers to the cold sweat of fear on the body, and then, by metaphor, to the terrors of the mind: cf. 1. 167.

221. **sacra,** 'awful'. **maior...humana,** sc. *imagine*, 'larger than life': cf. Suet. *Claud.* 1 *species barbarae mulieris humana amplior.*

223. Cf. Lucr. v. 1219 *cui non correpunt membra pavore,* | *fulminis horribili cum plaga torrida tellus* | *contremit et magnum percurrunt murmura caelum?* Lightning was admitted by the orthodox to be Jupiter's method of punishing perjury: see n. to III. 145.

224. **quoque,** 'even' the first.

225. **fortuitus:** either the *i* is here short, as in Stat. *Silv.* I. 6. 16 *largis gratuĭtum cadit rapinis*; or the two middle syllables coalesce as in *pītuïta.*

226. **iudicet,** 'chose its victims', not 'condemned the guilty'; *iudicare* often means 'to exercise deliberate choice', never 'to condemn': cf. Sen. *de Ben.* II. 28. 2 '*plura illis* (than this) *hominibus turpissimis data sunt.' quid ad rem? quam raro fortuna iudicat* (chooses the recipients of her gifts). The lightning really acts blindly, like fortune; but the guilty man cannot believe this. *iudicet* might mean, 'acted as judge': but the context (cf. *fortuitus,* l. 225; *ad sua corpora,* l. 230) is decisive in favour of the other interpretation.

227. **illa,** sc. *tempestas.* For the omission of **si,** cf. III. 100.

228. **sereno** is a noun, as VII. 179; so *tranquillum* Lucr. v. 12.

233. **cristam...galli** = *cristatum gallum*: cf. Mart. IX. 68. 3 *nondum cristati rupere silentia galli.* For such offerings, cf. XII. 96 and Plato, *Phaedo,* 118 A τῷ Ἀσκληπιῷ ὀφείλομεν ἀλεκτρυόνα.

236. **malorum** is masculine.

237. **superest constantia**, 'they have plenty of boldness'; cf. l. 109.

239. Perhaps a recollection of Horace, *Epp.* I. 10. 24 *naturam expelles furca, tamen usque recurret.*

240. **damnatos**, 'condemned by themselves': when the crime is committed, they see their guilt and condemn it, but go on sinning: cf. Ovid, *Met.* VII. 20 *video meliora proboque, | deteriora sequor. damnare* (when applied to things, not persons) is the exact opposite of *probare.*

241 ff. For the thought, cf. Cic. *in Verr.* II. 3. 177 *ita serpit illud insitum natura malum consuetudine peccandi libera, finem audaciae ut statuere ipse non possit. tenetur igitur aliquando, et in rebus cum maximis tum manifestis tenetur.*

242. **attrita de fronte**: blushing is a sign of modesty (cf. X. 300, XI. 54 and 154): when the face and forehead had lost the power to blush—had become, as we say, brazen—it was considered to be the consequence of rubbing: hence *perfricui frontem* 'I lay modesty aside': cf. also *frons durior*, VIII. 189. 'Shyness' is *mollitia frontis.*

244. **in laqueum**: a metaphor from field-sports.

245. **uncum**: see n. to X. 66.

246. For the punishment of *deportatio in insulam*, see n. to I. 73; for **scopuli** applied to such islands, see n. to X. 170: islands of as little amenity and convenience as possible were chosen for the purpose.

247. **magnis**, 'of importance': so *magnus amicus.*

249. **Teresian** = *caecum*: see n. to l. 93.

SATIRE XIV

The satire deals with two subjects, the influence of parental example on the young (ll. 1–106), and the vice of avarice (107–331). The connection between the two is but slight.

1–58. *Many vices are learnt by the young from the example of their own parents, such as gambling, gluttony, cruelty, and wantonness. A young man must have a fine nature indeed, if he is not tainted by a bad example at home. The moral is, that all parents should take heed to their ways. If your son imitates your vices and betters your instruction, you will have no right to get angry and cut him off with a shilling.*

1. Fuscinus is unknown. **et**, 'both'.

2. **nitidis...rebus,** 'upon an honourable name': the *res* would be *nitidae*, but for the stain.

3. **monstrant,** by example: see n. to x. 363.

5. **bullatus:** *praetextatus* is commoner in this sense: for the *bulla*, see n. to v. 164.

movet arma: for this common metaphor, see n. to I. 91.

6. **melius,** i.e. better than of the infant gambler.

7. **tubera terrae,** 'truffles': cf. v. 116. All the dishes mentioned are costly luxuries: for **boletum,** see n. to v. 147.

9. **mergere,** 'to swallow down': cf. xi. 40. As instances of *mergere* used absolutely in this sense are very rare, some translate here, 'to steep'. But this makes *natantes* superfluous; and, as Seneca uses *demittere* 'to swallow' (*Nat. Quaest.* iv. 13. 10; *Epp.* 72. 8), it seems likely that *mergere* also could be so used. The *e* in *ficedula* is elsewhere long, so that Lachmann wished to read *ficellas* here.

nebulone parente | et cana monstrante gula: in English the two nouns are not naturally co-ordinated, and we say, 'instructed by the hoary gluttony of his worthless parent'; but the hendiadys is common in Latin poetry: cf. viii. 251. Exactly the opposite is found xiii. 137 where our language requires the co-ordination of two nouns, of which the one governs the other in Latin.

10. Children lose their first teeth about their seventh year, and it was at this age, at latest, that formal education was generally begun in antiquity: cf. Quint. i. 1. 15 *quidam litteris instituendos, qui minores septem annis essent, non putaverunt.*

12. **barbatos...magistros,** i.e. philosophers, a long beard being a common attribute of a philosophic teacher: cf. Hor. *Sat.* ii. 3. 35 *sapientem pascere barbam* 'to grow a philosophic beard', i.e. to become a philosopher.

inde, 'on one side'; **hinc,** 'on the other': cf. i. 65.

14. **culina** is used παρὰ προσδοκίαν for some more reputable word: perhaps it means 'cookery' here, not 'kitchen'.

15. **mores...aequos,** 'gentleness'.

16. **nostra** is fem. sing., agreeing with *materia* 'of the same substance as we' the masters: cf. Quint. iii. 8. 31 (on the morality of enlisting slaves) *liberos enim natura omnes et eisdem constare elementis et fortasse antiquis etiam nobilibus ortos, dici potest.* The words *materia* and *elementa* may be taken from Lucretius, in whose poem, familiar to Juv., they constantly occur to denote the atoms. For this humane view of slaves, see n. to vi. 222.

19. **nullam Sirena flagellis | conparat,** i.e. thinks the sound of the lash sweeter than the song of any Siren: cf. XII. 21 for a constr. exactly similar, and XII. 121. For *flagellis*, see n. to VI. 479.

20. Antiphates and Polyphemus are typical names for cruel ogres: the former was king of the Laestrygones: both appear in the Odyssey. Cf. Ovid, *ex Pont.* II. 2. 113 *nec tamen Aetnaeus vasto Polyphemus in antro | accipiet voces Antiphatesve tuas: | sed placidus facilisque parens veniaeque paratus.*

21. **tortore:** see n. to VI. 480.

22. **lintea:** presumably the towels had been stolen at the bath (cf. III. 263) by the slave's negligence, as such thefts are often mentioned. The slave is punished for the loss, by branding.

23. **iuveni** = *filio*: cf. l. 121, and see n. to III. 158.

24. **inscripti,** 'branded slaves', στιγματίαι: the word occurs also in Mart. VIII. 95. 9 *quattuor inscripti portabant vile cadaver*; in a list of cruel punishments Seneca includes *ergastula* and *inscriptiones frontis* (*Dial.* V. 3. 6).

The MSS. have *inscripta ergastula* which Mr H. Richards (*Class. Rev.* II, 326) thus corrects. The phrase is hardly too bold for Juv.; the real stumbling-block is the want of either the copula or an epithet before *carcer*. For **ergastula,** see nn. to VI. 151, VIII. 180.

25. **rusticus expectas,** 'are you so dull as to expect...?': cf. VI. 239. Larga is not mentioned elsewhere.

28. **respiret,** 'take breath'.

conscia matri...fuit, 'she was in her mother's secrets, while still a maid'. The *cerae pusillae* are the same as the *tabellae* of VI. 233 and 277, love-letters being written on tablets of small size.

30. **cinaedis:** transl. 'go-betweens'.

32. **magnis auctoribus,** 'on high authority', i.e. that of our parents: an abl. absol.

35. **meliore luto,** 'of finer clay'. The Titan is Prometheus who, according to one legend, made the first men out of clay: see n. to VI. 12.

36. **reliquos:** for the quantity of this word, see n. to X. 260.

37. **orbita,** 'beaten track'.

38. **vel | una,** 'at least one': if there was no other reason for refraining, this alone would be strong enough.

41. **turpibus** and **pravis** are neuter. The thought may be taken from Seneca, *Epp.* 97. 10 *omne tempus Clodios, non omne Catones feret. ad deteriora faciles sumus.*

42. **axe,** lit. 'sky', hence 'part of the world': so *Gallicus axis*, VIII. 116.

43. The 'uncle of Brutus' is Cato of Utica, whose sister Servilia was mother of M. Brutus. So Caligula is described (VI. 615) as Nero's uncle.

45. **pater:** cf. XII. 98: this gives just the same meaning as the traditional *puer* which has very slender authority. The Schol. (*ubi filios habes*) seems to have had *pater es* before him.

procul...inde: a common form of πρόρρησις or warning to profane persons to begone before mysteries or sacred rites are performed: cf. Virg. *Aen.* VI. 258 *procul, o procul este, profani*, perhaps taken from Callimachus, *Hymn to Apollo*, l. 2 ἑκάς, ἑκάς, ὅστις ἀλιτρός. A number of similar examples are quoted by Bentley on Hor. *Epp.* II. 2. 199. Juv. implies that the innocence of a child is as sacred a thing as any mysteries.

puellae: these were the slaves of the *lenones*, who are therefore sometimes called *magistri*.

46. **pernoctantis parasiti,** 'that night-bird, the parasite', L.: cf. *pernox*, VIII. 10.

47. **puero** = τῷ παιδί, 'to your son': the context shows that this is not stated as a general truth but with reference to a particular instance.

48. **nec,** 'and do not...': cf. Cic. *de Fin.* I. 25 *nec mihi illud dixeris*.

tu is not emphatic, but is often added in exhortations: M. quotes VIII. 228, X. 342; Hor. *Sat.* II. 2. 20 *tu pulmentaria quaere | sudando*.

49. **peccaturo:** for the hiatus after this word, see n. to X. 281.

infans: emphatic: even the babe in the cradle.

50. There were no censors elected under the Empire; but their duties were performed by the emperors under various titles, such as *praefectus morum* (cf. *iudex morum*, IV. 12): only Claudius and Domitian took the title of *censor*.

51. **quandoque,** 'at some future time'; see n. to v. 172; as the sentence begins with *si*, *quando* would be more correct.

52. **filius,** i.e. heir. The copula which follows may be omitted in English: the subjunctive (*peccet*) is consec., *qui* being = *talis ut*. For an exactly similar constr., cf. VII. 211 and 212.

53. 'Following in your footsteps, commits all your sins and outdoes them.'

54. **nimirum,** 'of course', is sarcastic, as generally after Lucretius.

55. tabulas mutare, 'to alter your will', i.e. to disinherit him: cf. *delebit tabulas,* XII. 123.

56. unde tibi frontem, sc. *sumes?* The ellipse of the verb is common where a personal dative is expressed: cf. Hor. *Sat.* II. 7. 116 *unde mihi lapidem?* For *frontem* ('the face' to do so), see n. to XIII. 242.

libertatem parentis, 'the freedom of speech (παρρησία) a father should have', but which you have forfeited.

58. hoc, 'your head', the father's.

cucurbita, properly 'a <u>gourd</u>', is also 'a cupping-glass' shaped like a gourd, which was applied to draw blood from the head in disorders of the brain. The epithet *ventosa* shows a wrong conception of its action, as the blood is drawn by the want of air in the glass, not by a current of air. But it is clear that the epithet was a regular one, as a cupping-glass is called simply *ventosa* in late Latin, and so *ventouse* in French.

59–85. *When you expect visitors, you are very careful to have your house and furniture clean; you ought not to be less careful to remove all moral pollution from the house where your child lives. It depends upon the training you give him, whether he turns out a good or bad citizen. And he will do what he sees you do, as birds and beasts imitate and reproduce the habits of their parents.*

60. There are several scenes in Plautus, where a host gives instructions of this kind to his slaves, e.g. *Stichus,* 355 A. *ego hinc araneas de foribus deiciam et de pariete.* | B. *edepol rem negotiosam.* C. *quid sit nil etiam scio:* | *nisi forte hospites venturi sunt.*

61. 'Down with the withered spider, web and all': for *cum tota tela,* cf. VI. 171.

62. leve (or *purum,* X. 19) **argentum** is 'plain silver plate'. **vasa aspera** are silver cups with embossed work (*emblemata* or *sigilla*) upon them. The former are smooth to the touch, the latter rough.

64. The first part of a double question begins here, the second part at l. 68: in Greek the first would be introduced by μέν, the second by δέ, the purpose of the opposition being to show that the two actions are inconsistent. In translating, the first of such a pair of clauses may begin with 'if'.

68. illud non agitas, 'are you not anxious about something else...': *illud* (= ἐκεῖνο δέ) is defined by what follows.

sanctam: see n. to VIII. 24. **omni** = *ulla*.

70. gratum est, 'we thank you': for the view, general in antiquity, that to add to the population is a service to the state, cf. *The Vicar of Wakefield*, chap. 1. 'I was ever of opinion, that the honest man who married and brought up a large family, did more service than he who continued single and only talked of population...I considered them (my children) as a very valuable present made to my country.'

71. **patria:** so Büch. after P and the Schol.: the abl. after *idoneus* is rare and apparently archaic; but the similar constr. of *dignus* makes it possible that Juv. used *patria* here, to avoid repeating *patriae*. The other MSS. read *patriae*.

utilis agris, i.e. as a husbandman.

73. **plurimum enim intererit:** a prosaic phrase: cf. Quint. v. 4. 2 *plurimum intererit, quis et quem postulet* cet. There is only one other elision in Juv. similar to that of *plurimum* here, viz. *quantulum in*, vi. 151: see n. to *Pontica*, vi. 661. Virgil gives himself more licence in this point: in the Aeneid he often elides the last syll. of *Ilium*, also of *omnium* (i. 599), of *alterum* (ii. 667), of *audiam* (iv. 387), of *abluam* (iv. 684), and of *eruam* (xii. 569). He also uses hiatus after a cretic, e.g. after *insulae* (*Aen.* iii. 211), and after *Ilio* (v. 261). Another device of Virgil's is synizesis of such words as *alveo*, *aureis*, and *ferrei*.

76. **sumptis...pinnis,** 'when they have put on wings': cf. iii. 80.

77. **iumento et canibus** are dead animals. **crucibus** refers to the bodies of malefactors left hanging upon the cross: cf. Hor. *Epp.* i. 16. 48 *non pasces in cruce corvos*.

80. Friedl. notes that vultures really nest in rocks in spite of Juv. and Ovid, *Am.* i. 12. 20 [*illa arbor*] *vulturis in ramis...ova tulit*. The Roman poets in general are careless observers of such matters: thus Martial makes the swallow collect material for her nest with her claws.

81. **famulae Iovis,** i.e. eagles.

82. **hinc,** 'from these', i.e. hare or roe.

83. **inde,** 'from the nest'. **levavit:** P reads *levaret*: it seems probable that *levavit*, the reading of T, is right, as the perf. ind. is wanted to describe the repeated action: see Madv. on Cic. *de Fin.* v. 41.

84. **famē:** for the quantity, see n. to vi. 424.

86–106. *Other vices than those already mentioned may be inherited: thus, if a man is extravagant in building or has a sympathy for Judaism, his son will ruin himself in bricks and mortar or conform in every particular to the Jewish ritual.*

86. A mania for building (*insanus* and *insanire* are com-

monly used in this connection) was characteristic of wealthy
Romans under the Empire, and is often satirised by Horace,
e.g. *Carm.* II. 18. 17–22, III. 1. 33–6; *Epp.* I. 1. 83–7: cf. also
Juv. I. 94 and Mart. IX. 46. 1 *Gellius aedificat semper.* The
number of country-houses (*villae*) belonging to Romans of
moderate fortune, such as Cicero and the younger Pliny, strikes
us as very remarkable. Friedl. thinks that in this respect, more
than in any other, Roman extravagance was greater than that of
modern times.

87. **Caietae:** a town on the coast of Latium, supposed to
have been named after the nurse of Aeneas, who died there:
cf. Virg. *Aen.* VII. 1. *tu quoque litoribus nostris, Aeneia nutrix,* |
aeternam moriens famam, Caieta, dedisti.

For Tibur and Praeneste, see n. to III. 190.

89. **culmina villarum:** cf. Mart. IV. 64. 9 *puris leniter
admoventur astris* | *celsae culmina* (the roofs) *delicata villae.*

Graecis...marmoribus: the quarries of white marble at
Luna (Carrara) in Etruria were the chief source from which
marble was originally brought to Rome: Mamurra got the
pillars for his house from there 48 B.C., and Augustus built the
temple of the Palatine Apollo with this material. But in later
times the coloured marbles from other countries became more
popular—especially green serpentine from Laconia (cf. XI. 175),
white marble with purple veins from Phrygia, now known as
pavonazzetto, and a red-yellow marble from Numidia (cf. VII.
182), now known as *giallo antico*. And these were only the
favourites among very many varieties.

longe petitis suggests the cost of carriage: cf. Plaut. *Most.*
822 A. *quanti hosce* (*postes*) *emeras?* | B. *tris minas pro istis
duobus praeter vecturam dedi.*

90. **vincens,** 'surpassing': cf. II. 143 *vicit et hoc monstrum
tunicati fuscina Gracchi.* There was a famous temple of Fortuna
at Praeneste, and another of Hercules at Tibur: each of these
looked mean beside the palace of Cretonius at each place.

91. Posides was a favourite freedman of Claudius: cf. Suet.
Claud. 28 *libertorum praecipue suspexit Posiden spadonem, quem
etiam Britannico triumpho inter militares viros hasta pura donavit.*
This passage shows that he built himself a splendid house near
the temple of Jupiter Capitolinus.

nostra, i.e. at Rome, distinguished from Tibur and Prae-
neste: cf. *nostra infantia,* III. 84.

92. **dum** is equiv. to 'because': for the same constr., cf. l.
95, and 1. 60.

93. **fregit opes:** cf. Hor. *Sat.* II. 3. 18 *postquam omnis res mea Ianum | ad medium fractast.*

94. **turbavit:** *conturbare* is commonly used in this sense without an expressed acc.: see n. to VII. 129.

96. The obvious peculiarities of the Jewish ritual are here mentioned—their keeping of the Sabbath, the absence of images in their temples, their abstention from swine's flesh, and the practice of circumcision. Their peculiarities gave them the reputation of misanthropists; and other Roman authors, e.g. Tacitus and Pliny the Elder, speak of them with contempt and dislike. The present passage seems to show that Jewish proselytes were not uncommon at Rome. Even in Horace's time, his friend Fuscus pleads, no doubt in jest, as a reason for postponing business, *hodie tricesima sabbata (Sat.* I. 9. 69).

metuentem, here and l. 101, has probably a technical meaning: for a similar term was applied by the Jews themselves to those not of Jewish birth who were friendly to the Jewish religion: thus Paul addresses his audience in the synagogue of Antioch, ἄνδρες Ἰσραηλῖται καὶ οἱ φοβούμενοι τὸν Θεόν, i.e, Jews and Judaisers (Acts xiii. 16).

sabbata: cf. Pers. 5. 184 *recutitaque sabbata palles.*

97. **nubes:** the shadowy and unsubstantial object of Jewish worship is thus contrasted with the anthropomorphism of paganism. Thus in Aristophanes' comedy, the followers of Socrates, having put aside other deities, worship the clouds. Tacitus has little good to say of the Jews, but he is forced to admit the pure nature of their worship: *Hist.* v. 5 *Iudaei mente sola unumque numen intellegunt: profanos qui deum imagines mortalibus materiis in species hominum effingant; summum illud et aeternum neque imitabile neque interiturum.*

caeli numen, 'the holy sky': cf. VI. 545: the idea that the Jews worshipped the sky, perhaps arose from their unwillingness to utter the name of God, which led them to substitute 'heaven', as in 'the kingdom of heaven'.

98. They think pork-eating as bad as cannibalism: cf. VI. 160 and Tac. *l.l.* 4 *sue abstinent, memoria cladis, quod ipsos scabies quondam turpaverat, cui id animal obnoxium.*

99. **praeputia ponunt,** 'they are circumcised': cf. Tac. *l.l.* 5 *circumcidere genitalia instituerunt, ut diversitate noscantur.*

100. Ancient accounts agree in attributing to the Jews a hatred for all nations except their own: cf. Tac. *l.l.* 5 *adversus omnes alios hostile odium...trangressi in morem eorum idem*

*usurpant, nec quicquam prius imbuuntur quam contemnere deos,
exuere patriam*; Quint. III. 7. 21 *perniciosam ceteris gentem*.

101. Perhaps the line is meant to parody biblical language,
such as abounds in Ps. cxix. The intense devotion of the Jews to
their Law (*leges Solymarum*, VI. 544) is often mentioned.

102. **Moyses:** cf. Tac. *l.l.* 4 *Moyses, quo sibi in posterum
gentem firmaret, novos ritus contrariosque ceteris mortalibus
indidit*.

103, 104. The Jewish Law is misrepresented as consisting
solely of prohibitions against showing the commonest offices of
humanity (what Petronius calls *tralaticia humanitas*) to any but
co-religionists. These good offices are often mentioned as due
from all men to their kind: cf. Cic. *de Orat.* I. 203 *ut commonstrarem
tantum viam et, ut fieri solet, digitum ad fontis intenderem*;
Sen. *de Ben.* IV. 29. 1 *ergo...nec aquam haurire permittes, nec
viam erranti monstrabis ingrato?* (i.e. is the ungrateful man to be
considered as outside the pale of humanity?).

105. **pater in causa,** 'the father (of the proselyte) is to
blame': here, as in the other cases, a bad example has been
followed and made worse: the father was content with keeping
the sabbath and abstaining from pork, the son carries out to the
full the inhumanity of his sect.

106. Tacitus (*l.l.* 4) takes the same view of the Jewish Sabbath
and Jubilee, that they were prompted by indolence:
septimo die otium placuisse ferunt, quia is finem laborum (i.e. their
wandering in the wilderness) *tulerit; dein blandiente inertia
septimum quoque annum ignaviae datum*.

vitae is used for 'the business of life'. The subject to *attigit*
is *lux*, not *qui* supplied from *cui*: the proselyte was no doubt
active enough on the other days of the week.

107–255. *Other vices are readily imitated by the young, but
avarice they need to be taught. And this instruction is systematically
imparted by parents. Love of money and desire for
wealth grow apace. Injustice to others and consequent disrepute
are of no account to the man who is in haste to be rich, and for
whose wants the simplicity of an earlier age is quite insufficient.
When you urge your son to grow rich, to make money his first
object, you must not be surprised if he betters your instruction
and commits perjury and even murder for money. Your own life
will not be safe, if you persist in living on when your son, too apt a
pupil, is eager to inherit your wealth.*

108. **quoque,** 'even' against their will. The fact that there

is no caesura after the 3rd foot in the verse, is disguised by the elision: x. 358 is similar.

109. **umbra,** 'by a faint resemblance to a virtue': it would be more in accordance with usage to say that avarice is itself an *umbra* of a virtue: cf. Plaut. *Mil. Glor.* 624 *siquidem te quicquam quod faxis pudet, | nihil amas, umbra es amantum magis quam amator.*

110. **cum,** 'since'.

111. **nec:** the negative contained in this word belongs entirely to *dubie*, 'in no doubtful terms'.

112. **tutela,** 'guardian': the word is often applied by the poets to persons and things: cf. Hor. *Epp.* I. I. 103 (to Maecenas) *rerum tutela mearum | cum sis.* It is regularly used for the 'figurehead' of a ship, which generally represented some guardian deity.

113. **fortunas,** 'wealth': a constant sense of the plur.: *fortuna* of P, which Büch. keeps, is a mere slip due to the repeated *s*.

114. The apples of the Hesperides and the golden fleece of the Colchians in Pontus were both guarded by a sleepless dragon, according to the legend.

hunc de quo loquor is any *avarus*. The additional fact is that the miser is thought a good hand at increasing his wealth as well as keeping it.

116. **quippe,** 'for'.

fabris is either dat. of advantage or abl. absol.

117. **sed...modo** is a parenthesis, qualifying the previous statement by adding that such men are unscrupulous in the race for wealth: cf. Hor. *Epp.* I. I. 65 *rem facias, rem, | si possis, recte, si non, quocunque modo rem.*

118. The metaphor of *fabris* is continued: so we say 'to strike while the iron is hot'.

119. 'Well then, a father also (as well as the common people) thinks a miser a happy man.' **animi,** 'in their mind', a locative common with certain adjectives, such as *aeger* and *dubius,* and some verbs, e.g. *pendēre.*

If the text is sound, there is some carelessness in the expression here: it is difficult to determine whether the relative clauses which follow *avaros,* have *pater* as antecedent, or form the subject to *hortatur.* If *credit* were preceded by a relative, or if l. 119 were removed, there would be no difficulty. It seems necessary to the sense that *hortatur* should be a principal, and not a dependent, verb. The simplest remedy is to put a semi-

colon for a comma after *avaros* (so C. F. Hermann): but the
expression still remains abrupt. Madvig read *mirantur* and
putant, on the authority of the inferior MSS.

122. **pergant:** *peragant* of P, which Büch. keeps, is unin-
telligible. **sectae,** 'way of life': this word is generally used
together with, and governed by, the verb *sequi*.

123. **elementa,** 'rudiments', 'alphabet': so Quint. heads
his first chapter on education, *quemadmodum prima elementa
tradenda sint*; cf. Lucr. I. 81 (when beginning to set forth the
first doctrines of Epicurus) *vereor ne forte rearis | impia te
rationis inire elementa*.

124. **inbuit,** 'he begins their instruction': the word is
constantly used of making a beginning of any kind: so the Argo,
as the first ship that ever sailed, is said *imbuere Amphitriten*
(Catull. 64. 11); Perillus, who made the brazen bull for Phalaris,
and was the first to be burned in it, is said *imbuere taurum*
(Ovid, *Ars,* I. 654); a cup filled for the first time is said *imbui*,
'to be christened' (Mart. VIII. 51. 17).

minimas...sordes, 'to learn by heart petty meannesses':
so Quint. (I. 1. 36) recommends for very young children *dicta
clarorum virorum et electos ex poetis...locos ediscere inter lusum*.

125. **mox,** 'later', when preparatory studies have fitted
them for such an acquirement.

126–33 describes the *minimae sordes*.

126. **servorum ventres:** cf. III. 167, and see n. to III. 141.

modio iniquo: he cheats his slaves by measuring out less
than their due of corn: cf. Sen. *Epp.* 80. 7 *servus est: quinque
modios accipit et quinque denarios* (as a monthly wage).

127. **neque enim:** see n. to XI. 30.

sustinet, 'he has not the heart to'; cf. XV. 88.

128. **mucida...frusta:** cf. V. 68.

129. **hesternum,** 'from the previous day' and therefore
'stale', or (of flowers) 'faded': cf. Ovid, *Am.* III. 7. 66 *hesterna
languidiora rosa*: ἕωλος gets the same meaning in a similar way.
servare: cf. Mart. I. 103. 7 (of a miser also) *deque decem plures
semper servantur olivae*. September was the hottest and most
unhealthy month at Rome: see n. to IV. 56: but even the cer-
tainty that his meat will go bad cannot overcome the miser's
reluctance to eat it.

131. The articles of food mentioned are of the poorest: cf.
III. 293 and Mart. VII. 78. 1 *cum...ponatur cauda lacerti | et,
bene si cenas, conchis inuncta tibi; | sumen, aprum, leporem, boletos,
ostrea, mullos | mittis.*

aestivam: i.e. at a season when it will not keep; see n. to I. 28.

132. **signatum,** 'locked up', lit. 'sealed', the seal-ring being used for security: cf. Mart. XIV. 79 *flagra. ludite lascivi, sed tantum ludite, servi: | haec signata mihi quinque diebus erunt* (i.e. I shall lock up the scourge during the Saturnalia).

133. **numerata includere,** 'to count before he puts away'. For **porrum sectivum,** see n. to III. 293.

134. **de ponte** = *mendicus:* see n. to IV. 116, and cf. Sen. *Dial.* VII. 25. 1 *in sublicium pontem me transfer et inter egentes abige.*

negabit: P and the Schol. read *negavit,* and similarly *scrove* for *scrobe,* l. 170: for this confusion of *b* and *v,* see n. to *verbum,* I. 161. *negare* is regularly used in this sense: cf. Mart. II. 69. 7 *en rogat ad cenam Melior te, Classice, rectam. | grandia verba ubi sunt? si vir es, ecce, nega.*

135. **quo divitias:** see n. to VIII. 9.

139. **crevit,** the reading of P, which has been supposed (by Stephan *Rhein. Mus.* XL, 281) to be a mere slip, is confirmed by T, which reads the same quite clearly.

140. **non habet** = *caret,* being treated as a single word: cf. Mart. III. 8 *Thaida Quintus amat: quam Thaida? Thaida luscam. | unum oculum Thais non habet, ille duos.*

paratur tibi, 'you buy': see n. to III. 224.

141. **altera villa:** see n. to l. 86.

143. **vicina,** 'belonging to your neighbour'.

144. **arbusta** = 'a vineyard', this being the regular word for the trees on which vines were trained: cf. *ulmos* VI. 150 and Virg. *Georg.* II. 416 *iam vinctae vites, iam falcem arbusta reponunt.*

canet: the word expresses the grey-green colour of the olive-leaf: cf. Stat. *Theb.* III. 466 *canentis olivae | fronde;* Lucr. 5. 1373 *olearum caerula...plaga;* Pind. *Ol.* 3. 13 γλαυκόχροα κόσμον ἐλαίας; Tennyson's 'olive-hoary cape'; and Browning's more elaborate imagery 'the hills over-smoked...by the faint grey olive-trees'.

145. **quorum** is governed by *dominus.* If the owner refuses all offers for his land, his crops can be ruined so that he must sell.

146. **lasso...collo:** cf. VIII. 66: they have been hard worked in the shafts and are therefore likely to eat more. This malpractice seems to have been ancient, as there was a law of the Twelve Tables assigning penalties for it.

148. With **domum,** supply *abibunt* from *mittentur.*

saevos, 'ruthless', i.e. ravenous.

149. **falcibus actum,** 'the sickle had been at work'; for the fields have been cropped clean.

152. **qui sermones:** *erunt* must be supplied: cf. x. 88.

foedae: *foede* is read by P; but both words are spelt identically as a rule in MSS., and the adv. would require a different verb (e.g. *sonat*) to be supplied. For **bucina,** cf. Cic. *ad Fam.* XVI. 21. 2 *polliceris te bucinatorem fore existimationis meae.*

153. **tunicam lupini,** 'a pea-pod', i.e. something worthless: cf. Hor. *Epp.* I. 7. 23 *nec tamen ignorat quid distent aera lupinis.*

154. **toto pago** is abl. of place. For the rich man's contempt of public opinion, cf. Hor. *Sat.* I. 1. 65 *sordidus ac dives, populi contemnere voces | sic solitus, 'populus me sibilat, at mihi plaudo | ipse domi, simul ac nummos contemplor in arca.'*

156. **scilicet,** 'I suppose', shows that the sentence is ironical. **debilitate,** 'bodily injury', such as the loss or incapacity of a limb: see n. to x. 227.

159, 160. The accumulation of immense landed properties in single hands was, in Pliny's judgment, the ruin of Italy: cf. *Nat. Hist.* XVIII. 35 *latifundia perdidere Italiam, iam vero et provincias. sex domini semissem* (one half) *Africae possidebant, cum interfecit eos Nero princeps.* The system led to the substitution of slave labour for that of freemen and of pasture land for arable, so that Italy had to depend on foreign countries for her corn supply, and her free peasantry ceased to exist.

160. Titus Tatius was king of the Sabines when they joined with the Romans under Romulus to form one nation.

161. **mox,** 'at a later time'. The worn-out veterans of the Punic wars were content with 2 acres apiece.

162. **Molossos:** see n. to XII. 108.

166. **curta fides,** 'a breach of faith', is predicate.

saturabat, 'fed': see n. to VIII. 118. This meaning of **glaebula** survives in our 'glebe'.

167. **turbam casae,** 'all the inmates of the cottage': *turba* is often used in Latin of the whole of a number, not large in itself: cf. Ovid, *Met.* VI. 199 (Niobe speaking) *non tamen ad numerum redigar spoliata duorum, | Latonae turbam,* i.e. two are all whom L. can boast.

feta, 'in child-bed'.

168. Cf. XI. 98.

The fact that of four children only one is a slave, is mentioned as typical of old times. When the imperial writers speak of children as pets in a household, these are almost always *vernulae*, not *ingenui*: cf. Mart. III. 58. 22 *cingunt serenum lactei focum vernae*; XIV. 54 *si quis plorator pendet tibi vernula collo*; Sen. *Epp.* 12. 3 *ego sum Philositi vilici filius, deliciolum tuum* (your little pet); Petron. quoted on v. 27. The reason of this is stated by Juv. VI. 594.

169. **magnis,** 'grown up': cf. l. 79. For the picture of rustic life, cf. XI. 82–9. Those who have been working all day on the farm require a larger meal than those who have stayed in the house.

171. **pultibus,** 'porridge': see n. to XI. 58.

172. **horto,** 'a kitchen-garden' for growing vegetables: see n. to I. 75. Juv.'s contemporaries are not satisfied with two acres of kitchen-garden.

173. **inde,** 'from this', i.e. from desire for more land and more money.

174. **ferro grassatur:** cf. III. 305.

179. A similar figure is Ofellus, who gives the same kind of advice to his *pueri* (Hor. *Sat.* II. 2. fin. *quocirca vivite fortes*).

For **casulis,** cf. IX. 60 *rusticus infans | cum matre et casulis et conlusore catello*.

180. The Marsi (III. 169), Hernici, and Vestini were small nations of Central Italy, probably of Sabine origin, who, after a vain struggle against the growing power of Rome, became her most faithful allies and bravest soldiers.

182. **numina ruris:** cf. Virg. *Georg.* I. 7 *Liber et alma Ceres, vestro si munere tellus | Chaoniam pingui glandem mutavit arista*.

184. **veteris,** 'their food for long': cf. XIII. 57; and see n. to VI. 21.

fastidia quercus: cf. Lucr. V. 1416 *sic odium coepit glandis*.

185. **fecisse,** 'to be guilty of': see n. to IV. 12, and cf. Sen. *de Ben.* III. 7. 7 *de quibusdam et inperitus iudex dimittere tabellam potest* (can give his vote), *ubi fecisse aut non fecisse pronuntiandum est* (where a mere verdict of 'guilty' or 'not guilty' is required). So *fecisse* is not an imitation of the aor. inf. in Greek, though such an imitation is common in Latin poetry: compare Hor. *Carm.* III. 4. 51 *fratresque tendentes opaco | Pelion imposuisse Olympo*, with Hom. *Od.* XI. 315, Ὄσσαν ἐπ᾽ Οὐλύμπῳ μέμασαν θέμεν.

The meaning is, that a man who is content with rough clothing and a hardy life, is not likely to commit crime.

187. **pellibus inversis,** 'skins with the hair turned inwards': cf. Pind. *Pyth.* IV. 81 ἀμφὶ δὲ παρδαλέᾳ στέγετο φρίσσοντας ὄμβρους.

peregrina = *stlattaria*, VII. 134.

188. **quaecumque est** repeats the sense of *ignota*: cf. Virg. *Aen.* v. 83 *Ausonium, quicumque est, quaerere Thybrim* (to seek the unknown river of Ausonia). **purpura** stands for 'fine raiment', as opposed to the skins of beasts: cf. Lucr. v. 1423 *tunc igitur pelles, nunc aurum et purpura curis | exercent hominum vitam.*

189. With **praecepta** supply *dabant.*

190. **post finem autumni,** i.e. at the beginning of winter, when days become short.

media de nocte = *post sextam horam,* 'after midnight': so *de meridie,* 'in the afternoon'; *diem de die differre,* 'to put off day after day'.

That it was the custom of diligent Romans to anticipate the day in winter, we know from the practice of the elder Pliny, Pliny, *Epp.* III. 5. 8 *lucubrare Vulcanalibus* (23 August) *incipiebat...statim a nocte multa, hieme vero ab hora septima* (1 a.m.)...*ante lucem ibat ad Vespasianum imperatorem; nam ille quoque noctibus utebatur.*

191. **ceras,** 'tablets' for writing on; see n. to VII. 23. The professions indicated are those of a barrister (*causas age*), and of a jurisconsult. That Juv. in the 7th satire represents these professions as anything but lucrative, is an inconsistency which need not trouble us.

192. **rubras...leges:** the first words of a law were written in red, as also the first words or letters of medieval MSS.; hence our 'rubrics', the directions in the Prayer-book which were often printed in red. Cf. Petron. 46 *emi ergo nunc puero aliquot libra rubricata, quia volo illum...aliquid de iure gustare*: the whole chapter gives a lively picture of such a father as Juv. is describing here.

193. **vitem posce libello,** 'petition for a vine-staff', i.e. for a post of centurion, the *vitis* being the attribute of this officer: see n. to VIII. 247.

libellus is the regular word for 'petition': the imperial secretary who dealt with them was called *a libellis.*

Under the Empire, a man of equestrian birth, who wished to enter the civil service, had first to serve in the army in three successive grades, for which see n. to I. 58. Eventually, a fourth grade preliminary to the others was added, that of *centurio.* When these grades had been completed, the man was entitled *a quattuor militiis* and became eligible for civil appointments, especially those of *procurator* or *praefectus.* Candidates for these military offices were called *militae* (sc. *equestris*) *petitores*: cf. Suet. *Gramm.* 24 *M. Valerius Probus diu centuriatum petiit donec taedio ad studia se contulit.*

It appears, however, that the youth mentioned here is not of equestrian rank but a member of the *plebs*; cf. VIII. 47 and Livy there quoted. For in the case of an *eques*, this military service was not a serious profession, but a mere passport to more lucrative civil employments. But the context shows that the youth here spoken of is to remain in the army all his active life and retire as a *primipilaris* (see n. to l. 197). This post was the highest ambition of a common soldier, who worked his way up by degrees; and it must have been a rare piece of good fortune for such a man to begin his service in the army even at the bottom of the list of centurions. It cannot be supposed that a man, who, as an *eques*, could begin his service as a centurion, would spend his whole active life in that position.

194. **buxo** = the comb, this wood being used for the purpose: cf. Mart. XIV. 25 *pectines. quid faciet, nullos hic inventura capillos, | multifido buxus quae tibi dente datur?* For carefully dressed hair as a sign of effeminacy, and much hair on the body as a sign of manliness, cf. Sen. *Epp.* 115. 2 *nosti complures iuvenes barba et coma nitidos, de capsula totos* (exactly as if they came out of a band-box)*; nihil ab illis speraveris forte, nihil solidum*; Aristoph. *Lysistr.* 800 A. τὴν λόχμην πολλὴν φορεῖς. XO. καὶ Μυρωνίδης γὰρ ἦν |τραχὺς ἐντεῦθεν μελάμπυγός τε τοῖς ἐχθροῖς ἅπασιν, | ὡς δὲ καὶ Φορμίων.

195. **Laelius** is the general, rather than (as the Schol.) a centurion, as the recruit is himself to be a centurion. There may have been at the time a successful campaigner of this name, who is not elsewhere mentioned.

196. Two satires of this last book, the 13th and 15th, were written after A.D. 127: see nn. to XIII. 17, XV. 27. This l. may refer to a somewhat earlier time: Hadrian, who became emperor in 117, had to quell a rising in Mauretania in the beginning of his reign; and he was in Britain in 121, when he built his famous wall to protect the province against inroads from the north.

attegias: the word occurs only once again, in an inscription (Alsatian): the huts of the Africans are generally called *mapalia*. The **Brigantes** were a large tribe who occupied the chief part of northern England: cf. Tac. *Agr.* 17 *Brigantum civitatem, quae numerosissima provinciae totius perhibetur, aggressus.*

197. **locupletem aquilam** = *divitias et primipilatum*: the

eagle, or chief standard of a legion, was placed in the first line in
battle, and was in charge of the senior centurion of the legion
(*centurio primi pili*). Such a man when discharged was called
primipilaris; he might have gained considerable wealth and was,
at least in the *municipia*, a person of importance, and often
gained equestrian rank. One of Martial's chief friends, Aulus
Pudens, was a centurion for whom Martial predicts the prize of
his profession: VI. 58. 10 *et referes pili praemia clarus eques*.

sexagesimus annus: under the Empire the period of ser-
vice, after which a *honesta missio* was given, was generally 20
years (*vicena stipendia*); but men were often retained in the
service for a much longer period as *evocati*. It is not known that
there was a rule of superannuation at sixty; but that age is
mentioned elsewhere as a suitable period to retire from active
life: cf. Sen. *Dial.* x. 3. 5 *audies plerosque dicentes: 'quin-
quagesimo anno in otium secedam. sexagesimus me annus ab
officiis dimittet.'*

199. **solvunt...ventrem,** 'upset your stomach': the phrase
might be copiously illustrated from Aristophanes.

200. **pares,** 'you must buy': cf. l. 140. Retail trade is
another way to a fortune. This advice, certainly, could only be
given to a member of the *plebs*: see n. to l. 193.

201. **dimidio** is abl. of the amount of difference.

202. Evil-smelling businesses, such as tanning, had to be
carried on across (i.e. on the Janiculan side of) the river: cf.
Mart. VI. 93. 4 *non detracta cani transtiberina cutis (tam male
olet)*.

204. **unguenta,** 'perfumes'. What follows is probably a
reference to a familiar story of Vespasian: Suet. *Vesp.* 23
*reprehendenti filio Tito, quod etiam urinae vectigal commentus
esset, pecuniam ex prima pensione admovit ad nares, sciscitans num
odore offenderetur; et illo negante 'atquin' inquit 'e lotio est'*.

206. **Iove digna poeta,** lit. 'worthy of Jupiter as poet', i.e.
'a verse worthy of Jupiter himself'. 'Know thyself' is very well
as a motto, and may be worthy of Apollo (see n. to XI. 27); but
the present piece of wisdom is worthy of Apollo's father.

208, 209 are bracketed by Jahn. Though excellent in them-
selves, they are out of place here, where Juv. is insisting on
parental influence. If retained, they must form part of the
father's speech.

208. **monstrant:** see n. to x. 363. **repentibus,** i.e. before
they can walk.

It is remarkable that T agrees with P in reading *reppentibus*:

perhaps this indicates that the true reading is *reptantibus*: *repere* is rarely used of creeping children, while *reptare* is exceedingly common.

assae: the Schol. explains that *assa nutrix* is a 'dry-nurse' (*quae lac non praestat infantibus*) and is borne out by inscriptions.

210. **instantem:** cf. l. 63.

212. **meliorem...discipulum,** 'I warrant the pupil to outdo the teacher'; i.e. your son will prove more avaricious than yourself; you need have no fears on that head.

214. **Peleă:** for the quantity, see n. to III. 266.

215. Cf. Virg. *Georg.* II. 363 *parcendum teneris*. The whole passage seems to be in Juv.'s mind; see l. 230.

Büch., following P, reads *medullas: naturae mala nequitia est*. But what does this mean? Surely *naturae* is a natural error for **maturae,** which is read by some MSS. and gives a simple and appropriate meaning: 'the bane of full-grown iniquity has not yet infected their marrow', i.e. has not penetrated to their core. This reading is well illustrated, and defended, by Sen. *Dial.* III. 16. 2 *in te duriora remedia iam solida nequitia desiderat... perbibisti nequitiam et ita visceribus immiscuisti, ut nisi cum ipsis exire non possit.*

implere, like ἀναπιμπλάναι, sometimes means 'to infect'; cf. Livy, IV. 30. 8 *vulgati contactu in homines morbi. et primo in agrestes ingruerant servitiaque; urbs deinde impletur.*

216, 217. **cum...cultri** = when he is grown up. For the age at which the beard was first cut, and that when regular shaving began, see nn. to VIII. 166, VI. 105. *culter* is 'a knife' or 'shears', not a razor, as the allusion is not to shaving but clipping the bread: cf. Mart. IX. 76. 3 (of a *barbae depositio*) *creverat hic vultus bis denis fortior annis, | gaudebatque suas pingere barba genas, | et libata semel summos modo purpura* (the dark growth) *cultros | sparserat.*

219. **etCereris,** 'even of Ceres': an oath by deities who presided over mysteries was especially binding: cf. III. 144. The sinner is so bold that he touches, while swearing, not only the altar, but the foot of the goddess: for the custom, cf. XIII. 89 and Livy, XXI. 1. 4 *fama est Hannibalem annorum fere novem... altaribus admotum, tactis sacris, iure iurando adactum se, cum primum posset, hostem fore populo Romano.*

220. **elatam...nurum,** 'take it that your daughter-in-law is already carried to the grave,' i.e. is as good as dead: for *elatam*, cf. I. 72.

vestra is used, because the same house is shared by father and son.

221. **quibus...digitis,** 'whose fingers will strangle her in her sleep!' The sentence is exclamatory; of course her husband is meant.

223. **putas:** the emphatic pronoun is required in English.

brevior via: cf. Sen. *de Ben.* VII. 26. 4 *alius totus lucri est* (is entirely bent on gain), *cuius summam, non vias, spectat.*

226. **penes te:** cf. Livy, XXVIII. 27. 11 (Scipio addressing mutinous soldiers) *causa atque origo omnis furoris penes auctores est: vos contagione insanistis.* Though such a word as *potestas* is normally subject to *penes est*, yet *culpa* (or *noxia*) is thus used in all periods of Latin: cf. Terence, *Hec.* 535; Livy, III. 42. 2, IV. 53. 5, V. 36. 10, XXXV. 33. 3, XLV. 10. 10; Sen. *de Ben.* VII. 18. 2; Stat. *Theb.* XI. 189; Trajan, *ad Plin.* 30. 2.

227–9. Büch.'s punctuation of the whole passage is followed in the text: according to this, the first relative clause (*quisquis... amorem*) is followed by a principal clause (*et...avaros*); then follows a second relative clause in which *conduplicari* is governed by *praecepit.* The constr. is very awkward; and l. 229 is not found in some MSS. and is bracketed by Jahn. Perhaps *et laevo* is corrupt: *et* is strange, and the Schol.'s comment (*subtili monitu*) seems unsuitable to *laevo.*

228. **et,** 'also'. **laevo** = *sinistro.* **producit,** 'brings up': cf. VI. 241.

230. **effundit habenas:** he lets his team get the bit in their mouths and finds it too late to check them. The metaphor is common in Latin, e.g. Virg. *Georg.* II. 364 (of the vine) *palmes ...laxis per purum inmissus habenis*; Livy, XXXIV. 2. 13 (Cato speaking in defence of the *lex Oppia* 195 B.C.) *date frenos impotenti naturae et indomito animali* (i.e. women) *et sperate ipsas modum licentiae facturas.*

231. **quem si revoces,** 'and if you try to stop him', i.e. your son. Büch. takes *curriculum* as antecedent of *quem*, the Latin grammarians vouching for a form *curriculus*; but it seems likely that part of the evidence for the form was supplied by this passage misunderstood.

234. **adeo** here is corrective and means, 'rather', with a sense like *immo.* In particular phrases it has this meaning often: cf. Cic. *in Verr.* II. 3. 21 *tot annis atque adeo* (or rather) *saeculis*; *ibid.* 142 *nova lege atque adeo nulla lege.* See Palmer's n. to Plaut. *Amph.* 677 *quam omnium Thebis vir unam esse optumam diiudicat,* | *quamque adeo cives Thebani vero rumiferant probam*; also

Tyrrell's n. to Cic. *ad Att.* I. 17. 9 *ego princeps in adiutoribus atque adeo secundus.*

Transl. 'not a bit of it! they give themselves freer licence'.

For the sentiment, cf. Sen. *Dial.* III. 7. 4 (of anger) *quarumdam rerum initia in nostra potestate sunt, ulteriora nos vi rapiunt nec regressum relinquunt; ibid.* 8. 1 (anger, if once suffered to begin) *faciet de cetero quantum volet, non quantum permiseris.*

235. **stultum,** sc. *eum esse.* Juv. may have in mind the scenes, common in Plautus, where a prudent father dissuades his son from helping an extravagant friend: e.g. *Trinummus,* II. 2.

donet: this verb has generally an acc.; but cf. v. 111 *donandi gloria*; Mart. IV. 40. 7 *iam donare potes, iam perdere.*

237. The apodosis begins here. The first **et** is 'both'. **circumscribere:** see n. to XV. 136.

238. **amor,** sc. *tantus est.*

239. **Deciorum:** see n. to VIII. 254.

240. Menoeceus, acting on a prophecy of Teiresias, stabbed himself to secure victory for Thebes against the seven invaders (Stat. *Theb.* X. 628–782). Here again Juv. expresses distrust of the records of Greece, and with better reason than in X. 174.

241. What follows is satirical: Juv. represents the miraculous legend as a thing of common occurrence at Thebes. The legend ran that Cadmus sowed the teeth of the serpent he had killed, that an armed host grew up out of the soil and fought with each other, and that five who survived, formed, with Cadmus, the original inhabitants of Thebes.

quorum: the antecedent is *Thebani,* understood from *Thebas.*

dentibus, abl. of origin, as in *amplissima familia natus.*

246. **nec tibi** = *ne tibi quidem:* this has more point than to make the *et,* contained in *nec,* merely connect the sentences.

247. **fremitus,** used for any loud noise (cf. VI. 261), denotes especially the roaring of lions: cf. VIII. 37; Sen. *Dial.* III. 1. 6 *spumant apris ora...leones fremunt, inflantur inritatis colla serpentibus.*

248. For **mathematici** or *Chaldaei,* see n. to VI. 553; for **genesis,** see n. to VI. 579. It was regarded as a sign of the wickedness of the age for a son to enquire of the astrologers when his father would die, or a wife about her husband's death: cf. VI. 565 and Ovid there quoted. But in the present case the son goes further and poisons a father who persists in living on.

249. **stamine nondum abrupto,** i.e. before the span of life

allotted you by fate is run out: your son anticipates 'the blind
Fury with the abhorred shears'.

251. The stag, like the *cornix* (cf. X. 247), was erroneously
supposed to live to a very great age, 900 years, says the Schol.

252. **Archigenem** = *medicum*: see n. to VI. 236. For
'Mithridates' mixture', see n. to VI. 661.

253. **composuit**, 'compounded': the regular word used of
drugs: cf. Quint. I. 10. 6 *antidotos...ex multis...componi
videmus*; Sen. *de Ben.* IV. 28. 4 *compositiones remediorum salu-
tarium*. **si vis...rosas**, i.e. if you wish to see another autumn
and another spring.

255. **et pater et rex**, 'both a father and a king', i.e. a modern
father like the ancient king. There does not seem to be any
allusion to the fact that a son of Mithridates rose in rebellion
against him.

256–302. *It is as good as a play, nay better than any, to watch the
dangers that the avaricious man incurs in the struggle for wealth.
A rope-dancer on the stage is not so amusing to the observer, as
the merchant who spends his life on board ship, at the mercy of
the winds. There are many forms of madness: Orestes is mad in
one way, Ajax in another, and a third madness is that of the man
who for the sake of mere pieces of stamped metal risks his life in
storms at sea. Shipwreck and beggary are his probable fate.*

It will be noticed that Juv. has now finally dismissed the
topic of parental influence.

256–64. Cf. V. 157.

257. **aequare** has the sense of *comparare*; a rare use. For the
connection of the praetors with the public shows, see n. to
VIII. 194.

258. **spectes**, 'look on and see': the sight is a real *spectacu-
lum*. **capitis discrimine**, 'danger to life'.

259. **domus** = *rei domesticae* or *familiaris*: see n. to VI. 357.

260. It was common to deposit money in a temple for safe-
keeping; hence arises a question which often turns up in Quin-
tilian, whether theft of such money was sacrilege. It appears
from this passage, that burglars had entered the temple of
Mars Ultor and carried off not only the money deposited there,
but also what parts they could detach of the god's own statue:
cf. XIII. 150–3.

vigilem Castora: the temple of Castor in the Forum, of
which three columns are still standing, is meant. Castor is called
watchful, because a military guard was posted there (so M.).

262. **suas** is emphatic, 'even his own', far less other people's.

Florae: cf. VI. 250: the *ludi Florales* were held 28 April–3 May: they included plays and also beast-baitings: cf. Mart. I. 35. 8 *quis Floralia vestit et stolatum | permittit meretricibus pudorem?*; idem, VIII. 67. 4 *cum...Floralicias lasset harena feras.*

263. The other shows mentioned are the *ludi Ceriales* (12–19 April), and *Megalenses* (4–10 April), for which see n. to XI. 193. Plays formed a part of all these shows: hence the word **aulaea:** cf. VI. 67 (on the passion of women for plays and actors) *ast aliae, quotiens aulaea recondita cessant, | et vacuo clausoque sonant fora sola theatro, | atque a Plebeiis longe Megalesia* (i.e. in the interval between 17 November and 4 April), *tristes | personam thyrsumque tenent et subligar Acci.*

264. **humana negotia** is the subject: the verb *sunt* must be supplied: cf. Hor. *Epp.* II. 1. 197 *spectaret populum ludis attentius ipsis, | ut sibi praebentem nimio spectacula plura.*

265. **petauro,** 'from the spring-board': but there is much doubt as to the exact nature of this machine. The word is Greek and meant originally 'a perch for fowls'; in Latin writers it occurs frequently as an apparatus for acrobats on which to perform dangerous and delicate feats of skill: Martial speaks of *graciles vias petauri* (II. 86. 7). In some passages it seems to mean 'a wheel'.

266. **rectum,** 'tight'. **descendere:** the prefix seems to show that the performer (a *funambulus* or σχοινοβάτης, III. 77) came down from the roof of the theatre on a slanting rope: cf. Hor. *Epp.* II. 1. 210 *ille per extentum funem mihi posse videtur | ire poeta, meum qui pectus inaniter angit.*

267. **Corycia:** Corycus was a town on the coast of Cilicia famous for its cave (the Κιλίκιον πολυώνυμον ἄντρον of Pind. *Pyth.* I. 17), and for the excellent saffron (l. 269) that grew there.

268. **habitas,** 'make it your home'.

tollendus: the gerundive has the sense of a future passive participle, 'on the point of being....'.

269. **perditus,** 'reckless': cf. *audacia perdita*, III. 73. **sacci olentis:** saffron (*crocus*) was largely used to perfume the stage during *ludi*, in the form of spray: cf. Mart. v. 25. 7 *hoc, rogo, non melius quam rubro pulpita nimbo | spargere, et effuso permaduisse croco?*

The text cannot be considered satisfactory: **vilis** must be nom., as *ac* could not couple *perditus* and *mercator*, nor could the saffron be called worthless. Apparently *vilis* must have the sense of *sibi vilis*. The reading of the inferior MSS. (*a siculis*) does not mend matters. Housman's *similis* means 'yellow as his saffron'.

270. **antiquae,** because of its prominence in mythology and early history.

271. **municipes Iovis,** because Jupiter was born in Crete: cf. IV. 33 and Martial there quoted.

272. **hic,** the *funambulus*.

274. **tu,** '*but* you'. He risks danger for the sake of necessary food and clothing; but you do so to get mere superfluities.

275. **centum villas:** see n. to l. 86.

276. **plenum:** the same adj. must be supplied with *portus*: cf. VIII. 129 where *cunctos* must be supplied with *conventus* from the following *cuncta*. **plus hominum,** 'the majority of mankind': a considerable hyperbole, if we consider merely the proportionate number of the sexes.

278. The Carpathian sea stretches between Crete and Rhodes and is called after the island Carpathus. The sea that bounds the north coast of Africa is here called the Gaetulian sea; the chief article of commerce there was the purple-fish.

279. **Calpe** is the Rock of Gibraltar; the quantity shows that this is the abl. of a form *Calpis*, for which *Calpe* (fem.) is else-where found in Latin. The Schol. explains the name thus: *urnae similis mons.*

280. The sun is supposed to plunge, like a mass of molten metal, with a hissing sound into the waters of the Atlantic, which is called after Hercules because legend represented him as having explored Europe as far as the Atlantic, setting up his Pillars as a memorial of his adventures.

For **stridentem,** cf. Virg. *Aen.* VIII. 450 *alii stridentia tin-gunt | aera lacu.*

281. **grande operae pretium:** cf. XII. 127: the same might be expressed by *tanti est*, 'is a price worth paying': for the sights mentioned below are not regarded as interesting novelties but as horrible dangers.

283. The Ocean is distinguished from the Mediterranean, where it was not supposed that such monsters existed: cf. X. 14; Tac. *Ann.* II. 24 *miracula narrabant,. . .monstra maris, ambiguas hominum et beluarum formas.*

iuvenes marinos, 'the young mermen'.

284. **non unus:** see n. to VIII. 213. **ille** is Orestes: cf. Eur. *Orest.* 264 (Orestes speaking in delirium to Electra) μέθες, μί' οὖσα τῶν ἐμῶν Ἐρινύων.

286. **hic** is Ajax, whose madness is related in Sophocles' play.

287. **Ithacum,** i.e. Odysseus: cf. x. 257.

parcat = does not tear: for this sign of madness, cf. Luke viii. 35: 'they found the man, from whom the devils were gone out, *clothed and in his right mind.*'

288. **curatoris eget** = *insanus est*: cf. Hor. *Epp.* I. I. 101 *insanire putas sollemnia me neque rides,* | *nec medici credis nec curatoris egere* | *a praetore dati.* The natural feeling, that it is madness to go to sea, lingered long in the ancient world: witness the Greek verse quoted on VI. 30.

289. **tabula distinguitur unda,** 'is separated by a (single) plank from the wave': cf. XII. 58. Some MSS. have *una* for *unda,* but the former can be understood while the latter cannot.

290. Coins, then as now, bore the miniature image of the ruler and his 'superscription': for **tituli,** see n. to I. 130. The common Latin for 'coined money' is *argentum signatum,* while 'plate' is *argentum factum,* though the epithet is often omitted in the latter case: e.g. XII. 43.

292. The owner of the cargo hastens to carry back his goods to market, in defiance of threatening weather.

294. **fascia,** 'the wrack', is properly 'a strip', and here means a cloud of that shape.

295. **aestivum:** the adj. is used as an adverb: cf. I. 16, VI. 485.

297. **zonam** = his purse, money being carried thus by travellers: cf. Hor. *Epp.* II. 2. 40 (of a soldier) *ibit eo quo vis qui zonam perdidit:* so Plautus translates βαλλαντιοτόμος by *sector zonarius.*

laeva, because he is keeping himself up in the water with the right hand. **morsu** = *mordicus,* 'with his teeth'.

299. **Tagus:** see n. to III. 55. The Pactolus in Lydia was equally famous for the same properties.

300. With **sufficient,** supply *ei.*

frigida inguina, 'his cold and nakedness'.

302. **picta...tuetur,** 'gets a living by a picture of the storm'. It was common for a shipwrecked sailor to have a picture of his disaster painted, and to carry it about the streets, begging. Cf. Pers. I. 88 *cantet si naufragus, assem* | *protulerim? cantas, cum fracta te in trabe pictum* | *ex umero portes?* From this it appears that the picture was, if possible, painted on a fragment of the wreck, to refute the sceptical.

303–31. *If it is hard to get money, it is still harder to keep it. The rich man takes a world of trouble to keep his treasures from being burnt; Diogenes, in his tub, has no fears of this kind. So Alexander felt that the Cynic was a happier man than himself. The proper limit of wealth is what nature requires. If this seems a hard saying for our modern manners, fix the limit at a 'knight's fortune', or twice or three times as much, if you still insist; but if that will not satisfy you, nothing will.*

303, 304. Cf. Sen. *Epp.* 115. 16 *maiore tormento pecunia possidetur quam quaeritur.*

305. **praedives** was applied to Seneca, x. 16.

vigilare cohortem: the words suggest the public fire-brigade of Rome (*vigiles*), organised by Augustus A.D. 6, and consisting of seven cohorts, one to each pair of *regiones*. They were generally freedmen; their fire-extinguishing apparatus consisted of *centones, siphones, perticae, calae,* and *hamae* (or *amae*): cf. Pliny, *ad Trai.* 33. 2 (of Nicomedeia the capital of Bithynia) *nullus usquam in publico sipho, nulla hama, nullum denique instrumentum ad incendia conpescenda.*

306. **Licinus:** see n. to I. 109.

attonitus pro, 'terrified for', like δεδιὼς περί. In classical prose, such verbs as *timere* are not followed by *pro* but by the personal dat. or *de*: see n. to VI. 18.

307. **electro** is either (1) amber, or (2) an alloy made of $\frac{4}{5}$ gold and $\frac{1}{5}$ silver, which got this name from its colour; either would be valuable.

Phrygia columna, 'pillars of Phrygian marble': see n. to l. 89.

308. For **ebore,** see n. to XI. 124; and, for **testudine,** n. to XI. 94.

dolia: Diogenes, to show his αὐτάρκεια, took up his abode in a *dolium*, a large jar (not *tub*) of earthenware, intended to hold wine and larger than a *lagena*: cf. Sen. *Epp.* 90. 14 *Diogenes... qui se conplicuit in dolio et in eo cubitavit.* A gem, representing the philosopher looking out of his *dolium*, like a dog out of a barrel used as a kennel, is reproduced on p. 343 of King and Munro's *Horace.* He had historical precedent, as some of the population of Attica, when crowded into Athens during the Peloponnesian war, lived in similar jars: cf. Aristoph. *Knights*, 792 τοῦτον ὁρῶν οἰκοῦντ' ἐν ταῖς πιθάκναισι.

The context shows that *dolia* is a metrical equivalent for *dolium.*

nudi: see n. to XIII. 122.

310. **plumbo commissa**, 'with a rivet of lead': this meaning of the verb is seen in *commissura*, 'a joint'. *plumbata* would express the same thing more briefly but perhaps too technically for poetry: cf. Sen. *Nat. Quaest.* IV. 2. 18 *argentum replumbatur* (i.e. the heat in Aethiopia makes the lead fastenings in silver plate fall out). **manebit**, 'will serve'.

311. The reference is to the famous interview between Alexander and Diogenes, when the Cynic, on being told to ask any favour of the king, begged him to stand out of his light.

testa may be applied to any vessel of earthenware: cf. XV. 128.

312. The emphatic position of **magnum** is intended to suggest that this title belonged more truly to the philosopher than to the conquering king. Cf. Sen. *de Ben.* v. 4. 4 (*Diogenes*) *multo potentior, multo locupletior fuit omnia tunc possidente Alexandro; plus enim erat quod hic nollet accipere quam quod ille posset dare.*

313. **totum...orbem**: cf. x. 168. The verbs are in the subj., because the thought of Alexander is expressed.

314. **aequanda** has here a future sense rather than the correct sense of the gerundive: cf. *tollendus*, l. 268; *sumendas*, XII. 61.

315, 316. **nullum...deam** is here repeated from x. 365; and is decidedly irrelevant in this place.

318. **in quantum** does not differ in meaning from *quantum*, and is commonly used for it by prose-writers of the silver age, Seneca, Tacitus, and Pliny: in verse it is very rare.

319. **Epicure**: see n. to XIII. 123. Epicurus is constantly cited by ancient writers as a pattern of frugality: cf. Sen. *Epp.* 18. 9 (*Epicurus*) *gloriatur non toto asse pasci* (that his food costs less than a penny): *Metrodorum, qui non tantum profecerit, toto.* 'At first sight the garden of Epicurus presents the idea of a society of ascetics rather than of voluptuaries, and of dietetic reformers rather than philosophers' (Wallace, *Epicureanism*, p. 48).

parvis: see n. to XIII. 123: the epithet is added because *horti* suggests a rich man's park.

320. For the temperate life of Socrates, see the panegyric of Alcibiades in Plato's *Symposium*, 215 ff.

ceperunt, 'contained': see n. to X. 171.

321. Nature, i.e. the unsophisticated man, has as simple wants as these philosophers.

322. **acribus,** 'severe', that require too much of human nature.

323. **nostris** = modern: the pron. *hic* is often used in this sense.

324. For a knight's fortune, and the position in the theatre secured to him by the law of Roscius Otho, see nn. to I. 106, III. 154.

325. **rugam...labellum,** 'frowns and pouts': cf. Sen. *de Ben.* VI. 7. 1 *voltus tuus, cui regendum me tradidi, colligit rugas et trahit frontem, quasi longius exeam.* The expression of face, proper to the unsatisfied man, is attributed to the sum itself.

326. **duos equites** is a bold expression for *censum duorum equitum.*

328. The wealth of Croesus, king of Lydia, and of the Persian kings was proverbial.

329. The wealth of Narcissus forms the climax: he was one of the most influential of Claudius' freedmen, and held the office of secretary (*ab epistulis*). He is said by Pliny to have been richer than Crassus.

331. The story is somewhat differently told by Tacitus *Ann.* XI. 33–8: he says that when Claudius could not make up his mind, Narcissus gave the order for Messalina's execution, and had it carried out, without authority from the emperor. Cf. *ibid.* 35 *mirum inter haec silentium Claudii...omnia liberto oboediebant.*

For the subservience of emperors to their freedmen, cf. Pliny, *Paneg.* 88 *plerique principes, cum essent civium domini, libertorum erant servi.*

SATIRE XV

A CASE OF CANNIBALISM IN EGYPT

1–32. *The strange religious customs of Egypt are well known, how they worship animals of all kinds. There is another side to this: the people, who will not eat animals, do not shrink from eating their fellow-men. Yet of all the marvels told by Ulysses to Alcinous, cannibalism is the most staggering. But now I will tell you of a modern instance of it, the crime, not of an individual, but of a people.*

1. The beginning of the satire is a reminiscence of Cic. *Tusc. Disp.* V. 78 *Aegyptiorum morem quis ignorat? quorum im-*

butae mentes pravis erroribus quamvis (adj.) *carnificinam prius
subierint quam ibim aut aspidem aut canem aut croco-
dilum violent, quorum etiam si imprudentes quippiam fecerint,
poenam nullam recusent.*

Volusi Bithynice: an unknown friend of the poet.

2. **portenta:** cf. Virg. *Aen.* VII. 698 (of Cleopatra's gods at
Actium) *omnigenumque deum monstra et latrator Anubis.* **colat,**
'worships'.

crocodilon: cf. Herod. II. 69 τοῖσι μὲν δὴ τῶν Αἰγυπτίων ἱροί εἰσι
οἱ κροκόδειλοι, τοῖσι δὲ οὔ, ἀλλ' ἄτε πολεμίους περιέπουσι.

3. **pars haec,** 'one district'. The ibis was worshipped be-
cause it was useful in killing the flying serpents from Arabia
(Herod. II. 75).

4. **sacri** is emphatic: 'the long-tailed ape is sacred and its
golden image glitters...'. The mummies of these have often
been found.

5, 6. For the roundabout description of Egypt, cf. the similar
description of Palestine VI. 159 and 160.

5. The musical statue of Memnon at Thebes was one of the
chief curiosities of the ancient world and much visited by
travellers. The statue was for long in ruins (**dimidio:** see n. to
VIII. 4), the upper part having been overthrown by an earth-
quake. Musical sounds, as of a stringed instrument (**chordae**)
came from it every morning at sun-rise. These are now attri-
buted, not to magic or fraud, but to the vibration caused in the
loosened mass of stone by the rapid change of temperature.
When the statue was restored by Severus A.D. 202, the pheno-
mena ceased.

Memnone: the abl., whether local or instrumental, is out of
place here: and, considering that P originally read *Memnonie*,
I am inclined to believe that Juv. wrote *Memnoni*, the dat., with
the final syll. short as in Greek. For similar datives, cf. Catull.
64. 247 *Minoidĭ*; idem, 66. 70 *Tethyĭ*; Stat. *Achill.* I. 285 *Palladĭ*.
In Stat. *Theb.* III. 521 the MSS. read *Iasone*, but Bentley's
restoration of *Iasonĭ* is universally accepted; perhaps here too
the *i* of the Greek dat. led to a corruption of the text. The dat.
gives a simple and natural construction, 'the strings of Mem-
non'.

6. Another of the chief sights of Egypt was the ruins of
Thebes of the Hundred Gates: cf. Tac. *Ann.* II. 60 *mox visit
(Germanicus) veterum Thebarum magna vestigia.* The city,
which in Homeric times was the richest and greatest in the
world (*Il.* IX. 381 ff.), is called *vetus*, to distinguish it from

Thebes in Greece, of Seven Gates, for which see n. to XIII. 27.

7. **illic aeluros**: the reading of P is *illicaeruleos*, of other MSS. *illic caeruleos*: Büch. adopts the latter, saying that the antithesis requires it, *caeruleus* being defined in various collections of glosses as *bestia marina*: the sea-fish is thus opposed to the river-fish.

Against this, it is pointed out by Friedl., that this noun *caeruleus* probably owes its existence to this passage alone; that *illic caeruleos* is a most natural, though wrong, correction of *illicaeruleos*; and that it would be surprising that Juv. should omit here the most sacred of all Egyptian animals, the cat (αἴλουρος): cf. Herod. II. 66 and 67; Anaxandr. *ap. Athenaeum* 300 (a Greek addressing an Egyptian) τὸν αἰέλουρον κακὸν ἔχοντ' ἐὰν ἴδῃς, | κλαίεις ἐγὼ δ' ἥδιστ' ἀποκτείνας δέρω.

It is a singular fact that the 'harmless, necessary cat', while cherished in such numbers in Egypt, was unknown in Greek and Roman households: no skeletons have been found at Pompeii, and the mouse-catching animals, which are sometimes mentioned (γαλέη, *mustela, faelis*, etc.), were probably all of the weasel or marten kind. The domestication of the cat is a modern event, compared with that of the dog, and was first effected in Egypt.

fluminis, i.e. the Nile, the only river in Egypt. For the sacred fish, especially eels, cf. Herod. II. 72, and Anaxandr. *l.l.* 299 τὴν ἔγχελυν μέγιστον ἡγεῖ δαίμονα | ἡμεῖς δὲ τῶν ὄψων μέγιστον παρὰ πολύ.

8. The dog was another sacred animal: cf. Herod. II. 66 and 67; Cic. *de Leg.* 32 *qui canem et faelem ut deos colunt*: and see n. to *Anubis* VI. 532.

The point of the opposition is that Diana as Huntress, was mistress of the dog, and that dogs were often sacrificed on her altars.

9. **porrum et caepe**: cf. Hor. *Epp.* I. 12. 21 *seu porrum et caepe trucidas* (perhaps Iccius, whom he is addressing, was of Egyptian descent).

10. This refers to the belief that the gods revealed themselves only to innocent men in an innocent age: see n. to VI. 19, and cf. Catull. 64. 383 *praesentis namque ante domos invisere castas | heroum, et sese mortali ostendere coetu, | caelicolae nondum spreta pietate solebant*. The vegetables are 'home-made' gods (*domi nata*).

11. **lanatis**: cf. VIII. 155.

13. The sentence must begin with 'but' in English. **atto-**

nito is a sarcastic reference to *Od.* XIII. 2 κηληθμῷ δ' ἔσχοντο κατὰ μέγαρα σκιόεντα. The narrative of Odysseus was evidently proverbial for its length and its marvels: cf. Plato, *Republic*, 614 B ἀλλ' οὐ μέντοι σοι 'Αλκίνου γε ἀπόλογον ἐρῶ. It is whimsical to say that cannibalism was the most incredible detail in the narrative.

15. **bilem aut risum:** cf. Hor. *Epp.* I. 19. 19 *ut mihi saepe | bilem, saepe iocum vestri movere tumultus.*

16. **moverat:** it seems certain that the later Roman poets used the plpf. in the sense of an aorist: this is especially common in Martial, where see I. 107, 3, II. 41. 2, IV. 63. 4, V. 52. 4, VI. 10. 6, IX. 43. 9 and 10, IX. 70. 1, IX. 94. 4, X. 79. 9, XI. 39. 12, XI. 71. 1. None of these instances bears the true sense of the plpf.: *dixerat*, for example, has the sense of *olim dixit*.

aretālogus, 'story-teller': according to the usual derivation from ἀρετή, these professional entertainers told tales of an ethical kind; but here, and often elsewhere, the context makes this meaning impossible. Cf. Suet. *Aug.* 74 (at dinner) *aut acroamata et histriones aut etiam triviales ex circo ludios interponebat ac frequentius aretalogos; ibid.* 78 *fabulatoribus* (= *aretalogis*) *arcessitis resumebat somnum.* Hence there is much force in the suggestion (of Meister) that the word is not connected with ἀρετή but with ἀρετός, 'agreeable', the third syllable being lengthened only from metrical necessity.

17. **abicit:** '*will* no one throw', is our equivalent for the idiom: cf. III. 296, IV. 130. The quantity (*ăbicit*) is also to be noticed: Ovid seems to be the first to have this scansion, regarding the first letter of *iacere* as a vowel, not as a consonant: cf. *ex Pont.* II. 3. 37 *turpe putas abici, quia sit miserandus, amicum*; Stat. *Achill.* I. 545 *subicit gavisus Ulixes.* All good writers spell this word (and other compounds of *iacio*) with only one *i* (see Munro on Lucr. I. 34); but the earlier poets lengthen the first syll.

vera, as opposed to the fictitious Charybdis he tells us of (Hom. *Od.* XII. 101–10): cf. VIII. 188 *iudice me dignus vera cruce.*

18. **fingentem,** 'for inventing'. Antiphates, king of the Laestrygones, and Polyphemus (see n. to XIV. 20) each devoured some of the shipmates of Odysseus.

19. **citius,** 'sooner' than tales of cannibalism.

Scyllam: cf. *Od.* XII. 80–100.

concurrentia saxa, 'the clashing rocks', a transl. of Συμπληγάδες. The Schol. pertinently asks whether Odysseus passed these. Juv., following Homer, identifies the Symplegades at the mouth of Pontus through which the Argo passed, with the Πλαγκταί, an equally dangerous pair of rocks which stood in the way of Odysseus (*Od.* XII. 59–72).

20. **Cyaneis:** the Cyanean rocks are identical with the Symplegades: cf. Eur. *Med.* 2 Κυανέας Συμπληγάδας. Munro says that *Cyaneis* denotes the whole of which *saxa* form part, and is dat., as in *cadentia membra homini*: but the expression is very strange. Ruperti proposed *Cyaneas* (those clashing rocks, the Cyaneae); and it is possible that the acc. in apposition may have misled the copyists.

plenos governs *tempestatibus*: the reference is to the ἀσκός given by Aeolus to Odysseus (*Od.* X. 19–27).

21. **crediderim,** 'I could believe': the constr. is like the aor. optat. with ἄν in Greek: see nn. to VII. 140, VIII. 30.

For the enchantments of Circe, cf. *Od.* X. 233–40; and for the fate of Elpenor, *ibid.* 552–60.

24. **merito,** sc. *dixisset*: cf. VI. 642: such an interruption would have been quite justified. Odysseus told the whole story *super cenam* (l. 14), while the wine was going round.

25. **Corcyraea:** the fairy-land of Phaeacia was generally identified by the ancients with Corcyra (Corfu).

duxerat, 'had drunk'. *deduxerat*, the reading of the inferior MSS. including the Bobio palimpsest (see Introduction, p. xlvi), is a corruption apparently due to ignorance of the scansion *tēmētum*.

26. **Ithacus:** cf. XIV. 287.

nullo sub teste, 'with none to bear him out', all his company having lost their lives before he reached Phaeacia. For the idiomatic use of *sub*, cf. Livy, II. 37. 8 *consules cum ad patres rem dubiam sub auctore certo* (on good authority) *detulissent. nullo teste* would mean the same: cf. X. 70.

27. **nuper:** consequently the tale, though marvellous, can be tested. An inscription proves that Aemilius Juncus and Julius Severus were consuls in the year A.D. 127. The satire was there-fore written after that date, though how long after, is uncertain, as *nuper* is a word of elastic meaning: cf. VIII. 120, where it refers to a period probably twenty years earlier; also Pliny, *Paneg.* 8, where *nuper* refers to an event of thirty-one years before.

28. **super,** 'beyond', i.e. south of: cf. *rursus*, X. 150.

Coptos was a city on the Nile, in the Upper Thebaid, of considerable importance: it lay between Ombi and Tentyra.

29. **cothurnis,** 'tragedies': cf. VI. 634.

30. **a Pyrrha** = from the time of the Flood: cf. I. 80–6.

syrmata = tragedies: see n. to VIII. 229.

volvas = *evolvas*, 'you read': see n. to X. 126: for the subj. after *quamquam*, see n. to VII. 14: but here the subj. may very well be potential, or else due to the generic notion of the 2nd person.

31. **accipe** = I will tell you: cf. VII. 36. **nostro** is emphatic: such horrors might be expected in tales of 'old, unhappy, far-off things'.

33–92. The people of Ombi worshipped different gods from the people of Tentyra; mutual hatred was the result. The occasion of a religious festival was seized for an attack; and after a bloody fight, the men of Tentyra turned tail. One of the fugitives slipped and fell, and was promptly torn in pieces and eaten by the crowd of pursuers.

33. **finitimos:** cf. Tac. *Hist.* I. 65 *veterem inter Lugdunenses et Viennenses discordiam proximum bellum accenderat...unde aemulatio et invidia et uno amne discretis conexum odium.*

The distance between Tentyra and Ombos is considerably over a hundred miles; but modern Egyptologists (cf. Maspero's *Dawn of Civilisation*, p. 202, n. 4) believe that Juv.'s Ombi is not Ombos but the town called Pampanis by the Romans and Pâ-Nubit by the Egyptians (now Negadeh). This is much nearer Tentyra.

vetus and **antiqua** are not mere synonyms: see n. to VI. 21.

35. **Ombos et Tentyra:** the names of the towns are in the acc., being in apposition with *finitimos* above.

36. **inde** = *ab eo*. The crocodile was probably the *causa belli*; it was worshipped at Ombi and persecuted at Tentyra: cf. Herod. quoted on l. 2; and Sen. *Nat. Quaest.* IV. 2. 15 *nec illos* (i.e. crocodiles) *Tentyritae generis aut sanguinis proprietate superant, sed contemptu et temeritate. ultro enim insequuntur fugientesque iniecto trahunt laqueo.*

39. **alterius populi:** it is remarkable that Juv. does not say which people began the attack; if *super* in l. 28 is to be taken strictly, the scene of the outrage was Ombi, both Pampanis and Ombos being to the south of Coptos; and therefore the Tentyrites were the aggressors.

42. **sentirent**, 'enjoy': the men of Ombi are the subject.

43. **pervigili toro,** 'night-long feasting', *torus* being the couches on which they reclined to eat. For *pervigili*, cf. VIII. 158.

quem...invenit, i.e. these festivals often last for seven days and nights.

44. **sane,** 'it is true'. This sentence is a parenthetical comment on the revelry.

45. The context shows that **luxuria** here denotes rather the will than the means to practise excessive indulgence.

quantum ipse notavi, 'as I myself have observed': for Juv.'s personal knowledge of Egypt, see Introduction, pp. xviii and xix.

notare is common in the silver-age writers in the sense of *animadvertere*: cf. XVI. 35.

46. **famoso,** 'notorious': for the reputation of Canopus, see n. to VI. 84: Seneca mentions it, together with Baiae, as a *diversorium vitiorum* which the virtuous man will, if possible, avoid. The people are here contrasted, as Greeks, with the natives of Egypt.

47. **adde quod,** 'besides': cf. XIV. 114: hence **et** before *facilis* is superfluous. Their choice of a time was partly due to spite, partly to knowledge of their enemy's helpless condition.

victoria de madidis: cf. Mart. IV. 23. 4 *palmam Callimachus ...de se | facundo dedit ipse Brutiano* (where *de* depends entirely on *palmam*, not on *dedit*). So τρόπαιον ἀπό τινος.

48. **blaesis:** the word is used to express the lisping of intoxication and also that of infancy: cf. Mart. IX. 87. 2 *denso cum iaceam triente blaesus*; idem, V. 34. 8 (of a child) *et nomen blaeso garriat ore meum.*

inde, 'on the one side', is opposed to *hinc*, 'on the other' l. 51. **virorum** is emphatic, the Romans thinking dancing an effeminate pastime.

49. **nigro tibicine:** abl. absol. **qualiacumque** is here an indefinite pronoun, and indicates that the perfumes were not choice.

50. For this description, cf. VI. 297 *coronatum et petulans madidumque Tarentum.*

51. **ieiunum odium,** 'hatred and an empty stomach': cf. *locupletem aquilam,* XIV. 197.

For **iurgia** as the beginning of a *rixa*, cf. V. 26. **prima** = at first.

52. **tuba,** lit. 'trumpet', i.e. signal: the phrase is applied to a person by Cicero, *ad Fam.* VI. 12. 3 *tibi, quem illi appellant tubam belli civilis.*

55. **aut** is corrective, 'or rather'.

toto certamine = *ex omnibus qui certabant.*

57. **dimidios,** 'mutilated': cf. l. 5 and VIII. 4.

alias, 'disfigured'.

59. **ipsi:** any bystander would have thought matters had gone far enough, but the combatants themselves are of a different opinion.

60. **calcent** is subj., because it expresses the reason in their minds. The emphasis falls on *cadavera*.

63. **inclinatis** ('back-bent') **lacertis** is to be taken with *torquere*, not with *quaesita*.

64. **torquere:** see n. to VI. 449. **domestica,** 'ready-made'.

65. The reference is to the heroes of Homer and Virgil, who use huge stones as missiles in war: cf. *Il.* XII. 380, V. 302; *Aen.* XII. 896.

nec, 'but not'. **qualis** is object of *torquebant* understood.

69. In each of the above cases, Homer, enlarging on the size of the stone, says that his own generation (οἷοι νῦν βροτοί εἰσι) could not rival the hero's feat: and to this Juv. is alluding satirically. The belief that the stature and strength of the race has been gradually diminishing, was general in antiquity, and stoutly maintained by Lord Monboddo at the end of last century. 'Ah, doctor,' he said to Johnson, 'poor creatures are we of this eighteenth century; our fathers were better men than we!' Johnson replied, 'Oh no, my lord; we are quite as strong as our forefathers and a great deal wiser.' On the theory of the gradual degradation, both physical and moral, of mankind, De Quincey remarks: 'as men ought physically to have dwindled long ago into pygmies, so, on the other hand, morally they must by this time have left Sodom and Gomorrah far behind. What a strange animal must man upon this scheme offer to our contemplation; shrinking in size, by graduated process, through every century, until at last he would not rise an inch from the ground; and on the other hand, as regards villainy, towering ever more and more up to the heavens. What a dwarf! what a giant! Why the very crows would combine to destroy such a little monster' (*Collected Writings*, I, 97).

71. He laughs at men, because they are pygmies (cf. XIII. 170–3), and hates them because they are scoundrels.

72. **a deverticulo,** 'after the digression' about the size of the stones: the apology is less necessary here than in many other places where it does not occur.

73. **aucti,** sc. *sunt*; the apodosis begins at *pars*.

74. **pugnam instaurare,** 'to renew the battle': see n. to VIII. 158.

75. The MSS. are in great confusion here. In P all is erased after *praestant*: other MSS. have *fuga* and various endings such

as *praestantibus omnibus instant*, or *praestant instantibus orbes*. The emendation given in the text is highly probable.

fugae is dat. after *terga praestant*: cf. Propert. IV(v). 2. 54 *turpi terga dedisse fugae* with Postgate's note: *terga dare* has the same meaning without the addition of *fugae*. **Ombis** (used for *Ombitanis*) **instantibus** is abl. absol.; and the subject to *praestant* is the whole of l. 76.

77. **hic**, 'hereupon', 'at this point': cf. III. 21. P reads *hinc*, but is not followed even by Büch.

78. **ast illum**: cf. III. 264 and n. there.

80. **corrosis ossibus edit**, 'devoured him and gnawed his bones': the nature of the case shows that the action described in the participle cannot, in spite of the tense, precede the action of the verb: see n. to v. 68.

81. **decoxit**: some word for 'roasting' must be supplied out of this to go with *veribus*.

83. **crudo**: the epithet is sometimes applied, by an extension of meaning, to the eater of raw flesh and especially human flesh: cf. Mart. IV. 49. 4 *cenam, crude Thyesta, tuam*; Ovid, *Heroid*. 9. 67 *crudi Diomedis imago,* | *efferus humana qui dape pavit equas*.

84. **hic**: cf. l. 77. Fire was brought down from heaven and would have been profaned by being used for such a purpose. The subject to *violaverit* is *victrix turba*.

86. Some edd. bracket from *elemento* to *reor*: but here as so often elsewhere there is no proof of interpolation. The difficulty lies in **te,** which is taken by some to refer to fire, by others to Volusius, the friend to whom the satire is addressed. In the first case, the sudden apostrophe to fire is very awkward; in the second, the reference to Volusius, who has been dropped since the first line and does not appear after this, is not less so. On the whole, considering that several of the satires are addressed to a perhaps imaginary friend who figures only in the first line (so Fuscinus, *Sat*. 14, Gallius, *Sat*. 16, and see n. to VI. 28), it seems more likely that fire itself is apostrophised.

88. **sustinuit**: cf. VI. 105, XIV. 127.

89–92. The sense is: the guilt being so great, you may well ask whether even the first to eat found pleasure in the taste: but you need not, for the last was as eager as the first.

89. **ne quaeras** and **ne dubites** are, if the text be sound, both prohibitions, and not final clauses. There are no other instances of this constr. in Juv.

90. **autem** offers another difficulty: some word like *nam* or *enim* or *immo* is wanted. In earlier Latin *autem* often means 'in

turn', but only of the second of a pair: Plautus, Lucretius, and Cicero use it so: here this meaning is inadmissible. The word is probably corrupt. If so, from *scelere* to *senserit* may be a parenthesis, containing a final sentence which introduces an indirect question.

93–131. *Under stress of war and siege, men have kept death at bay by eating human flesh. For such we may have pardon and pity; but the Egyptians have no such excuse. They are no wild and warlike race, from whom one might expect such atrocities.*

93. The Vascones were a Spanish tribe: after the death of Sertorius (72 B.C.) their chief town Calagurris was besieged by the Roman army under Afranius; and the sufferings which they endured from hunger became proverbial.

94. **produxere animas,** 'prolonged their lives'.

res diversa, 'the case is different': cf. VIII. 215.

95. **ultima** is neut. plur.: cf. *discriminis ultima*, XII. 55.

97, 98. 'For the instance, now in question, of this food (i.e. human flesh) deserves pity, inasmuch as the people I have just spoken of (i.e. the Vascones)....'. The antecedent of *quod* is *exemplum*, not *huius*.

There is great awkwardness in the expression; and also the sense, which **sicut** must bear, is doubtful. There are instances in Plautus, where *sicut* is apparently = *siquidem*. Housman reads *tibi, si cui*, with a full stop after *gens*: i.e. 'you, if anyone, ought to pity...'.

Büch. explains as follows: *huius in Aegypto facinoris miserabile exemplum usus ciborum talis....*, viz. 'if an instance of cannibalism is to excite pity, there must be such a diet as in the case of the Vascones'. Thus *sicut* secures its right meaning: but greater difficulties are raised. He takes *cibi* as nom. plur.: how then can the verb (*debet*) be singular? And would not *cibi* require some such epithet as *tales*?

102. **famē,** 'because of hunger': for the quantity, see n. to VI. 424.

104. **ventribus** is much more forcible and picturesque than *urbibus*, the reading of P, but cannot be called certain: *viribus* of the inferior MSS. is impossible.

106. **nos,** 'us moderns': cf. *nostris*, XIV. 323.

107. Zeno was the founder of the Stoic philosophy, and taught that moral virtue (*honestum*) was the only good, and that a man should rather die than do wrong. But it is a remarkable fact that the Stoics rather favoured, than condemned, the

practice which Juv. is attacking. For the evidence, see M. *ad loc.*

omnia quidam: so P: *quidam* are Zeno and his followers. Cf. Sen. *Epp.* 70. 7 *non omni pretio vita emenda est.* The inferior MSS. have *omnia, quaedam*, 'not all things but only some things'; in that case the subject to *putant* is *praecepta.*

108. **sed...stoicus,** 'but how could a Cantabrian be a Stoic philosopher...': the Vascones were not actually Cantabrians, but the name is used to denote the most savage tribes of Spain.

109. **Metelli:** Q. Caecilius Metellus Pius is meant, who carried on the war against Sertorius 79–72 B.C.

110. **nunc,** 'but now' as opposed to the rude age of Metellus. There is therefore less excuse for such a crime.

Graias nostrasque Athenas, 'the culture of Greece and Rome': 550 years before Pericles had called Athens παίδευσις τῆς Ἑλλάδος, and her political extinction had extended, rather than diminished, her sway over the minds of men.

111. Tacitus throws some light on the beginnings of British oratory: *Agric.* 21 (*Agricola*) *principum filios liberalibus artibus erudire, et ingenia Britannorum studiis Gallorum anteferre, ut qui modo linguam Romanam abnuebant, eloquentiam concupiscerent.* The passage refers to a date about A.D. 80.

112. For this pleasant l., see Introduction, p. xli.

rhetore, 'a professor of rhetoric': see n. to I. 16. The ancients often refer to *Ultima Thule*, as we might to Spitzbergen, as the most distant land to the North; but it is uncertain whether they meant Iceland, a Shetland, or some other island, by the name.

113. **ille populus,** the Vascones.

114. **virtute atque fide,** 'in valour and loyalty'.

Zacynthos: this name is common in the poets for Saguntum, which was supposed to have been founded by colonists from the Greek Zacynthus. The town lay in Hispania Tarraconensis and suffered great hardships when besieged and taken by Hannibal 218 B.C. That the inhabitants were driven to cannibalism is not mentioned by the historians, and is perhaps an invention of a later time. Cf. Petron. 141 *Saguntini oppressi ab Hannibale humanas edēre carnes.*

115. **Maeotide ara,** the altar of the Tauric Artemis, where strangers were sacrificed as victims: cf. Eur. *I. T.* 384 αὐτὴ δὲ θυσίαις ἥδεται βροτοκτόνοις. The *palus Maeotis* (Sea of Azov) is immediately to the North of the Tauric Chersonnese (Crimea).

116. **quippe,** 'for': see n. to XIII. 26.

117. **iam** has the force which it usually has after *si* or *ut*: it suggests that the legends, though *not* credible, may pass as such *for the moment*: cf. Lucr. v. 195 *quod si iam ignorem rerum primordia quae sint*; idem, III. 843 *et si iam nostro sentit de corpore postquam | distractast animi natura animaeque potestas*. Lucr. is convinced that he is *not* ignorant of the composition of matter, and equally convinced that sensation vanishes when body and soul are separated; but he is willing in both cases to assume, for the purposes of argument, what he believes to be untrue.

carmina refers to the tragedy of Euripides and other sources of the legend.

119. **cultro**: abl. after the comparatives. The victim need not fear being eaten as well.

modo, 'lately': cf. *nuper*, l. 27.

120. **hos**, i.e. the Egyptians, is opposed to *illa*, l. 116.

vallo, abl., goes closely with *infesta*, and the phrase = *infesto vallo*: cf. *infestis sagittis* l. 74.

122. **aliam**, transl. 'in any other way'.

123. The fertility of Egypt depends entirely, in modern as in ancient times, upon the autumn rising of the Nile.

invidiam facerent Nilo, 'put the Nile to shame'. Thus when Ino was driven out to sea by Juno, the Theban women by their lamentations *invidiam fecere deae* (Ovid, *Met.* IV. 547). The river is regarded as a divinity; and it was a regular practice of ancient religions, if the gods failed to do what was expected of them, to commit outrages in order to put them in the wrong and make them ashamed: cf. Suet. *Calig.* 5 (on the death of Germanicus) *lapidata sunt templa…partus coniugum expositi*; and Herod. I. 159, where Aristodicus, in order to bring Apollo to a sense of his guilt in ordering the surrender of a suppliant, began to rob the birds' nests round the temple. Angry children often act in this way; and the savages, from whom the ancient religions were inherited, were like children in their mental processes.

124. **Cimbri**: see n. to VIII. 249: the verb understood is *saevierunt*.

Brittonĕs (cf. *Vasconĕs* l. 93) = *Britanni*: Mart. XI. 21. 9 has the form *Brĭtōnis* (gen.). Our ancestors had a bad reputation: cf. Hor. *Carm.* III. 4. 33 *Britannos hospitibus feros*.

125. The Sauromatae or Sarmatians lived on the site of modern Poland and much of Russia: they were fierce savages and were constantly at war with Rome in the 2nd cent. A.D. The Agathyrsi belong more to legend: they are placed by Hero-

dotus in what is now Transylvania. Virgil (*Aen.* IV. 146) calls them *picti*.

126. **rabies,** as distinguished from *ira*, is the rage of a wild beast and hence is often used of cannibalism; cf. Livy, XXII. 51. 9 *cum...in rabiem irā versā laniando dentibus hostem expirasset.* For the hiatus after *rabie*, see n. to X. 281.

127, 128. The epithets **parvula** and **brevibus** are meant to suggest the feeble powers of the Egyptians. For **testae**, see n. to XIV. 311: *testae* is gen., governed by *remis*. The want of wood forced them to make boats of unusual materials.

131–74. *Man is distinguished from animals by the power to weep: pity is the noblest manifestation of human nature, and human civilisation is based upon sympathy. But man has fallen until he is even lower than the animals: lions do not prey on lions, but man attacks his fellow and even devours him.*

133. **haec,** 'this', i.e. 'sympathy', refers to *mollissima corda*: 'of all our feelings sympathy is the noblest'.

134. **iubet:** the subject is *natura*. The constr. is, *natura iubet nos plorare squalorem amici causam dicentis et rei*; but there is some objection to this, as *causam dicentis* and *rei* are identical in meaning; hence Kiaer's somewhat improbable conjecture, *squalorem atque rei*, by which *rei* becomes a noun and is distinct from *amici*.

135. **squalorem** refers especially to the untrimmed beard and hair by which men on trial endeavoured to excite pity: cf. Mart. II. 24. 1 *si det iniqua tibi tristem fortuna reatum,* | *squalidus haerebo pallidiorque reo.*

ad iura, 'into court': generally *in ius*: cf. X. 87.

136. **circumscriptorem:** cf. X. 222: the ward, whose guardians have robbed him of his property, brings them into court: *circumscriptio* is used of legal chicane of all kinds.

137. **incerta,** i.e. it is difficult to tell whether he is a boy or a girl: Roman boys wore their hair long and are hence called *capillati* (Mart. X. 62. 2), and *cirrhati* (idem, IX. 29. 7): they must not be confused with the class of slaves to whom the same names were given.

138. **adultae,** i.e. *nubilis*, death at that age being considered especially tragic: cf. Lucretius quoted on XII. 118.

140. **minor igne rogi,** 'too small for the funeral fire': it was the custom, as still among Hindus, to bury, not to burn, infants: cf. Pliny, *Nat. Hist.* VII. 72 *hominem prius quam genito dente cremari mos gentium non est.* Cremation, though general,

was not universal at Rome even for adults: thus it is mentioned that Sulla was the first member of the *gens Cornelia* to be burnt: all before him had been buried.

For the abl., see n. to IV. 66. **face dignus arcana**, 'worthy of the mystic torch', i.e. worthy to bear a part in the Mysteries of Demeter at Eleusis. An imitation of the πρόρρησις at Eleusis, or warning to the wicked to depart, is given in Aristoph. *Frogs*, 354–71. The Hierophant 'wishes men to be' of a certain character, by proclaiming the exclusion of those who fall short of the standard.

142. Perhaps an allusion to the famous verse of Terence (*Haut. Tim.* 77) *homo sum; humani nil a me alienum puto*.

143. **mutorum**: cf. *animalia muta*, VIII. 56.

ideo, 'for that reason'. The logic is dubious: because we have the power of sympathy, therefore we have it (l. 146).

venerabile, 'worthy of reverence': some explain 'reverential', which certainly harmonises with *divinorum capaces*: but it seems doubtful whether Juv. would use an adj. of this form with an active sense: see n. to XII. 73. They are common in earlier poets: see Munro's n. to *genitabilis* Lucr. I. 11. Martial (IV. 19. 9) has *frigus penetrabile* (= *quod penetrat*); but that may be a reminiscence of Virgil.

146. As Juv. cannot mean to deny that animals have 'sensation', **sensus** must here mean much the same as *communis sensus*, 'sympathy', as we use 'feeling': see n. to VIII. 73.

147. A reminiscence of Ovid, *Met.* I. 84 *pronaque cum spectent animalia cetera terram,* | *os homini sublime dedit caelumque videre* | *iussit et erectos ad sidera tollere vultus.*

148. **communis conditor**, 'the creator of both us and them'.

149. Where **anima** and **animus** are contrasted, the former is 'the vital principle,' ψυχή, while the latter is 'the rational principle', ψυχή: cf. Sen. *Epp.* 58. 14 *quaedam* (sc. *animantia*) *animum habent, quaedam tantum animam.*

150. **adfectus**, 'friendly feeling': the word is used by classical writers to denote only 'a feeling'; it has the sense of *studium* or *gratia* occasionally in silver-age Latin: see n. to VI. 214.

iuberet is final, not consecutive. The Creator gave mankind an instinct intended to produce civilisation.

153. **laribus**, 'home': a single household is properly called either *lar* or *penates*, each household having one of the former and two of the latter: cf. VIII. 14, XIV. 20: but *lares* is often used

of a single house, e.g. Mart. I. 70. I *ire iuberis | ad Proculi nitidos, officiose, lares.*

155. **collata fiducia,** 'confidence born of union': it is = *collatorum fiducia.*

157. **defendier** is to be noticed as one of the very few archaisms in Juv.: *duelli* (I. 169) is another, but perhaps forms part of a proverb. *defendier* has an epic sound: cf. Virg. *Aen.* VIII. 493. *ingens* also, as an epithet of *vulnus*, is epic: cf. Virg. *Aen.* X. 842, XII. 640.

159. **iam,** 'nowadays': the Creator's intentions have been frustrated by our wickedness.

Moralists in all ages have pointed to the behaviour of animals to their own kind as an example to man; but the facts are not quite as the moralists have stated them.

160. Two constructions are possible: (1) 'the wild beast, alike in spots, spares its kin'; (2) 'a wild beast of the same kind spares kindred spots', i.e. a spotted creature of its kind. Of these the first is preferred by Friedl., but the second seems more natural and more like Juv. The meaning is the same in either case—that panther will not prey on panther.

165. **incude produxisse** = *procudisse*, which Lucr. (v. 1265) uses for working metal into swords on an anvil.

166. **parum est,** i.e. they are not content with killing, but must also eat, their victims.

cum, 'though'.

167. **marrae, sarcula** and **vomer** all occur III. 311. The statement sounds conventional and untrue when compared with Lucretius's pictures of primitive man: he says that men first fought with teeth and nails, or stones (*domestica tela*, l. 64) and pieces of wood, and then with swords as soon as they discovered the use of metal (v. 1283–96).

168. **gladios extendere,** 'to forge long swords': cf. Lucan, IV. 417 *carinas | extendunt* (they build long keels).

169. **aspicimus,** '*for* we see'.

171. **crediderint:** the 1st pers. sing. regularly means 'I am inclined to believe': but there the subj. seems not to be potential but consecutive after *qui* (= *tales ut*) which is to be supplied from *quorum*: the omission of a repeated relative pronoun is common in Latin: see n. to XI. 25.

174. **tamquam homine:** cf. XIV. 98.

Pythagoras ate no flesh but only vegetables, and not every kind of vegetable: beans in particular were excluded from his diet, though there is some uncertainty as to his reasons: see n.

to III. 229, and cf. Hor. *Sat.* II. 6. 63 *faba Pythagorae cognata*
with Palmer's n.

SATIRE XVI

THE ADVANTAGES OF A MILITARY CAREER

The Scholiast says of this fragmentary Satire that many held
it to be spurious and refused to allow that Juvenal wrote it.
Most modern critics believe in its authenticity. Not only is it
found in all MSS. and attributed to Juvenal by Servius and
Priscian, who quote from it; but, even if it were anonymous, a
reader familiar with Juvenal's style would surely pronounce it
to be his. The belief in fatalism, the banter at the expense of
the gods, the sustained irony which, under a mask of praise,
points out the unjust privileges of the soldier over the civilian—
all these are characteristic.

1–6. *A soldier's life is a happy one. I would enlist myself, if I were*
sure of luck, as luck is all-important in the army.

1. **praemia,** 'advantages': it is not used here though often
elsewhere of the pecuniary rewards given to soldiers on their
discharge (*missio*): cf. Tac. *Ann.* I. 26 *de praemiis finitae militiae*;
Suet. *Iul.* 70 *missionem et praemia flagitantes.* Under Augustus a
legionary received 12,000 sesteces (£120) on his discharge:
later the sum was probably less.

Galli: the name does not recur in the satire (see n. to xv. 86),
and nothing is known of him.

2. **militiae:** the details which follow show that service in the
praetorian guard is meant. This force was recruited, as far as
possible, in Italy; and the men received double the pay of the
legionaries, and could retire at an earlier age: cf. Tac. *Ann.* I. 17.

nam, etc., explains *felicis*: luck is necessary to success.

3. **excipiat** is optative.

pora, sc. *castrorum.*

4. A lucky star does more for a soldier than a letter of recom-
mendation addressed to the God of War by his mistress, Venus,
and his mother, Juno. The significance of **hora** is shown by VI.
581: the luck is attached to a particular time.

5. For the influence of Venus over Mars, cf. Lucr. I. 31 ff.
A recruit may often have provided himself with an *epistula*
commendaticia to the commanding officer, such as Cicero often
wrote to solicit the good offices of important personages for his
friends: the 13th book of his letters consists entirely of these.

6. **Samia...harena** is a periphrasis in Juv.'s manner (see n. to III. 25) for Juno, the mother of Mars; she was supposed to show special favour to the island of Samos.

7–34. *In the first place a civilian, when assaulted by a soldier, has to claim redress before a court of centurions and is likely to be disappointed. He has the whole regiment against him and cannot persuade his witnesses to appear.*

7. **communia**, 'common to all soldiers': the advantages of individuals must have been treated in the missing part of the satire.

8. **illud...ne:** *ut non* would be correct, as an *ut* clause defining *illud* is properly consecutive, not final. Perhaps the distinction was becoming obliterated: cf. Mart. IV. 64. 19 *essedo tacente,* | *ne blando rota sit molesta somno*, where the last clause ought to be consecutive in meaning.

togatus, 'a civilian': so *paganus* l. 33: cf. VIII. 49.

9. **dissimulet,** 'conceals it', i.e. pretends he has not been beaten.

10. It appears from this passage, that if an *actio iniuriarum* against a soldier was brought before the praetor, he was obliged or allowed to appoint centurions to try the case, and that the trial took place in the camp. But this may be a satirical exaggeration of the actual procedure.

12. **medico nil promittente** = *de quo medicus nil promittit*, 'about which the doctor gives no certain promise', i.e. can only hope for the best: cf. Pliny, *Epp.* I. 22. 11 *medici secunda nobis pollicentur: superest ut promissis deus adnuat.*

relictum: he has not actually lost his eye, like his teeth: it is left in his head but is in a bad way. M. explains 'given over, abandoned': for this sense of *relictus*, cf. Sen. *Epp.* 78. 14 *quotiens deploratus sum a meis, quotiens a medicis relictus!* But the rest of the l. seems inconsistent with such a desperate condition of the eye.

13. **Bardaicus calceus** is a soldier's boot, so named from the Bardaei (or Vardaei), a tribe of Illyria: cf. Mart. IV. 4. 5 *lassi vardaicus...evocati*. The boot is used for its wearer, a centurion, just as *caligae* is used below for 'common soldiers': cf. our 'red-coats', and the French *pantalons rouges*. **punire** = to get redress for.

14. **grandes surae** are the other centurions, who need big benches to sit on as *iudices*, being big, strong men, like the *magni centuriones* of Horace (*Sat.* I. 6. 73).

15. **Camilli:** his name is mentioned as the originator of a standing army; we need not suppose that he laid down rules to this effect.

17–19. **iustissima...querellae** is a reflection attributed to some injured person; it is followed by a reply which shows that it is not very satisfactory to seek redress in a court of centurions.

18. **cognitio:** see n. to VI. 485.

20. **cohors:** the praetorian guard was divided into ten cohorts, each 1,000 strong; each cohort had three maniples.

21. **curabilis** = *curanda*, 'requiring medical treatment': so M.: the redress you get is of such a kind that you must go to the doctor again with worse injuries. This seems better than the explanation of Friedl.: 'that the punishment is slight (lit. easily cured) and therefore more painful (to the plaintiff) than the original assault': for it is difficult to supply a different person with the second adjective.

Would it be possible to take **magno** with *curabilis* in the same sense as III. 166 and XI. 148? **consensu** is better without the epithet; but the order of the words is certainly an obstacle to this explanation.

23. **mulino corde** = stupidity. For Vagellius, see n. to XIII. 119.

24. **duo,** '*only* a single pair'.

caligas, tot is a necessary correction for *caligatos* of all MSS. See n. to l. 13. The hobnails (*clavi*) of soldiers' boots were mentioned III. 248.

26. **praeterea:** another difficulty is that you cannot produce witnesses: your friends are frightened of the soldiers, and say they can't go out of town so far as the camp.

tam Pylades = *tam amicus*.

molem aggeris ultra: see n. to VIII. 43. The praetorian camp, a fortified barrack (see n. to X. 94), was really close to the city, outside the *agger*, and between the Colline and Viminal gates.

28. **se excusaturos,** 'for they are sure to give excuses': so *periturae*, I. 18. For the fut. particip. expressing a concessive clause, cf. VI. 39.

29. **da testem,** 'produce your witnesses'.

audeat...et credam = *si audebit, credam*.

30. **pugnos,** 'the fisticuffs'. **vidi:** cf. VII. 13.

31. Such straightforwardness is worthy of our long-haired ancestors: see n. to IV. 103.

34. 'To attack the pocket and the honour of a man in uni-

form': for this sense of **fortuna,** cf. XIV. 328; and for **pudorem,** cf. VIII. 83.

35–50. *When a civilian goes to law, he suffers by the law's delay; but the soldier can get summary justice.*

35. **notemus:** see n. to *notavi,* XV. 45.

36. **sacramentorum** = *militiae.*

37. **campum,** like *convallem,* governs *ruris.*

38. **sacrum saxum** is the *terminus* or boundary-stone between two properties: these were worshipped as statues of the god Terminus, though they were merely posts or rough stones: cf. Ovid, *Fasti,* II. 641 *Termine, sive lapis sive es defossus in agro* | *stipes,... te duo diversa domini pro parte coronant,* | *binaque serta tibi binaque liba ferunt*; and see Munro's n. to Lucr. V. 1199.

39. **annua,** i.e. on the feast of the Terminalia, 23 Feb.

41 is an almost exact repetition of XIII. 137, where see nn.

42. 'I must wait for a year (i.e. a long time) before the hearing of the suits of the whole people begins', and consequently longer still for my own case. Civil cases were heard in the order in which application was made to the praetor. So M.

Others, following Servius on Virg. *Aen.* II. 102, take *annus qui lites inchoet* as = *annus litium,* and explain this as equivalent to *rerum actus,* the part of the year during which legal business was taken. So the speaker means, 'I must wait till vacation is over and the flood of litigation begins.' But it is remarkable that there is no authority except Servius for this phrase *annus litium.*

44. **subsellia...sternuntur,** 'the benches (for the *iudices* and *causidici*) are prepared and no more'. All is ready for the trial: the barrister on each side is making his preparations to begin his speech, when the case is suddenly adjourned.

45. **ponente lacernas,** because he had to speak in the toga: cf. Suetonius quoted on XI. 204.

46. **Caedicio:** the name occurred XIII. 197, and **Fuscus,** XII. 45; but there is no certainty that they are the same persons.

47. The emphasis falls on **lenta,** which may be translated as an adv.

48. **balteus** is a leather belt, worn over the shoulder, from which the sword was hung: cf. VI. 256.

50. 'Nor is their substance worn away by the everlasting drag of their suit.' For **res atteritur,** cf. *deteret,* III. 24. **sufflamen** is properly the drag on the wheel of a carriage: cf. VIII. 148: and is here metaphorically applied to the suit which drags

on and so wears away the litigant's wealth. The verb *sufflaminare* is also used metaphorically: thus Augustus said of a too rapid orator, *aliquando sufflaminandus est*, 'he needs the drag sometimes', words which were applied later by Ben Jonson to Shak.

For the length of lawsuits at Rome, cf. Mart. VII. 65. 1 *lis te bis decimae numerantem frigora brumae | conterit una tribus, Gargiliane, foris.* The lawyer's point of view, that business should not be taken too fast, will be found stated by Pliny, *Epp.* VI. 2. 5 ff.

51–60. *A soldier too has the peculiar privilege that he can own, and bequeath, property, though his father is still alive. None but the brave deserve the rewards of bravery.*

51. Under the Empire a soldier, though still *in manu patris*, could dispose freely of the money he had gained by service in the army: this was called *castrense peculium*. Civilians had no similar privilege until a late date.

53. **placuit,** 'it has been settled'.

in corpore census, 'included in the property'.

54. **omne regimen,** 'absolute power of disposal'.

Coranum: though this name occurs in Hor. *Sat.* II. 5. 57 in a connection somewhat similar, this must be a coincidence; Juv. must be referring to some notorious incident of his own time.

56. **iam tremulus,** 'quite palsied', i.e. very old: cf. x. 198.

For the practice of *captatio* (legacy-hunting), see nn. to III. 129, IV. 19, V. 98.

favor is Ruperti's necessary correction of *labor* which all MSS. read, but which the epithet *aequus* makes impossible: the origin of the corruption may have been that *favor* became *fabor*: see n. to *negabit* XIV. 134.

57. **provehit,** 'promotes'.

sua dona, 'the right rewards'.

58. **ducis...referre:** this constr. of *refert* is not classical and is used chiefly by Sallust: *refert* can be used with the adjectives *meā, tuā*, etc.; but with the gen. of the person, *interest* is used by classical writers.

60. **phaleris et torquibus:** cf. Cic. *in Verr.* II. 3. 185. *Q. Rubrium...corona et phaleris et torque donasti*; Gellius (II. 11) tells of a famous soldier, L. Sicinius Dentatus, tribune of the plebs 454 B.C., that he had received eighty-three *torques* and *phalerae* twenty-five times, besides many other distinctions.

phalerae were worn, not only by horses (cf. XI. 103), but also by soldiers as medals for distinguished service: they were thin plates of precious metal worn upon the breastplate. They were often given with *torques* and *armillae*; and the three together were called *dona*.

The sentence and the satire here break off abruptly. The satire is obviously unfinished. There is also external evidence which proves this: see Introduction, pp. xlvii and xlviii. It follows that part of the book is lost, though the amount lost is a matter of conjecture. A 'book' (*volumen*) of Latin poetry may contain as many as 1,000 lines: few, except those of Lucretius, exceed this number. This mutilated book, contains 814 lines; there are 990 lines in the 1st book, 661 in the 2nd, 668 in the 3rd, and 704 in the 4th. Thus it is already longer than three of the books, but considerably shorter than the 1st.

ADDITIONAL NOTES

I. 149. **omne in praecipiti vitium stetit:** Mr Richards' explanation of this phrase is strikingly confirmed by an expression of Seneca (*Dial.* III. 7. 4), *vitiorum natura proclivis*, where the context shows the meaning to be that many vices are in a state of unstable equilibrium: they cannot remain where they are but must go on till they reach their extreme point. This is exactly what Juv. means.

III. 216. **conferat inpensas:** as the meaning, which *inpensas* apparently has here, is not common, another instance may be quoted from the book of portents (*Prodigiorum Liber*) compiled from Livy by Julius Obsequens: 28 [87] *Tarracinae, sereno, navis velum fulmine* † *exanimatum in aquam deiectum, et* † *impensas omnes quae ibi erant, ignis absumpsit*: part of the text is corrupt, but the meaning of *impensae* is clearly 'materials' of some kind which formed the ship's cargo.

VI. 295. **paupertas Romana:** this striking phrase may take its origin from a very similar passage in Seneca's *Epistles* (87. 41: the subject is the futility, for practical purposes, of the syllogisms of the philosophical schools which prove that wealth is not a 'good'): *his* [*interrogationibus sumus*] *effecturi, ut populus Romanus paupertatem, fundamentum et causam imperii sui, requirat ac laudet, divitias autem suas timeat? ut cogitet has se apud victos reperisse, hinc ambitum et largitiones et tumultus in urbem sanctissimam temperantissimamque inrupisse?*

INDEX

References to the Introduction are by pages with Roman
lower case numerals, to the Notes by satire and line

INDEX
TO NEW INTRODUCTION

REFERENCES TO
INDIVIDUAL SATIRES AND TO
PASSAGES OF JUVENAL